ALLE ZEIT WACH 1842

Springer Series in Wood Science

Editor: T. E. Timell

M. H. Zimmermann
Xylem Structure and the Ascent of Sap (1983)

J. F. Siau
Transport Processes in Wood (1984)

R. R. Archer
Growth Stresses and Strains in Trees (1986)

W. E. Hillis
Heartwood and Tree Exudates (1987)

S. Carlquist
Comparative Wood Anatomy (1988)

L. W. Roberts / P. B. Gahan / R. Aloni
Vascular Differentiation and Plant Growth Regulators (1988)

C. Skaar
Wood-Water Relations (1988)

J. M. Harris
Spiral Grain and Wave Phenomena in Wood Formation (1989)

B. J. Zobel / J. P. van Buijtenen
Wood Variation (1989)

P. Hakkila
Utilization of Residual Forest Biomass (1989)

J. W. Rowe
Natural Products of Woody Plants, (1989)

P. Hakkila

Utilization of Residual Forest Biomass

With 240 Figures

Springer-Verlag Berlin Heidelberg NewYork
London Paris Tokyo Hong Kong

Dr. PENTTI HAKKILA
Finnish Forest Research Institute, Unioninkatu 40A, SF-00170 Helsinki, Finland

Series Editor:

TORE E. TIMELL
State University of New York, College of Environmental, Science and Forestry,
Syracuse, NY 13210, USA

Cover: Transverse section of *Pinus lambertiana* wood. Courtesy of Dr. Carl de Zeeuw, SUNY College of Environmental Science and Forestry, Syracuse, New York

ISBN-13:978-3-642-74074-9 e-ISBN-13:978-3-642-74072-5
DOI: 10.1007/978-3-642-74072-5

Softcover reprint of the hardcover 1st edition 1989

Typesetting: K+V Fotosatz GmbH, Beerfelden;

2131/3020-543210 – Printed on acid-free paper

To Pirkko

Preface

An increase in the demand for wood results in improved recovery and less residual biomass in the forests. Paradoxically, interest in forest residue as a renewable source of raw material seems to be in a reverse ratio to its availability in a certain area.

Finland and Sweden are probably more dependent on forestry and forest industries than any other developed countries in the world. A sufficiency of raw material for integrated forest industries is vital for the national economy of both countries, and a great deal of attention is being paid to the long-term potential of unutilized biomass left behind in logging operations. Furthermore, since these countries possess no reserves of fossil fuels, and since their per-capita consumption of primary energy is exceptionally high, they also consider unmerchantable forest biomass a realistic source of indigenous energy.

A joint Nordic research project on harvesting and utilization of logging residue was carried out in 1969–1976 under the auspices of the Nordic Research Council on Forest Operations. This fruitful cooperation soon gave rise to related national projects in Sweden, Finland, Norway, and Denmark, stimulating further research and producing practical applications. Concurrently, particularly after the worldwide energy crisis in 1973, research on all aspects of utilization of forest biomass mushroomed in the United States, Canada, and the Soviet Union. An explosive increase occurred in both the number and diversity of biomass studies. According to Young (1977), world literature on forest biomass quadrupled in 6 years in the mid-1970s. This ever-accelerating trend seems to continue.

The Editor of the then new Springer Series in Wood Science, Professor T.E. Timell of the State University of New York, surprised me in May, 1982, with a letter offering an opportunity to produce a monograph on the utilization of forest biomass. I took his letter as a great honor, and recognized the need for a holistic presentation of the problem area. However, I personally knew many distinguished scientists with much more suitable background and experience to write the book in English.

Gradually, after a thorough examination of this opportunity I understood it as recognition of the joint Nordic efforts for the improved utilization of all parts of trees. Since only a small fraction of the Nordic literature is reported in the major world languages, these results are not well known internationally. When I realized that the work could now be reviewed in an international forum and integrated in a monograph of world literature, with some hesitance I decided to accept the offer.

This book grew out of forest technology research at the Finnish Forest Research Institute (FFRI) and a 3-year project on Complete Utilization of Forest Biomass, financed by the National Research Council for Agriculture and Forestry of the Academy of Finland. The research program was financially supported by SITRA, the Finnish National Fund for Research and Development. The exchange of ideas and information with distinguished foreign scientists has been an essential part of past work. Among the many colleagues who over a period of 20 years contributed to the research programs referred to above, I wish to express my particular thanks to Harold E. Young, John R. Erickson, Peter Koch, and Rodger A. Arola of the United States of America; Imants Ievin of the Soviet Union; P.O. Nilsson of Sweden; Olav Gislerud of Norway; Wolfgang Knigge of the Federal Republic of Germany; and Veijo Heiskanen, Bror-Anton Granvik, Matti Kärkkäinen, Ilpo Palenius, and N.E. Virkola of Finland.

The object of this monograph is to provide an interdisciplinary but still simple reference to various aspects of utilization of the presently unmerchantable parts of trees. The proportion of residue left behind in logging operations is largest in the virgin forests in the north and in the tropics where very few scientific data are available. Furthermore, since long transport distances and other serious constraints will probably prevent the recovery of forest residue from remote areas until some future time, this book is primarily a review of forest residue potential in countries where forestry has a long tradition, and where the forests are under regular silvicultural management. Because most of the scientific data and practical experience is from northern Europe, *Pinus sylvestris* (Scots pine) and *Picea abies* (Norway spruce) are the species most frequently mentioned in the text. Had the book been written by another author from another geographic region, the emphasis would most probably have been different.

The goal is a balanced review of scientific literature and recent practical development in the field of forest biomass utilization. The scope covers aspects of forest management, mensuration, ecology, logging and transport, wood technology, processing, and utilization.

Although the use of residual biomass is, to a great extent, an economic problem, with some minor exceptions this review does not include an economic evaluation of alternative methods of biomass utilization. Due to differences in the cost of labor and capital, logging technical and socioeconomic conditions, and varying exchange rates between countries, the inclusion of economic comparisons was not considered feasible. Because the intention is a brief, but comprehensive presentation, a cautious generalization of complex phenomena has been required.

The production of this book would not have been possible without financial support from the Foundation for Research of Natural Resources in Finland and the Finnish Cultural Foundation. Their generous contribution provided me with the much needed opportunity to concentrate exclusively on writing during the final phase of the work. I was fortunate to enjoy the steadfast support of the Finnish Forest Research Institute and its Director, Professor Aarne Nyyssönen. The Institute kindly permitted me the use of its published and unpublished scientific material in writing the book.

Many staff members of the Department of Forest Technology, Finnish Forest Research Institute, contributed to the book. Hannu Kalaja was extremely effective and helpful in the collection of field material over many years. Hannu Aaltio was responsible for the data processing of the several unpublished Finnish biomass studies used in Chapter 2. Maija Tuuri, with the assistance of Heidi Koskinen, patiently, carefully and with dedication typed the manuscript, not only once, but several times, as the text underwent new revisions. Leena Muronranta made most of the line drawings, always with care and skill. Pentti Sairanen briefed several Russian publications on biomass utilization. Timo Heikka, Risto Helkiö, Pirkko Kinanen, Tapio Nevalainen, Raija Siekkinen, Aune Suinuniemi, and Jaakko Sumanen assisted in many ways. Special appreciation is expressed to Kaija Kanninen who always knew how to make things happen. Without her encouragement, thoughtfulness, and unselfish sacrifices, far beyond the call of duty, I would not have been able to complete the work.

Several colleagues reviewed parts of the manuscript. Their constructive criticism and advice were most useful in pointing out inaccuracies, enabling me to remove many imperfections. For their part in reviewing one or more chapters, my warm thanks are due to Aaro Alestalo, Olli Eeronheimo, Rihko Haarlaa, Osmo Liiri, Bruno Lönnberg, Eino Mälkönen, Seppo Kellomäki, Raili Voipio, and especially to Harold E. Young and Matti Kärkkäinen. I am also indebted to Anne Johnson, Marita Näsman-Repo, and Valerie Vainonen, who checked and improved the English language.

Illustrations were contributed from many sources, which are mentioned in the figure legends. I am particularly indebted to T.E. Timell and W.A. Côté of the SUNY College of Environmental Science and Forestry, Syracuse, NY USA, for the illustrations of the structure of wood in stem, branches, and roots. Photographic material was submitted also by Rodger A. Arola of the Forestry Sciences Laboratory at Houghton Mich., Allan Bruks of Bruks Mekaniska AB, Ken Harrison of Koehring Canada, Imants Ievin and V. Lazdan of Scientific and Industrial Research Association SILAVA, Hannu Kalaja and Erkki Oksanen of the Finnish Forest Research Institute, Peter Koch of Wood Science Laboratory Inc., Timo Kyttälä of Lokomo Forest Oy, V. Lauren of Mackmyra Cellulosaflis AB, Göran Lönner, Jonas Palm, and Kerstin Tordmar of the Swedish University of Agricultural Sciences, Arthur Riedacker of Agence Française Pour la Maitrise de l'Énergie, Pekka Saranpää of the University of Helsinki, and many others. John R. Erickson and Andy Baker, U.S. Forest Products Laboratory, kindly supplied me with many important reference publications. I am also grateful to the authors and publishers who kindly gave me permission to reproduce previously published figures and tables.

It gives me great pleasure to express my gratitude to the publisher, Springer-Verlag, and the Editor, T.E. Timell, for the opportunity to produce this book in the Springer Series in Wood Science. Their patience, flexibility, advice, assistance, and encouragement from the beginning to the very end of the project made the experience inspiring and pleasant.

Finally, I wish to express my deep gratitude to my mother and my family for their unceasing understanding and support. My wife, Pirkko, had to sacrifice a great deal during the years I prepared the book. Not only did she give her support and encouragement for the work in all possible ways, but she also gave me linguistic advice and even made the attractive operational drawings which illustrate biomass harvesting systems much more clearly than would have been possible to express in words. Without her dedicated support the book would not be a reality.

Puuksanharju Peninsula, Finland
May 1989

P. HAKKILA

Contents

Chapter 1

Introduction

Plants of an ecosystem produce carbohydrates and many minor compounds from carbon dioxide and water through photosynthesis. The total amount of organic matter produced by photosynthesis is called *gross primary production.* Part of these photosynthates is consumed by the plants in respiration. The remainder is incorporated into the plants and referred to in plant physiology as *net primary production.* However, organic matter falls from trees as litter, is sloughed as dead roots, or is grazed by insects and herbivores. Therefore, in applied forestry, and for the purposes of the present review, production of a forest stand is defined as the change in the accumulated biomass between two measurements. *Forest biomass* can be defined as the accumulated mass, above and below ground, of the wood, bark, and leaves of woody shrub and tree species both living and dead. As such it is a significant component of the total biomass of the plant kingdom (Young 1980b).

The world population expansion and the rising standard of living place increasing pressure on forests and the demand for wood-derived products. Fortunately, the world still possesses enormous reserves of largely untouched resources. There is a vast pool of interchangeable materials for an almost infinite variety of uses. Backed by this hypothetical world pool of resources, forests could play a much greater role than they are accorded. Wood has always been and remains a vital element for mankind, essentially for the following three reasons (Glesinger 1977):

1. Wood is *versatile,* which makes it an almost universal raw material;
2. Wood is *renewable* and, in contrast to minerals and fossil fuels, not threatened by exhaustion;
3. Wood offers almost unlimited *possibilities of expansion.*

The wood-processing industries procure their raw material very selectively. While all important species of the boreal coniferous zone are merchantable, only a small fraction of the tropical tree species is accepted. Even among the preferred species, industrial use is limited to the stem, or actually only to its best part, which meets given quality and dimensional requirements, whereas few of the large variety of biochemical products potentially available in the tree crown are utilized. However, the concept of *merchantable stem* has changed over the decades and may presently include timber which was once culled because of species, size, or defects.

In forest resource utilization, yesterday's residual is today's raw material. The first loggers generally take only the best logs from the best trees, and leave the rest. In North-American pulp mills, for example, as recently as in the early 1940s, workers carefully drilled the knots out of hemlock cants and then fed the cants directly into the chipper. The use of anything but the cleanest knot-free hemlock was as strongly resisted then, as the use of today's forest residuals is resisted now.

But the potential for making today's residual into tomorrow's furnish is too great to ignore (Trawers 1976).

The basis of the forest products industry yesterday was the merchantable stem, whereas tomorrow, it may be the forest biomass (Keays 1975). Against the background of the well-established merchantable stem concept, at the beginning of the 1960s, Young (1964) presented the *complete-tree concept* embracing both biological and technological investigation of the entire tree from root tips to leaf tips. The tree stem is traditionally measured by volume because of the simplicity and relatively high accuracy of the scaling. Both of these features disappear in the face of the odd and small size of branches, leaves, and roots. Hence, mass, either fresh or dry, becomes a more appealing unit to quantify all components of a tree. Furthermore, in the production of pulp, chemicals, and energy, mass is the most feasible measure for value estimation from the yield and processing point of view. For practical reasons, particularly when volume measurement is used also for the end product, it must be, and still is, used as an alternative to weight scaling of forest biomass.

With the inclusion of successional species, small trees, and shrubs, Young (1968, 1975, 1978b, 1981) later extended the complete-tree concept into the *complete-forest concept*. This concept is defined as the biological and technological investigation of all woody shrub and tree species, from root tips to leaf tips, for intensive management and utilization of a portion of the forest for solid products such as boards and planks, reconstituted products such as paper and particleboard, solid and liquid fuel, food and animal fodder, and feedstocks for the chemical industries. To this must be added the basic requirement that all forests must be managed for clean air, clean water, and recreation whether or not products are removed. The complete forest concept is incorporated more often into the management and utilization of hardwoods than softwoods (Young 1977a). However, although intensive crop production and utilization of everything is almost standard practice in the broad field of agriculture, this expanded view of biomass utilization is not yet widely accepted in forestry.

The traditional forest inventory often recognizes only that part of a tree which is considered usable by the forest industries. Although the inventory data thus do not directly cover unmerchantable tree components, they can still be used for rough biomass estimates. Accordingly, compared to the inventoried stem mass, the actual tree biomass stock of exploitable, closed forests in Canada is 41% higher, in the Soviet Union 46% higher, and in the United States 56% higher (United Nations 1985). The corresponding figure for Europe is 49%, as shown at page 3 (European Timber ... 1986).

Forest industries not only leave tree biomass in the forests, but also fail to use the raw material at the mill completely. As long as primary and secondary *industrial residues* are left unutilized, large-scale recovery of *forest residue* is hardly feasible. Industrial residues such as bark, saw dust, slabs, cores, and lignin-based waste liquor are much more homogeneous than forest residue, and the cost of delivery is often negligible. In integrated forest industry enterprises the residue of one mill is often valuable raw material for another, and if it cannot be used for processing, it is still of value for the production of energy. Therefore, they are often called by-products rather than residues.

	Forest biomass stock of Europe	
	Million tons oven dry	Percent of inventoried material
Inventoried material:		
Wood	6575	89
Bark	796	11
Total	7371	100
Other above-ground biomass	2127	29
Total above-ground biomass	9499	129
Stumps and roots	1482	20
Total tree biomass	10982	149

Careful planning, modern processing technology, and integration can make utilization of industrial residues almost complete. Economic preconditions for the processing or conversion of forest residue in a given area are improved when this stage of raw material utilization is approached. For example, when practically the entire allowable annual cut of stem wood and all forest industry residues were needed and used for industrial, rural, and energy purposes in Finland for the first time in the early 1970s, and many pulp mills seemed to have been threatened by a raw material shortage, only then was serious attention given to forest residue as the last, untapped, raw material reserve readily available for expanding the forest industries.

Utilization of forest residue plays a central role in the complete-tree concept. The primary aim is to supply industry, power plants, and other consumers with additional biomass. Simultaneously, the recovery and utilization of forest residue offers other important benefits as well. In fact, even in a surplus situation where the raw material balance does not require utilization of undersized trees, tops, branches, stumps, or roots, other advantages connected with residue recovery may make utilization feasible. The most obvious of these are listed below.

Utilization of logging residue promotes *rationalization of small-tree harvesting* and thus improves the productivity of work. Fast adoption of whole-tree chipping methods in the United States at the beginning of the 1970s in fact was partly a result of an acute shortage of manpower for forest work. Despite the increase in biomass yield and bulkiness of the raw material, integration of crown recovery with harvesting of merchantable stem may reduce the cost of logging of conventional timber substantially. This is mainly because delimbing, which is a significant cost factor, can either be totally avoided or transferred from the woods to a centralized processing yard or mill, where single-tree treatment is usually replaced by multiple-tree treatment.

Both naturally and artificially regenerated forests all over the world produce large quantities of silviculturally undesirable, small-sized trees. Forest management programs promote removal of unnecessary stocking, initially in early precommercial, and later in commercial thinnings, to reduce suppression mortality and to make way for the best trees selected for growth. But if there are no markets for low-quality timber and small trees, even later thinnings may remain uncommercial. Then only too often the trees are left standing. Resulting overstocking, in terms of producing high-quality trees, may seriously retard the development of the stand. Because whole-tree logging essentially lowers the harvesting costs of small-sized trees, it may *promote implementation of silvicultural projects* and, consequently, improve the possibility for production of high-quality timber. In addition to thinnings, whole-tree logging may also stimulate conversion of poorly productive stands of inferior species to natural stands of desired species or to plantations.

Utilization of forest residue may also offer advantages to *environmental hygiene*. Intensive recovery of biomass reduces fire hazards in cutovers. Slash burning, which is restricted by law because of the danger of air pollution in many regions of North America, becomes unneccessary under effective biomass recovery on regeneration areas. Moreover, if forest biomass is substituted for fossil fuels on a large scale, sulfur emissions are reduced and the long-term carbon dioxide balance of the atmosphere is improved.

Although whole-tree logging increases the productivity of forest work, it does not necessarily result in fewer jobs in forestry or elsewhere. On the contrary, increasing the profitability of small-tree operations creates *new job opportunities*. For example, in Finland, it is estimated that the use of undersized trees and logging slash as fuel for 1- to 10-MW chip-fired heating plants creates sorely needed jobs in rural areas as follows:

1. The building phase of the chip-fired heating plant temporarily creates a demand for manpower to manufacture and install the equipment and to construct the plant.
2. The heating plant offers permanent jobs. The average manpower requirement is 1 man-year per 2000 m^3 solid chips or 385 tons of oil equivalent (Toropainen 1982).
3. Chip procurement offers permanent jobs in the woods and on-highway transport. The manpower requirement varies greatly with the tree size, logging conditions, professional skill of the workers, degree of mechanization, and the logging method. An average figure in Finland, covering the operations of professional forest workers as well as those of self-employed forest owners, is 1 man-year for delivering from stump to heating plant 600 m^3 solid fuel chips from small-sized trees (Hakkila 1984).
4. Procurement of fuel chips and operation of the heating plant creates indirect jobs. These are estimated at 90% of the direct jobs for the building phase, and 130% of the number of permanent jobs (Toropainen 1982).

In forest-rich countries residual biomass forms *an indigenous energy reserve*. It increases the energy self-sufficiency of forest industry enterprises and the entire country, thus creating a secure supply against times of crisis, but this requires, of course, adequate technical preparedness.

Substituting unmerchantable biomass for imported fuels is beneficial for the local and national economy. Table 1.1 shows the monetary net benefits, by beneficiary groups, due to improved employment, increased stumpage, and better

Table 1.1. Average net benefits to the national economy from chip-fired district heating plants in Finland at the beginning of the 1980s, by beneficiary groups, in US$ ($ 1 = 4.50 Finnish marks) per MWh of produced heat (Hankala and Toropainen 1985)

Beneficiary group	Net benefits (US $/MWh)			
	Before completion of heat distribution network		After completion of heat distribution network	
	Plant size (MW)			
	≦1	>1	≦1	>1
Heat consumers in own municipality	−7.69	−5.24	−2.46	−1.49
Own municipality and its other population	+17.28	+10.22	+17.16	+12.93
Other municipalities and their population	+0.09	+0.93	+0.11	+1.27
State finances	+4.36	+0.84	+4.57	+1.92
Total	+14.04	+6.75	+19.38	+14.63

silviculture which result from the use of fuel chips in local district heating plants in Finland. Although substitution of chips for oil, if not subsidized, actually increased the expenses of private heat consumers, it simultaneously produced significant savings for other beneficiaries. However, full benefits were not achieved until the heat distribution network was sufficiently large to utilize the whole capacity of the plant.

Despite its many benefits, intensive recovery of forest biomass for industrial processing or large-scale energy production is still an exception. Various economic barriers and the risk of reducing the production capacity of forest soil due to accelerated nutrient loss are some of the factors constraining utilization of residual forest biomass. The roles of traditions, attitudes, and lack of information are also significant.

In setting policy and strategic objectives for biomass recovery and utilization, the following factors must be taken into consideration in every country. They all add up to the selection of an appropriate forest management pattern and harvesting technology (Spiers 1984).

1. Economics. The value of various components of biomass and the relative costs of labor and machines differ widely in various parts of the world. An understanding of these differences must be a prime consideration when selecting systems or components.

2. Environment. Not only do the physical factors of the terrain in any forest area affect man and machine performance in the short term, but the ecological relationship between them and the harvesting system influence soil and water, as well as recreational and aesthetic values. A balance must be achieved between profitable productivity in the present, and desirable land management in the long run.

3. Human Resource. The availability, level of knowledge, and training of the work force must influence the selection of a system, and its health and well-being must be considered in selecting the components.

Thus, the intensity of biomass recovery as well as the appropriate harvesting and processing technology are the result of many ecological, socioeconomic, geographic, and technical factors, which vary from country to country, and frequently within a country. Technology should therefore be applied with caution, and generalization of biomass data must be avoided when possible.

The use of research results from biomass utilization studies is usually restricted to specific local applications, but a few distinguished scientists deserve special credit for stimulation of development worldwide. *Harold E. Young* (Fig. 1.1) started his work for complete-tree utilization at the University of Maine at Orono, Maine, in 1959. In addition to his pioneering, personal research on the potential and inventory of forest biomass, he has promoted international understanding and exchange of utilization-oriented information in a unique way, especially within the framework of IUFRO and through personal contacts. The late *John L. Keays* (Fig. 1.2) analyzed the utilization possibilities of unmerchantable top, foliage, branches, crown, stump, and roots in a landmark series of five review publications at the Western Forest Products Laboratory in Vancouver, British Columbia, in the early 1970s (Keays 1971a–e). This thorough literature review of complete tree utilization continues to be a useful source of biomass data. *Imants Ievin* (Fig. 1.3) has conducted a unique, comprehensive biomass utilization program in the Latvian Scientific Research Institute in Riga, Latvia, since the 1960s. The achievements are especially significant and encouraging in

Fig. 1.1. Harold E. Young. Professor Emeritus, University of Maine, Orono, Maine

Fig. 1.2. John L. Keays (1914–1976). Director, Pulping Division, Western Forest Products Laboratory, Canadian Forestry Service, Vancouver, B.C. Canada

Fig. 1.3. Imants Ievin. Director of the Latvian Scientific Research Institute, Riga, Latvia

Fig. 1.4. John R. Erickson. Director of the U.S. Forest Products Laboratory, Madison, Wisconsin

Fig. 1.5. Peter Koch. President of the Wood Science Laboratory Inc., Corvallis, Montana

Fig. 1.6. P.O. Nilsson. Professor of Forest Technology, The Swedish University of Agricultural Sciences, Garpenberg, Sweden

the field of foliage utilization. *John R. Erickson*, together with his colleagues (Fig. 1.4), developed harvesting, comminution, and upgrading of biomass from low-value trees at the Forestry Sciences Laboratory at Houghton, Michigan, in the 1970s. As a research manager, he continues to be a strong supporter of biomass research and international cooperation. *Peter Koch* (Fig. 1.5) designed and developed entirely new, highly mechanized systems for recovery of both above-ground and below-ground biomass at the Southern Forest Experiment Station at Pineville, Louisiana, in the 1970s. All aspects of the complete-tree utilization of southern pines and hardwoods growing on southern pine sites are discussed thoroughly in his two monumental handbooks (Koch 1972a, b, 1985a, b, c). *P.O. Nilsson* (Fig. 1.6) has headed many successful national and international research and development projects on improved biomass utilization at the Swedish University of Agricultural Sciences at Garpenberg, Sweden, since the mid-1970s. As Chairman of the IEA forestry energy harvesting program, and as Coordinator of IUFRO Division 3, he has contributed in an exceptional way to forest energy research and smooth exchange of information between scientists all over the world. Recently, *T.E. Timell*, State University of New York, published a massive three-volume book on compression wood which also includes a comprehensive review of the technical properties of coniferous branch and top woods and their suitability for pulp and paper production. This much needed book, *Compression Wood in Gymnosperms* (Timell 1986a, b, c), is an inspiring encyclopedia for both research and practical planning for improved utilization of residual forest biomass.

Despite the recent progress in wood and forest science, forestry, forest industries and forest machine manufacturing, available knowledge is still quite limited on many central aspects of forest biomass utilization. This state-of-the-art report is an attempt to review the general basis, to present possibilities, and to discuss the long-term potential of recovery and utilization of residual forest biomass. Answers are sought to the following questions, mainly for European and North American forestry conditions.

1. What is the potential of residual forest biomass as a renewable raw material reserve (Chap. 2)?
2. What are the technical properties of residual forest biomass (Chap. 3)?
3. How can residual forest biomass be recovered and transported to the point of utilization (Chaps. 4 to 7)?
4. How can residual forest biomass be upgraded and used for industrial raw material, fuel, or fodder (Chap. 8)?
5. What kind of ecological consequences may result from intensive recovery of forest biomass, and how can possible risks be reduced (Chap. 9)?

Chapter 2

Residual Forest Biomass as a Raw Material Reserve

A biomass study may by undertaken for several reasons. These include collection of basic information for purposes of botany, wood technology, tree mechanics, ecology, nutrient cycling, timber measurement, value estimation, wage calculation, logging mechanization, fire hazard prediction, and utilization of forest biomass for energy, pulp, composite boards, chemicals, or fodder. The end use of the data definitely affects research methodology, sampling requirements, data processing, and conclusions. Therefore, a biomass study is directed according to the ultimate aim.

Biomass measurements have long been of principal interest to ecologists, whereas studies directed to recovery and utilization are relatively new to forestry. The bulk of the world literature on forest biomass is concentrated in ecological publications with emphasis on the annual primary productivity and nutrient cycling (Hitchcock and McDonnell 1979). However, efficient management, planning, and development of harvesting and processing machines and systems also presuppose comprehensive knowledge of forest biomass and its distribution among tree components. A key question concerning the utilization potential of unmerchantable biomass is its availability and distribution into components.

A clear, consistent, and uniform nomenclature is of utmost importance for reporting on biomass studies. However, it seems almost impossible to agree on a common, detailed classification of tree components due to differences in tree species, tree sizes, measurement methods, and application of biomass data. A suitable and widely accepted frame is offered by Young et al. (1964). The term *whole or full tree* includes all the components of the tree biomass above the stump cross-section, but excludes the stump and roots. The entire tree is referred to as the *complete tree* (Fig. 2.1).

Because the biomass components of a tree are difficult to define, any classification is bound to be arbitrary to a considerable extent. The biomass component nomenclature recommended by Keays (1968), with some minor modifications, is the most appropriate for the purposes of this review. Unfortunately, due to differences in the procedure for selecting and measuring the components in the reference studies, its strict application is not possible throughout the review. If not otherwise noted, the following classification is used:

1. *Unmerchantable top* of stem, henceforth usually simply *top*, is defined by local logging practice. For example, the bottom diameter of the top may vary from as much as 15–20 cm in some areas of British Columbia to 5–8 cm in Europe.
2. *Branch mass* includes all wood and bark of live and dead branches but is free of leaves, shoots, and reproductive organs of a tree. Branch mass is often divided into size classes by diameter, but class division varies from study to study according to conditions and needs.

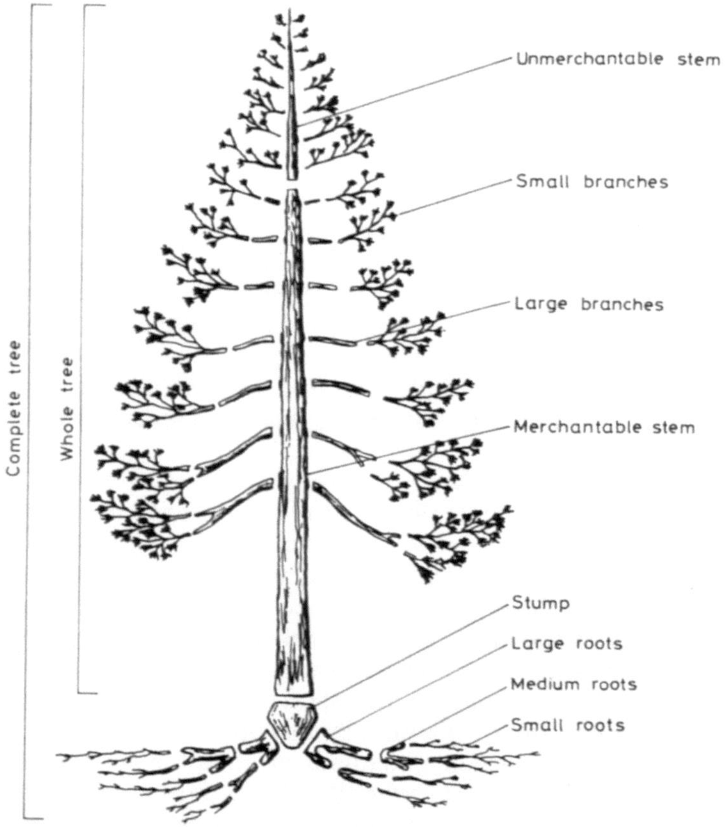

Fig. 2.1. The components of complete tree and whole tree (redrawn from Young et al. 1964)

3. *Foliage* includes all leaves and new shoots of branches. Reproductive organs are normally also included in foliage mass. In a few studies they are segregated from foliage and considered a separate biomass component. *Commercial or technical foliage*, a commonly used term in Russian biomass literature, includes not only the foliage but also twigs and branches of less than a specific diameter. A standard limit in the Soviet Union is 0.6 cm.
4. *Crown* is defined in the present review as all live and dead branches plus all foliage and reproductive organs. However, in many reference studies dead branches are excluded.
5. *Logging slash* in the cutover areas is defined in practice as not only the unmerchantable tops and crowns, but also undersized trees, culls, defective and broken stem parts, and any other tree biomass left unutilized on the ground in timber harvesting, except the stumps and roots.
6. *Stump* is the unutilized above-ground biomass below the bottom of the merchantable stem, and its under-ground projection, excluding the lateral roots. Of all tree components, the stump is the most difficult to define and stump biomass data the most laborious to measure and interpret.

7. *Roots* include all side or lateral roots but exclude the taproot, which is a part of the stump as a natural elongation of the stem. Like branch mass, root mass may also be divided into subclasses by diameter.
8. *Stump-root system* is comprised of the stump and roots.

Volume, with or without bark, is the traditional unit of measure for standing stems and conventional timber assortments. When tree components such as branches, foliage, stumps, and roots are considered, the standard methods for volume determination become impractical and almost impossible due to peculiar shapes and variable basic densities of the pieces. Therefore, mass rather than volume, is the most feasible unit of measure for tree components, a complete tree, or a complete forest.

On the average, roughly one-half of the total fresh mass of a living tree consists of water. However, moisture content varies widely from species to species, from tree to tree within a species, among tree components within a tree, and from week to week or even day to day depending on season, weather conditions, and storage of biomass. Although the fresh mass may be an appropriate basis for determination of, e.g., transport rates for biomass, it is not constant and is, therefore, insufficient for most scientific studies or for value estimation of raw material. Instead, the dry mass of a tree component or entire tree is a precise and stable unit of measurement.

Conversion of fresh weight to dry weight requires an estimation of moisture content. Collection of a representative and reliable sample from biomass components for moisture content determination is a difficult, time-consuming, and error-prone task even for an experienced scientist. The procedure for selection of trees and subsampling for moisture content within these trees is therefore of great importance. Using a large sample improves accuracy but increases costs, and may result in unacceptable disturbance of a forest stand.

In wood science, *moisture content* is customarily expressed *on the basis of dry mass*, i.e., the mass of water in wood in percent of the dry mass of the wood. In practical forestry and forest industries it is common to calculate the moisture content *on the basis of fresh mass* as received, i.e., the mass of water in wood in percent of the total mass of water and dry wood. Since the present review to a large extent is practically oriented, the moisture content of biomass is given on the basis of total mass of dry matter and water as received, unless otherwise indicated.

Comparisons between biomass data gathered from various tree species, tree sizes, and geographic areas are often best made by defining component mass in percent rather than in absolute mass. Relationships used most frequently in biomass literature are:

1. As a percentage of merchantable stem mass (with or without bark);
2. As a percentage of full stem mass (with or without bark);
3. As a percentage of whole-tree mass;
4. As a percentage of complete-tree mass;
5. As a percentage of complete-forest mass.

Keays (1971a) expressed the component mass as a percentage of stem mass, on a bark-free and dry-mass basis, and recommended that the resulting ratio be used as a standard basis for component comparisons. Full stem was preferred to merchantable stem because it is a fundamental characteristic of a tree, whereas the merchantable stem is dependent on a particular choice of minimum top diameter of timber. Keays also expected that tree-length, whole-tree and complete-tree logging systems would become established practice in the future, whereupon the present concept of merchantable stem would lose its meaning in the long run. Although this standard procedure would facilitate comparisons of research results, the proposal has not been commonly accepted, mainly because of the difficulties in determining bark-free component masses. Furthermore, while measurement of bark-free biomass might be desirable from the point of view of pulp industries, it is not feasible for energy calculations. Consequently, biomass components in the present review are expressed primarily as percentages of stem mass or whole-tree mass, all components including bark.

A clear understanding of the interactions and relationships of biomass components to the complete tree and complete forest will help to manage forests and optimally utilize the resultant products. *Forest Biomass* by Satoo and Madgwick (1982) is a valuable source of information on biomass production and methods of estimating forest biomass. The quantities and utilization properties of unmerchantable tops, foliage, branches, crowns, stumps, and roots were analyzed in five separate volumes by Keays (1971 a–e). The analysis, in which Russian biomass literature is also extensively covered, was revised and updated by Weldwood (1979 a, b) and now has a strong emphasis on biomass recovery as well.

The primary objective of Chapter 2 is to provide background information for harvesting and utilization rather than for planning silvicultural treatments. The aim is to explain the distribution of biomass in trees and stands of certain characteristics, passing over growth and yield questions. Due to differences in longevity between the tree components, tree and stand biomass data at the time of harvesting are in this conjunction more relevant than data on trees and stands which are to be cut in the far future.

2.1 Biomass Components of a Tree

Models of stem development in individual trees and entire stands are an important tool for forest research and forest management planning. However, for a deeper understanding of the behavior of forest trees, knowledge is required not only of the stem, but also of the crown and root system. Therefore, the models should be constructed for the whole or complete tree rather than the stem. These *tree models* could form a cornerstone for larger planning systems of stand management and biomass utilization.

Knowledge of the distribution of biomass into its main components in an individual tree is the basis for quantitative evaluation of forest biomass for a stand, logging site, or larger area. To establish relationships between the biomass components of standing trees and conventional measurable tree characteristics, the first phase of biomass studies usually concentrates on per-tree data. To reduce unnecessary conflicts to a minimum, and to encourage uniformity in approach and in the expression of results, Alemdag (1980) has made a proposal for a standard method of collecting field data and processing them in the laboratory.

The tree components dealt with are stem, merchantable stem, top, foliage, branches, crown, stump, roots, stump-root system, whole tree, and complete tree. The components may be broken down further to wood, bark, and foliage fractions, or to size classes. Biomass data of individual trees may be expressed either in the form of *mass or weight tables* or as *regression functions*. The strength of the functions is in their compatibility and ease of use with computers. Young (1976) published a summary and analysis of major weight table studies prior to 1976. Hitchcock and McDonnell (1979) later updated the list. Stanek and State (1978) compiled a listing of regression equations for calculation of the biomass of several, mainly Canadian, tree and shrub species.

The masses and outer dimensions of tree components are strongly correlated with one another. Due to this *allometric structure* of trees, the mass of a given tree component, such as crown or stump-root system, can be indirectly predicted from independent stem, tree, or stand variables. To ensure smooth application of weight tables and functions in conjunction with conventional volumetric inventories, the variables should be easily measurable. The most commonly used independent predictors can be grouped as follows:

1. Stem growth and form indicators:
   Breast height diameter/age
   Tree height/age
   Tree height/breast height diameter
   Increment of the last growth rings
2. Stem size indicators:
   Breast height diameter
   Tree height
   Stem volume
   Tree age
3. Crown size indicators:
   Crown length
   Crown ratio
   Stem diameter at the base of the live crown
   Crown surface area
   Crown class
4. Stand and site indicators:
   Stand age
   Stand density
   Stand history
   Site quality
   Soil type.

These variables can be used alone, combined, or transformed. The relations between a biomass component and the independent variables may be in natural, semi-log, or log-log form. Although the equations should be as accurate as possible, they should also be easy to use. Additivity between component equations is also desirable but is often achieved at the expense of accuracy of single-component functions. Methods for ensuring additivity of biomass components by regression were published by Kozak (1970).

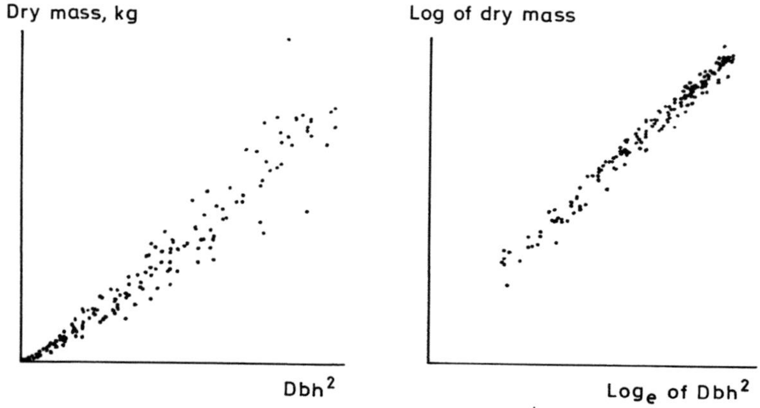

Fig. 2.2. Dry mass of whole Appalachian hardwood trees related to the square of breast height diameter (dbh) and its logarithmic transformation (redrawn from Wiant 1979)

Generally, variation in component masses is explained better by tree-level information than stand-level information. The *breast height diameter* (dbh) is the most appropriate independent variable for predicting green or dry masses of tree components. The relationship between component mass (y) and breast height diameter (D) is exponential. Consequently, the following allometric or power function is a widely accepted curve form:

$$y = b_1 D^{b_2} .$$

A frequently used convenient b_2 value is 2.0, presupposing biomass to be directly proportional to the basal area of a tree. However, this is rarely the exact case. For example, in a study of nine Appalachian hardwoods, b_2 values for green and dry whole-tree mass ranged from 2.4 to 2.7 (Wiant 1979).

The variance of component mass around its mean increases with the diameter. The model can be linearized by means of logarithmic transformation, which also results in a stabilization of the variance (Fig. 2.2). According to Madgwick (1976), the following models are used most commonly in world literature, when tree mass is related to breast height diameter:

$$y = b_0 + b_1 D^2$$
$$y = b_1 D^{b_2}$$
$$\log y = b_0 + b_1 \log D .$$

Marklund (1987) tested several analytical expressions of the breast height diameter in order to find one that would fit best to the above-ground components of *Picea abies* along a wide diameter range. Using log D was not sufficient and caused trouble when D was close to zero. The best equations for stem, foliage, and crown mass were obtained by using various kinds of diameter transformations which included a component formed by the sum of the diameter and a constant. For example, the best single variable explaining foliage or crown mass proved to be of the following form:

$$y = D/(D + b_0) .$$

Another stem size indicator used to predict total or component masses is *the tree height* (H) or height to a given top diameter. However, the breast height diameter generally explains the tree-to-tree variation of biomass components with essentially higher accuracy than does the height of the tree. The incorporation of height in the function, in addition to dbh, often does little to improve predictions. Although the crown mass of a tree naturally increases with stem diameter, height, and volume, it also tends to increase with an increasing taper within a given dbh class, i.e., with decreasing height and stem volume. This fact notwithstanding, tree height may improve biomass functions significantly under certain conditions. For example, the closest correlation for data on 12 tree species in Québec was obtained by applying the following nonlinear model (Quellet 1983):

$$y = b_1 D^{b_2} H^{b_3} .$$

A slightly closer correlation is obtained when a crown size indicator, rather than tree height, is used as an additional independent variable to predict component masses. Stem diameter at the base of the crown and *crown ratio* (C) are more effective than crown height in explaining component masses (Loomis et al. 1966; Hakkila 1971). A common equation type is the following:

$$y = b_0 + b_1 D^2 C .$$

If the purpose is to predict relative component proportions instead of absolute component masses, then the functions tend to improve when a combination of diameter and crown ratio is replaced by a combination of tree height and crown ratio. However, the deviation from the regression equation remains great.

Knowledge of component relationships is needed in order to work out biomass estimates for specific cases and to plan the collection of material for further biomass studies. The tree-to-tree variation in crown mass is wide, erratic, and a result of stand history. Limitations in the precision of component and tree mass estimates derived from weight tables and biomass equations are also partly a function of the representativeness and accuracy of the field sampling. Due to laborious fieldwork, the sample used in most biomass studies is rather small, often comprising only a few trees (Cunia 1979a). But even when a study is based on an extensive, representative sample, the confidence limits are still obviously bound to remain essentially wider than those for stem mass or stem volume, respectively. Crow and Schlaegel (1988) published a guide to help managers select the most appropriate equation available for a given situation for estimating tree biomass.

2.1.1 Tree Crown

The crown has many definitions in the literature. In addition to the branch mass and foliage, many publications also include the unmerchantable top in the crown mass. This makes comparisons difficult, since the minimum top diameter of merchantable timber varies from one geographic area to another, from species to species, and in the course of time. Dead branches are often excluded from the crown mass, particularly in ecological studies, because they have no important role for the functions of a tree. The present review defines *crown* as comprising all wood, bark, foliage, and reproductive organs in the live and dead branches of a tree.

Quantitative data on crown mass may be used for several purposes, but in the present review primarily to calculate the amount of residual forest biomass as a raw material reserve. Crown data are also useful in various ecological studies, and, for example, in conjunction with the following estimation tasks:

1. Effect of crown on the mass of whole tree, center of gravity of a tree, and delimbing difficulty during logging operations;
2. Additional mass and load volume caused by crown material during off-road and on-road haulage;
3. Effect of logging slash from crown material on difficulty of manual and mechanized regeneration operations in cutover areas;
4. Nutrient loss from forest soil caused by whole-tree logging and residue recovery;
5. Effect of biomass removal on nutrient leaching and watershed management;
6. Forest fire potential in cutover areas.

2.1.1.1 Development of the Crown

A forest may be defined as a collection of trees occupying a specific ground area and forming an ecosystem together with many other living and dead organisms in an inorganic environment, including mineral soil and atmosphere. The components of such an ecosystem interact with one another. An indispensable condition for a definition of a forest is that the trees are not independent as in a park or savanna but influence one another. The crowns of the trees in a forest ecosystem form a layer called a *crown canopy*. If the *forest is open* and the trees grow in isolation, foliage is produced almost to ground level, and as such trees mature, this foliage increases exponentially. When the canopy is continuous and the *forest is closed*, tree crowns are shaded by neighboring trees and the amount of foliage in a tree of a given diameter is greatly reduced compared with open-grown trees of the same species. Hence, trees growing in a closed forest and forming a crown canopy have a different shape from isolated, independent trees (Satoo and Madgwick 1982).

The competition between individual trees in a forest affects *tree architecture*. The extension growth of the primary axis of a branch in any one year is usually greatest at the top of the tree and decreases as one progresses downward in the crown. In a given branch it decreases as the tree grows in height while the location of the branch remains unchanged. For example, in a fast-growing *Pinus radiata* plantation in New Zealand, the maximum foliage production and maximum mass occurred in the second yearly growth zone from the tree apex up to the age of 6 years. In older plantations, the canopies were deeper and maximum foliage production occurred in the fourth annual growth zone or below. Maximum foliage and branch mass per zone occurred lower in the canopy. These changes in the vertical distribution of crown mass are a result of the transition from a shallow canopy comprising many densely spaced, intensely competing individuals to a less competitive distribution of similar foliage mass over fewer and taller trees (Madgwick et al. 1977b). In slow-growing boreal forests, maximum growth and concentration of crown mass occur in annual branch zones farther from the apex, but in principle the development of the canopy structure essentially follows the same pattern.

As leaf mass accumulates in a plant community, consumption of photosynthate by leaf respiration increases in proportion to leaf mass, whereas total photosynthesis by leaves reaches an asymptotic upper limit, dependent on radiation intensity. Hence, there is an optimum leaf mass for maximum production. Foliage mass in a stand increases with total biomass as long as the latter is small but gradually becomes independent of it (Monsi and Saeki 1953).

In a closed forest, competition is severe among individual tree crowns and even among branches within a tree. Total crown biomass will build up to an equilibrium value where production is balanced by loss through death and shedding. Due to differences in longevity of components, marked changes occur in the relative distribution of biomass. The equilibrium is dependent on stocking (Satoo and Madgwick 1982).

When a tree grows, older branches are left in the shade of the new upper crown of the same tree and the neighboring trees. At the base of the crown, consumption of photosynthate starts to exceed production and gradually leads to the death of lower branches. The dead branches are attacked and destroyed by fungi and insects. Some of the dead branch material falls to the ground but some remains on the crown. The pattern of branch fall varies among tree species. In hardwoods an entire dead branch usually breaks off at once, whereas in many softwoods the process advances more slowly, piece by piece. Adherence of dead branch stumps to the stem depends on tree species, spacing of trees, and the size and shape of the branch (Knigge and Schulz 1966).

The branchiness ratio of a tree, defined as the ratio of crown mass to stem mass, is a function of stand density. The accumulation of stem and crown mass is related to the growth space as indicated by the crown projection area of a tree. The above-ground biomass of a *Pinus sylvestris* tree in a closed 15- to 20-year-old stand in Finland was halved when the number of trees per hectare was doubled (Kellomäki and Väisänen 1986). If a plantation is dense at the outset, the canopy closes, branches remain thin, and the reduction of the live crown starts at an early phase of stand development. Figure 2.3 shows the effect of spacing on the crown form development of young *Larix leptolepis* trees.

In addition to stand density, other factors also affect the branchiness of trees. For example, the positive correlation between site fertility and average branch diameter is commonly known, although all species do not react in the same way. In a large sample of 17- to 26-year-old *Pinus sylvestris* trees in Finland, only 20% of variation in the average branch thickness of a tree was attributed to stand density. Inclusion of site fertility in the regression model, as indicated by the nitrogen content of the foliage, increased the coefficient of multiple determination to 74%. A further improvement to 83% was achieved by adding the calcium content of the foliage to the model. Within certain limits, the branchiness development of Scots pine may actually be more dependent on the site fertility than on stand density (Turkia and Kellomäki 1987).

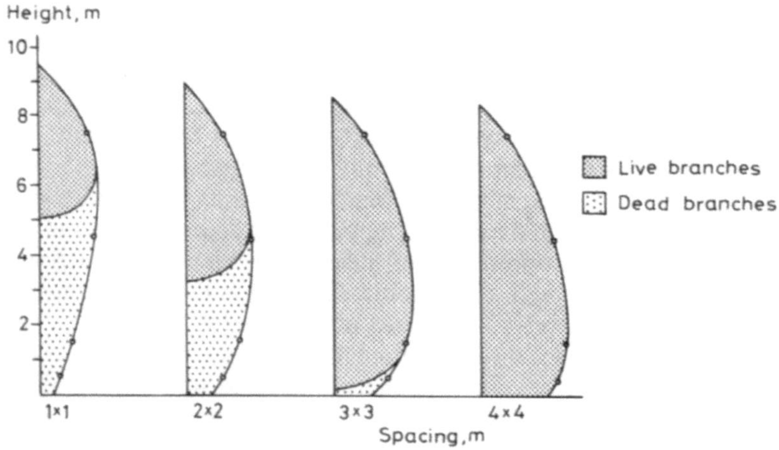

Fig. 2.3. Effect of spacing on the crown development of *Larix leptolepis* (Mayer-Wegelin 1955)

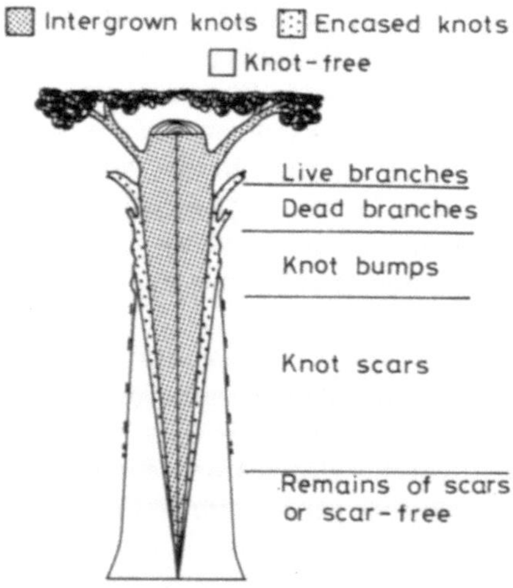

Fig. 2.4. Outer stem and branch characteristics as an indicator of knottiness inside the stem (redrawn from Schulz 1959)

A former branch section that is included in the wood of a stem by growth around its base is called a *knot.* Knots are undoubtedly the most common natural defect in timber. As long as the branch is living, the cambium of the stem and branch is continuous and the resulting knot is *intergrown or tight.* Once the branch has died, the continuity of the cambium is interrupted and the knot produced is *encased or loose.* The appearance of a knot in a piece of lumber depends on the direction of cut through the included branch. When the cut is made along the axis of the original branch, the knot appears as a spike knot on the surface of the wood. When a transverse cut is made through the branch, the knot appears round or oval (Kollmann and Côté 1968).

When a branch is pruned naturally or artificially, the branch end is grown over. Any stem wood produced subsequently will be knot-free, but *knot bumps*

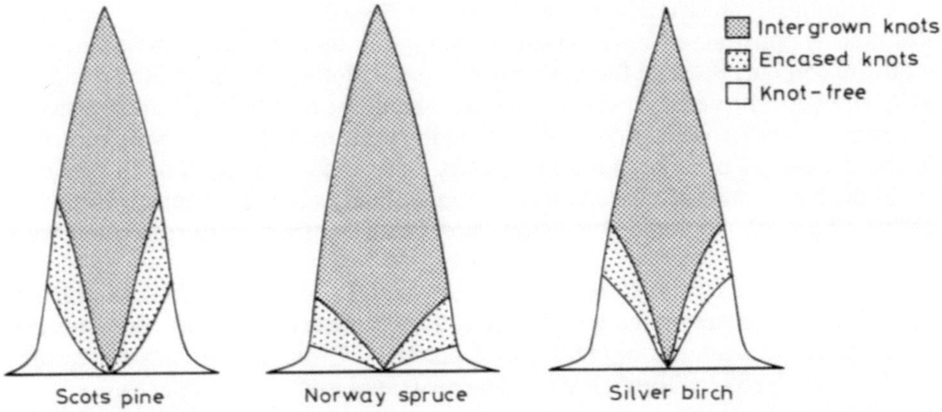

Fig. 2.5. Longitudinal stem section of *Pinus sylvestris*, *Picea abies*, and *Betula pendula* divided into knot zones. Average sawtimber trees from southern Finland, dbh 27 cm (redrawn from Kärkkäinen 1986)

and scars will still develop on the stem (Fig. 2.4). Intergrown knots appear in a stem zone with the form of a cone with the vertex down, surrounded by a mantle of encased knots. The base of this cone is at the height of the lower limit of the live crown. The quantity, quality, and angle of knots are specific to each tree species.

Computer models have been developed for predicting *knottiness of stem wood*, i.e., the frequency, location, and quality of knots, in order to estimate the value of saw timber trees. The occurrence of intergrown and encased knots can be estimated by simulating crown development. The lower limits of live and dead branches in a tree are used in simulation models as external tree characteristics for division of a stem into zones according to knot properties. Figure 2.5 (Kärkkäinen 1986) shows an example of simulation results. *Pinus sylvestris* and *Betula pendula* represent light-demanding species, whereas *Picea abies* is a shade-tolerant species with a long, live crown. The zone of intergrown knots is by far the largest in spruce stems.

2.1.1.2 Foliage Mass

Foliage is the assimilation organ of a tree. Therefore, it plays a crucial role in the tree's vital functions. Knowledge of its mass, moisture content, chemical composition, behavior, and relation to other tree components could provide clues for rational management of forests. Foliage mass should also be known when estimating nutrient loss in conjunction with residue recovery, when determining biomass composition in whole-tree logging, and when estimating the raw material potential for chemical or fodder industries. At the present time, foliage is usually considered a nuisance, but in the long run it may become one of the most valuable products of the forests.

Quantitative measurement of foliage is based on destructive methods. Research is constrained by problems of sampling and further subsampling for moisture content. Because the moisture content of foliage is high and varies diurnally and with the weather, research results should primarily be given in terms of dry mass. Another difficulty is the seasonal variation of foliage mass, not only in the deciduous trees but also in conifers with long-lived needles. In order to avoid the inconvenience caused by continuous changes in foliage mass during the period of growth, biomass studies on hardwoods are frequently concerned with leafless trees.

Considering its importance to the production of forests, information in silvicultural literature on the quantity of foliage is meager. It is even more meager in technical literature dealing with forest products. Utilization-oriented literature has been analyzed by Tomčuk and Tomčuk (1966) and Keays (1971a).

2.1.1.2.1 Moisture Content of Foliage

A tree needs water for assimilation and transport of nutrients and photosynthate. Water is taken from the soil by the fine roots and conducted upward in the outer sapwood of the stem by tension through transpiration in the foliage (Zimmermann 1983). Water moves in the leaves from wet cells into the intercellular spaces and further by diffusion of water vapor to the outside air. The amount of water transpired by trees and forests is large. For example, a *Picea abies* tree in Switzerland uses about 165 kg water per each kg of dry needle mass annually (Burger 1939a).

Since the moisture content of foliage is related to the transpiration function, it varies accordingly. Leaves dehydrate in the daytime, but during the night the moisture content increases again. The moisture content of *Quercus rubra* leaves in Wisconsin was 60% at 5 a.m. but only 56% at 2 p.m. (Chaney and Kozlowski 1969). A majority of the moisture content measurements for forest biomass presumably represent daytime conditions.

The transpiration function of the foliage results in moisture content differences within the crown of a tree. For example, the moisture content of *Picea abies* needles in Switzerland was 52% in the upper sun crown but 55% in the lower shade crown (Burger 1939a). In a *Fagus silvatica* stand, the average moisture content of the leaves was 61% in dominant trees, 62% in intermediate trees, and 63% in shaded, suppressed trees (Burger 1939a, b). On the other hand, although the old needles are partly shaded by younger ones, within a given branch they may have a slightly lower moisture content. In a study on *Pinus taeda* in the southeastern United States, the moisture content was 51% in the needles older than 1 year, and 52% in those younger than 1 year (Metz and Wells 1965).

Changes in the transpiration activity and structure of leaves during the growth period result in distinctive seasonal variations in the moisture content of the foliage. The moisture content of *Querqus rubra* leaves in Wisconsin was 75% at the end of May, 60% in mid-July, and 55% in early September (Kozlowski and Clausen 1965). Horntvedt (1983) found considerable seasonal variation in the moisture content of *Picea abies* needles. The lowest values were measured in the spring when the transpiration was high but the soil was probably still frozen. The low moisture content was thus partly a result of desiccation, but it was assumed to be affected also by a temporary accumulation of starch and soluble carbohydrates in the foliage.

Foliage moisture content is also affected by the weather, soil moisture content, vitality of a tree, and tree species. During June–August in Finland, the moisture content of the leaves of young, vigorous, 1–12-cm-thick trees was 70% in *Betula pubescens* and *Populus tremula* and 67% in *Alnus incana* (Simola 1977). In the United States, the moisture content of the needles of southern pines ranges from 50 to 75% with average values near 60% (Koch 1972a). In large material of *Pinus contorta* var. *latifolia* in western North America, trees with 76, 152, or 228 mm dbh had a moisture content of 54.2, 52.3, and 51.7%, respectively (Koch 1987).

2.1.1.2.2 Vertical Distribution of Foliage Mass

Tree crown can be divided into two parts on the basis of the light conditions in the foliage. The upper part is called *sun or light crown*, and the lower part *shade crown*. The hypothetical boundary is determined by the largest diameter of the crown (Fig. 2.6).

As a tree grows, the light conditions become less favorable in the lower crown parts. Defoliation starts in the inner shade crown, resulting in gradual death of lower branches and formation of a *leafless crown heart*. In a 35-year-old *Picea abies* stand in Switzerland, the leafless heart comprised 33% of the crown volume in the shade crown, but only 1% in the sun crown. The shade crown is thus char-

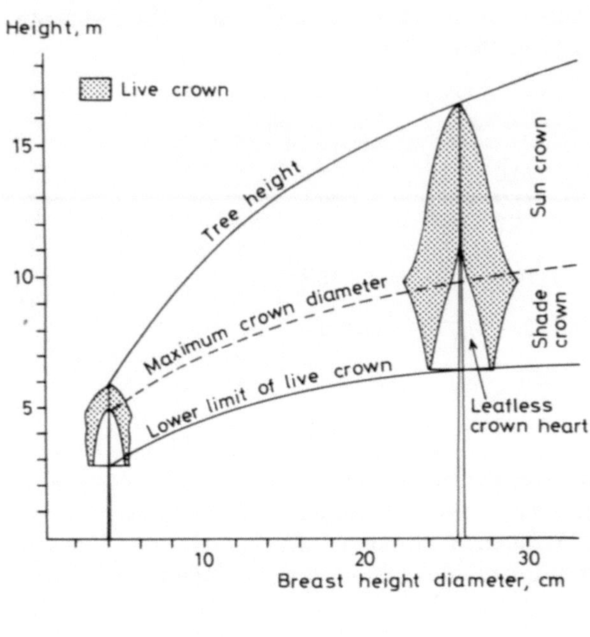

Fig. 2.6. A schematic presentation of sun crown and shade crown as a function of breast height diameter in a *Picea abies* stand (redrawn from Burger 1939a)

Table 2.1. The fresh mass of 1000 needles by whorls in seven *Pinus sylvestris* trees of various ages in Poland (Lemke 1978)

Sequence of whorl	Fresh mass of 1000 needles (g) Tree age (years)						
	18	26	36	46	56	70	90
1	28.1	37.1	73.3	59.5	48.8	31.4	44.1
2	33.0	41.4	65.1	44.8	53.3	30.6	45.9
4	17.7	28.5	51.0	50.4	37.4	23.9	26.4
6	12.1	23.3	39.4	42.7	36.0	21.4	32.5
8		12.7	37.3	32.6	25.3	21.8	28.9
10			18.9	21.5	26.3	18.0	33.0
12			15.9	24.1	23.9	17.0	29.3
14				18.8	18.3	17.2	29.3
16						17.5	29.6
18						13.2	27.1
20						12.9	23.3
22						13.1	19.0

acterized, not only by shaded light conditions, but also by less leaf mass per unit volume of crown (Burger 1939a).

This development is reflected further in the size of the needles. The heaviest needles are found just below the apex of the tree. The lower the location of a whorl in the crown, the lighter are its needles. Table 2.1 shows the fresh mass of 1000 needles by whorls in seven *Pinus sylvestris* trees of various ages in Poland. Although the average mass of a needle varies from tree to tree, the trend within a tree is clear. The fresh mass per needle in withering branches at the base of the live crown is only one-third of the mass of the uppermost needles. The sun leaves are thus larger in size than the shade leaves. This is also shown by the following data from Burger's (1939a) study on *Picea abies:*

	Sun needles	Shade needles
Needle length, mm	14.0	13.6
Surface area, mm^2/needle	47.9	39.1
Specific leaf area m^2/kg fresh needle mass	5.7	5.9
Specific leaf area m^2/kg dry needle mass	11.7	13.1
Number of needles/kg fresh needle mass	119000	151000
Number of needles/kg dry needle mass	246000	334000

Specific leaf area, i.e., leaf surface area per unit mass of leaves, thus increases with decreasing radiation intensity in the canopy. The sun leaves in the upper layer of the canopy have lower specific leaf areas compared with shade leaves. Another example is a *Metasequoia glyptostroboides* plantation in Japan, where the specific leaf area increased from 12 m^2/kg dry needle mass at the top of the canopy to 30 m^2/kg at the base (Satoo and Madgwick 1982). In a dominant aspen tree in Wisconsin, the specific leaf area was 12.0 m^2/kg in the upper third of the crown and 14.7 m^2/kg in the lower third of the crown (Zavitkowski 1971). The vertical change in specific leaf area parallels changes in the structure and function of the leaves.

A decision to favor or restrict the recovery of foliage is dependent on its utilization possibilities for chemical, fodder, or energy production but also on the fertility and sensitivity of the site under consideration. If only partial harvesting of foliage is considered feasible, the removal of foliage can be controlled by setting the directions for delimbing and topping accordingly. Information on the vertical distribution of leaf mass is thus of importance for a holistic system development for whole-tree logging.

Figure 2.7 shows the vertical distribution of dry needle mass in *Pinus sylvestris* and *Picea abies* trees of four height categories in a large sample collected from logging sites in southern Finland. The category of short trees primarily represents trees removed in early commercial thinnings. The second and the third categories represent later selective thinnings from below, and the third and the fourth categories represent final cuttings of repeatedly thinned stands. Each of the hypothetical *foliage mass profiles* are drawn from averages of 10–46 trees. Compared to single trees, the use of averages compacts the profile, making it somewhat too slim and high, but provides a telling characterization of the distribution of needle mass standwise.

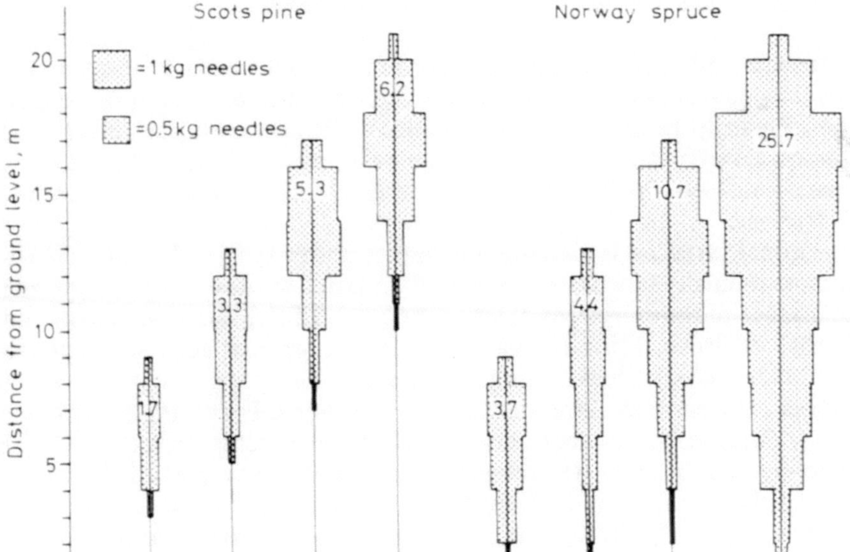

Fig. 2.7. Vertical distribution of dry mass of foliage in 9-, 13-, 17-, and 21-m-high *Pinus sylvestris* and *Picea abies* trees at the phase of removal in selective thinnings and final cuttings in southern Finland. Average dry mass of foliage indicated in kg for each height category

Table 2.2. The effect of topping on needle removal when harvesting *Pinus sylvestris* and *Picea abies* in southern Finland unlimbed

Tree height (m)	Reduction in needle removal (%)			
	Top length 3 m		Top length 5 m	
	Scots pine	Norway spruce	Scots pine	Norway spruce
9	56.0	47.3	94.7	79.4
13	44.7	30.2	80.1	57.9
17	40.0	23.0	76.4	49.2
21	26.9	10.9	64.2	28.9

The same data are used in Table 2.2 to show how needle removal from a forest may be reduced through topping of unlimbed trees. If a 3-m top is left on site and the rest of the tree is harvested unlimbed, needle removal can be reduced by 47–56% in early thinnings, 23–45% in later thinnings, and 11–27% in final cuttings, depending on tree species and size. The loss of woody branch mass would be essentially smaller, and no merchantable stem mass would be left behind.

2.1.1.2.3 Seasonal and Annual Variation in Foliage Mass

Both systematic and sporadic changes occur in the foliage mass of a tree in the course of time. During each growing season the mass increases up to the annual maximum. Typically, in mixed hardwood stands in Tennessee the peak is achieved at midseason (Harris et al. 1973). After the peak, some studies report a static foliar mass until leaf fall in the autumn, whereas others show an immediate gradual decrease.

The seasonal variation is naturally the largest in tree species shedding the entire leaf mass annually but, from the utilization point of view, the phenomenon is of importance in the evergreens as well. The range of the variation depends on the longevity of leaves, being the narrowest for tree species such as *Thujopsis dolobrata*, whose needles have a long life span.

Evergreens also carry the lowest foliage mass in winter. For example, like many other pines, *Pinus echinata* sheds one-third of its needle mass each year, and its summer foliage can be estimated roughly by multiplying the mass of the winter foliage by 1.5. However, the ratio is not constant due to erratic year-to-year variation. Weather, especially during the growing season, influences the quantity of needle mass produced during a year. Weather also affects the annual foliage fall through drought, high winds, ice storms, hail, or extreme winter cold. Favorable or adverse growing conditions in a particular year may thus influence both needle mass on the trees and needle fall for as long as 5 years (Loomis et al. 1966). Furthermore, a high seed crop may result in a decrease in needle production.

Table 2.3. The dry needle mass of a 25-year-old *Picea abies* tree in Germany by whorls and age classes 3 years after fertilization (Beemtsma 1979)

Whorl	Dry mass (g) Age of needles (years)							Total
	1	2	3	4	5	6	7	
1	7							7
2	77	6						83
3	254	70	11					335
4	134	53	55	5				246
5	170	85	146	38	3			442
6	124	73	214	101	28			540
7	97	84	181	118	39	8		527
8	112	74	176	131	62	36	0	591
9	19	15	68	64	27	12	–	205
10	24	24	79	56	24	11	0	217
Total	1018	484	930	513	183	67	0	3195

Table 2.3 is an example of year-to-year variation of needle production. The needle mass of a 25-year-old *Picea abies* tree, growing on poor soil in Germany, was analyzed from the ten uppermost whorls 3 years after fertilization. All age classes of needles were separated from whorl 1 to whorl 10, including the internodes. Large variations were caused by small whorls, the low mass of 2-year-old needles in Table 2.3 reflecting a very dry year. Although the variation in needle mass and nutrient concentration in the crown is basically fairly systematic by age and whorl, annual fluctuation in needle production may result in unpredictable variations.

Year-to-year variation in needle mass may also be caused by insects such as the spruce budworm (*Choristoneura fumiferana*) in the northeastern United States and eastern Canada. This defoliator feeds primarily on the current year's shoots, buds, and needles of *Abies balsamea* and spruces. In order to create an index for estimating a 5-year annual defoliation history of a damaged stand, Kleinschmidt et al. (1980) developed a model to predict the foliage mass at each age on a branch for a given whorl from undefoliated balsam firs.

Systematic seasonal and sporadic annual variations of needle mass create problems in biomass studies. The effect of these sources of error is seldom known. In order to avoid the sampling problems resulting from seasonal variations in foliage mass, biomass materials of hardwoods are frequently collected from naked trees during winter, but this results in loss of essential information.

2.1.1.2.4 Effect of Tree Size on Foliage Mass

The quantity of foliage is affected by a highly complex network of variables. Tree size is perhaps the most important factor for practical forestry and biomass utilization, but foliage mass is also strongly correlated to the current growth. Because tree size and age are closely interrelated, their individual effects are not easy to separate. This section is restricted to tree size.

Foliage mass increases rapidly with age and tree size as long as the tree grows on an open site without competition. The rate of increase depends on a variety of factors, including species, site, and stocking. For closed forests, many different trends and patterns are reported. Some studies show a buildup to a maximum stable total value, while others show slight decreases in the foliage mass of a stand after a certain maximum phase (Satoo and Madgwick 1982).

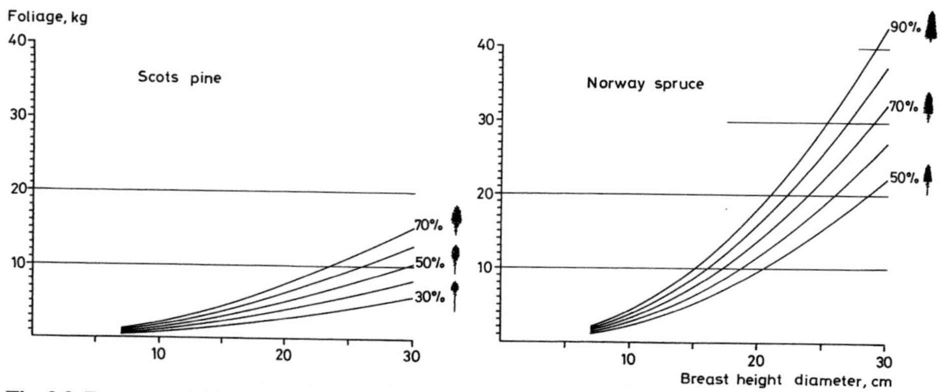

Fig. 2.8. Dry mass of foliage in *Pinus sylvestris* and *Picea abies* trees at the time of removal in selective thinnings and final cuttings in southern Finland, as a function of dbh and crown ratio

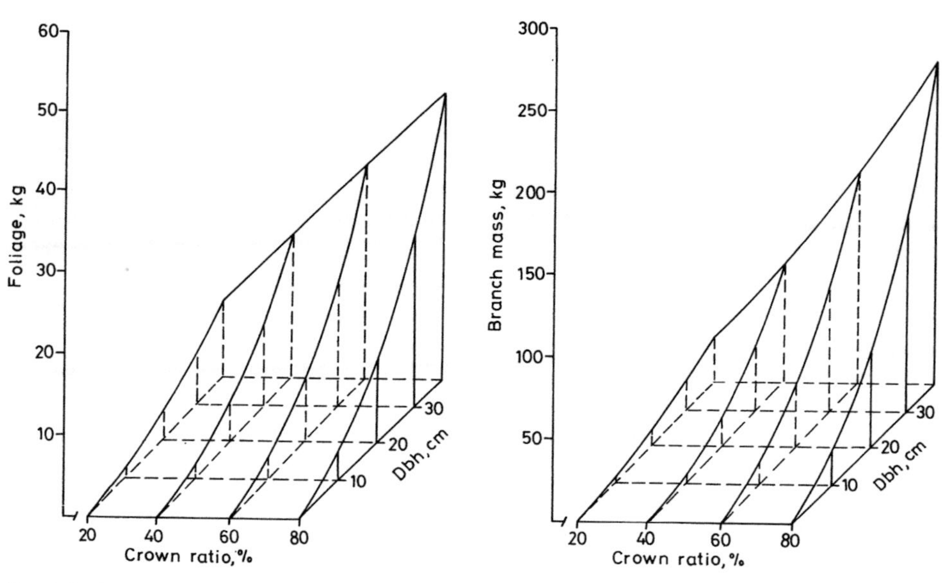

Fig. 2.9. Dry mass of foliage and branches in *Pinus echinata* in winter in Missouri as a function of dbh and crown ratio (redrawn from Loomis et al. 1966)

Figure 2.8 shows the dry mass of needles as a function of the breast height diameter and crown ratio of *Pinus sylvestris* and *Picea abies* at the time of removal during selective thinnings from below and final cuttings in managed forests in southern Finland. Because the trees in thinnings are primarily selected from suppressed and intermediate crown classes, the needle mass is probably higher in the remaining trees of a stand than in those removed. Scots pine trees with a breast height diameter of 10 cm had an average dry needle mass of 1.5 kg, and Norway spruce about 3 kg. The needle mass for 30-cm trees was 10 and 30 kg, respectively. The significant difference between pine and spruce reflects the tolerance of the species. In southern Finland the crown ratio is typically 50–55% in Scots pine but as much as 70–80% in Norway spruce. Moreover, the life span of needles in spruce is roughly twice as long as in pine.

Another example of the effect of breast height diameter and crown ratio on foliage mass is presented in Fig. 2.9. When used alone, both breast height diameter and crown height were poor estimators for foliage mass of *Pinus*

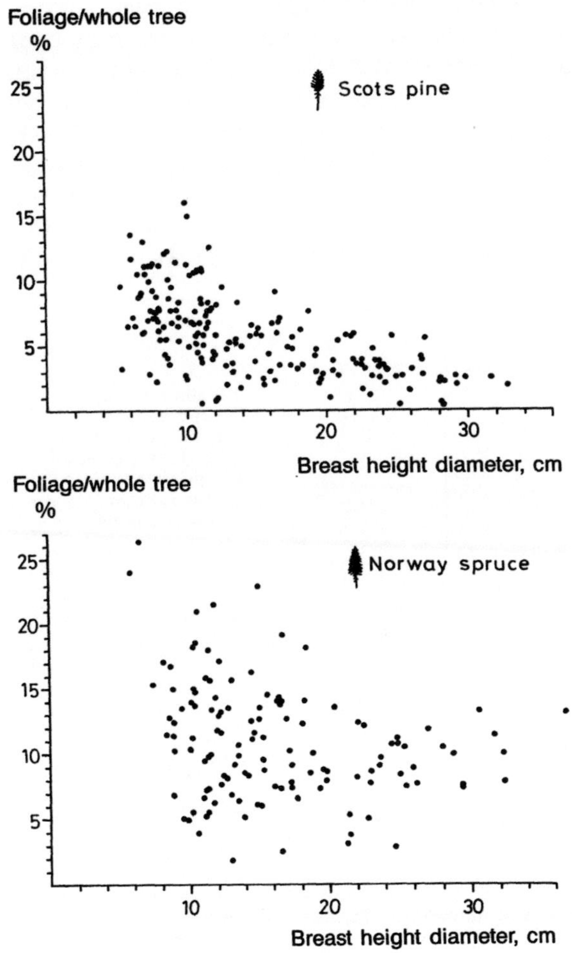

Fig. 2.10. Dry mass of foliage in percent of whole-tree biomass in individual *Pinus sylvestris* and *Picea abies* trees at the time of removal in selective thinnings and final cuttings in southern Finland, plotted against dbh

echinata. But when crown height was replaced by crown ratio and combined with breast height diameter to a single independent variable, the standard deviation from the regression line was significantly reduced (Loomis et al. 1966).

Quantitative data on the effect of tree size on foliage mass are available for many tree species. Even though a majority of the studies are based on a small number of samples, the information can be used for rough estimations of raw material availability, nutrient loss, etc. Comparisons between results from different studies are often easier if the data are given as foliage percentage. Figure 2.10 shows the variations in foliage percentage in whole-tree biomass on a dry mass basis in the study referred to in Fig. 2.8. An unweighted average value is 5.8% for *Pinus sylvestris* and 10.9% for *Picea abies.* The smaller the trees, the higher is the percentage of foliage. However, the variation is wide, especially for small trees. As noted at the beginning of Section 2.1, breast height diameter explains quite well the tree-to-tree variation in the absolute mass of tree components, but loses its power when the relative mass of the components is explained. The relative component masses are correlated more closely with tree height.

Figure 2.11 (Young and Carpenter 1967) shows the dry mass of foliage in percent of the dry mass of the stem or whole tree as a function of tree height for small trees of seven species in Maine. Percentage of foliage for trees shorter than 10 m decreases rapidly with increasing tree height. In 2-m-high saplings it is between 29 and 130% of the stem mass, and in 11-m-high trees between 6 and 27%, depending on species. When related to whole-tree mass, the proportion of foliage is 20–45% in 2-m-high trees and 5–20% in 11-m-high trees. In the study under consideration, there are two distinctly different groups of tree species in this respect. Shade-tolerant *Abies balsamea, Picea rubens,* and *Thuja occidentalis*

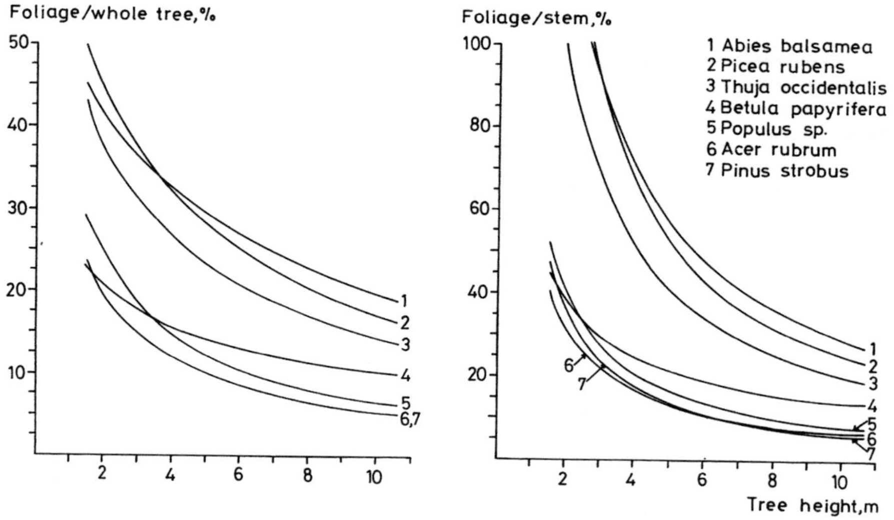

Fig. 2.11. Dry mass of foliage in percent of stem mass and whole-tree mass in small trees of seven species in the natural ecosystems in Maine, as a function of tree height (drawn from data in Young and Carpenter 1967)

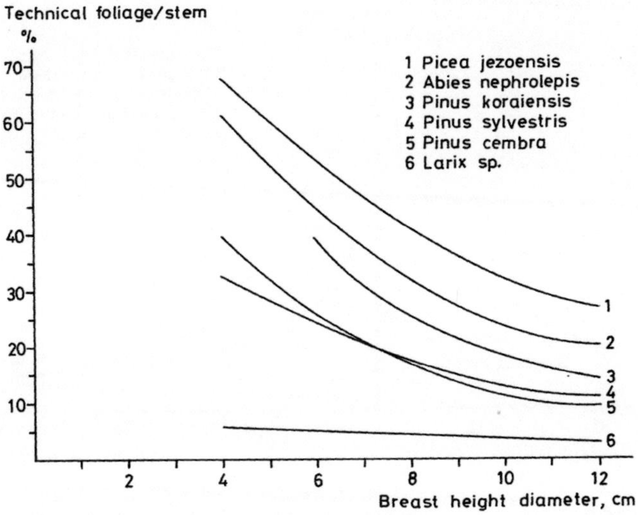

Fig. 2.12. Dry mass of technical foliage in percent of stem mass in small trees of six Siberian species (drawn from Tomčuk and Tomčuk 1966)

carry a heavy foliage mass. *Tsuga canadensis* belongs to the same group of tolerant species but is not included in Fig. 2.11 because it follows the curves of *Abies balsamea* and *Picea rubens* very closely. Another group is comprised of four less tolerant species; three hardwoods and *Pinus strobus*. Despite large differences at the quantitative level, the same function pattern fits both groups.

Figure 2.12 is drawn from studies on the dry mass of technical foliage of a number of Siberian conifers, consisting of all leaves, shoots, twigs and branch sections up to 0.6 cm diameter. The mass of technical foliage is naturally greater than that of the foliage proper, but it follows the same variation pattern when related to stem mass and plotted against breast height diameter. It is largest in tolerant *Picea jezoensis* and *Abies nephrolepis*. By far the lowest percentage of technical foliage was measured in *Larix* sp., an intolerant conifer which sheds its needles annually.

Figure 2.13 illustrates the dependence of fresh and dry leaf mass of small hardwood trees in Maine on breast height diameter. The curves represent an average for the following 11 species: *Acer rubrum*, *A. saccharum*, *Alnus rugosa*, *Betula alleghaniensis*, *B. papyrifera*, *B. populifolia*, *Fagus grandifolia*, *Populus tremuloides*, *Prunus pensylvanica*, *P. virginiana*, and *Salix* spp. These species form large areas of dense *puckerbrush* stands in the northeastern United States. These stands often contain considerable amounts of biomass but are very poorly utilized because of their small size. The foliage of a 5-cm-diameter puckerbrush tree has a dry mass of 0.4 kg, and a fresh mass of 1.0 kg. The foliage mass of a 15-cm-diameter tree is 2.6 and 6.5 kg, respectively. Again, the percentage of foliage decreases with increasing tree size. In a 15-cm-diameter tree, the percentage of foliage is 3.6 when compared with stem on a dry mass basis, and 3.2 when compared with whole-tree mass. Because of the high moisture content of the

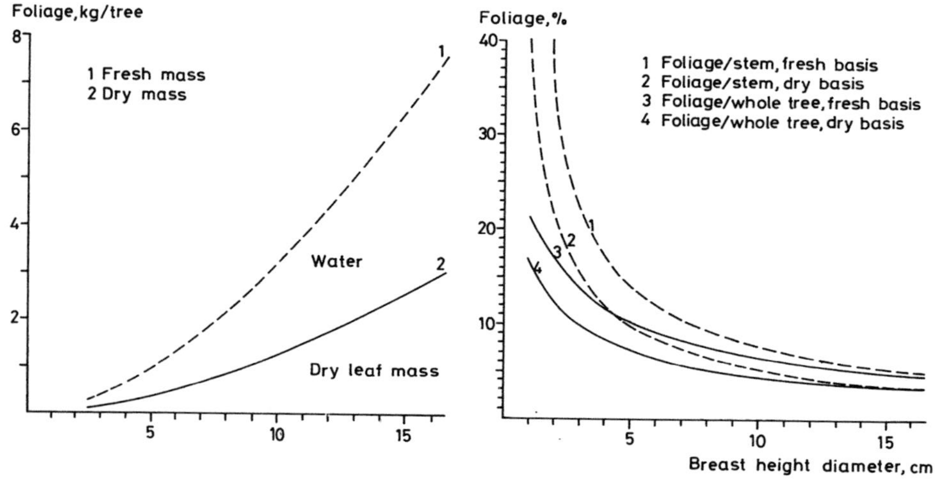

Fig. 2.13. Fresh and dry mass of foliage in kg per tree, in percent of stem mass, and in percent of whole-tree mass as average values of 11 puckerbrush hardwood species in Maine, as a function of dbh (drawn from data in Ribe 1973)

leaves, the percentage of foliage is higher on a fresh mass basis, particularly in very small trees.

The diameter of sapwood at breast height or crown base is an interesting though impractical tree characteristic for prediction of foliage mass. High correlations have been reported for *Abies procera*, *Pinus ponderosa*, and *Pseudotsuga menziesii* (Grier and Waring 1974).

2.1.1.2.5 Effect of Crown Class on Foliage Mass

As long as trees grow in the open, development of the crown and foliage is unaffected by competition between individuals. However, even in single trees lower branches die as a result of shading by the upper crown. When a stand closes, foliage mass per unit volume of crown decreases in the lower shade crown, and less competitive branches die. The higher the number of trees per hectare, the smaller are the crown projection area and the crown space available for a tree in a stand. The reduction in crown size as a result of competition is rapid in suppressed trees of intolerant species down to the point where the tree dies.

Table 2.4 shows the percentage of foliage in the dry whole-tree mass of *Populus tremuloides* and *P. grandidentata* in Wisconsin, as a function of tree age and crown class. The percentage of foliage decreases steeply with increasing age. For example, in 21- to 40-year-old trees it is less than half of that measured in 11- to 20-year-old trees. Within any age group, the crown class of the tree affects the percentage of foliage strongly. It was 78% in the codominant, 63% in the intermediate, and only 43% in the suppressed trees compared with the dominant trees.

From the biomass utilization point of view, it is important to know the percentage of foliage, particularly at the time of removal of a tree. The selective thin-

Table 2.4. Percentage of foliage in dry whole-tree mass of *Populus grandidentata* and *P. tremuloides* as a function of tree age and crown class in Wisconsin (Zavitkovski 1971)

Age (years)	Foliage/whole tree (%) Crown class				All trees
	Dominant	Codominant	Intermediate	Suppressed	
1–10	12.4	11.9	9.3	4.9	9.9
11–20	11.2	6.4	4.7	5.0	6.8
21–40	3.7	2.6	2.3	1.9	2.4
41+	2.8	2.5	1.8		2.4

Table 2.5. The proportion of foliage in the dry crown and whole-tree mass of *Pinus sylvestris* and *Picea abies* trees removed in successive cuttings from managed forests in southern Finland

Tree species	Treatment	Stem (m^3)	Foliage (kg/tree)	Foliage		
				Live crown (%)	Total crown (%)	Whole tree (%)
Pine	Early thinning	0.032	1.7	41.0	32.3	8.0
	Late thinning	0.140	3.0	32.9	26.0	4.7
	Final cutting	0.406	6.1	26.1	23.0	3.7
Spruce	Early thinning	0.051	4.1	41.7	38.2	12.6
	Late thinning	0.116	7.4	39.6	36.2	10.5
	Final cutting	0.364	24.3	37.6	35.4	10.4

nings from below favor vital dominant trees, directing the removal to suppressed, intermediate, and codominant crown classes. The trees with the largest foliage mass are left standing and harvested in final cuttings. In systematic thinnings, removal of a tree is determined by the location in the stand, rather than vitality or crown class.

Table 2.5 shows the proportion of foliage in the above-ground biomass of *Pinus sylvestris* and *Picea abies* trees removed in successive cuttings from managed forests in Finland. The foliage percentage in the dry whole-tree mass of Scots pine from early thinnings is 8.0, and in that from final cuttings only 3.7%. The influence of crown class is smaller than that of tree size. The differences due to cutting sequence are less conspicuous for the more tolerant Norway spruce, the foliage percentage falling from 12.6 to 10.4, respectively. If foliage is to be recovered from managed forests for fodder or the chemical industries, all cuttings of spruce stands and early thinnings of pine stands are attractive harvesting objects. The reduction in foliage percentage with increasing tree age and size is obviously much faster in unmanaged, dense forests.

2.1.1.2.6 Effect of Tree Species on Foliage Mass

The absolute and relative amounts of foliage in an individual tree of any species are affected by a large number of interacting tree variables, site factors, climatic conditions, and stand history. It would be meaningless and misleading to present mean values for percentage foliage of a specific species.

Sections 2.1.1.2.2–2.1.1.2.5 show great differences between softwoods and hardwoods, but the available data do not entitle researchers to estimate generalized averages by species. For preliminary calculations, Keays and Barton (1975) suggested, as a rule of thumb, the following rough averages for young and mature trees growing in well-stocked stands on average to good sites. They emphasized, however, that for detailed economic and engineering studies the quantity of biomass components should be determined by direct species-by-species measurement.

	Foliage/stem (%)	
	Softwoods	Hardwoods
Young trees	25	15
Mature trees	10	5

The previous sections indicate high percentages of foliage in tolerant softwood species. Light-demanding softwoods, such as many species of pine, actually resemble hardwoods rather than spruces, firs, and other tolerant softwoods. An example of foliage percentage in hardwoods is given in the following table representing average values for 15-cm-diameter trees of 22 species growing on pine sites in the southeastern United States (Koch 1985a).

	Foliage in percent		
	of stem	of whole tree	of complete tree
Green mass basis	5.9	4.3	3.4
Dry mass basis	4.9	3.7	2.9

Comparison of foliage mass of various tree species is more feasible on an areal basis in tonnes per hectare. The dry foliage mass of nine coniferous species in Japan is plotted in Fig. 2.14 against the mean life span of needles. Foliage mass increased among species linearly with average life span but seemed to be rather independent of annual foliage production, the average of which for all nine species was 3.5 t/ha. The life span of the needles is usually the longest in tolerant species (Satoo and Madgwick 1982).

Tadaki (1976) considers the leaf mass of deciduous forests as a basic unit of foliage mass of closed forests. In an extensive study in Japan it was 2–4 t/ha on a dry mass basis in most stands of deciduous trees (Table 2.6). For a large number of tree species in evergreen stands the foliage mass per hectare seemed to be

Table 2.6. Dry mass of foliage and leaf area index in stands formed by various tree species in Japan (Tadaki 1976)

Tree species	Dry mass of foliage (t/ha)			Leaf area index (ha/ha)		
	Mean	S.D.	No. of stands	Mean	S.D.	No. of stands
Deciduous, broad-leaved:						
Fagus crenata	3.8	1.6	58	5.7	1.0	11
Betula ermanii	2.9	0.9	23	4.5	1.0	4
B. platyphylla	1.1	0.3	13	2.8	0.7	12
B. maximowicziana	2.3	0.5	3			
(*Betula* spp.)	2.2	1.1	39	3.2	1.1	16
Alnus, 5 spp.	2.8	1.1	20	4.6	5.2	2
Populus, 4 spp.	3.8	1.8	9	5.1	2.5	9
Others, 11 spp.	2.4	1.0	15	5.4	1.9	10
Deciduous, needle-leaved:						
Larix leptolepis	3.0	1.0	30	4.2		1
Metasequoia glyptostroboides	5.0		2			
Evergreen, broad-leaved:						
Quercus phillyraeoides	8.6	2.0	7	6.7	1.7	7
Q. glauca	6.3	0.5	3	7.1	0.3	3
Q. myrsinaefolia	8.8	1.9	10	8.4	0.5	3
(*Quercus* spp.)	8.4	2.0	20	7.1	1.5	13
Castanopsis cuspidata	8.3	2.3	9	8.6	2.1	7
Machilus thunbergii	12.0	1.2	4	8.9	0.9	4
Camellia japonica	8.0	2.7	5	5.7	1.0	4
Acacia, 2 spp.	5.7	2.3	11	7.1	2.0	6
Evergreen, needle-leaved:						
Pinus densiflora	6.4	1.3	120			
P. thunbergii	7.7	2.1	15			
P. taeda	8.8	2.7	6			
P. elliottii	10.8	1.2	5			
P. strobus	6.1	2.5	6			
P. pumila	21.7	3.0	4			
Cryptomeria japonica	19.6	4.4	126	6.0	1.2	12
Chamaecyparis obtusa	14.0	2.5	26	5.1		2
Thujopsis dolabrata var. *hondai*	19.4	4.6	7			
Abies firma	16.5	2.6	4			
A. veitchii (-*A. mariesii*)	16.1	4.7	38	9.9	1.9	7
A. sachalinensis	22.8	3.7	5			
A. spp.	16.8	4.9	47	9.9	1.9	7
Tsuga sieboldii	7.8		2			
Picea glehnii	17.0	6.3	5			
Picea abies	18.0	3.9	5			

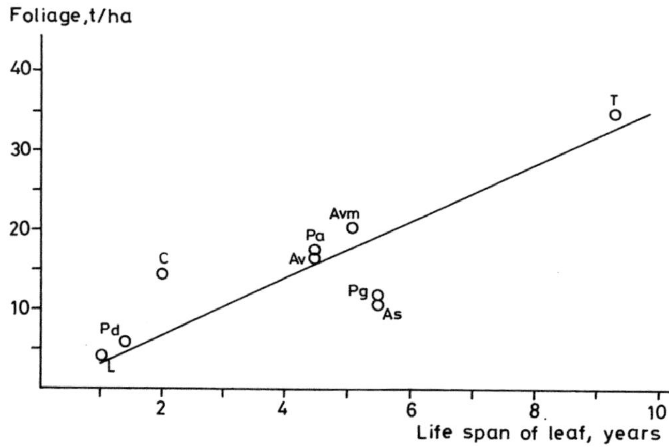

Fig. 2.14. The relationship between dry mass of foliage per unit area and life span of needles in nine coniferous species in Japan. *As* = *Abies sachalinensis* (3 stands); *Av* = *A. veitchii* (7); *Avm* = mixed *A. veitchii* and *A. mariesii* (1); *L* = *Larix leptolepis* (4); *Pg* = *Picea glehnii* (1); *Pa* = *Picea abies* (6); *Pd* = *Pinus densiflora* (37); *C* = *Chamaecyparis obtusa* (1); *T* = *Thujopsis dolabrata* (3) (redrawn from Satoo and Madgwick 1982)

broadly related to the annual production of foliage. Table 2.6 also shows the *leaf area index*, which is defined as the leaf surface area of forest per unit of land surface area. Only one side of the leaf of a broad-leaved tree is measured, whereas the total leaf area is used for coniferous trees.

2.1.1.2.7 Effect of Atmospheric Pollution on Foliage Mass

Atmospheric pollution causes damage to the foliage of forest trees in Europe and North America. This alarming forest-decline development, called *Waldsterben* in German, together with other causes of damage, results in a gradual loss of foliage mass. The extent of the damage is monitored in many European countries by classifying individual trees and stands according to crown condition.

The crown of *Pinus sylvestris* usually encompasses three or four, and that of *Picea abies* up to seven needle years. The number of full needle years in tree crowns by species is used as the bio-indicator for evaluation of crown damage. Unfortunately, the classification systems used in different countries are not fully comparable. In the Federal Republic of Germany, depending on the estimated loss of foliage, five classes are differentiated. If 25–60% of the foliage shows yellowing symptoms, the damage class is increased by one. If more than 60% exhibits yellowing color, the class is increased by two (Breloh and Kirch 1984):

Damage class	Loss of foliage (%)	Extent of damage
0	0–10	None
1	11–25	Slight
2	26–60	Moderate
3	61+	Severe
4	Dead	Dead

According to Table 2.7, as much as 51.9% of the forests in the Federal Republic of Germany was classified as damaged in 1985, and 19.2% of the forests had lost more than one-fourth of their foliage mass. The greatest degree of crown damage was observed in mature stands, at high altitudes, and in fir stands. However, in all tree species the damage was considered extensive enough to lead to serious ecological consequences and to affect the biomass composition in conjunction with whole-tree utilization (Bundesministerium ... 1985). The proportion of damaged forests was 53.7% in 1986 and 52.3% in 1987 (Waldschäden ... 1987).

Inventories in northern Europe show significantly less damage. For example, in 88% of the coniferous trees in Finland the loss of foliage mass was less than 10%. The loss was highest in old and large trees, at higher altitudes, in cold climate, and at the edge of the forest (Jukola-Sulonen et al. 1987).

Studies of damaged trees indicate that stem and crown dimensions, the growing space of a tree, and the specific competitive environment evidently can outweigh the influence of foliage loss on growth. On the other hand, the magnitude of the crown surface area has proved to be particularly indicative of the incremental performance of *Picea abies*. Increment per m^2 crown surface area appears, therefore, to be a better criterion for growth loss than the absolute growth increment of a tree. This relative increment value decreases with increasing needle loss (Kramer 1986).

Table 2.7. Forest damage in the Federal Republic of Germany by species in 1985 (Bundesministerium ... 1985)

Tree species	Total area (million ha)	Percent of area Damage class				Total
		0	1	2	3+4	
Picea abies	2.9	47.8	28.2	21.4	2.6	100.0
Pinus sylvestris	1.5	42.5	40.5	15.3	1.7	100.0
Abies alba	0.2	12.8	20.5	50.3	16.4	100.0
Fagus sylvatica	1.3	45.5	40.1	13.1	1.3	100.0
Quercus robur, Q. petrea	0.6	44.7	38.9	15.7	0.7	100.0
Others	1.0	69.5	22.7	6.7	1.1	100.0
Total	7.4	48.1	32.7	17.0	2.2	100.0

2.1.1.3 Composition of Branch Mass

Branch mass consists of the wood and bark of live and dead branches of a tree. Foliage is thus not included. The total mass of the foliage and branches is called crown.

The branching habit and tree form of many hardwoods are difficult to predict, while the branches of conifers are arranged in a more regular pattern about the central stem. Tree form, which determines the distribution of the above-ground wood and bark mass between stem and branches, is commonly interpreted in terms of terminal bud dominance. In trees with *excurrent branching* the terminal bud and main stem outgrow the lateral branches, resulting in a cone-shaped crown with a clearly defined central stem. Trees with weak terminal bud dominance exhibit *decurrent or deliquescent branching*, with lateral branches growing almost as fast as the stem and the central stem becoming lost in the upper crown. For example, in the southeastern United States *Liriodendron tulipifera*, *Liquidambar styraciflua*, and *Nyssa aquatica* typically have excurrent branching, while the oaks, elms, maples, and hickories represent the more common deliquescent hardwoods (Clark 1978).

Keays (1971 c) recommended that branch mass be determined for purposes of utilization-oriented biomass studies for at least two size categories, as the raw material properties and processing techniques are strongly dependent on branch diameter. Young and Chase (1965) defined three categories of diameter: less than 6.4 mm, 6.4–25 mm, and over 25 mm. A standard procedure in the Soviet Union is to form a special category, technical foliage, by combining branch sections thinner than 6 mm with foliage.

Branch mass of an individual tree of a specific species can be estimated from various external characteristics indicating stem size and form, growth, or crown development. Branch mass is also influenced by genetic factors, site quality, and stand characteristics, such as spacing. The present section deals with the variations in branch mass only from the standpoint of tree size and crown class, but the matter is touched upon again in Section 2.1.1.4 in conjunction with crown mass.

2.1.1.3.1 Moisture Content in Branches

The proportions of live tissues and sapwood in a tree biomass at a given stem height increase from stem base to top, resulting in an increasing moisture content of the wood. In many tree species, the base of larger branches contains inactive heartwood and, due to extraneous materials, has a high basic density. This lowers the moisture content expressed as a percentage of the fresh mass. As water passes into transpiring leaves via increasingly scanty vascular tissue, the moisture content tends to increase in a branch from base to tip. However, since the transition obviously depends on weather conditions and transpiration, this trend does not always prevail.

Table 2.8 shows the average moisture content of wood and bark in branch sections of different diameters in saw-timber-sized *Pinus sylvestris* and *Picea abies*

Table 2.8. Moisture content of wood and bark in *Pinus sylvestris* and *Picea abies* branches in the summertime in Finland on a fresh mass basis by diameter classes (Hakkila 1969)

Tree species	Component	Moisture content (%) Branch diameter (mm)				
		10–20	20–40	40–60	60–80	10–80
Pine	Wood	55.8	54.8	52.2	48.2	54.1
	Bark	58.8	59.5	60.2	60.3	59.4
	Wood + bark	56.7	55.6	53.1	49.2	55.1
Spruce	Wood	43.8	41.5	36.7		42.3
	Bark	52.6	52.6	55.2		52.6
	Wood + bark	46.5	44.4	40.5		45.4

trees in the summertime in Finland. The moisture content of Scots pine is significantly higher than that of Norway spruce. This difference between species is partly a result of differences in the basic density of the branch wood. In both species, the moisture content of branch wood decreases and that of branch bark increases with diameter. The moisture content is particularly high in bark, probably because the bark in branches is mainly composed of phloem. Moisture content relationships within and between branches vary seasonally (Phillips et al. 1976).

Examples of average moisture content (m.c.) of wood and bark in leafless branches of small understory hardwoods in North Carolina and Georgia are presented below. The high moisture content in the stems and branches of *Liriodendron tulipifera* and *Liquidambar styraciflua* reflect a rather low, 400 kg/m^3 basic density of wood, while the corresponding basic density values for the other species under consideration are 500–600 kg/m^3 and even more for *Carya* (Phillips 1977).

Understory species	Average m.c. of wood and bark (%)	
	Stem	Branches
Liriodendron tulipifera	55.0	55.4
Liquidambar styraciflua	57.7	57.5
Acer rubrum	45.8	46.8
Quercus alba	41.3	42.0
Q. prinus	40.8	42.5
Q. falcata	41.2	40.8
Cornus florida	44.1	44.9
Carya sp.	37.1	39.4

2.1.1.3.2 Percentage of Bark in Branches

The bark percentage shows the proportion of bark mass of the total mass of wood and bark on a dry or fresh mass basis. Because bark differs from wood in its anatomical, physical, and chemical properties, the percentage of bark is a critical raw

Table 2.9. Percentage of bark and distribution of bark mass by diameter categories in live branches of sawtimber-sized *Pinus sylvestris* and *Picea abies* trees in Finland, on a dry mass basis (Hakkila 1969)

Diameter (mm)	Percentage of bark		Distribution of bark (%)	
	Scots pine	Norway spruce	Scots pine	Norway spruce
−10	55.0	57.0	61.0	58.8
10−20	26.9	28.1	13.0	18.8
20−40	15.7	20.9	19.0	22.0
40−60	10.8	14.7	6.5	0.4
60−80	7.9		0.5	
All	40.8	43.5	100.0	100.0

material characteristic from the utilization point of view. For example, excessive bark from branches causes difficulties in the pulping processes and lowers both yield and product quality (Sect. 8.2). Branch bark consists chiefly of phloem and contains a large amount of extractives (Sect. 3.3.2.1).

The percentage of bark on branches is highly variable, depending on tree species and more particularly on branch diameter. The double-bark thickness on 5-mm-diameter *Pinus sylvestris* and *Picea abies* twigs is about 2 mm. The double-bark thickness on 50-mm branches is 4 and 7 mm, respectively. Since bark thickness increases rather slowly in relation to branch diameter, the percentage of bark decreases sharply as the diameter increases. On thin twigs or branch tips of less than 1 cm in diameter, the mass of bark exceeds that of wood. If the thin parts are removed and only the thickest parts are directed to industrial use, the bark content of the branch mass declines.

Table 2.9 shows the percentage of bark in branch sections of various diameters in large Scots pine and Norway spruce trees. About 60% of all branch bark in these species is concentrated in branch parts of less than 10 mm diameter. By diameter categories, these two tree species differ only slightly, Norway spruce branches having an average bark content of 43% versus 41% in Scots pine. The percentage of bark in the dry mass of branches was significantly lower in another Finnish study for small-sized trees of the same species: 30% for Scots pine, 31% for Norway spruce, and 32% for *Betula* spp. (Kärkkäinen 1976a).

The relationships of the percentage of bark to branch diameter in the four major species of southern pine agree fairly well with the data on Scots pine and Norway spruce, although southern pine branches within a certain diameter category have a lower content of bark. The percentage of bark is highest in *Pinus palustris* and lowest in *P. taeda* (Phillips et al. 1976).

On the average, the proportion of bark in the branches of large coniferous trees varies from 20 to 45%, although values lower than 20% have been reported. Most hardwood species fit within the same limits (Keays 1971c).

Table 2.10. Percentage of bark in branch and stem mass of eight understory hardwood species in North Carolina and Georgia, on a dry mass basis (data from Phillips 1977)

Understory species	Percentage of bark		
	Stem	Branches	Whole tree
Liriodendron tulipifera	20.2	31.2	21.8
Liquidambar styraciflua	22.7	29.3	23.8
Acer rubrum	16.7	23.4	18.1
Quercus alba	24.3	35.2	25.6
Q. prinus	23.9	31.0	24.8
Q. falcata	29.7	29.9	29.7
Cornus florida	15.4	17.7	16.1
Carya sp.	27.5	30.9	28.0

Average branch diameter increases with tree size, resulting in a decline of the bark percentage. The following table shows the bark content in the stem and branches of the four major species of southern pine, as a function of breast height diameter. The influence of tree size on the bark content is strong, but in all diameter categories the average bark content of the branches is roughly 2.5 times as high as that of the stem (Clark and Taras 1976).

	Breast height diameter (cm)						
	15	20	25	30	35	40	45
	Percentage of bark						
Stem mass	15	14	12	11	10	10	9
Branch mass	36	34	30	28	26	25	24

A study by Phillips et al. (1976), concerning the whole-tree mass of small understory hardwoods in North Carolina and Georgia, indicates wide species-to-species variations in bark content (Table 2.10). While the percentage of bark in branches was only 17.7 in *Cornus florida*, it was 35.2 in *Quercus falcata.* According to the authors, the percentage of bark in understory trees less than 13 cm in diameter at breast height ranged from 5 to 10 percentage units higher than most reported values for mature stems. The bark percentages were even higher in branches. However, the difference between stem and branches appears to be smaller in hardwoods than in softwoods.

2.1.1.3.3 Dead Branches as a Biomass Component

The erratic tree-to-tree variation in branch mass is partly due to differential natural losses of dead branches. Although dead branches are of minor significance from a biological point of view, their presence in the stem is extremely important for the quality development of sawtimber trees. They are not an accept-

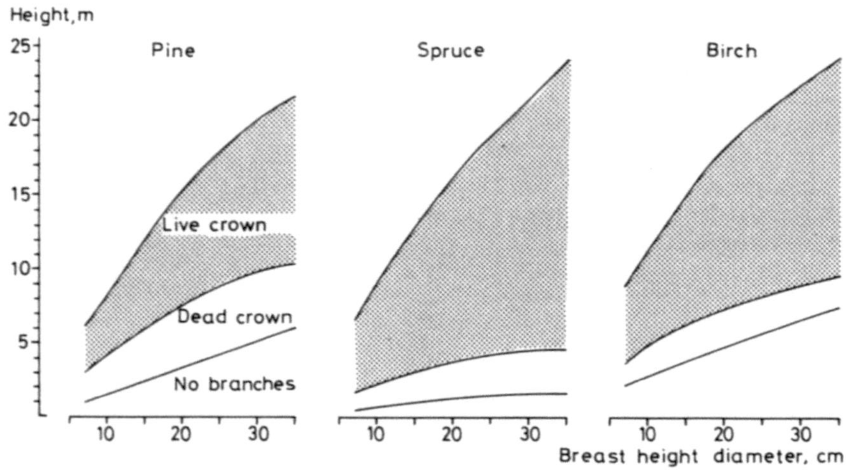

Fig. 2.15. Height of total tree, live crown, and dead crown in *Pinus sylvestris*, *Picea abies*, and *Betula* spp. in southern Finland as a function of dbh (redrawn from Hakkila et al. 1972)

able raw material for pulp and composite board production, although they contain very little bark. On the other hand, they are very suitable for fuel because of their low moisture content.

Since dead branches do not belong to the functioning organs of a tree, they are litter from the ecological point of view. The quantity of dead-branch mass is very difficult to explain with conventional tree variables. Many biomass studies, particularly when the approach is primarily ecological, are restricted to the living components of a tree, and dead branches are often ignored as a component of branches, crown, whole tree, or complete tree. However, depending primarily on tree species, stand density, and site quality, a considerable portion of the stem may retain its dead branches. For example, when logging wages are calculated on the basis of piece rates, the presence of dead branches must be taken into account as a work difficulty factor.

Figure 2.15 shows the height development of the total tree, live crown, and dead crown as plotted against breast height diameter in a large sample collected from the survey tracts during the Finnish national forest inventory. The results represent naturally regenerated, managed stands in southern Finland. The stem section with dead branches is short in *Picea abies*, because it is a tolerant species, and in *Betula* spp., because the dead branches fall off readily. In *Pinus sylvestris*, on the other hand, the role of dead branches is more conspicuous. According to Ager (1972), the dead-branch section is longer in thinning stands than in final cuttings. Marklund (1987) found the mass of dead branches more closely correlated to the tree height than to breast height diameter.

This review recognizes dead branches as a potential source of biomass, and as basically included in the mass of branches and crown, although they are ignored in several reference studies. At least in some cases the role of dead branches may be noteworthy, as shown by the following results from a 27-year-old *Pinus resinosa* plantation in Ontario (Alemdag and Stiell 1982):

Spacing (m)	Dry mass of dead branches (t/ha)	Dead branches, percent of dry crown mass
1.5×1.5	15.4	28.5
1.8×1.8	15.1	27.8
2.1×2.1	13.9	26.1
3.1×3.1	12.4	25.7
4.3×4.3	11.6	26.3

Table 2.11 presents the mass of dead branches as a proportion of the total dry or fresh mass of all branches, stem, and whole tree in *Pinus sylvestris*, *Picea abies*, and *Betula* spp. trees removed in selective thinnings and final cuttings in southern Finland. Great differences may be observed between tree species and various types of cuttings. Compared with the two conifers, birch has a relatively small mass of dead branches, reflecting weak durability and fast breaking of dead material, which is typical of many hardwoods. Among the conifers, Scots pine has more biomass in dead branches, although the total mass of branches is larger in Norway spruce. For all species, the proportion of dead branches of the total biomass is largest in the thinning phase of stand development, and decreases toward the end of the rotation. For example, during early thinnings of pine, spruce, and birch, 5.4, 2.7, and 2.2% of the whole-tree mass of felled trees consisted of dead branches. For final cuttings, the corresponding figures were 2.0, 1.8, and 0.4%, respectively. In a large Swedish sample, representing all sizes and crown classes of Norway spruce trees, the proportion of dead branches was 1.6% and that of cones 0.2% of the whole-tree mass (Marklund 1987). In an Australian study of unthinned, 12-year-old *Pinus radiata* plantations, dead crown mass consisted of 8.8% of the whole-tree and as much as 33.2% of the crown on a dry mass basis (Williams 1976).

Table 2.11. The mass of dead branches in relation to the mass of all branches, stem, and whole tree in *Pinus sylvestris*, *Picea abies*, and *Betula* spp. trees removed during selective thinnings and final cuttings in Finland

Tree species	Treatment	Dead-branch mass in percent of			
		branch mass (fresh basis)	branch mass (dry basis)	stem mass (dry basis)	whole-tree mass (dry basis)
Pine	Early thinning	16.4	22.0	7.3	5.4
	Late thinning	18.0	24.7	5.3	4.2
	Final cutting	8.0	12.0	2.4	2.0
Spruce	Early thinning	5.6	8.2	4.3	2.7
	Late thinning	7.0	10.2	4.3	2.9
	Final cutting	3.9	5.6	2.9	1.8
Birch	Early thinning	11.5	12.8	2.7	2.2
	Late thinning	4.9	5.4	1.0	0.7
	Final cutting	2.6	3.1	0.5	0.4

Dead branches should be taken into consideration especially when forest biomass is used for production of energy. However, in many whole-tree harvesting systems most of the dead branches break off and are lost during the felling, off-road transport, and chipping phases. On the other hand, in conjunction with certain logging systems much of the dead branch mass can be recovered, for instance, when a terrain chipper reduces residue from slash piles left by logging machines in clear-cutting areas.

2.1.1.4 Crown Mass

In many biomass studies, the unmerchantable top of the stem is included in the crown mass. The present review does not follow this practice, for the base diameter of the top is not constant. The *crown* is here the sum of all live and dead branches plus foliage. For utilization purposes, it is important to know not only the total crown mass but also its distribution into components.

Crown mass may be described as a product of the radius, length, form factor, and density of the crown. Crown form factor and crown density are very difficult to measure, but the former can be described by the social status of the tree in combination with other variables, and crown density may depend on site index, tree age, etc. (Marklund 1987).

Tree species, tree size as indicated by the height or the breast height diameter, diameter at crown base, age, dominance, stand density and history, genetic factors, season, etc. are the major variables explaining the absolute and relative quantity and composition of the crown mass. Tree-to-tree variation is usually smaller in even-aged plantations than in naturally regenerated stands.

Regression equations may be constructed to predict fresh or dry crown mass in trees with a regularly formed crown. However, trees with abnormal crowns are found in most stands. One-sided crowns are common, and equations for a full crown will undoubtedly overestimate crown mass for such individual trees. But even though the error for an occasional tree may be fairly large, the irregularities only slightly affect the estimate for an entire stand. Therefore, the accuracy of regression equations is acceptable for most purposes. Overmature trees with flattened crowns may also cause problems, but such trees are a minor factor in managed forests. Moreover, seasonal differences in foliage mass must be considered when the mass and composition of the crown is estimated (Loomis et al. 1966).

2.1.1.4.1 Crown Ratio

The branching habits and shade tolerance of a species, the degree of mutual shading, and the longevity of the foliage determine the amount of crown mass over a unit of ground area in a closed forest. Solar radiation, when it passes through the canopy, is absorbed and reflected by leaves and, to a lesser extent, by the branches and stems. Radiation intensity at a given level in the crown canopy depends on the leaf area above. Shade leaves consequently, although they are better adapted to shaded light conditions, always show lower photosynthetic rates than sun leaves in the upper crown. If consumption of organic matter by respiration exceeds the daily production by photosynthesis in shade leaves, and if the net photosynthesis is negative over a period of time, leaves cannot produce enough material for their own support. The minimum intensity of radiation to ensure survival and reproduction of leaves and branches is relatively constant for a specific tree species, growing stock, and site. If the foliage mass in a canopy exceeds a certain limit, the leaves in the lowermost crown do not receive enough radiation to replace the photosynthate they and their supporting branches consume and use for production of new leaves. As a result, the branch dies. As a tree loses the lower part of the live crown and new growth is added to the upper crown, the canopy as a whole moves upward (Satoo and Madgwick 1982). Attacks by fungi in the lower crown may occasionally cause this development to proceed stepwise.

The proportion of the live crown as a percent of the overall height of the tree is called *crown ratio*. This tree characteristic is an indication of the development of competition and need of thinning in a stand. Crown mass is positively correlated with the crown ratio. Because the crown ratio can be determined fairly easily, it is frequently used as an additional supporting, independent variable, together with a tree size indicator such as dbh, in prediction equations for crown mass.

In closed *Pinus radiata* plantations in New Zealand, the height to the lowest live whorl increased at approximately the same rate as the total tree height in the absence of thinnings (Siemon et al. 1976). The crown length thus remained nearly constant, resulting in a decrease of the crown ratio over time. After thinning, branches once suppressed might live indefinitely even though they rarely regained active growth. Consequently, thinning increased the crown length and crown ratio. Development of the crown ratio of an individual tree is determined largely by its position in the stand.

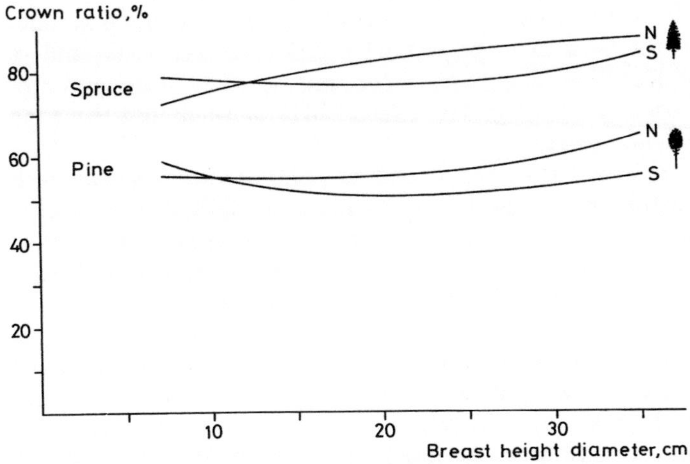

Fig. 2.16. Crown ratio of *Pinus sylvestris* and *Picea abies* in southern (*S*) and northern (*N*) Finland, as a function of dbh (redrawn from Hakkila et al. 1972)

Figure 2.16 shows the crown ratio of *Pinus sylvestris* and *Picea abies* for managed forests in southern and northern Finland. Because trees in the suppressed and intermediate crown classes are removed in thinnings from below, the average crown ratio stays rather constant or increases slightly over the years. The crown ratio is heritably higher in the north, which is an indirect result of the severe growth conditions, since risk of snow damage forces the trees to develop narrow and long crowns. Moreover, the spacing is wider and stocking essentially lower in the north.

Crown ratio is largest in the dominant and codominant trees that are important for the future production of a stand. On the other hand, the crown ratio of the suppressed trees removed in thinning cuttings may be of special interest from the utilization point of view. The crown ratio of trees harvested during thinnings is lower than that of trees in average stocking, as shown by the following data from southern Finland:

	Crown ratio (%)		
	Pinus sylvestris	*Picea abies*	*Betula* spp.
Trees removed in early thinnings	56.7	78.1	60.9
Trees removed in late thinnings	42.8	66.5	55.1
Trees removed in final cuttings	44.7	75.2	55.6
Trees of average stocking	53.9	78.8	60.1

2.1.1.4.2 Vertical Distribution of Crown Mass

Knowledge of the vertical distribution of the crown mass is needed in the design of logging machines and harvesting systems. The location of the center of gravity from the base of a tree is important to equipment designers and machine operators because it reflects tree stability and the dynamic forces required in moving a whole tree from the vertical to horizontal position or vice versa and in loading and transport activities. The mass, form, and position of crown components strongly affect the center of gravity (Keen 1963; Steinhilb and Erickson 1970, 1972; Dunfield et al. 1974).

Dry mass data are important from the utilization point of view, whereas fresh mass data are more useful for transport cost calculations and machine design. The present section deals mainly with dry mass. It can be roughly converted to fresh mass by multiplying by two, depending on tree species, proportion of dead branches in the crown, site, geographic area, season, and weather.

The following figures are examples of the average moisture content of crown mass during the removal phase in thinnings or final cuttings in southern and northern Finland. The conifers were felled in the summer and *Betula* spp. either in early spring or after defoliation in the fall. The moisture content seems to be lower under the extreme conditions in the north near the Arctic circle. The difference between the live crown and total crown is greatest in *Pinus sylvestris*, since it has the largest amount of dead branches.

	Moisture content (%)			
	Live crown		Total crown	
	South	North	South	North
Pinus sylvestris	55.4	54.5	52.4	51.5
Picea abies	52.0	48.0	50.6	46.9
Betula spp. (leafless)	46.7	44.6	46.2	44.2

The base diameter of a branch may be used to predict foliage and total mass when destructive sampling of whole trees is not possible (Cummins 1941). Correlations between branch diameter and the biomass components of the branch are characteristically high. However, other factors, apart from branch size, also affect the relationship between component mass and branch diameter. In a study of *Pinus radiata* in New Zealand, where branch diameter alone or the product of the branch basal area and branch height were used as predictor variables, branch mass at the top and base of the crown was overestimated and the mass in midcrown was underestimated. Overall, this led to an underestimate of total foliage mass. Use of the relative height of branch location as an additional variable improved predictions for individual branches, but overcompensation gave total foliage mass predictions greater than the actual mass. Shading effect was believed to be the underlying biological factor causing vertical variation in the relationship, depending on crown form and crowding within the stand canopy (Madgwick and Jackson 1974).

Compared with foliage, crown mass is more evenly distributed in the vertical direction of a tree. The proportion of foliage is highest in the uppermost crown, whereas the lower crown parts consist of thicker branches with fewer leaves, a

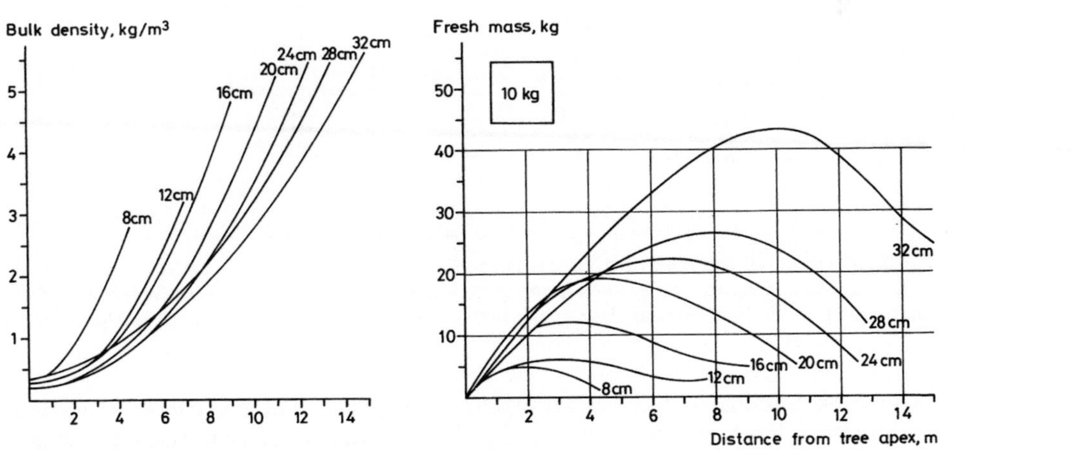

Fig. 2.17. Bulk density and vertical distribution of fresh mass in the live crown of *Picea abies* trees of various diameters in the natural forests of the northern Soviet Union. Dbh in cm indicated in the figure (redrawn from Korotjaev 1984)

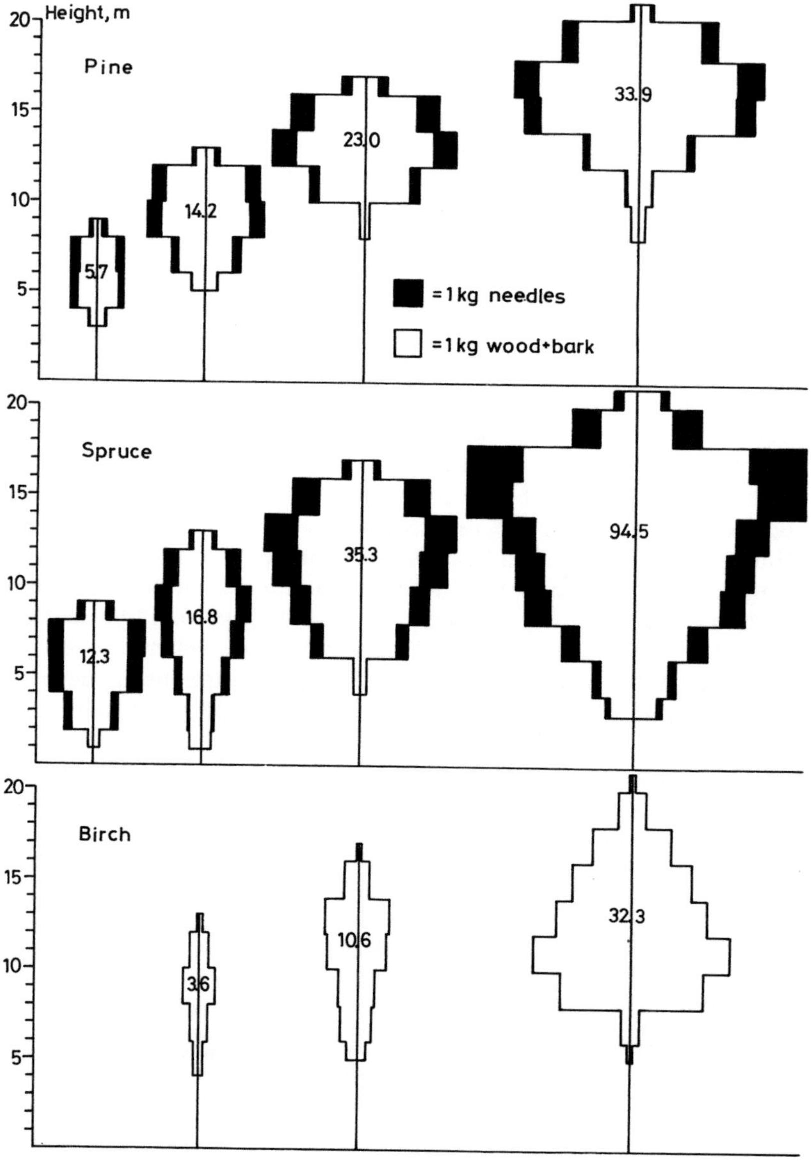

Fig. 2.18. Vertical distribution of dry mass in the live crown of 9-, 13-, 17-, and 21-m-high *Pinus sylvestris*, *Picea abies*, and leafless *Betula* spp. trees at the phase of removal in selective thinnings and final cuttings in southern Finland. Average dry mass of crown indicated in kg for each height category

lower bark content, and higher basic density of wood. The percentage of foliage in the crown mass of *Pinus echinata* in Missouri was 63 in 1-cm-thick branches, 21 in 5-cm-thick branches, and only 11 in 10-cm-thick branches (Loomis et al. 1966). In a detailed analysis of crown dry mass of a typical 15-year-old, 15-m *Pinus taeda* tree in the Carolinas, the percentage of foliage was 27 at the base of the crown 7 m from the ground level, but increased steadily toward the tip of the tree, exceeding 50% at 13 m height (Hepp and Brister 1982).

This change in the biomass composition within a tree crown is reflected in the *bulk density of crown*, i.e., kg fresh mass per m^3 crown volume. In *Picea abies*, the bulk density increases from the top to the base of the live crown (Fig. 2.17). In a large sample from natural forests in the Archangel and Vologda regions in the northern Soviet Union, the bulk density of the fresh crown mass was only 0.3 kg/m^3 near the tree apex but up to 5 kg/m^3 at the base of the crown. Mean crown bulk density of a tree increased linearly with breast height diameter from 0.9 kg/m^3 in small trees to 1.3 kg/m^3 in large sawtimber trees. The differences in bulk density are reflected in the vertical distribution of crown mass. Compared with foliage mass, the center of gravity of the total crown mass is lower in the crown. In a 20-cm tree in northern natural spruce forests, most of the crown mass is located about 2–8 m from the apex. In a 32-cm tree the main concentration of crown mass is 4–14 m from the apex (Korotjaev 1984). However, the vertical distribution of crown mass varies from tree to tree depending on growth conditions, genetic factors, etc. In a single tree the majority of the crown mass may be concentrated in the upper or lower crown, or it may be evenly distributed throughout the crown height (Kärkkäinen 1972).

The crown mass profiles in Fig. 2.18 show the average vertical distribution of live crown mass at the time of removal in small trees during selective thinnings from below, and in larger trees in final cuttings in southern Finland. In *Pinus sylvestris* the majority of crown mass occurs 1–7 m from the apex, independently of tree size. In *Picea abies* the live crown is considerably longer, the peak mass of foliage and branches being 3–7 m from the apex. In both species the content of foliage is highest in the uppermost crown. The crown mass of defoliated *Betula* spp. is significantly lower, in large trees it is concentrated in a few heavy branches at the base of the crown.

Partial recovery of crown mass through reduced delimbing, topping, etc. may be considered as a compromise between stem-only logging and whole-tree utilization. The crown mass profiles are a useful tool for system planning and method evaluation for this purpose.

2.1.1.4.3 Effect of Tree Size on Crown Mass

As the crown mass is a result of a large number of interacting site, stand, climatic, and genetic variables reflecting past rather than present spacing and other growth conditions of a tree, its tree-to-tree variation is wide and difficult to predict. Nevertheless, the mass or volume of any biomass component is related to tree size.

As a single tree characteristic, stem diameter is commonly the strongest independent variable for explaining crown mass variation. In some cases stem diameter at the crown base or sapwood basal area may explain the variation better than diameter or basal area at breast height, but the slight improvement in precision by the use of sapwood area over breast height diameter does not pay for the extra labor and impracticality of its measurement (Snell and Brown 1978). As mentioned at the beginning of Section 2.1, the relationship between a component mass and breast height diameter is exponential.

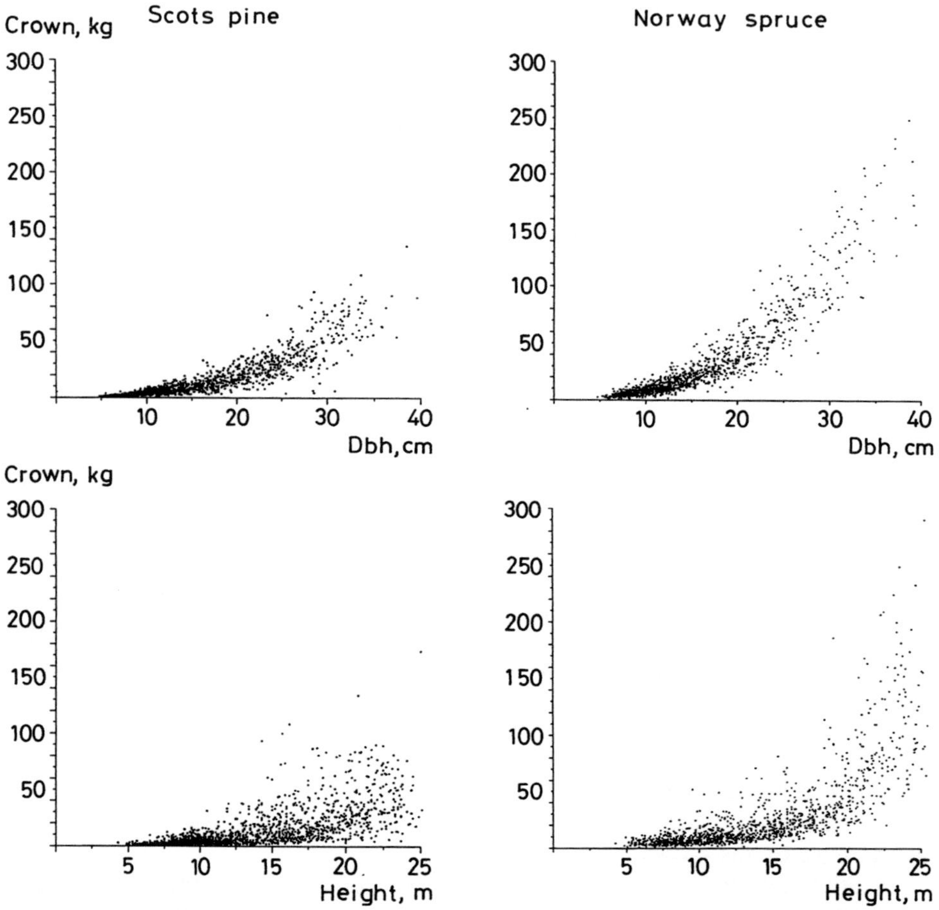

Fig. 2.19. Tree-to-tree variation of dry crown mass of *Pinus sylvestris* and *Picea abies* at the phase of removal in selective thinnings from below and final cuttings in southern Finland, plotted against dbh and tree height

Variation of dry crown mass in kilogram per tree in *Pinus sylvestris* and *Picea abies* is plotted in Fig. 2.19 against breast height diameter and height. The variance increases with tree size. The relationship follows the same pattern in both species, although spruce has an essentially larger crown mass. The correlation with diameter is closer.

There is a general trend toward a decrease in the percentage of crown mass, related to both stem mass and whole-tree mass, with increasing tree age and size (Fig. 2.20). Compared with the crown mass in kilogram per tree, the variation in percentage of crown mass is even larger and more erratic, being largest for small trees. As explained in Section 2.1, tree height predicts the percentage of crown mass with better accuracy than does the breast height diameter.

When the goal is utilization of forest biomass and the approach to the problem is technical, crown mass is of much more importance than crown volume.

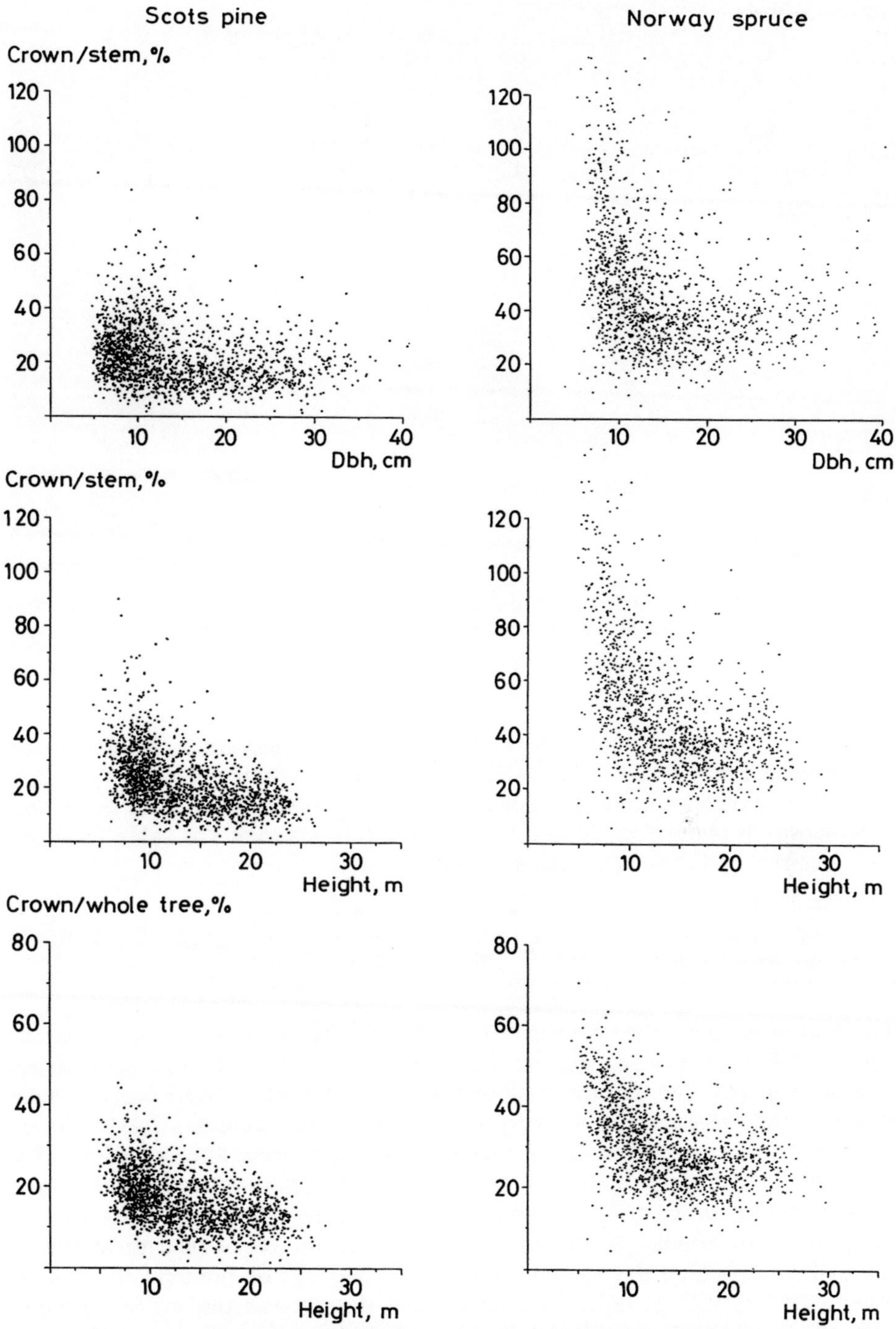

Fig. 2.20. Tree-to-tree variation of percentage crown mass as related to stem mass and whole-tree mass of *Pinus sylvestris* and *Picea abies* at the phase of removal in selective thinnings from below and final cuttings in southern Finland, plotted against dbh and tree height, on a dry mass basis

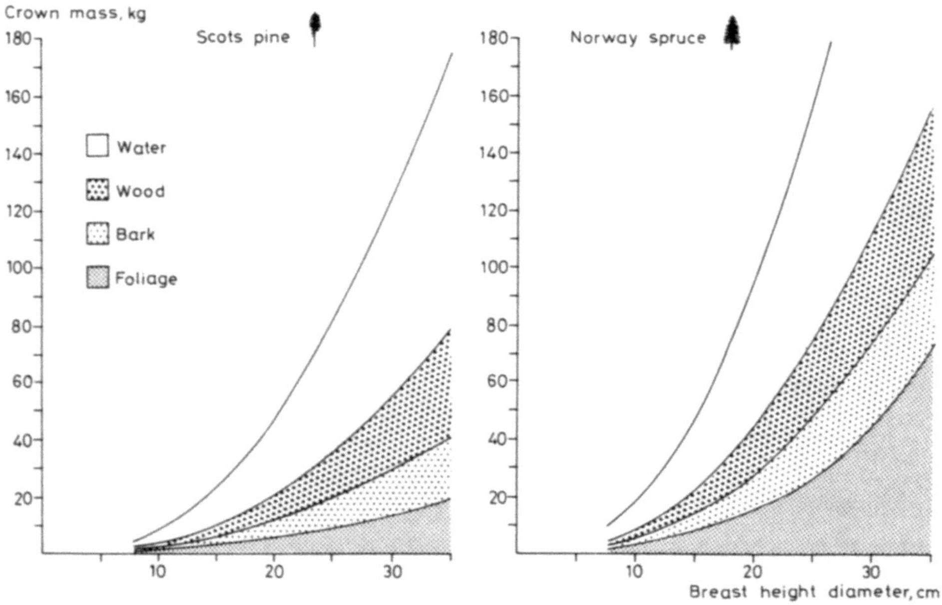

Fig. 2.21. The quantity and composition of fresh and dry crown mass in *Pinus sylvestris* and *Picea abies* at the phase of removal in selective thinnings from below and final cuttings in southern Finland as a function of dbh

On the other hand, for purposes of practical forest management, in some cases crown volume may be more indicative. The conversion factor between these two variables is crown bulk density. For example, in northern *Picea abies* the crown bulk density is about 1 kg fresh mass per m^3 crown volume. The value increases with tree size as follows (Korotjaev 1984):

Breast height diameter (cm)	8	16	24	32
Crown bulk density (kg/m^3)	0.82	0.98	1.14	1.30

The utilization properties of crown mass depend on the distribution of wood, bark, and foliage. The proportions vary with tree species, tree size, stand structure, season, etc. The amount and composition of fresh crown mass in *Pinus sylvestris* and *Picea abies* in southern Finland are illustrated in Fig. 2.21. The percentage of foliage in crown mass is highest in Norway spruce due to the longevity of its needles. Consequently, the total amount of crown mass available in a cutover area of spruce or other tolerant softwood species is largely dependent on the recovery of foliage. If whole trees or branches are let to shed their foliage through transpiration drying before removal, the yield of biomass for fuel or other purposes will be radically reduced. For deciduous trees, especially large trees, the foliage factor is of minor importance, because the percentage of foliage is much lower. Deciduous trees are leafless during the dormant season, and even in summer harvesting their foliage is usually lost gradually in succeeding opera-

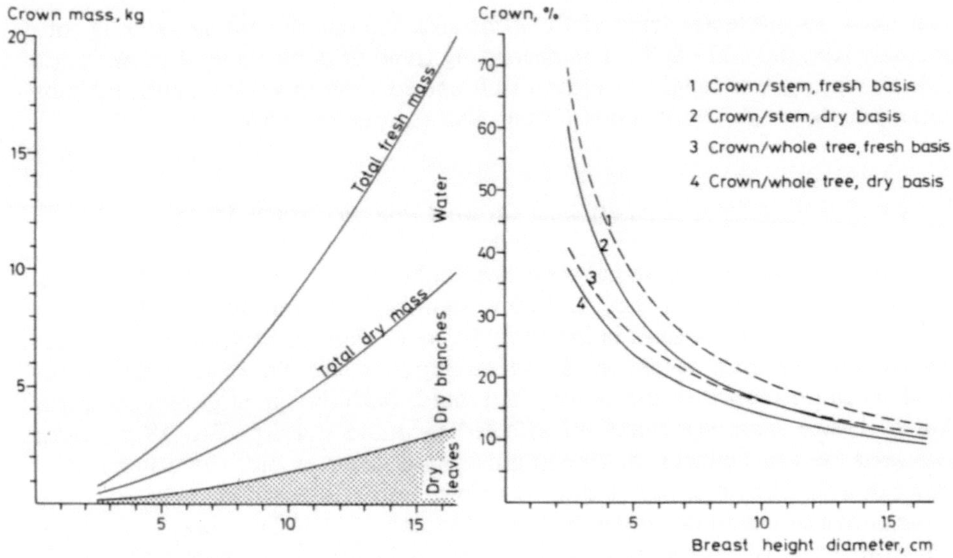

Fig. 2.22. Fresh and dry mass of crown in kg per tree, in percent of stem mass, and in percent of whole-tree mass. Average values of 11 puckerbrush hardwood species in Maine, as a function of dbh (drawn from Ribe 1973)

tions of the recovery system, unless special care is taken to protect the foliage component.

If a high wood content is desirable, crown mass from large trees is the best alternative. If the primary use is to be fodder or feedstock for the chemical industries and a high foliage content is desired, crown mass from small trees is more attractive. The same decreasing trend in the proportion of foliage was demonstrated by Hepp and Brister (1982) in an extensive study based on 182 sample plots of *Pinus taeda* plantations in the lower coastal plain of the Carolinas. The foliage percentage in the dry mass of crown was 58, 48, 37, and 29 in 5-, 10-, 20-, and 30-cm-diameter trees, respectively. Considering the danger of nutrient loss from forest soil, removal of crown material from large trees appears to be less risky.

Figure 2.22 represents puckerbrush stands in the northeastern United States. Small-sized hardwoods from these stands cannot be harvested economically using conventional stem-only methods. Nevertheless, they are an important potential source of whole-tree chips for energy or other purposes. According to Ribe (1973), the crown mass of 11 puckerbrush species was on the average 31% of the stem mass and 24% of the whole-tree mass of a 5-cm tree, on a dry mass basis. For a 15-cm tree the respective values were only 12 and 10%. As a rule of thumb, for small puckerbrush trees the crown-mass percentage of the whole-tree mass roughly doubles when tree height is reduced to one-half. Simultaneously, the proportion of foliage mass increases.

Independent of tree size, the percentage of foliage and that of supporting branches are highest in tolerant species, such as spruce, fir, and hemlock. In 2-m-high trees of tolerant species, the crown mass may be many times larger than the stem mass. In intolerant trees of the same size, the corresponding value is considerably less, only 60–80%. The decreasing trend in percentage of crown mass with increasing tree height is clear in all species, the inversion being perhaps somewhat slower for hardwoods (Young and Carpenter 1967).

2.1.1.4.4 Effect of Stand Density and Crown Class on Crown Mass

The growth increment in stem wood accumulates in its entirety into the tree, whereas leaves, whole branches, and outer stem bark die and fall off when a tree grows. Due to the differences in longevity between biomass components, the percentage of crown mass decreases with increasing stem size. The development is accelerated by the closing of the canopy in a stand. Both the quantity and composition of crown mass are therefore affected by stand density. Two 19-m *Abies balsamea* trees in Ontario, one open-grown and 42 years old, the other forest-grown and 62 years old, are compared below. Half of the wood and bark mass in the crown of the forest-grown tree is composed of dead branches, and all live branch material is less than 25 mm in diameter. In the open-grown tree, branch sections exceeding 25 mm in diameter are an important crown component (Honer 1971).

	Distribution of dry crown mass (%)	
	Open-grown balsam fir	Forest-grown balsam fir
Live branches:		
– 6 mm	20.0	24.0
6–25 mm	21.1	13.2
25–70 mm	18.5	–
Dead branches	14.2	38.0
Foliage	26.2	24.8
Total crown	100.0	100.0

Stand density generally has a negligible effect on tree height growth, number of whorls per meter of crown length, and number of branches per whorl. Factors contributing to nodal elongation and axial bud formation are under strong genetic control and independent of thinning treatment (Kärkkäinen 1984). In a 23-year-old *Pinus radiata* plantation in New Zealand, branches above 80% of the total tree height seemed to be free from competition for light irrespective of stand density. In the lower crown, at a certain stand density, light level became critical for growth and longevity of branches. Consequently, the following crown characteristics were dependent on thinning: height to the lowest live whorl, crown length, crown ratio, number of live whorls, number of live branches, mean branch diameter, and branch cross-sectional area (Siemon et al. 1976). Figure 2.23 shows the close relationship between the total branch cross-sectional area of the crown and the stem cross-sectional area at the crown base. Both variables are smallest in unthinned stands and highest in radically thinned stands.

Figure 2.24 shows the ratio of crown mass to stem mass, or *branchiness ratio*, in *Pinus sylvestris* stands of various densities at the age of 19–25 years in southern Finland. Although the stands are relatively young, the influence of spacing may be seen clearly. The percentage of crown mass is smaller on poorer *Vaccinium* sites. On the other hand, the branchiness ratio of suppressed trees is smaller than that of dominant trees (Kellomäki and Väisänen 1986).

Compared with shaded, suppressed trees, the dominant trees of a stand have a larger proportion of their crown mass in the upper sun crown. This results in a lower crown bulk density in the suppressed trees. The following example illustrates the crown bulk density in trees of various sizes in a *Picea abies* stand in Switzerland (Burger 1939a):

Dbh (cm)	Fresh mass, kg/m^3 crown volume		
	Foliage	Branches	Total
5	0.21	0.15	0.36
10	0.67	0.47	1.14
20	0.99	0.79	1.78
30	0.91	0.95	1.86

Figures 2.25 and 2.26 show the percentage of crown mass in trees removed in selective thinnings and final cuttings in Finland. The potential raw material reserve in the crown could theoretically supply additional biomass of as much as 21–34% in *Pinus sylvestris*, 48–54% in *Picea abies*, and 16–21% in *Betula* spp., compared with stem mass. However, it is not possible to recover crown mass completely by any harvesting system in use today. Dead branches, in particular, break off during logging. The percentage of crown mass in all three tree species is highest in trees harvested from early thinnings, mainly due to the small tree size. In intolerant pine and birch stands, the percentage of crown mass declines in successive treatments. Despite their taller size, the dominant and codominant spruce trees removed in final cuttings seem to have a higher percentage of crown mass than the suppressed spruce trees in late thinnings.

2.1.1.4.5 Effect of Tree Species on Crown Mass

Tree age and size, stand density, crown class, and site have different effects on different species in various geographic areas. Two trees of the same species and age, one growing under optimum conditions and the other at the extreme limit of its geographic range, may differ greatly in mass and composition of the crown. Trees growing in dense, unmanaged forests will also differ in the average percentage of crown mass from those in managed, repeatedly thinned forests. This is not only due to differences in stand density but also to the fact that a large part of the suppressed trees were removed earlier from the thinned stands. Consequently, only a rough comparison of percentage of crown mass between tree species is possible.

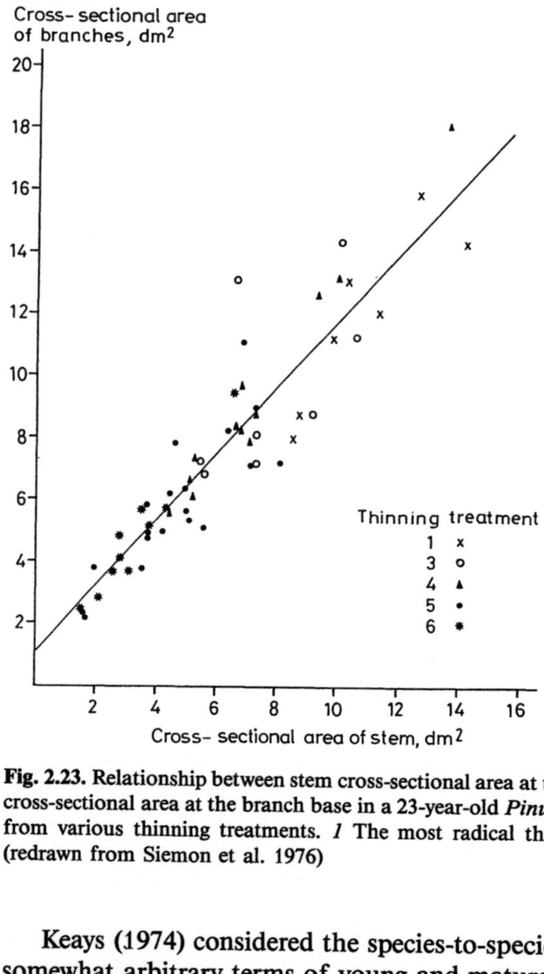

Fig. 2.23. Relationship between stem cross-sectional area at the base of the live crown and total branch cross-sectional area at the branch base in a 23-year-old *Pinus radiata* stand in New Zealand, resulting from various thinning treatments. *1* The most radical thinning and *6* an unthinned control plot (redrawn from Siemon et al. 1976)

Keays (1974) considered the species-to-species variation in crown mass in the somewhat arbitrary terms of young and mature trees, young trees being defined as those having a breast height diameter smaller than 7.5 cm, while trees with a breast height diameter of 15 cm or more were classified as mature. An analysis of the available literature provided the following summary table:

	Mass percent of stem mass		
	Branches	Foliage	Crown
Softwoods, young	25	25	50
Softwoods, mature	15	10	25
Hardwoods, young	30	15	45
Hardwoods, mature	25	5	30

According to Keays, the above values are applicable primarily to most *Abies, Picea, Pinus, Pseudotsuga, Thuja,* and *Tsuga* species among softwoods, and to most *Acer, Betula, Fagus, Populus,* and *Quercus* species among hardwoods. The

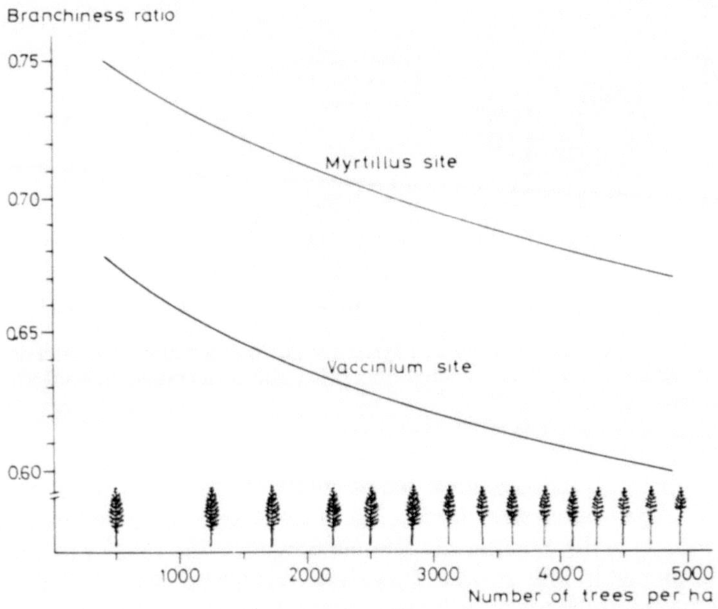

Fig. 2.24. Branchiness ratio (branch mass/stem mass) of *Pinus sylvestris* growing on dry *Vaccinium* sites and moist *Myrtillus* sites in southern Finland, as a function of stand density as indicated by the number of trees per ha (redrawn from Kellomäki and Väisänen 1986)

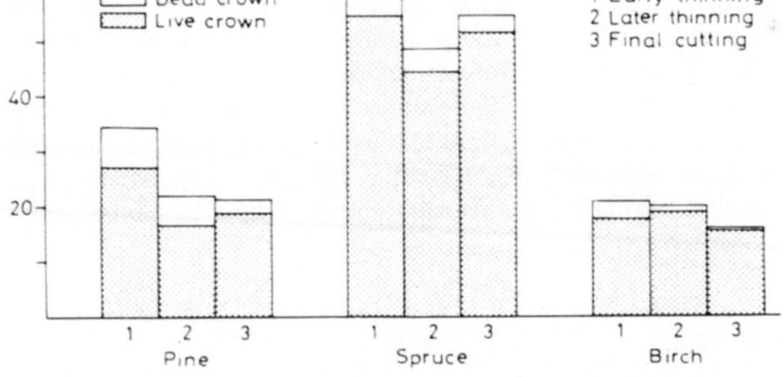

Fig. 2.25. Crown mass in percent of stem mass in *Pinus sylvestris*, *Picea abies*, and leafless *Betula* spp. at the phase of removal in selective thinnings from below and final cuttings in southern Finland, dry mass basis

majority of individual values for mature trees were based on data for trees with a breast height diameter of close to 20 cm. The reliability was greater for mature trees.

These general values are not applicable, however, to the estimation of crown mass for a particular species. The effect of shade tolerance and longevity of

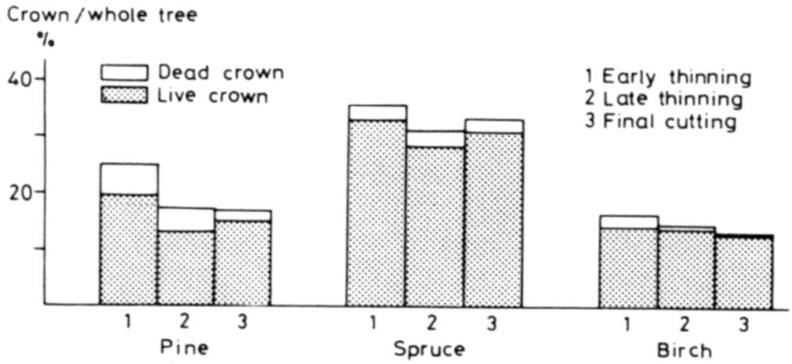

Fig. 2.26. Crown mass in percent of whole-tree mass in *Pinus sylvestris*, *Picea abies*, and leafless *Betula* spp. at the phase of removal in selective thinnings from below and final cuttings in southern Finland, dry mass basis

foliage is clear. Tree species shedding their leaves annually generally have the smallest foliage mass. Large branch size means a low percentage of bark and often a low percentage of foliage in the crown mass. Branch longevity and tenuity affect the mass of dead branches in the crown. *Tsuga* and *Thuja* are very tolerant species. *Abies* and *Picea* are tolerant as well but usually demand somewhat more light. *Pseudotsuga* is an intermediate species. Most *Pinus* species are intolerant, but *P. strobus* may be included in intermediate trees. *Larix* species are all intolerant (Sarvas 1964).

Table 2.12 summarizes data on the dry mass of crown from two North American studies, one from Maine (Young et al. 1980) and the other from British Columbia (Standish et al. 1985). These studies were selected to illustrate variations between tree species, because both of them represent comprehensive softwood and hardwood samples collected under comparable conditions. Both are from naturally regenerated, unmanaged forests, and in both cases the crown mass data are given for trees of about 20 cm in diameter. Because biomass is affected by many variables, and due to possible differences in sampling, individual species information must be considered only as suggestive. Hardwoods particularly are variable in stem form and branching habits and, consequently, in percentage of crown mass.

2.1.2 Unmerchantable Top

The merchantable stem is the main product of forestry. Since it is not a component of the residual forest biomass, it is outside the scope of this review. When timber is harvested by conventional methods, the stem is usually topped on the site or at the landing at some fixed minimum diameter or at an arbitrary stem position governed by the quality and length requirements of the timber. The raw material potential of the resulting unmerchantable top is analyzed briefly in the present section.

Equations have been developed to predict the biomass of tree components, but these equations seldom estimate the mass of the unmerchantable tops separately.

Table 2.12. Crown mass in percent of stem mass in trees of about 20 cm in diameter of several softwood and hardwood species in naturally regenerated unmanaged forests in Maine (Young et al. 1980) and British Columbia (Standish et al. 1985), on a dry mass basis

Tree species	Mean dbh (cm)	Mass percent of stem mass		
		Branches	Foliage	Crown
Softwoods in Maine:				
Abies balsamea	21	8.5	14.8	23.3
Picea rubens	21	16.1	18.3	34.4
Pinus strobus	21	16.8	7.8	24.6
P. resinosa	21	18.1	12.4	30.5
Tsuga canadensis	21	26.4	20.5	46.9
Thuja occidentalis	21	29.8	16.1	45.9
Larix laricina	21	21.2	10.4	31.6
Softwoods in British Columbia:				
Abies amabilis	16	23.4	25.8	49.2
A. grandis	23	17.0	15.8	32.8
A. lasiocarpa	16	37.7	20.0	57.7
Picea sitchensis	17	30.8	26.4	57.2
P. glauca	19	19.3	13.2	32.5
P. engelmannii	24	18.5	9.9	28.4
P. mariana	17	18.7	13.2	31.9
Pinus ponderosa	28	24.6	7.7	32.3
P. monticola	19	19.4	10.1	29.5
P. contorta var. *latifolia*	19	11.8	5.5	17.3
P. contorta var. *contorta*	16	26.5	14.1	40.6
Pseudotsuga menziesii	21	11.7	6.5	18.2
P. menziesii var. *glauca*	26	23.9	8.8	32.7
Tsuga heterophylla	19	14.5	6.5	21.0
T. mertensiana	20	12.7	7.9	20.6
Thuja plicata	22	22.4	12.7	35.1
Chamaecyparis nootkatensis	18	11.1	10.4	21.5
Larix occidentalis	24	12.1	2.4	14.5
Hardwoods in Maine:				
Betula alleghaniensis	21	22.2	4.7	26.9
B. papyrifera	21	10.2	4.7	14.9
B. populifolia	21	5.2	2.1	7.3
Fagus grandifolia	21	21.1	3.7	24.8
Acer saccharum	21	14.4	3.0	17.4
A. rubrum	21	11.9	4.3	16.2
Populus spp.	21	11.8	4.0	15.8
Prunus pensylvanica	21	13.4	9.6	23.0
Salix spp.	21	12.5	6.5	19.0
Hardwoods in British Columbia:				
Alnus rubra	13	29.8	6.0	35.8
Betula papyrifera	15	22.8	6.0	28.8
Populus tremuloides	18	17.0	3.2	20.2
P. trichocarpa	16	34.5	7.6	42.1

Most of the information on tops is available on a volume basis due to the fairly regular form of the tree tops and the ease of measurement, particularly in softwoods. However, because the basic density of wood and bark varies in the longitudinal direction of a tree stem (Sect. 3.4.2), the quantitative measurements should be based on mass rather than volume, preferably on a dry mass basis. The top mass is expressed either in kilogram per tree or in percent of the total dry mass of a stem.

The top mass is primarily dependent on the minimum top diameter of merchantable timber. It is also a function of stem geometry, the basic density and moisture content of wood and bark, and percentage of bark in tops. Because these variables are not consistent, caution must be exercised when converting percentage of volume to percentage of mass.

2.1.2.1 Percentage of Bark in the Unmerchantable Top

In many tree species the unmerchantable top is composed partly of juvenile wood (Sect. 3.1.3). In spite of this, the top is still the component of residual forest biomass which in many respects differs least from normal, mature stem wood. If tops are used for manufacturing of pulp or composite boards, the most important technical difference between the merchantable stem mass and unmerchantable top mass is obviously their bark content.

Tree bark consists of two parts: outer dead bark and inner live bark. The function of *the outer bark* is to protect the tissues of the inner bark, cambium, and xylem against drying and injuries caused by animals, fungi, fire, etc. The outer bark of all tree species is thickest at the butt of the stem because the need for protection is greatest near ground, and because the bark is oldest at the stem base. The primary function of *the inner bark* is to transport assimilation products from the tree crown and, therefore, its thickness is usually fairly constant from butt to top. Consequently, the composition of bark varies in the longitudinal direction of a tree. The proportion of the inner bark is highest in the unmerchantable top (Fig. 2.27). This is important from the pulping point of view, because the inner bark is less harmful in pulp manufacturing than the thick outer bark.

As the outer bark becomes thinner toward the top of a tree, the total bark thickness changes accordingly, following a species-specific pattern. Figure 2.28 gives an example of bark thickness variations in 15-m stems of *Pinus sylvestris*, *Picea abies*, and *Betula* spp. in southern Finland. The change from the butt upward is greater in Scots pine and birch, both having typically rough butt bark. In Norway spruce the change is less rapid, and of the three species spruce actually has the thickest bark in the top.

The dry mass of bark at a certain stem height is a result of stem diameter, stem form, bark thickness, and the basic density of the bark. Figure 2.28 shows the dry mass of bark in grams per linear millimeter of stem. Bark mass in kilogram per linear meter of stem has the same numerical value and is a more meaningful unit for practical forestry. The figure thus also shows the dry mass of bark in a 1-m bolt from any height of a stem.

The longitudinal variation in bark and wood mass of a stem affects the percentage of bark. Figure 2.29 illustrates the variation pattern of the percentage of

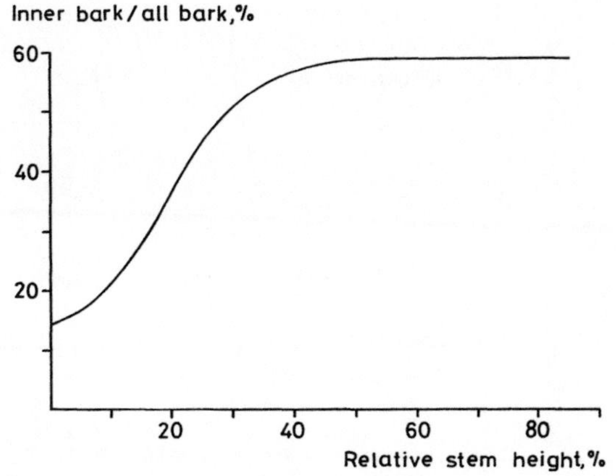

Fig. 2.27. The proportion of the inner bark in the total bark mass at different relative heights of 10–20-m-high *Pinus sylvestris* trees in southern Finland, dry mass basis (Hakkila 1967)

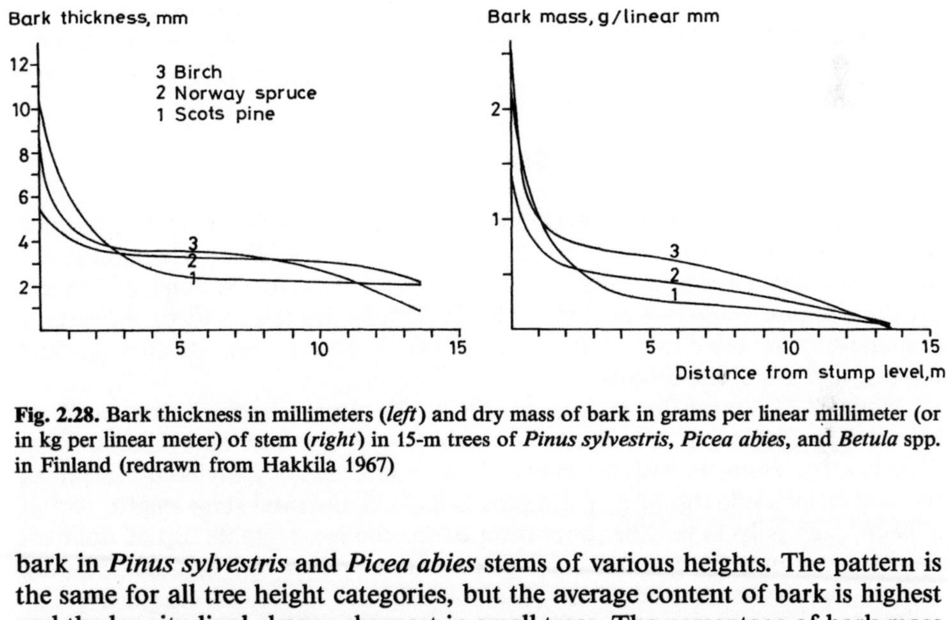

Fig. 2.28. Bark thickness in millimeters (*left*) and dry mass of bark in grams per linear millimeter (or in kg per linear meter) of stem (*right*) in 15-m trees of *Pinus sylvestris*, *Picea abies*, and *Betula* spp. in Finland (redrawn from Hakkila 1967)

bark in *Pinus sylvestris* and *Picea abies* stems of various heights. The pattern is the same for all tree height categories, but the average content of bark is highest and the longitudinal change sharpest in small trees. The percentage of bark mass decreases in Scots pine from ground level to about 40% of the height, and in Norway spruce to about 20% of the height. Thereafter, the percentage starts rising with increasing speed toward the top, reaching 12–14 in Scots pine and 15–17 in Norway spruce at top diameter of 7 cm. The proportion of bark in the uppermost top may exceed 50%. These very high values have been measured for the uppermost top in other tree species as well, for example, in *Populus grandidentata* and *P. tremuloides* (Zavitkovski 1971).

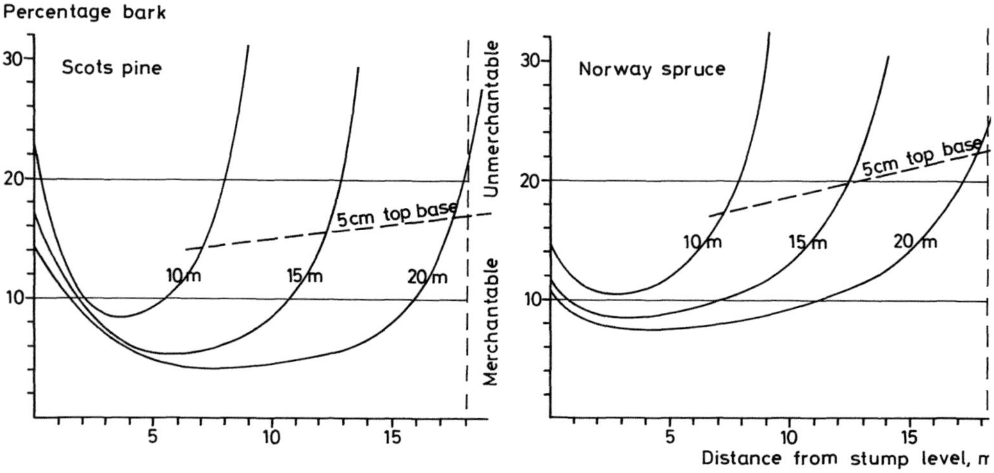

Fig. 2.29. Vertical variation of percentage of bark in 10-, 15-, and 20-m stems of *Pinus sylvestris* and *Picea abies* in Finland, dry mass basis. The *dashed line* indicates the base of 5-cm unmerchantable top (redrawn from Hakkila 1967)

2.1.2.2 Mass of Unmerchantable Top

Most of the quantitative information on the unmerchantable top is on a volume basis. Stem volume is accumulated at the base of the tree. While a 20% butt section of the stem length contains about 40% of the total stem volume, a 20% top section contains only 2–3% (Fig. 2.30). Variations are the result of differences in stem form and taper in the top of a tree depending on species, site quality, stand density, and wind conditions.

Figure 2.31 shows the average length of the unmerchantable top in *Pinus sylvestris* in Finland as a function of breast height diameter and minimum top diameter. For example, with a 5-cm minimum diameter for merchantable timber, the unmerchantable top of a 10-cm pine is 33% of the total stem length, and of a 30-cm pine only 11%. The proportion of the unmerchantable top of the total stem volume may be calculated by placing these values in Fig. 2.30. The top volume then is 7% for the 10-cm pine but only 1% for the 30-cm pine.

Both the moisture content and the basic density of the unmerchantable top differ from the average values for the stem (Sect. 3.4.2). The average moisture content usually increases toward the top, partly because the top does not contain heartwood. The trend, for basic density particularly, is species-specific. The basic density of pine wood decreases from butt to top, where the proportion of thin-walled juvenile wood is high. The same pattern is typical of many other conifers with a relatively short live crown. The percentage of top material is less for these species on a dry mass basis than on a volume basis. The variation pattern is different for spruces and many other tolerant conifers with a long-live crown. A slight decrease in basic density from the butt upward changes to an increase in the upper half of the crown (Fig. 3.41). The variation is smaller and more irregular in hardwoods (Fig. 3.42).

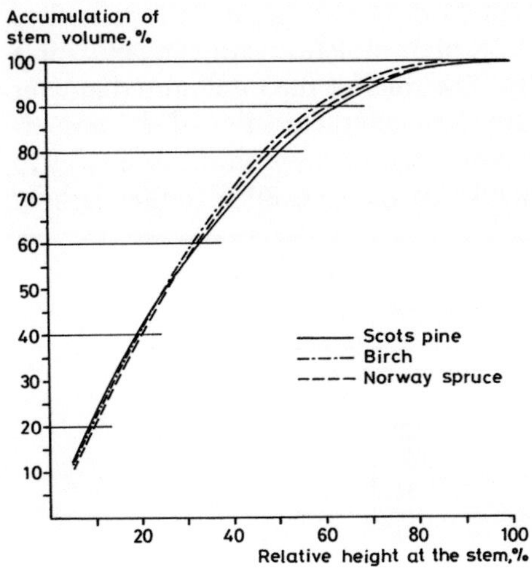

Fig. 2.30. Accumulation of stem volume in *Pinus sylvestris*, *Picea abies*, and *Betula* spp. in Finland as a function of relative height at the stem (redrawn from Laasasenaho 1975)

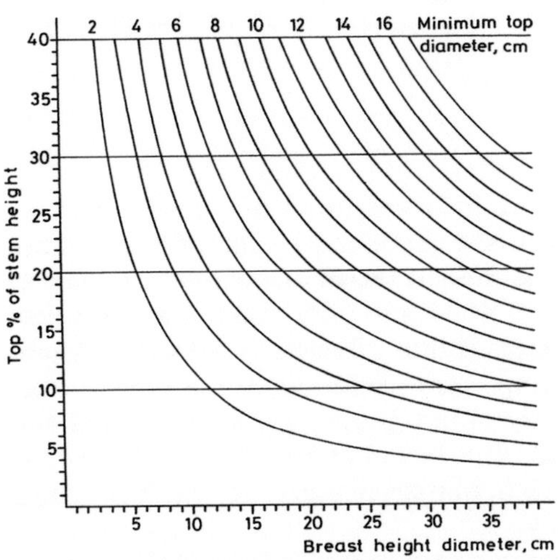

Fig. 2.31. The length of the unmerchantable top in percent of the total tree height in *Pinus sylvestris* in Finland, as a function of breast height diameter (over bark) and the minimum top diameter of merchantable timber (under bark) (redrawn from Laasasenaho 1975)

Alemdag (1982a) related the percentage of merchantable stem mass (M) to the ratio of the minimum top diameter (Dm) to breast height diameter (dbh). The best coefficients of determination were obtained by functions containing a ratio variable in the form of Dm/dbh. The introduction of tree height to these equations did not result in any considerable improvement over the Dm/dbh ratio for estimating percentages of merchantable stem mass. Since the merchantable portion of a stem may in some cases be defined by its height instead of its top diameter, the percentage of merchantable mass could also be estimated using the ratio of the merchantable height (Hm) to total height (Ht):

$$M,\ \% = b_0 + b_1\,(Dm/dbh) + b_2\,(Dm/dbh)^2$$

$$M,\ \% = b_0 + b_1\,(Hm/Ht) + b_2\,(Hm/Ht)^2\ .$$

The potential yield of additional biomass from unmerchantable tops is thus characterized by a high percentage of bark, a high moisture content, and for a number of species, a low basic wood density. The smaller the minimum diameter of merchantable timber, the more critical are these characteristics of the unmerchantable top. The following example represents tops of large sawtimber trees in southern Finland when the minimum diameter of merchantable timber is only 6 cm (Hakkila 1969):

	Pinus sylvestris	*Picea abies*
Moisture content in top (%):		
Top wood	60.5	58.7
Top bark	63.0	58.0
Average	60.8	58.5
Bark, percent of dry top mass	13.2	19.6
Basic density (kg/m^3):		
Top wood	372	393
Merchantable stem	417	392

2.1.3 Stump and Roots

Of all biomass components of a tree, the stump, roots, and stump-root system are least clearly defined and standardized in the literature. In fact, the terms are misleadingly used interchangeably. In the present review the *stump-root system* is defined as all below- and above-ground wood and bark mass of a tree below the stump cross-section (Fig. 2.32). Its mass is thus affected by the stump height which refers to the distance of the stump cross-section from the ground level (Sect. 4.4.1.1). *Stump mass* consists of the stump proper above the ground and its direct continuation below the ground, thus including the taproot. *Root mass* consists of the side roots or laterals. The minimum diameter to which the roots are measured varies from study to study.

Three distinct parts may be separated in a stump-root system: the root crown at the base of the stem; the rapid-taper section, comprising the root parts which provide lateral support and anchorage; and the network, where thinner roots meander and intermingle and may extend several meters laterally without noticeable change in diameter. The area occupied by the roots of eastern hardwoods in the United States is usually a little more than twice the diameter of the crown (Stout 1968).

In spite of the massive amount of high-quality fiber hidden in the stump-root systems, the problems of harvesting and processing prevent their large-scale utilization in the forest industries. In the field of wood science and forest engineering, the interest in laborious stump and root mass studies has until recently been rather limited. Since the roots play a vital role in life functions, such as nutrient and water uptake of a tree, much more attention has been paid to them in ecological studies. Nevertheless, the estimate of roots is always an ecologist's

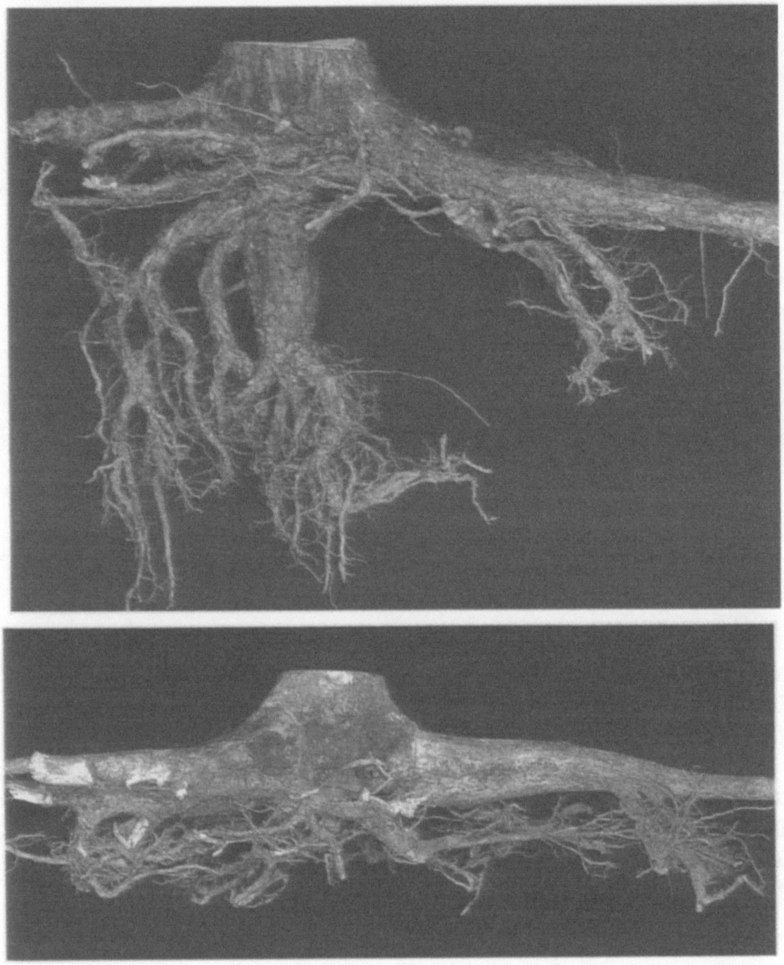

Fig. 2.32. Stump-root system of *Pinus sylvestris* (*above*) and *Picea abies* (*below*) (courtesy of Lalli Laine)

nightmare, as stated by Harpner (1967), and in many experimental stands the disturbance caused by root studies is unacceptable. The following root classification for ecological purposes by Böhm (1979) indicates that in these studies the emphasis is largely on the thinnest root parts if not, even more, on root hairs and mycorrhizas.

From the utilization point of view, the interest is in the stump and large sections of roots. Since studies of the below-ground biomass are slow and laborious, field procedures often require compromises which reduce comparability. The following factors frequently cause differences among the meager studies on stump and root wood:

Root category	Diameter over bark (mm)
Very fine	−0.5
Fine	0.5−2
Small	2−5
Medium	5−10
Large	10−20
Very large	20−

1. Stump height varies depending on logging practice. For instance, winter logging in deep snow results in high stumps. On the other hand, severing with shears produces lower stumps than motor-manual cutting with a chain saw.
2. The minimum diameter of root sections actually included in the total mass of the stump-root system varies from study to study depending on the goal and method of extraction.
3. Independent of the minimum diameter, part of the roots is always lost and left in the ground, depending on the extraction method.
4. Grafted unions between neighboring trees may sometimes make it impossible to determine which roots belong to each partner.

According to the limited data available, stumps and larger roots comprise a potential biomass reserve that, with some allowance for losses in the harvesting phase, is roughly 20% compared with the stem mass. The solution model for a recovery system depends on which part of this reserve is to be utilized. Recovery of stump-root systems complete with fine roots is neither ecologically advisable nor technically possible. More practical goals may be just lowering the stump height, recovery of the whole above-ground part of the stump, recovery of the stump and its below-ground continuation in their entirety, recovery of the central stump and root mass within a certain radius around the stump center, or recovery of the entire stump and root mass to a certain minimum diameter, e.g., 10 or 5 cm (Fig. 2.33).

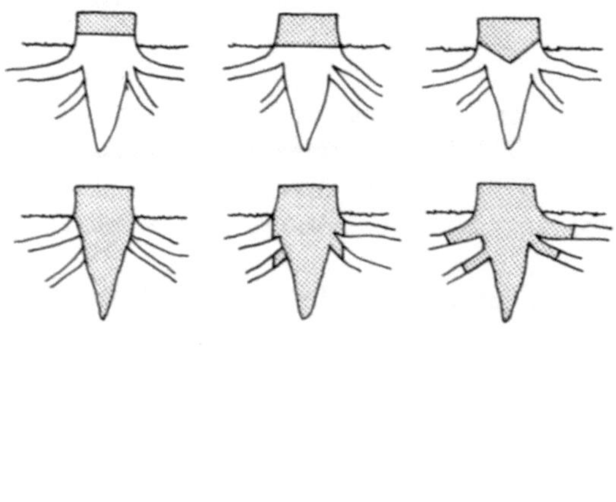

Fig. 2.33. Alternative degrees of stump and root mass recovery

2.1.3.1 Distribution of Stump-Root System Components

Machine design, cost calculations, and systems analyses require data not only of the total mass in stump-root systems but also of the distribution of its components. Machine performance, efficiency of recovery, and feasibility of alternative harvesting methods under different stand conditions may be compared only when these variables are known. Unfortunately, quantitative information on the components is less consistent than that for the whole stump-root systems.

Methods of stump and root mass recovery are being developed along two different lines. Either only the stump mass, i.e., the stump and its below-ground continuation is harvested, or the stump-root system is recovered with the laterals within a given radius or to a given diameter. The suitability of a method depends on the structure of a root system, which varies greatly depending on tree species and soil type. Stump-root systems of 20 hardwood species in the southern United States were characterized from the recovery point of view by Koch (1985a).

Figure 2.34 shows the distribution of the dry mass of over 5-cm components in the stump-root systems of sawtimber-sized *Pinus sylvestris* and *Picea abies* in a large sample from southern Finland. The former represents a species with a distinct taproot, although it does not develop in a cold climate as well as under southern conditions. More than one-half of the mass of the Scots pine stump-root system is contained in the central stump. An exception is in wet swamps where Scots pine does not develop a taproot, and only one-third of the dry mass is in the central stump. Norway spruce has a very superficial root system without any taproot at all. The central stump contains only one-third of the stump and root mass on mineral soils, and even less in swamps, whereas thick lateral roots are an important mass component, particularly in drained peatlands. Taking into account only root parts over 5 cm, the following figures are obtained in an average sawtimber stand in southern Finland when different recovery alternatives are applied.

Portion harvested	Cumulative efficiency of recovery (%)	
	Scots pine	Norway spruce
Above-ground portion of stump	15	15
Central stump incl. taproot	58	35
Central stump + over 20-cm roots	66	60
Central stump + over 10-cm roots	85	85
Central stump + over 5-cm roots	100	100

Roots thinner than 5 cm are not included in this efficiency evaluation, since their utilization seems unrealistic. They break easily when stump-root systems are extracted and are, therefore, largely left in the ground. They are difficult and expensive to clean and the thinnest roots give low-quality pulp and lower yield than the stump and larger roots. Since a considerable part of the nutrient supply in root systems is located in the thinnest roots, there are also ecological constraints against their removal (Sect. 9.1.3).

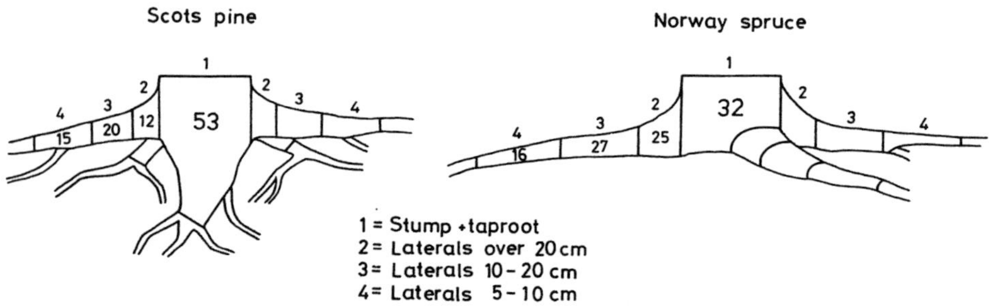

Fig. 2.34. Relative distribution of dry mass in stump-root systems of sawtimber-sized *Pinus sylvestris* and *Picea abies* in southern Finland. Under 5-cm root sections are excluded (Hakkila 1975)

Despite the many reasons why root parts under 5 cm should not be utilized, for ecological reasons it is useful to know their biomass content in relation to the recoverable stump and root mass. According to Mälkönen (1972), 1–5 cm roots would add one-third to the mass of the thicker stump and root components. In Sweden, practical harvesting operations have demonstrated that if the minimum diameter is reduced from 5 to 2 cm, the recovery of raw material from stump-root systems of *Pinus sylvestris* and *Picea abies* may be increased by 10–12% when excavator-mounted stump harvesters are used (Magnusson and Nylinder 1977).

The proportion of roots thinner than 5 mm is particularly high in the early development of a tree but significant also in mature trees. Heth and Donald (1978) estimated that roots under 5 mm comprise 18% of the total root mass in 27-year-old trees and 6% of that in 36- to 39-year-old trees of *Pinus radiata* in South Africa. In a *Pinus sylvestris* plantation, roots under 5 mm comprised 87% of the total root mass at an age of 7 years, and 37% at 55 years (Ovington 1957). In a 5-year-old *Pinus elliottii* plantation in Florida, fertilizer treatment of 392 kg/ha diammonium phosphate increased the total root mass 58% and the mass of roots under 5 mm by as much as 145% (White et al. 1971).

2.1.3.2 Percentage of Bark in Stump and Root Mass

Variation in stem bark thickness is largest at the butt of a tree. This results in a wide within-stand and between-stand variation in the percentage of bark in the stump proper. On the average, the percentage of bark in the stump and in its below-ground continuation does not differ significantly from that in the stem. In roots the percentage of bark increases with increasing distance from the root crown (Fig. 2.35).

In practice part of the bark falls off during presplitting, haulage, and storage of stump and root pieces. The percentage of bark will thus be lower when the material arrives at a pulp mill. A decisive factor is the length of the storage time. In long-time storage of over a year, weathering and insect activity may result in the disappearance of a majority of the bark. Concurrently, dry matter losses may occur also in the wood proper.

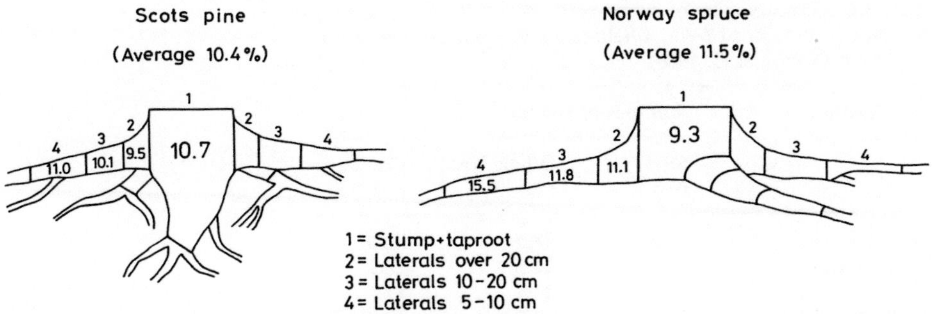

Fig. 2.35. Percentage bark in various components of the stump-root systems of sawtimber-sized *Pinus sylvestris* and *Picea abies* in southern Finland (Hakkila 1975)

2.1.3.3 Dry Mass of the Stump-Root System

The dry mass of a stump-root system can be estimated by species from the stump diameter. However, the confidence limits for a single stump-root system are broad. Stand density, wind conditions, topography, and soil are among the influencing factors. Of all tree components, the roots are most sensitive to site conditions such as soil structure, moisture availability, soil temperature, and nutrient supply. Care taken in felling affects the relationship as well, for increasing stump height means a smaller stump diameter but additional residual mass, thus disturbing prediction functions based on stump diameter alone.

Figure 2.36 shows the dry mass of *Pinus sylvestris* and *Picea abies* stump-root systems in mineral soils of southern Finland, with root parts thinner than 5 cm excluded. For example, if the stump diameter under normal logging conditions on mineral soils is 30 cm, the dry mass of a stump-root system is 30 to 40 kg. In a

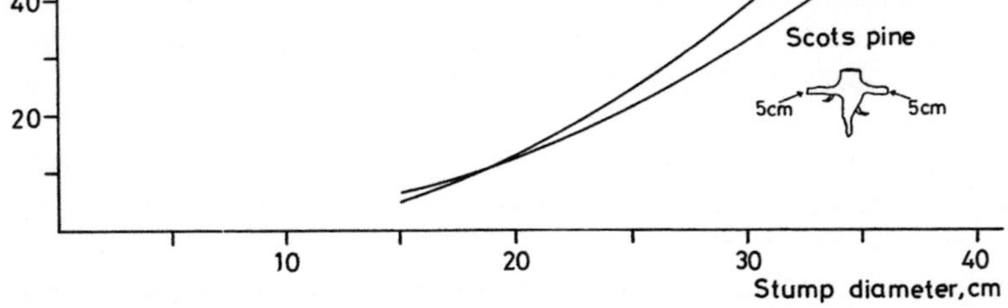

Fig. 2.36. Dry mass of the stump-root system of *Pinus sylvestris* and *Picea abies* as a function of dbh in southern Finland (Hakkila 1972)

Table 2.13. The dry mass of stump-root system of 13 tree species in Maine in percent of stem, whole-tree and complete-tree mass; dbh 30 cm (Young et al. 1980)

Tree species	Stump-root system		
	Stem (%)	Whole tree (%)	Complete tree (%)
Abies balsamea	35.0	28.3	22.0
Picea rubens	35.0	26.3	20.9
Pinus resinosa	27.7	21.4	17.6
P. strobus	28.5	23.2	18.8
Tsuga canadensis	40.9	27.5	21.6
Thuja occidentalis	29.3	21.0	17.3
Larix laricina	37.1	29.1	22.5
Softwoods	33.4	25.3	20.1
Acer rubrum	27.2	23.9	19.3
Acer saccharum	28.6	24.3	19.5
Betula alleghaniensis	33.4	26.3	20.8
B. populifolia	23.2	20.8	17.2
Fagus grandifolia	29.6	23.7	19.2
Populus spp.	23.4	20.4	16.9
Hardwoods	27.6	23.2	18.8

restricted sample from drained peatlands, the dry mass of Norway spruce stump-root systems was double that in mineral soils due to the marked butt swell and thick lateral roots. As shown in Section 2.1.3.1, thin roots would add considerably to these values, but their harvesting and utilization does not seem possible.

For various planning purposes, information is often needed on the dry mass of stump-root systems in percent of stem mass. These data are especially useful in calculations of the amount of stump and root wood potentially available to an industrial plant from a certain geographical area. It can be used also to estimate machine performance and the need for machines.

Table 2.13 shows the percentage of stump and root mass in seven coniferous and six hardwood species in Maine in relation to the dry mass of stem, whole tree, and complete tree. Although the sample is not adequate for general comparisons between conifers and hardwoods, the following indicative rough rule of thumb for northern growth conditions can be drawn from Table 2.13: the dry mass of a stump-root system is about 30% of stem mass, 25% of whole-tree mass, and 20% of complete-tree mass. Taras (1978), on the basis of findings of several investigations on southern pine species in southern United States, considered 18±3% of the complete-tree biomass a reasonable estimate for southern pine root wood.

The percentage of stump and root mass is generally between 25 and 35 compared with the stem. Although the tree-to-tree and stand-to-stand variations are wide, they are far less extreme than for the crown components. In sawtimber-sized trees the variation in the percentage of stump and root mass with stump diameter is actually small. Prediction equations based on the breast height diameter, instead of stump diameter, have only slightly more power. The effect of soil and site is significant.

For the estimation of utilizable stump and root mass for industrial purposes, thinner roots must be deducted. In the Soviet Union the technically harvestable portion of stump-root systems of sawtimber-sized *Pinus sylvestris* was 22–23% (Kiščenko et al. 1962; Komšilov et al. 1963) and in Finland 23–24% (Hakkila 1972) as related to the stem on a dry mass basis. Comparison on a volume basis in Poland and Finland gave somewhat lower values, 15–22%, partly because the higher basic density of stump and root wood was not recognized (Kamin'ski 1961; Mäkinen 1965; Mälkönen 1972). These results are in agreement with studies of southern pine in the United States, where the following percentages of recoverable stump and root mass have been reported: 19% for *Pinus taeda* and 22% for *P. elliottii* and *P. palustris* (Koch and Coughran 1975); 22% for *P. elliottii* (Howard 1973c); and 22% for *P. elliottii* and *P. taeda* (Clark 1978). The proportion of over 5-cm stump and root mass of *Picea abies* was 22% in southern Finland and 30% in northern Finland not far from the northern limit of the natural range of the

Table 2.14. The dry mass of harvestable stump-root system in percent of stem and complete-tree mass of 22 pine-site hardwood species in the southern United States, when lateral roots are severed at 31- or 92-cm radius from the tree center (Koch 1985a)

Tree species	Central stump-root system with 31-cm radius		Central stump-root system with 92-cm radius	
	Stem (%)	Complete tree (%)	Stem (%)	Complete tree (%)
Fraxinus pennsylvanica	15.0	11.9	20.3	15.4
F. americana	18.6	12.7	27.0	17.4
Ulmus americana	28.0	17.7	37.6	22.5
U. alata	20.9	13.3	34.2	20.0
Celtis spp.	26.8	16.6	38.1	22.1
Carya spp.	30.0	18.7	38.1	22.7
Acer rubrum	27.9	12.2	49.2	19.6
Quercus velutina	22.4	15.7	31.8	20.8
Q. marilandica	34.1	14.2	58.5	23.4
Q. falcata var. *pagodaefolia*	27.4	16.7	37.7	21.6
Q. falcata var. *falcata*	23.5	14.8	33.6	20.0
Q. prinus	23.4	15.8	31.4	20.0
Q. laurifolia	22.9	15.0	33.5	20.4
Q. rubra	21.9	14.5	31.5	19.5
Q. stellata	27.0	18.6	34.9	22.8
Q. coccinea	25.3	17.2	31.8	20.8
Q. nigra	20.1	12.4	30.8	17.3
Q. alba	26.0	16.1	36.6	21.0
Magnolia virginiana	25.2	14.5	40.2	20.9
Liquidambar styraciflua	28.4	18.0	39.0	23.2
Nyssa sylvatica	32.9	19.1	46.6	25.1
Liriodendron tulipifera	15.9	11.7	25.6	17.6
Average of 22 hardwoods	24.7	15.3	35.8	20.6

species, where steep taper reduces the mass of a stem. Considerably higher values were observed on drained peatlands characterized by trees with marked butt swell and strong lateral roots. In some sawtimber Norway spruce trees on drained peatlands, the percentage of stump and root mass exceeds 50% (Hakkila 1972).

Koch (1985 a) studied stump-root systems of 22 pine-site hardwood species in the southern United States, the breast height diameter of the trees averaging 15 cm (Table 2.14). To confine mass data to material realistically available for utilization by existing or foreseeable equipment, the measurements were limited to the portion of the stump-root system that is within a radius of either 31 or 92 cm from the tree center. Since the sample consisted of only three trees per species and within-species variation was considerable even in a single diameter class, the data must be accepted with caution. However, Table 2.14 gives a clear indication of the proportion of utilizable stump and root mass in relation to the mass of the stem or complete tree. If the radius of the acceptable mass is 31 cm around the tree center, almost 25% additional mass is available compared with the stem. With a 92-cm radius, the proportion is as much as 36%. In a study on *Picea mariana* on organic soils in Ontario the proportion of recovered stump mass was 10–15% of stem mass, when a circular shear was hypothetically placed around the base of the stem and then forced downward through the root collar to sever all laterals from the root crown (Drolet 1973).

2.2 Composition of Whole-Tree Biomass

According to definition, whole-tree biomass consists of the stem and crown. The majority of the whole-tree mass of a mature tree is in the stem. Because external stem-size characteristics accurately predict not only the volume but also the mass of a stem, the whole-tree mass can be estimated from the breast height diameter and height of a tree. For rough orienting calculations, the breast height diameter alone is often sufficient to predict whole-tree mass. Within a certain diameter class the stem mass decreases with increasing stem taper, but this is compensated by a simultaneous increase in crown mass. Figure 2.37 shows *the effect of tree size* on the whole-tree mass of *Pinus sylvestris* and *Picea abies* in southern Finland. In the same breast height diameter class, these species differ only slightly in stem mass, but Norway spruce tends to have more whole-tree mass due to its heavier crown.

From the utilization point of view, data on the composition of whole-tree mass are often even more important than the quantitative per-tree data. It is the composition which determines the suitability of whole-tree mass for various purposes and the need to upgrade the raw material before industrial processing. Table 2.15 gives an example of the composition of the fresh and dry whole-tree mass in naturally regenerated, dense puckerbrush stands in Maine. Representing average values of 11 puckerbrush species, the data reflect the general pattern of increasing concentration of whole-tree mass in the stem with increasing tree size. In a 1-cm-diameter tree the foliage contains as much as 17% but in a 15-cm tree only 3% of the dry whole-tree mass. The proportion of branch mass drops from 42 to 7%, while the stem mass increases from 41 to 90%, respectively. Because

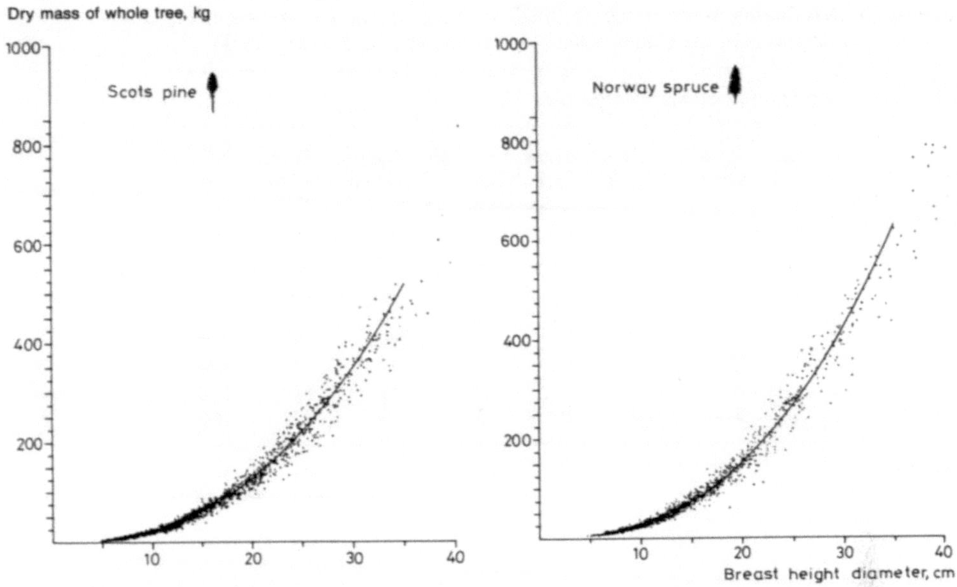

Fig. 2.37. Dry whole-tree mass in *Pinus sylvestris* and *Picea abies* at the phase of removal in selective thinnings from below and final cuttings in southern Finland as a function of dbh

Table 2.15. Average proportion of foliage, branches, and stem in fresh and dry whole-tree mass of 11 hardwood species in dense puckerbrush stands in Maine. For species, see Fig. 2.13 (data from Ribe 1973)

Dbh (cm)	Distribution of fresh mass (%)				Distribution of dry mass (%)			
	Foliage	Branches	Stem	Whole tree	Foliage	Branches	Stem	Whole tree
1.0	22.3	39.0	38.7	100.0	16.8	42.3	40.9	100.0
2.5	15.4	25.4	59.2	100.0	11.2	26.5	62.3	100.0
5.0	10.4	16.4	73.2	100.0	7.3	16.5	76.2	100.0
7.5	7.9	12.2	79.9	100.0	5.6	12.0	82.4	100.0
10.0	6.5	9.8	83.7	100.0	4.5	9.5	86.0	100.0
12.5	5.5	8.2	86.3	100.0	3.8	7.8	88.4	100.0
15.0	4.8	7.1	88.1	100.0	3.3	6.7	90.0	100.0

the moisture content is highest in the leaves, their proportion is larger in fresh mass than in dry mass of a whole tree.

The changes in component proportions with increasing tree age and size are more predictable in conifers. Table 2.16 shows the development of whole-tree mass in natural stands of southern pine in the southern United States when the breast height diameter of a tree grows from 15 to 50 cm. The proportion of stem wood increases from 63 to 79%, and, although the proportion of branch wood decreases simultaneously, the total content of stem and branch wood increases

Table 2.16. Distribution of dry whole-tree mass of southern pine into foliage, bark, and wood components in the United States, as a function of dbh (Taras 1978)

Dbh (cm)	Distribution of whole-trees mass (%)							
	Foliage	Branch bark	Stem bark	Total bark	Branch wood	Stem wood	Total wood	Whole tree
15	5	5	12	17	15	63	78	100
20	5	4	11	15	13	67	80	100
25	4	4	10	14	11	71	82	100
30	4	3	9	12	11	73	84	100
35	4	3	8	11	10	75	85	100
40	3	3	8	11	9	77	86	100
45	3	3	7	10	9	78	87	100
50	3	3	7	10	8	79	87	100
Average	4	3	9	12	11	73	84	100

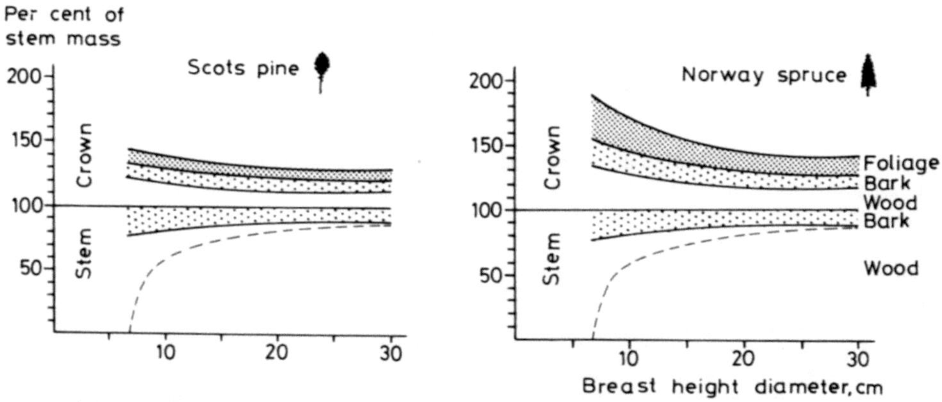

Fig. 2.38. Composition of dry whole-tree mass of *Pinus sylvestris* and *Picea abies* in Finland as a function of dbh. Stem wood divided with a *dashed line* into merchantable (*lower*) and under 6 cm unmerchantable (*upper*) portions. Relative quantity of stem mass 100 (redrawn from Hakkila et al. 1975)

from 75 to 86%. The same change is seen in *Pinus sylvestris* and *Picea abies* in Fig. 2.38 which illustrates the potential yield increase and the change in raw material composition, when stem-only logging is replaced by whole-tree logging. The harvesting system may thus have a remarkable effect, particularly on small-tree operations. In 7-cm-diameter *Pinus sylvestris* not more than 69% of whole-tree mass is composed of stem and branch wood, and in *Picea abies* the corresponding figure is only 60%.

In addition to tree size, *the effect of spacing* or stand density on the composition of whole-tree mass is of great significance. The early development of a stand is especially important. An extreme result of stand density is seen in Fig. 2.39

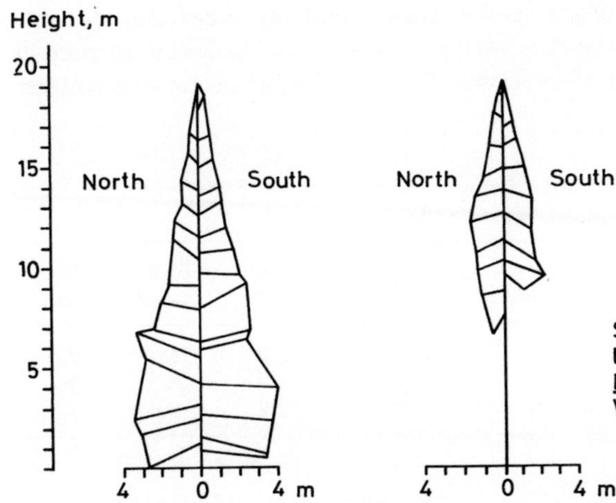

	Open-grown fir	Forest-grown fir
	Dry mass, kg	
Stem	333	141
Branches	194	20
Foliage	86	14
Whole tree	613	175

Fig. 2.39. Comparison of the quantity and composition of dry whole-tree mass of two *Abies balsamea* trees of equal height in Ontario. Dbh 40 cm in the open-grown (*left*) and 24 cm in the forest-grown (*right*) tree (redrawn from Honer 1971)

showing two *Abies balsamea* trees of equal height. An open-grown balsam fir is then defined as growing free of competition from the aerial components of neighboring trees and having a full crown extending to approximately 95% of the total tree height, whereas a forest-grown balsam fir is defined as having a crown sufficient to sustain tree life and growth in competition with surrounding trees. When the variation of whole-tree mass and its components was represented by log-log equations with dbh and tree height as independent variables, the equations differed significantly for forest-grown and open-grown trees. The dependence on dbh was strong, while the height factor as an additional variable was insignificant in whole-tree equations but significant in component equations of forest-grown balsam fir. For open-grown trees the height variable was significant only in explaining the variation of stem mass. When the equations were tested under differing climatic, soil, and stand-history conditions in New Brunswick 800 km from the original study area in Ontario, component estimates were seriously biased. Stem wood estimates showed the smallest error and branch estimates the greatest (Honer 1971).

A study by Alemdag and Stiell (1982) shows the effect of stand density on the mass of a whole tree and its components in high-survival, unthinned, 27- to 54-year-old *Pinus resinosa* plantations in Ontario with spacings from 1.5×1.5 m to 4.3×4.3 m. Stepwise regression analysis was applied to the whole-tree and component mass data, first using a combined form of dbh and height, then assessing the additional contribution of spacing, and finally that of age. The addition of spacing and age showed successive improvements in the accuracy of the estimates over those provided by the combined tree-size variable alone. Although the improvements were minimal for the whole tree and stem bark, markedly better statistics resulted for stem wood and crown components when spacing and age were introduced. As long as the effect of spacing was not recognized, stem wood was underestimated and crown mass overestimated at close spacings and vice versa at wide spacings. The basic tree-size variables were thus sufficient to predict the total mass of a whole tree, but spacing was needed for more accurate estimates of its composition:

Independent variable	Coefficient of variation (%)					
	Techn. foliage	Live branches	Dead branches	Stem bark	Stem wood	Whole tree
$Dbh^2 \times$ height	56.6	60.4	86.2	13.0	13.8	9.8
$Dbh^2 \times$ height + spacing	37.2	39.2	57.6	13.0	10.3	9.7
$Dbh^2 \times$ height + spacing + age	31.5	36.6	53.8	12.4	9.6	9.7

Table 2.17 shows *the effect of crown class* on the composition of whole-tree biomass in *Populus tremuloides* and *P. grandidentata* in Wisconsin. The percentage of stem wood is lowest in the dominant trees, and this is not fully compensated by a relatively high amount of branch wood. The highest percentage of stem wood with the least amount of stem bark was found in codominant trees followed by intermediate and suppressed trees. Codominant trees are thus the most desirable for many purposes of wood utilization. Forest management should perhaps aim at production of a large proportion of codominant trees, which may be controlled to some degree with spacing (Zavitkovski 1971). In young stands of *Pinus sylvestris*, a suppressed tree position favors foliage growth over the growth of other tree components, whereas under unsuppressed conditions branch growth increases substantially. The share of stem mass in the total growth is greatest under moderate shade conditions, i.e., in codominant trees (Kellomäki 1981).

Barclay et al. (1986) studied *the effect of thinning and fertilization* on the distribution of whole-tree mass in a 34-year-old *Pseudotsuga menziesii* stand in British Columbia. Four treatment combinations were applied: no thinning and no fertilization (T_0F_0); no thinning and fertilization with 448 kg urea per ha (T_0F_1);

Table 2.17. Composition of dry whole-tree mass in 6- to 45-year-old *Populus tremuloides* and *P. grandidentata* in Wisconsin by crown classes (Zavitkovski 1971)

Crown class	Distribution of whole-tree mass (%)			
	Foliage	Branch mass	Stem bark	Stem wood
Dominant	7.5	16.7	14.1	61.7
Codominant	5.8	11.2	12.8	70.2
Intermediate	4.6	9.9	16.8	69.3
Suppressed	4.0	10.7	17.3	67.9
Average	5.3	12.0	14.8	67.9

Table 2.18. Composition of whole-tree mass 9 years after thinning (T) and fertilization (F) in a 30-year-old *Pseudotsuga menziesii* stand in British Columbia. Explanation of the treatments, see text. Component masses are nonadditive (Barclay et al. 1986)

Component	Distribution of whole-tree mass (%) Treatment			
	T_0F_0	T_0F_1	T_1F_0	T_1F_1
New foliage	1.5	1.7	2.1	2.0
Old foliage	5.2	6.0	7.9	8.2
New twigs	0.36	0.41	0.42	0.44
Live branches	6.4	8.2	11.2	12.4
Dead branches	4.9	4.8	2.2	1.9
Stem bark	11.7	11.3	10.8	10.2
Stem wood	68.1	66.5	64.6	64.0

thinning by removal of two-thirds of the basal area and no fertilization (T_1F_0); thinning by removal of two-thirds of the basal area and fertilization with 448 kg urea per ha (T_1F_1). Nine years later the biomass components were distributed differently among the various treatments (Table 2.18). Thinning increased significantly the proportion of new and old foliage, new twigs, and live branches but decreased the proportion of stem mass and dead branches. The action of fertilization was very similar in direction, although not in magnitude. Both treatments increased the crown mass and total photosynthetic base of the tree at the expense of the structural components. Stimulation of foliage growth and improved light penetration resulted in further retardation of lower branch mortality and, consequently, in smaller mass of dead branches.

Tree species differ both in quantity and composition of whole-tree mass. Variation between species is caused by differences in shade tolerance, longevity of foliage, thickness of branches, durability of dead branches, stem form, thickness of bark, basic density of biomass components etc. Reaction to aging, size increase, spacing, crown class, and silvicultural treatments is specific to species. Furthermore, the same species may behave differently in various parts of its natural range or when introduced in a new geographic area. For example, *Pinus sylvestris* has rather thin branches and good natural self-pruning in northern Europe, but tends to become branchy in Central Europe and when grown as an exotic in North America.

Table 2.19 gives examples of between-species variation in the composition of whole-tree mass in British Columbia. Conifers tend to have a larger proportion of the biomass in foliage, thin twigs, and dead branches, while hardwoods seem to have a higher percentage of thicker branches and stem bark. *Larix occidentalis*, which sheds its foliage annually, resembles in those respects hardwoods. As regards the proportion of stem wood, it is difficult to find regular trends and differences between the northern softwoods and hardwoods.

Table 2.19. Composition of dry whole-tree mass in 22 tree species in British Columbia (data from Standish et al. 1985)

Tree species	Number of trees sampled	Mean dbh (cm)	Distribution of whole-tree mass (%)							
			Foliage	Live branches (cm)			Dead branches	Stem bark	Stem wood	Whole tree
				−0.5	0.5−2.5	2.5+				
Abies amabilis	45	16	17.1	3.6	10.4	1.6	1.1	8.2	58.0	100.0
A. grandis	40	23	11.6	3.2	6.9	2.4	2.3	9.8	63.8	100.0
A. lasiocarpa	89	16	12.3	6.7	12.0	4.4	3.3	8.7	52.6	100.0
Picea sitchensis	40	17	16.6	4.7	11.2	3.4	1.4	7.3	55.4	100.0
P. glauca	105	19		4.2	7.6	2.9	1.7	7.7	67.9	100.0
P. engelmannii	43	24	7.6	3.7	7.1	3.4	1.7	7.0	69.5	100.0
P. mariana	60	17	9.7	5.2	8.8	0.8	2.2	9.2	64.1	100.0
Pinus ponderosa	42	28	5.7	0.1	9.3	8.7	2.5	9.9	63.8	100.0
P. monticola	40	19	7.7	2.6	8.3	3.8	1.7	6.9	69.0	100.0
P. contorta var. *latifolia*	98	19	4.6	1.4	6.5	2.0	1.9	6.1	77.5	100.0
P. contorta var. *contorta*	40	16	9.5	3.1	11.9	2.8	5.5	6.6	60.6	100.0
Pseudotsuga menziesii	49	21	5.5	1.7	3.9	4.2	0.9	10.8	73.0	100.0
P. menziesii var. *glauca*	41	26	6.5	3.8	6.7	7.3	1.6	12.2	61.9	100.0
Tsuga heterophylla	70	19	5.3	3.0	5.2	3.7	1.4	10.0	71.4	100.0
T. mertensiana	39	20	6.5	2.7	5.5	2.3	0.4	13.1	69.5	100.0
Thuja plicata	70	22	9.2	2.4	8.0	5.8	2.1	8.9/	63.6	100.0
Chamaecyparis nootkatensis	41	18	7.5	1.3	5.5	1.7	0.7	6.0	77.3	100.0
Larix occidentalis	41	24	2.1	2.5	4.9	3.1	0.3	10.1	76.9	100.0
Average of 18 conifers		20	8.5	3.1	7.8	3.6	1.8	8.8	66.4	100.0
Alnus rubra	41	13	4.3	2.4	13.0	6.1	2.2	8.8	63.2	100.0
Betula papyrifera	40	15	4.6	3.3	11.2	3.2	0.5	10.6	66.7	100.0
Populus tremuloides	40	18	2.6	1.7	8.5	3.8	1.3	14.4	67.7	100.0
P. trichocarpa	40	16	5.3	1.3	18.0	4.6	1.4	10.4	59.0	100.0
Average of 4 hardwoods		16	4.2	2.2	12.7	4.4	1.4	11.0	64.1	100.0

2.3 Biomass of an Entire Stand

Many alternative methods can be used to convert per-tree biomass data to a land area basis. Differences in approaches to biomass estimation hinder comparisons between studies. According to Art and Marks (1971), the following methods are being used for estimation of standwise biomass stocking or production. Regardless of the method selected, a certain number of trees must be felled for destructive sampling to determine component mass:

1. *Mean Tree Method.* Biomass stocking and production of an entire stand are estimated from the analysis of a tree or trees of mean dimension. Sample data are converted to a land area basis by multiplying the mass and production of sample trees by the number of trees per unit area. This method is used most widely in plantations and other even-aged stands, but is thought to be inappropriate for uneven-aged stands.
2. *Stratified Tree Method.* This technique is a variation of the mean tree method for uneven-aged stands that are stratified by diameter classes. Each class is sampled by a tree of mean dimensions for that particular stratum.
3. *Regression Estimation Method.* This technique involves three phases: Trees are first tallied by species and diameters at breast height on a quadrate basis; for sample trees, difficultly measured allometric biomass or growth parameters are then related by regression analysis to more easily measured parameters, ultimately to dbh; finally, the conversion to a land area basis is accomplished by applying the regressions of phase 2 to the quadrate measurements of phase 1.
4. *Unit Area Method.* All individuals within small randomly located plots are completely harvested. Alternatively, all individuals are harvested in a large plot representative of a larger vegetational complex.
5. *Canopy Area Method.* Sample tree data are converted to a land area basis by dividing its biomass by the effective canopy projection of the tree.
6. *Basal Area Proportion Method.* Biomass of a stand is estimated from sample trees by the ratio of stand basal area to sample tree basal area.

In order to offer a guide to the literature and a rapid means of comparing data accumulated from different forest stands, Art and Marks (1971) published a simple working table listing by species, location, and stand age, the dry mass and net annual primary production as well as the leaf area index for over 280 stands. In the majority of the citations, data are restricted to the above-ground biomass of living trees.

Cannell (1984) explained the variation of above-ground woody stand biomass, foliage excluded, in published data on 640 stands in a total of 23 countries around the world. Figure 2.40 shows the individual stand data and a log-log regression line with the product of stand basal area (G) and mean height (H) as an independent variable. The smallest values are for young or stunted forests and the largest for stands of *Sequoia sempervirens* in northern California. The regression equations differed significantly between species. For example, for the same GH values hardwood stands had generally more above-ground woody biomass per hectare mainly because of their higher basic wood density (B).

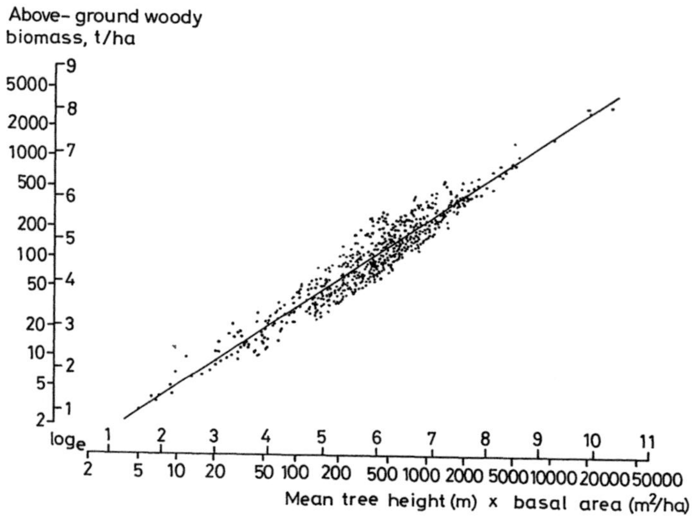

Fig. 2.40. Relationship between the above-ground woody dry mass and the product of mean tree height and stand basal area in 640 stands of all kinds of forests and woodlands throughout the world (redrawn from Cannell 1984)

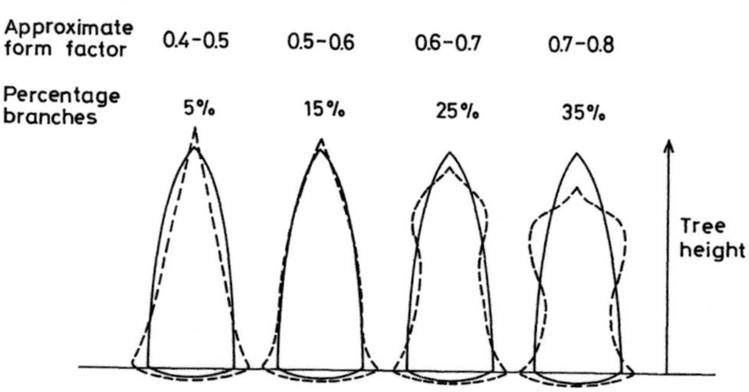

Fig. 2.41. Hypothetical profiles of the above-ground woody components of trees having different percentages of branch mass, shown relative to a quadratic paraboloid. *Solid lines* represent paraboloids, and *dashed lines* illustrate shapes of woody mass accumulation as if a ring had been passed up the stems from the base (Cannell 1984)

Variation from regression in Fig. 2.40 is caused also by differences in the *form factor for above-ground woody biomass.* For any given stand basal area and mean tree height, the greatest woody biomass will be contained in those stands with a large proportion of branch mass. Figure 2.41 illustrates hypothetical profiles of above-ground woody components of trees having different percentages of branches and, accordingly, different biomass form factors (F). When the form factor is known, the total above-ground woody biomass (M_T) in a stand can be calculated with the following equation:

$$M_T = F \times HG \times B \ .$$

Although the biomass form factor represents basically the same function of percentage of branch mass irrespective of whether the trees are conifers or hardwoods, on a world scale it is generally higher for hardwoods due to their branchiness, particularly in the tropics. According to Cannell, the biomass form factor is 0.60 ± 0.05 for the majority of conifer stands but is considerably greater for tropical hardwoods. An extreme case is *Hevea braziliensis* with a form factor exceeding 1. The biomass form factor is related to the percentage of branch mass in the total above-ground woody biomass, foliage excluded, as follows:

Percentage of branch mass	Form factor for above-ground woody biomass
5	0.4 – 0.5
15	0.5 – 0.6
35	0.7 – 0.8

Stand biomass is a net value rather than the total production up to the time of sampling. The stocking of a forest ecosystem is primarily the result of the annual accumulation of dry matter production in stems. On the other hand, foliage, branches, fine roots, and bark die and are renewed, and trees may die as well. Under natural conditions in unmanaged forests, production and decay will tend to come into balance.

The productive structure of a forest ecosystem can be illustrated by diagrams showing the vertical distribution of radiation intensity, the dry mass of non-photosynthetic biomass components, and the distribution of foliage (Monsi and Saeki 1953). Hardly any plants can grow under dense stands of tree species with a large foliage mass, while smaller understory trees, shrubs, and ground vegetation thrive under a canopy of trees with a small foliage mass (Fig. 2.42). Tree foliage is thus complemented by the leaf mass of the undergrowth. Although the differences among individual stands are large when canopy and undergrowth are compared separately, the total sum of the leaf area in forest ecosystems with hardwood species as the top layer is, under given conditions, approximately constant (Satoo and Madgwick 1982).

An analysis of 21 *Pinus sylvestris* stands forming an age succession in central Sweden showed that the masses of foliage and branches grow rapidly in the initial phase of stand development, but soon reach a constant level. After 40 years of growth the branch mass may even decrease. Thereafter the increase in stand biomass with age is mostly a result of accumulation of stem, stump, and coarse root mass. During stand development, the relation of foliage and branches to stem mass decreased from approximately 100% for both to 5 and 15%, respectively. The proportions of stumps and coarse roots over 30 mm remained rather constant, 7 and 10%. In addition to 12.4 t dry mass of coarse roots per hectare, the roots included also 4.2 t of 2–30 mm material, and about 2.0 t of under 2 mm material per hectare (Albrektson 1980).

The development of the proportional distribution in the above-ground biomass of *Pinus sylvestris* and *Picea abies* stands is illustrated in Fig. 2.43 from drained, fertilized peatland sites in Norway. During the 30-year-study period, stem wood became progressively more dominant as stand density increased. The addi-

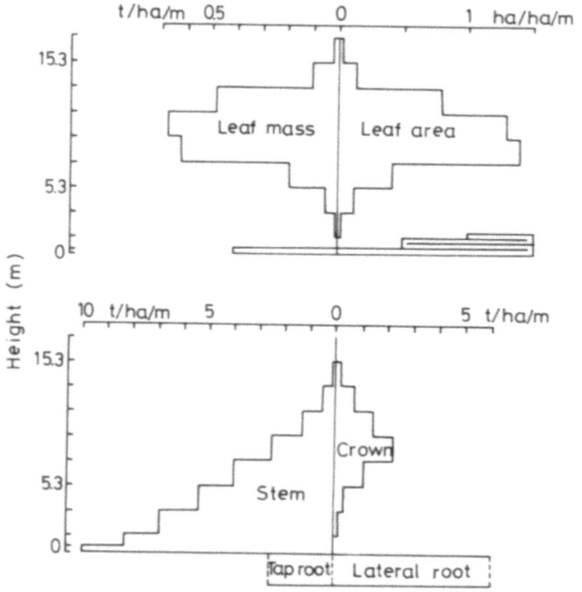

Fig. 2.42. Vertical distribution of leaf mass, leaf area, and complete-tree mass in a 17-year-old, 14.7-m-high *Metasequoia glyptostroboides* stand in Japan (redrawn from Satoo 1974)

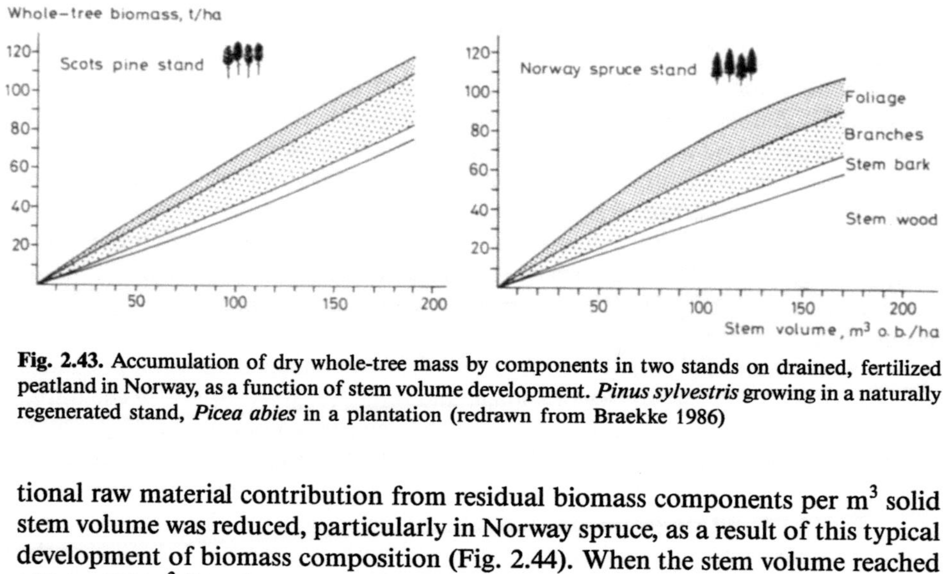

Fig. 2.43. Accumulation of dry whole-tree mass by components in two stands on drained, fertilized peatland in Norway, as a function of stem volume development. *Pinus sylvestris* growing in a naturally regenerated stand, *Picea abies* in a plantation (redrawn from Braekke 1986)

tional raw material contribution from residual biomass components per m^3 solid stem volume was reduced, particularly in Norway spruce, as a result of this typical development of biomass composition (Fig. 2.44). When the stem volume reached 100 to 120 m^3/ha, the mass of foliage stabilized. As a tolerant species, Norway spruce retained much more foliage than Scots pine.

Table 2.20 (Alemdag and Stiell 1982) illustrates the general *effects of spacing* on the amount and composition of the whole-tree biomass in a 27-year-old, unthinned *Pinus resinosa* plantation in Ontario. Low total biomass is clearly associated with wider spacing, and the same is true for all biomass components except large branches. Proportionally, percentages of stem wood and stem bark decline with greater spacing, whereas the percentage of large live branches increases, and

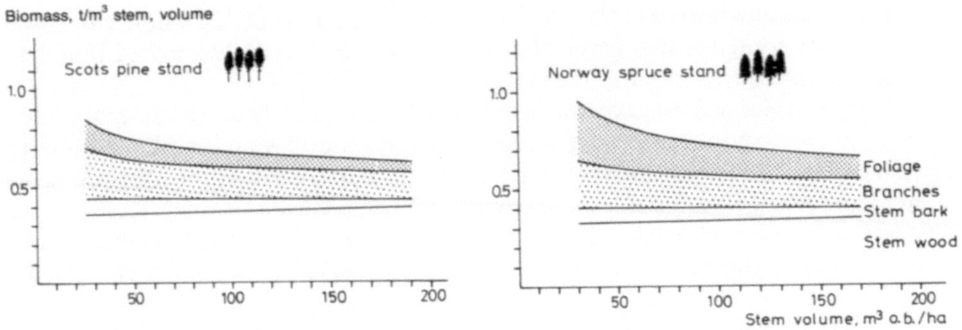

Fig. 2.44. Composition of whole-tree mass of *Pinus sylvestris* and *Picea abies* in two peatland stands in Norway. Total amount of dry above-ground biomass in t/m³ solid of unbarked stem volume. Stand data given in Fig. 2.43 (redrawn from Braekke 1986)

Table 2.20. The effect of spacing on the quantity and composition of dry whole-tree mass in a 27-year-old *Pinus resinosa* plantation in Ontario (Alemdag and Stiell 1982)

Biomass component	Spacing (m)				
	1.52	1.83	2.13	3.05	4.27
Whole-tree	Above-ground biomass (t/ha)				
	206.0	186.8	181.1	139.8	104.4
	Distribution into components (%)				
Stem wood	66.4	63.8	63.8	59.2	51.8
Stem bark	7.4	7.1	6.7	6.3	6.0
Large live branches	6.7	8.7	10.2	13.4	17.1
Live twigs and foliage	12.0	12.3	11.6	12.2	14.0
Dead branches	7.5	8.1	7.7	8.9	11.1
Whole tree	100.0	100.0	100.0	100.0	100.0

that of twigs and needles shows an upward trend only at the widest spacing. These results receive support from similar findings with *Abies balsamea* by Baskerville (1966).

Baskerville (1966) studied the effect of stand density on the biomass composition of naturally regenerated *Abies balsamea* stands in New Brunswick 43 years after release. When the number of trees per hectare increased from 1750 to 12500, the dry mass of stem components increased from 31 to 52 t/ha and that of over 2-mm root mass from 13 to 21 t/ha, whereas only minor changes occurred in the crown mass. Consequently, the proportion of stem mass increased significantly at the cost of crown components. Another example of biomass composition in dense stands is given by Johnstone (1971) from *Pinus contorta* var. *latifolia* in Alberta. In a 100-year-old stand with over 2500 trees/ha and an average dbh of

16 cm, the proportion of stem mass was as much as 87% of the whole-tree mass and 79% of complete-tree mass. Needles and branches both comprised only 5% of the complete-tree mass in the stand.

The examples demonstrate the influence of stand density on the biomass composition. However, the effects of stand density are complex and in thinned forests often reflect the stand history rather than the current situation. According to Satoo and Madgwick (1982), the quantity and composition of stand biomass are, in fact, a function of *relative stand density*, which is the ratio of the actual number of trees to the maximum number of trees expected under given conditions.

2.4 Forest Biomass Inventory

A biomass inventory may be carried out separately or integrated with a conventional forest inventory. Information required of a conventional inventory can be divided into two broad groups. *Area information* describes the size of forest and other land-use classes and the distribution of forest area according to site quality, forest type, stand structure and condition, ownership, etc. *Area-related tree information* describes in quantitative figures the stem volume of growing stock and its composition by species, diameter, quality classes, etc. For both information groups, data collection is based on the joint use of remote sensing and ground measurements. In biomass inventories, the collection of information on forest area and its division into classes is equally important. Tree enumeration may be carried out on the sample plots in much the same way as in conventional inventories, but additional work is required in order to include noncommercial species, small trees, shrubs, and possibly other biomass (Kuusela and Nyyssönen 1981).

In temperate and boreal zones the trees are characterized by a central stem extending from ground to top. The common inventory target is then to assess the total stem volume or the volume meeting a certain minimum stem diameter. In tropical and subtropical zones, including monsoon climates, the trees are often characterized by a large proportion of branch mass. Although the overall consumption of fuel wood in these zones is much greater than that of industrial wood, the forest inventories are presently designed to estimate only the volume of clear stem of the commercial species (Kuusela and Nyyssönen 1981).

The intensity and accuracy of forest inventory, the process by which reliable and satisfactory information on the forest is obtained, are directly related to the intensity and sophistication of forest management or demand for timber. To increase the use of trees as feedstock for diversified chemical industries, fodder, and production of energy, information restricted only to stem volume is inadequate. Although present forest management is largely concerned with the volume of the merchantable stem, the target of the inventories should be modified to feasibly meet the requirements of developing forest management and utilization that is already taking place (Young and Tryon 1978).

For complete forest utilization there must be complete forest management to ensure healthy, viable forests. The first step, and the key to complete forest management, is a forest biomass inventory of all woody shrub and tree species. Such an inventory provides versatile information on the standing forest, permits the estimation of site productivity, and when coupled with nutrient sampling, can be used to assess periodically the forest vigor. It also makes it possible to work out realistic cutting allocation and utilization plans (Young 1978a). The need for expansion of traditional forest inventory from stem volume to tree biomass is particularly urgent in areas of the world with a fuel wood scarcity.

Forest biomass inventories may still be premature by many standards because of present harvesting techniques and product utilization, and most resource managers view it only as an option with some merit for the future. However, the sudden, unexpected change in the interest of residual forest biomass as a renewable source of energy represents dramatic evidence of the need for more comprehensive information on forest resources. As the demands of forest resources multiply, the biomass inventory will be indispensable (Tryon and Edson 1979). Aldred and Alemdag (1988) published a manual describing procedures for including forest biomass assessment in forest inventory. The procedures cover the collection of data in the field, laboratory analysis, development of single-tree biomass equations, and applications of the equations to both current and proposed new inventories.

Although there is a great deal of similarity between a biomass inventory and a traditional forest inventory, certain fundamental differences may be distinguished. Three important features characterize the biomass inventories. First, all trees and shrubs are included, independently of species, size, and commercial value. Secondly, in addition to the stem, other biomass components of a tree are also estimated. And thirdly, instead of volume, the unit of measurement is mass, either fresh or dry.

Mass tables and regression functions in terms of individual tree components, groups of tree components, complete tree, and shrubs are essential to a biomass inventory. However, there is a current need for more and better tables setting forth mass by species and tree components. In fact, this is presently the weakest link of the biomass inventory proposal and will remain so until mass studies are made throughout the forest areas of the world (Young and Tryon 1978).

Mass tables are developed in much the same way as stem volume tables. Generally, trees are destructively sampled and the mass of each component is determined and related by regression to standing tree dimensions. However, mass is more difficult to predict from external characteristics since the dry mass per unit volume may vary with geographic location, age, size, growth rate, and species. To permit conversion from fresh to dry mass, determination of mass per unit volume, and estimation of percentage of bark, the field procedure should include subsampling for moisture content, basic density, and separation of wood and bark by components.

According to Hitchcock (1979), over 80% of the total cost of a biomass inventory would be associated with destructive sampling. Therefore, the least-cost method would be that which minimized destructive biomass sampling. Clark (1979) listed the following basic methods for *determining the whole-tree mass:*

1. Harvest and weigh whole trees on portable scales. The method is accurate, can be used in rough terrain, requires low equipment investment, and allows subsampling. However, it is labor-intensive and slow when sampling large trees.

2. Harvest and weigh whole trees with mechanized equipment, and obtain component masses by removing components and reweighing the remaining mass. Advantages of this method are accuracy, small manpower demand, and possibility for subsampling. However, large equipment and open areas are required for weighing.
3. Harvest and chip whole trees and weigh the reduced mass. This method is basically accurate, fast, and needs minimum manpower. On the other hand, it requires large equipment and does not allow component separation. Mass loss may occur in conjunction with haulage, particularly if skidding is used, and with chipping.
4. Harvest the whole tree and weigh the crown by hand, measure the stem volume, and estimate stem mass from the measured volume factors based on subsamples. The method is accurate, has lower manpower requirement than weighing a whole tree, and allows subsampling. Furthermore, stem mass can easily be estimated for various combinations of products. A prerequisite for the use of this method is adequate subsampling.
5. Measure stem and crown volume of a standing tree and estimate the mass based on volume and assumed mass per unit volume factors. The obvious advantages of this method are small manpower requirements, low costs, and the fact that the trees are not destroyed. However, the crown estimates are inaccurate and thus hardly sufficient for the construction of mass tables. In this alternative subsampling is not possible.

The mass data thus obtained from individual trees are then related to easily measurable external tree characteristics for the construction of regression functions and tables for use in biomass inventories. Three main criteria determine the *method of sampling*. First, the sample should be representative of the target population. Secondly, the selection procedure must be such that a sufficiently valid regression analysis is possible. And finally, the selection procedure must be cost-effective. Since the crown biomass is more, rather than less variable, than the stem volume, larger samples of trees are needed for reliable biomass tables. Cunia (1979a, b), when analyzing requirements for selecting sample trees, sample size, and regression methodology as applied to biomass tables, suggested two sampling procedures. One is for the construction of *specific biomass tables* applicable to a specific, restricted forest area and the other for *general biomass tables* applicable to forest inventories from large areas such as states and regions.

As shown for various components and component combinations of residual forest biomass in Section 2.1, by far the most important mass predictor for single trees is some kind of transformation of the breast height diameter, which is also simple to measure under all conditions. The estimation of stem mass may be improved with height and taper variables. Crown mass estimation, which is much more difficult, could be improved by measuring the diameter at the base of the live crown, but its application in practical inventories is limited because of difficulties in measurement. However, in many cases crown ratio has almost the same predictive power and is relatively simple and fast to determine with adequate accuracy. For stump and root mass, stump height and soil type may improve the functions considerably.

Five alternative methods to convert per-tree biomass data into a land area basis are presented in Section 2.3. Of these the regression estimation method is most suitable for inventory purposes. Appropriate single-tree equations, relating mass of tree components and groups of components to dbh, tree height, and possibly other variables, such as crown ratio, are applied to each species either by individual tree measurements or by dbh class. The total biomass of a stand by tree components is then the sum of the corresponding masses of single trees or dbh classes.

If single-tree measurements are lacking and only stem or stand volumes are available, as may be the case with some conventional forest inventory data, approaches permitting direct mass estimates on a stand volume basis are needed. In this case, previously established ratios between stand mass and stand volume of a given stand type by age, spacing, and site quality classes must be provided. Once the stand stem volume is converted into stand stem mass, the mass of other components can be calculated by employing component percentages. The following figures from even-aged, naturally regenerated, untreated *Pinus banksiana* stands in Ontario are examples of such stand mass conversion factors, when stem volume is measured in all cases without bark (Alemdag 1982).

	Conversion factor (kg/m^3)
Stem wood mass excl. bark/stem volume	412
Stem mass incl. bark/stem volume	459
Whole-tree mass/stem volume	554
Stem mass incl. bark/merchantable stem volume	518
Whole-tree mass/merchantable stem volume	626

Table 2.21 gives an example of the results of a combined volume and biomass inventory on public lots in Maine. All trees larger than 11.7 cm dbh were tallied by species and product for conventional stem volume inventory in variable-radius plots. Mass data on trees below commercial size were collected on fixed-radius, 13.5-m^2 plots, and all trees between 1.3 and 11.7 cm dbh were sampled and tallied by species and dbh class. Trees and shrubs 1.3 cm dbh and smaller were recorded by species and height class to a minimum height of 15 cm on a 4-m^2 plot.

It is a common experience in conventional forest inventories that mixed tropical forests, representing much variation, are much more difficult than temperate forests in sampling design. This is obviously the case for forest biomass inventories as well. In addition, in mixed tropical forests the clear stem of merchantable tree species often comprises only a small percentage of the complete tree and a still smaller percentage of the total forest biomass, including nonindustrial species. The stem volume does not, therefore, have the same value as a key characteristic as in temperate forests. Furthermore, while forests and other wooded areas in the tropics vary widely by their nature, the needs and pressure for biomass inventories on marginal woodlands are actually more urgent than in better stocked forests (Kuusela and Nyyssönen 1981).

Table 2.21. An example of the results of a forest biomass inventory on a 25000 ha public lot in Maine (Young 1978a)

Species category	Dbh (cm)	Forest component			
		Stems	Crowns	Stumps and roots	Total
		Percent of complete forest			
Industrial tree species	Over 11.7	48.1	21.8	17.5	87.4
	Under 11.7	5.4	2.4	1.9	9.7
Nonindustrial tree species	Over 11.7	0.8	0.3	0.3	1.4
	Under 11.7	0.6	0.2	0.2	1.0
Shrubs		0.3	0.1	0.1	0.5
Complete forest		55.2	24.8	20.0	100.0
		Percent of merchantable stem mass			
Industrial tree species	Over 11.7	100.0	45.4	36.4	181.8
	Under 11.7	11.1	5.0	4.0	20.1
Nonindustrial tree species	Over 11.7	1.6	0.7	0.5	2.8
	Under 11.7	1.1	0.5	0.4	2.0
Shrubs		0.6	0.3	0.2	1.1
Complete forest		114.4	51.9	41.5	207.8

2.5 Quantity of Residue in Practical Logging Operations

Full utilization of forest biomass may be defined as the extraction and use of all above-ground components of all trees, including hardwoods and softwoods, preferred and weed species, large and small trees, trees dead because of decay, disease, insect attack, old age, crowding, or accident, and trees conventionally culled because of decay, crook, sweep, or any other cause. It would include all standing timber, all logging debris, and such down timber which might have some utilization value (Keays and Hatton 1975).

Unmerchantable above-ground biomass left behind in a cutover area is called *logging residue*. Two major components of logging residue may be differentiated. One is *logging slash*, defined as all above-ground biomass left on the ground after a timber harvesting operation. The other is trees bypassed because of their species, small size, or inferior quality. These living remnants of old stands, if not destroyed during regeneration operations, will compete for growing space with valuable trees without contributing to timber production. Such remnants, although included as part of the timber resource, interfere with the new stand and therefore have a negative value from a forest management point of view. They are referred to as *living residue* (Welch 1980).

Overmature, unmanaged forests typically contain large amounts of unmerchantable biomass in rough, rotten, and dead trees, whereas young sapling stands yield large amounts of undersized trees during precommercial or early commercial thinnings. In repeatedly thinned managed forests, dead, defective, and undersized trees are of little importance. Consequently, in such stands the majority of the residual biomass is composed of crowns and tops of harvested trees.

The physical presence of logging residue affects the trafficability of terrain, work difficulty of timber harvesting, and efficiency of subsequent activities of stand establishment, such as site preparation, manual or mechanized planting, and thinning of young plantations (Sect. 9.2). Logging residue is a fire hazard and an eyesore (Sect. 9.3.4), for which disposal is laborious and costly. On the other hand, it is a source of nutrients (Sect. 9.1), a protective cover preventing erosion and rutting (Sect. 9.3.1), and a reserve of raw material and energy. By considering such factors as optimum mass, size, and arrangement of logging slash, management can select the most appropriate harvesting methods and *residue management* alternatives.

Viewed from a long-term perspective, residue is an essential part of the forest ecosystem. Residue management must therefore be accepted as an integral part of forestry. It affects mechanization of timber harvesting and transport, work organization, the productivity of forests, industrial processes, and energy utilization. Treatment of residue may have both detrimental and beneficial consequences. Hence, in addition to biomass utilization aspects, residue management must take into consideration protection of the soil and its nutrient supply, establishment of a new stand, and water quality (Ruth and Harris 1975).

A great deal of attention has been focused on *national and regional residue assessment* during the past decade. The purpose of these surveys has usually been the estimation of the potential of logging residue as an indigenous impact on energy supply in a certain country or area. The prospect of new or expanded markets has gradually changed the emphasis of analysis of logging residue. What was earlier a general concern for national or regional reserves has become a need for specific information on residue in more restricted, defined areas. Questions now concern also characteristics such as piece size, number of pieces per unit area, dispersion, topography, and technical properties (Howard 1981). Although the gross statistics are valuable for broad national calculations, they are seldom specific enough for residue management planning or feasibility studies. Before feasibility of residue utilization can be analyzed, it is first necessary to have detailed information on the amount, composition, technical logging characteristics, and accessability of the residual biomass.

The mass and properties of logging slash play an important role in wildfires and prescribed burns. Appraising the potential fire hazard is a process of interpreting the fuel in terms of fire behavior, including characteristics such as rate of spread, fireline intensity, and flame length. This survey of logging slash is called *fuel appraisal.*

2.5.1 Estimating the Quantity of Logging Residue

If a majority of the logging residue is composed of crown material and tops, as is the case in managed forests, the quantity and composition may be *predicted from timber cruise data* with biomass functions and tables. The total mass of the residue is computed by multiplying the number of trees in each dbh class by the expected crown and top mass for that class. Figure 2.45 illustrates the fresh mass of residue from crown and under 7.5-cm top of *Pinus sylvestris* and *Picea abies* in a nomogram form as a function of tree height, crown base height, and dbh. If only the total stem volume for each dbh class is available, the quantity of residual biomass may be predicted from stem volume on a percentage basis.

In unmanaged, overmature forests the role of residue from dead and defective trees, breakage, and down material may be even more important than that from crowns and tops. The handbook for predicting residue mass of Pacific Northwest conifers proposes the following four-step procedure (Snell and Brown 1980):

1. Summarize inventory data of trees to be cut, by number of trees per hectare by species and dbh, or by basal area per hectare by species and dbh. In addition to trees to be cut, include also an estimate of trees that will be lost from *trampling*, i.e., the pushing over of small trees by logging equipment.
2. Select the desired mass table and multiply either the number of trees or the basal area per hectare by the appropriate table values for *crown and top mass* for each dbh group by species. Sum all groups to obtain the total crown and top mass per hectare.
3. Estimate residue mass from tree *defect and breakage*. This estimate cannot be obtained from mass tables but must be based on experience and knowledge of local harvesting operations. Defect percentages must represent the stem material actually remaining on site. In the Pacific Northwest, breakage in old-growth stands is commonly estimated at 15% and in second-growth stands at 5% of the cruised merchantable stem volume. The basic density of unbarked wood is used to convert the volume to mass.
4. If relevant, estimate the mass of *dead and down trees* as instructed by the Handbook for Inventoring Downed Woody Material (Brown 1974).

A traditional way to estimate the quantity of residue is to select a certain number of fixed-area sample plots and *weigh all residual biomass within each plot*. Since the method is laborious it permits only a small plot size and is best suited for areas where the residue is small in size, evenly distributed, and not very abundant. For timber balance calculations, stem residue is sometimes estimated in conjunction with National Forest Inventories. For example, in Finland Mikkola (1969, 1972) recorded all stumps from trees cut less than 1 year previously within a 5-m-wide belt of inventory tracts. The stem residue from each of these trees was measured by volume, whether occurring inside or outside the 5-m belt. In addition, residual stem material which could not be assigned to any particular stump was measured from 10-m belts. In Sweden, the stem volume left as logging residue was estimated in conjunction with the National Forest Inventory from 1500 tracts, each containing 32 to 40 sample plots with a radius of 10 m (Wallden 1982).

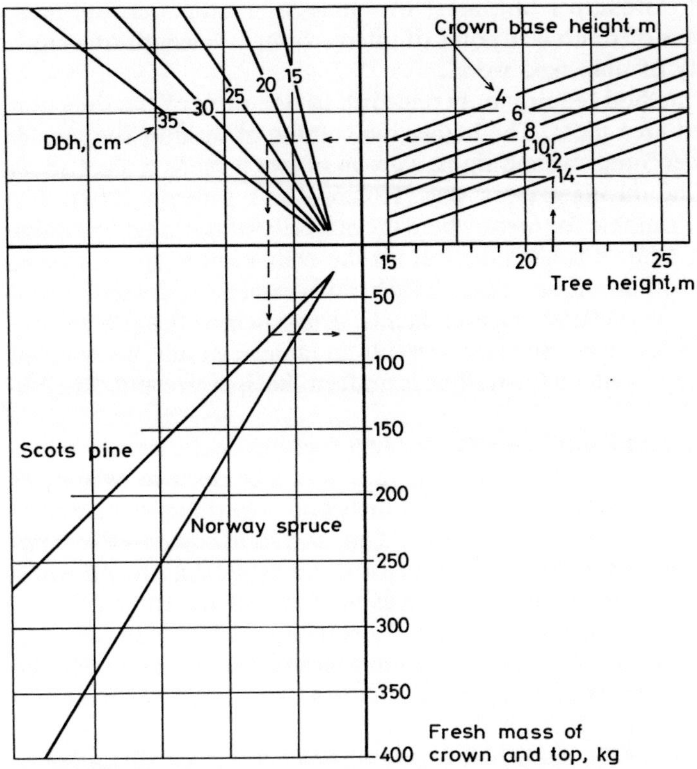

Fig. 2.45. Nomogram for estimating the total fresh mass of crown and under 7.5-cm top of *Pinus sylvestris* and *Picea abies* in Sweden as a function of tree height, crown base height, and dbh (Gunnarfelt 1973)

A more economical way of estimating the quantity and characteristics of on-ground logging residue is offered by the *line intersect method.* It was first used by Warren and Olsen (1964) in New Zealand, then retested and further refined in the United States, Canada, and Sweden by van Wagner (1968), Bailey (1969, 1970), Howard and Ward (1972), Martin (1976), Holappa and Jonsson (1987), etc. The method uses a sample line without width, or a transect that may be located systematically or randomly in the cutover area. To avoid the effect of bias in orientation of residue pieces, the direction of a continuous line is usually changed in either a fixed or random manner after a predetermined distance. If the line is not continuous, its segments may be oriented unidirectionally, randomly, paired at right angles to each other, etc. Whenever a piece of residue crosses the line, its diameter is measured at the point of intersection. The volume and mass of residue is then estimated by using the following basic formulas which can be modified to fit any set of units:

$$V = \frac{\pi^2 \sum D^2}{8L}, \quad M = \frac{\pi^2 B \sum D^2}{8L};$$

where V = volume of residue per unit area; M = mass of residue per unit area; D = diameter of a piece of residue at point of intersection; L = length of sample line; B = basic density of unbarked wood.

The line intersect method requires only one-fifth to one-third of the time normally spent with fixed-area plots. The higher the volume of residue, the greater is this advantage. Under Swedish conditions, a group of two persons cruises about 300 m of inventory line during a work day (Holappa and Jonsson 1987). The method is particularly suitable for areas where the majority of the logging residue consists of stem mass, while it is less practical for the evaluation of crown masses from small trees. If information on defect, length, and diameter is desired, a subsample may be taken for detailed measurements. Areas where the pieces vary greatly in size and where the residue is noticeably in clumps should be sampled more intensively than areas where the residue is uniform both in size and distribution (Martin 1976).

Clear-cut areas are sometimes prepared for regeneration operations by windrowing the residue using a crawler tractor equipped with a brush rake. Windrows may contain considerable amounts of potential fuelwood. The quantity of woody material is a result of the length, cross-sectional area, and *solid content of a windrow*. The solid content values are naturally specific to the type of slash and windrowing method. On the pine-site, clear-cut areas in the southern United States, they vary from 0.14 to 0.22. Rough values for the solid content may be obtained with two easily estimated variables, viz. maximum windrow height and an average diameter of over 7.5-cm material in the windrow, and used to assess the energy value of windrow residue (McNab 1980).

Maxwell and Ward (1976) published a *photo series for quantifying forest residue* in the coastal *Pseudotsuga-Tsuga* and *Pseudotsuga-hardwood* forests in the Pacific Northwest of the United States. It is used to evaluate impacts of residue on various aspects of forest management, to identify areas of unacceptable residue loading, to estimate the amount of utilizable material, and to predict fire behavior of logging residue. These series are arrays of photos, with each array showing different residue loading levels generated from similar timber types and cutting practices. Each photo is supplemented with information on measured residue quantities according to size classes, average depth, ground area covered, fuel ratings, and logging data. Users can estimate any of the residue characteristics on an area being inventoried by comparing them with the photos as follows. These steps are repeated for each characteristic desired:

1. Observe each characteristic of the residue on the ground, for example, diameter classes 0.6–2.6, 2.7–7.7, 7.8–22.9 cm, etc.
2. Select a photo which nearly matches, or photos that bracket, the observed characteristic.
3. Obtain the quantitative value for the characteristic being estimated from the data sheet accompanying the selected photo.

2.5.2 Examples of the Quantity of Logging Residue

The mass, composition, condition, and arrangement of logging residue are affected by many factors: age and size of the trees harvested, history and structure of the stand, species composition, topography, market conditions, stumpage prices, utilization standards, logging equipment and methods, climatic conditions, and the attitude of the persons involved. Wide variation in tree size, characteristic of climax or near-climax stands, contributes to the quantity of residue because felling of large trees tends to break small ones, and heavy equipment needed to extract large logs tends to break small timber (Ruth and Harris 1975). Keays and Hatton (1975) differentiated three levels of additional raw material recovery from harvesting of all above-ground biomass, compared to the yield from conventional stem-only logging. The first level represents managed forests that have been thinned before, whereas the others refer to unmanaged forests:

1. For an even-aged stand on an average to good site, fully stocked without weed species and containing trees of uniform dbh, cull and defective trees, but no dead standing timber, the residual biomass may be at least 30% for conifers and 35% for hardwoods.
2. For a mature, unmanaged stand growing on an average site at high stand density and containing small trees, some dead standing timber, and some weed species, the residual biomass may be about 60% for conifers and 100% for hardwoods.
3. For unmanaged, mixed stands growing on poor sites, with a wide range of tree sizes, many tree species, and a high percentage of defective trees and dead standing timber, the residual biomass for predominantly coniferous stands may be 100 to 200%; for predominantly hardwood stands 100 to 300%.

These rough figures reflect the very high raw material and energy potential of residual forest biomass, particularly in mature, unmanaged stands. However, the averages ignore the effect of tree size and species, both of which strongly affect the percentage of residual biomass. The following table, although representing single trees instead of stands, and stem wood instead of all biomass, shows the percentage of stem wood recovered from various types of stands in tree-length logging to 10 cm top diameter inside bark in the northeastern United States. Decurrent hardwood species such as oaks differ significantly from excurrent tree species (Clark 1978):

Stand type	Recovery of stem wood (%) Breast height diameter (cm)		
	15	20	25
Quercus rubra	70	79	77
Liriondendron tulipifera	66	84	93
Natural pine stands	67	86	92
Pine plantations	72	84	

Table 2.22. Distribution of stem volume between live trees, standing dead trees, and down material in mature stands in the northern Rocky Mountains (Benson and Schlieter 1979)

Stem component	Distribution of stem volume (%)				
	Pinus contorta	*Larix occidentalis*	*Abies grandis*	*Pseudotsuga menziesii*	
				Moist site	Dry site
Live trees:					
Merchantable logs	44.3	49.8	53.5	54.3	56.1
Culls	2.3	3.3	7.1	7.1	1.8
Tops	9.1	1.9	2.6	2.3	2.5
Small trees	4.9	9.7	2.0	11.3	10.2
Subtotal	60.6	64.7	65.2	75.0	70.6
Standing dead trees:					
No defect	8.7	1.3	–	3.8	–
Sound defect	5.8	0.4	0.3	1.0	2.7
Solid rot	2.7	7.2	–	0.8	0.7
Crumbly rot	–	4.5	3.2	1.2	–
Subtotal	17.2	13.4	3.5	6.8	3.4
Down material:					
No defect	7.1	1.6	3.5	5.7	1.5
Sound defect	6.2	1.0	0.1	1.1	0.6
Solid rot	4.2	1.8	3.9	2.9	6.1
Crumbly rot	4.7	17.5	23.8	8.5	17.8
Subtotal	22.2	21.9	31.3	18.2	26.0
Total stem volume (%)	100.0	100.0	100.0	100.0	100.0
Total stem volume (m^3/ha)	351	477	559	328	207

In managed stands, the bulk of the above-ground residual biomass is composed of crowns and tops. Unmanaged, overmature stands typically hold, in addition to large live trees, some small suppressed trees, defective trees, standing dead trees of various sizes, and an accumulation of down material. If the stand is decadent, the amount of dead and rotten material may be very high. Table 2.22 shows the importance of standing dead and down material *in overmature conifer forests* in the northern Rocky Mountains according to a study by Benson and Schlieter (1979). *Sound defect* includes crook, sweep, fork, splits, and drying checks that prevent the use of the log for solid products but not for fiber. *Solid rot* includes pieces with rot that can be handled in logging, whereas *crumbly rot* is material that will not hold together in logging.

Clear-cutting such stands leaves behind large amounts of residue, the condition of which varies from stand to stand. In the study under consideration, residue volume of over 7.6-cm stem material ranged from 90 m^3/ha in dry-site *Pseudotsuga menziesii* stands to 260 m^3/ha in *Abies grandis* stands. In addition, there was 22 to 56 t/ha smaller stem and crown mass, including foliage. In general, no close relationships were found among the volume of various residue components and stand conditions. Mature stands are subject to a number of interacting factors, such as fire, insects, disease, and storms that determine residue conditions, and these may occur at irregular intervals. However, the condition of dead material appears to be related to decay conditions in the stand. In *Abies grandis* and *Larix occidentalis* stands on moist warm sites, almost all dead material was rotten, while in *Pinus contorta* stands at high elevations with dry cool conditions that do not favor rapid decay, much of the residue consisted of sound, dead material (Benson and Schlieter 1980b).

Increased stand age and overmaturity are associated with abundant residue and a large percentage of rotten material. Ruth and Harris (1975) reported an extreme case on the Olympic Peninsula in Washington, where clear-cutting of a *Thuja plicata*-dominated, old-growth stand left behind 1302 m^3 or 509 t of dry mass per hectare of residue. The effect of age is clearly indicated by the following data from the Douglas fir region on the west coast of the United States (Howard 1973c):

Stand age	Residue after logging (m^3/ha)	Percentage of rotten material in residue
–100	87	12
101–200	137	18
201–300	225	26
301–	334	31

The amount of residue also depends on cutting prescription, the level of utilization, and on the extent to which utilization specifications are followed during logging. For example, when an overmature, 511 m^3/ha *Pseudotsuga-Larix* stand in Montana was skyline logged to 15 cm sawlog top diameter, 206 m^3 residual biomass was left per hectare. When all trees down to 2.5 cm dbh were harvested, only

Table 2.23. The average amount of stem material left as residue in forest in final cuttings and selective thinnings from below in Sweden (Wallden 1982)

Stem component left in forest	m^3/ha		Percent of stem volume	
	Final cutting	Commercial thinning	Final cutting	Commercial thinning
Tops	3.2	2.0	2.2	3.7
Butt-offs	0.3	0.3	0.2	0.5
Overhigh stumps	0.1	0.0	0.1	0.1
Other stem sections	0.5	0.4	0.4	0.7
Processed timber	0.5	0.3	0.3	0.6
Whole stems	2.5	2.8	1.7	5.1
Total stem material	7.1	5.8	5.0	10.8

12 m^3 sound material and 91 m^3 unsound material was left (Benson and Schlieter 1980a).

The examples above represent the residue potential in overmature, old-growth forests on the west coast and in the Rocky Mountains in the United States. The situation is quite different *in managed forests* in Europe and other areas, where the demand for wood and stumpage prices are higher, and repeated selection thinnings from below have long traditions. This is illustrated in Table 2.23 by Sweden, where the National Forest Inventory in 1975 and 1980 was extended to estimate the residual stem volume from trees larger than 5 cm at breast height. In the final cuttings only 7.1 m^3/ha or 5.0%, and in thinnings 5.8 m^3/ha or 10.8% of stem volume in total felling was left in the forests. In Finland, 6% softwood and 14% hardwood stem volume was left in the forests in the late 1960s, when the residue from early precommercial thinnings was also counted (Mikkola 1969, 1972). If only stem sections meeting the quality and 6–7-cm minimum diameter requirement of industrial timber are included, the proportion of residual stem volume in Finland is not more than 1.4% for softwoods and 2.3% for hardwoods, 70% of which consists of excessively high stumps and over 6-cm tops (Halinen 1986).

As stated earlier, interest in utilization of residual forest biomass in a given area often seems to be inversely proportional to the amount of biomass left in the forests. Finland and Sweden are countries where great efforts are being placed into the development of forest residue utilization, although relatively little stem mass is left unutilized and the residue is restricted primarily to trees smaller than 10 cm at breast height, crowns, stumps, and roots. Even under these conditions remarkable amounts of low-quality biomass suitable for energy and other purposes may exist in the forests. For example, the total potential forest energy reserve in Sweden's forests amounts to 752 million tonnes of dry matter, consisting of the following sources (Marklund 1981):

Source	Potential reserve, million tonnes of dry mass
Whole-tree mass from under 10-cm unmerchantable trees	136.6
Stem mass from over 5-cm dead trees	23.9
Top and crown mass from over 10-cm merchantable trees	334.4
Over 5-cm stump and root mass from over 10-cm merchantable trees	257.4

Table 2.24 (Danielsson and Nilsson 1977) is based on the stem volume statistics of the National Forest Inventory in Sweden and on a 75 million m^3/a cutting level. Corresponding amounts of annual crown, stump, and root residue were estimated using biomass functions. According to these calculations, 49 million m^3 solid of residual forest biomass is produced and potentially available annually in Sweden. However, availability is radically reduced in practice by factors such as difficult terrain, poor and sensitive sites, loss of biomass in recovery operations, etc. Consequently, only 17 million m^3/a is estimated to be readily available for

Table 2.24. The amount of residual biomass potentially and technically available at the current 75 million m^3/a cutting level in Sweden (Danielsson and Nilsson 1977)

Source of residual biomass	Million m^3 solid/a				
	Softwood		Hardwood		Total
	Final cuttings	Commercial thinnings	Final cuttings	Commercial thinnings	
Biomass potentially available:					
Stumps + roots	10.30	–	–	–	10.30
Tops + crowns	16.60	8.10	–	3.00	27.70
Small trees	1.50	4.90	2.50	0.35	9.25
Oversized top sections	0.80	*	–	0.16	0.96
Other stem residues	0.53	0.23	–	0.18	0.94
Subtotal	29.73	13.23	2.50	3.69	49.15
Biomass technically available:					
Stumps + roots	5.70	–	–	–	5.70
Tops + crowns	4.70	2.50	–	0.60	7.80
Small trees	0.42	1.50	0.70	0.07	2.69
Oversized top sections	0.80	*	–	0.16	0.96
Other stem residues	0.24	0.10	–	0.08	0.42
Subtotal	11.86	4.10	0.70	0.91	17.57
Wood mass technically available:					
Stumps + roots	4.50	–	–	–	4.50
Tops + crowns	2.80	1.75	–	0.35	4.90
Small trees	0.32	1.05	0.50	0.05	1.92
Oversized top sections	0.64	*	–	0.13	0.77
Other stem residues	0.20	0.08	–	0.06	0.34
Subtotal	8.46	2.88	0.50	0.59	12.43

* Included in top and crown mass.

harvest without major technical or environmental problems. On the other hand, in addition to the logging residue from current cutting activity, large amounts of raw material are also available in Sweden in young stands that are presently outside cutting activity due to the high cost of logging.

The Swedish results are in fairly good agreement with a corresponding Finnish calculation, according to which 15 million m^3 solid of residual biomass is technically available at a 55 million m^3 annual cutting level. Recovery of this reserve could yield an annual 7.1 million m^3 solid of whole-tree chips from small-sized trees, 4.8 million m^3 solid of chips made of crowns and tops, 2.7 million m^3 solid of split stump and root wood, and 0.3 million m^3 solid of aged wood material mined as a by-product from fuel peat bogs (Puuvaraselvitys . . . 1976).

In the United States, the Forest Service completed its first tree-biomass compilation in the early 1980s. The principal finding was that the country's almost 200 million ha of commercial forest land supports 32.6 billion tonnes of fresh

Table 2.25. Total above-ground tree biomass on commercial forest land in the United States by class of timber and geographic region, on a fresh mass basis (Tree biomass ... 1981)

Biomass category	Million green tonnes				%
	Geographic region			Total USA	Total USA
	North	South	West		
Commercial trees:					
Merchantable stem	5461	7281	6717	19459	59.6
Under 10 cm top and crown	1867	1489	1341	4697	14.4
Defective trees:					
Merchantable stem	1165	1346	275	2786	8.5
Under 10 cm top and crown	471	333	96	900	2.8
Small trees	1593	2564	642	4799	14.7
Total above-ground mass	10557	13013	9071	32641	100.0

above-ground biomass (Table 2.25). Material not recognized in conventional forest inventories, i.e., defective trees, small trees, tops, and crowns, add 43% to the merchantable stem mass of the softwood growing stock and 99% to that of hardwood growing stock. The largest potential increase through biomass recovery, 93%, occurs in the northern region of the country due to the high proportion of hardwoods. The gain is rather evenly distributed among tops and crowns, small-sized trees, and defective trees. In the southern region the potential biomass gain is greatest in small-sized trees and in the western region by recovering tops and branches of merchantable trees. Nationwide, if all the potential residual biomass reserves were recovered, there would be a two-thirds increase in the forest biomass base (Tree biomass ... 1981).

Using conventional inventory data, Carpenter (1980) estimated the potential annual fuel reserve of unmerchantable forest biomass from conventional logging operations in the eastern United States as 118 million tonnes of dry matter. This potential reserve would be available annually from logging areas if all cull sections, tops, and crowns from trees cut for various products were salvaged, and if all rough, rotten, salvable dead, and mortality trees in logging areas were removed at the time of cutting. The calculation did not include noncommercial stand improvement operations. According to Groton et al. (1985), the potential fuelwood supply from logging residue and noncommercial trees for a geographic region covering 83 million ha of commercial forests in 13 southeastern states of the United States represents an annual 2.2 tonnes of fresh biomass per hectare, or a total of 177 million tonnes.

The U.S. Forest Service estimated the amount of unutilized forest biomass potentially available as 540 million dry metric tonnes per annum, excluding stumps and roots. Of this total amount 144 million dry tonnes were classified as logging residue. If the total reserve could be used for energy, the share of wood in the energy budget of the country could be increased from 1.9% to as high as 12.9%. However, the calculation refers to a theoretical maximum potential, which could be recovered in its entirety under no circumstances (The nations ... 1981):

	Potentially available, unutilized forest biomass in USA (million dry tonnes/a)
Excess growth and small trees	195
Mortality	86
Logging residue	146
Salvable rough, rotten, and dead trees	18
Wood processing residue	18
Residue from land clearing	18
Urban wood residue	63
Total	540

Chapter 3

Technical Properties of Residual Tree Components

Tree crowns, unmerchantable tops, stumps, and roots form a huge reserve of renewable raw material. Compared with merchantable stems, some 50% additional biomass is potentially available from these unutilized tree components in silviculturally managed forests and, if undersized trees and defected stem parts are included, a considerably higher percentage in overmature, unmanaged forests.

Utilization of the residual biomass is constrained and frequently totally prevented by many economic, technical, and ecological barriers. A critical factor is the high cost of recovery due to the small unit size, low areal yield, low bulk density, and need for special equipment. In the conventional use of raw material, high content and irregular variation of bark and foliage are serious drawbacks as well. This problematic situation is reflected in the perspective of wood technology research, which is traditionally aimed at improved utilization of stem wood. Only a very small fraction of the efforts is directed toward the utilization of other components of the tree.

Presently, development of effective machines and methods for harvesting (Chaps. 4 to 7) and segregation of wood proper from bark and foliage (Chap. 8) are key questions. In the long term, utilization of residual forest biomass may be facilitated by reduction of recovery costs and emerging technology for segregation of biomass components from each other. Knowledge of the qualitative properties of crown, top, stump, and root mass will then be of great importance. In fact, lack of detailed information on the quality of residual tree components is a serious drawback for the feasible use of forest biomass in its entirety. If the quality and variation were better known, the use and processing could be planned more effectively.

In forestry and industrial planning, it is of primary importance to know not only the quantity but also the technical properties of biomass inventories. Only when the quality is known can biomass be sorted and utilized to the best advantage, and only then is there a firm basis for forecasts of raw material behavior and consumption in an industrial process. Quality, in turn, is the result of physical and chemical characteristics possessed by a tree or part of a tree that enables it to meet the property requirements for different end products (Mitchell 1961). Quality is thus a concept combined with the use of biomass. As there is no unit by which the biomass quality can be expressed directly, that end must be attained by a roundabout route. One or several properties indicative of the suitability of biomass for a certain use must be determined, depending on the intended use. These properties are called biomass *quality indicators*. To be serviceable, a biomass quality indicator should meet the following requirements:

1. The indicator must give an idea of the suitability of the biomass for various uses.

2. The indicator must be a sensitive measure of the variation in biomass quality.
3. The indicator must be easy to interpret and use, not only in research but also in practical forestry and forest industries.
4. The indicator must be applicable to all tree species and all components of biomass.

Utilization of residual tree parts as industrial raw material seems most possible for fiber, chemical, and biochemical products, whereas use in the mechanical forest industries, composite boards excluded, is generally out of the question. Accordingly, the quality of various components of residual biomass is reflected best by the cell dimensions (Sect. 3.2), chemical composition (Sect. 3.3), and basic density of wood and bark (Sect. 3.4). For deeper understanding of the variation of these biomass quality indicators, a short presentation of the anatomical structure of residual tree components is also included (Sect. 3.1). Since, in addition to industrial processing, residual forest biomass is actually used in considerably larger quantities for the production of energy, special attention is also paid to the heating value of various tree components (Sect. 3.5). Bark percentage is discussed in Sections 2.1.1.3.2, 2.1.2.1 and 2.1.3.2, as it is closely linked to the questions of the quantity and composition of tree biomass.

3.1 Anatomical Structure of Residual Tree Components

Under normal growth conditions forest trees produce an erect, vertical stem. Wood or xylem of such stems is characterized by regular concentric annual increments or *growth rings*, which can be distinguished from one another because of differences in the structure and proportions of various types of cells produced in the beginning and at the end of the period of annual growth. The thin-walled *earlywood* cells formed at the beginning of a growth period function primarily as water transportation tissue, whereas thick-walled *latewood* cells formed at the end of the season give mechanical strength to stem, branches, and roots. Cells from earlywood and latewood also differ in their paper-making properties. In softwoods, earlywood tracheids have a larger lumen and thinner walls, resulting in lighter color and lower wood density. Latewood tracheids, on the other hand, are smaller in radial diameter and have thicker walls.

According to Mork's (1928) definition, latewood in softwoods includes tracheids in which the common cell wall in the tangential direction between two cells is exactly half or over half of the radial width of the lumen. If the joint width of two cell walls is less than this, the tracheids are considered to belong to earlywood. Mork's definition was intended originally for spruce, but is used commonly for other softwoods as well. The *transition from earlywood to latewood* in a conifer may be gradual or abrupt (Figs. 3.1, 3.2), depending on the species, age from the pith, and, to a lesser extent, on environmental conditions. The proportion of latewood volume to earlywood volume, denoted as *percentage of latewood*, is an important wood quality indicator when wood is used for pulping or in the mechanical forest industries. The percentage of latewood denotes, however, only the proportion of the xylem that meets the minimum cell-wall thickness

Fig. 3.1. Scanning electron micrograph of normal softwood with gradual transition from earlywood to latewood. Stem wood of *Pinus strobus* (courtesy of W. A. Côté, N. C. Brown Center for Ultrastructure Studies)

as defined above and gives no indication of the variation in wall thickness within the earlywood and latewood zones. Its use is also limited by its unsuitability for hardwoods.

Hardwoods are classified into two groups on the basis of vessel arrangement within a growth ring. A vessel is formed from a vertical row of cells, vessel elements, from which the end walls have disappeared. In *diffuse-porous hardwoods*, the vessels or pores are evenly distributed and rather uniform in size (Fig. 3.3). In *ring-porous hardwoods*, the earlywood vessels are significantly larger and arranged in a ringlike fashion (Fig. 3.4).

Fig. 3.2. Scanning electron micrograph of normal softwood with abrupt transition from earlywood to latewood. Stem wood of *Pinus resinosa* (courtesy of W. A. Côté, N. C. Brown Center for Ultrastructure Studies)

Wood is composed of various types of cells which are either parenchymatous or prosenchymatous. The *parenchymatous cells* function as storage tissue. They are typically short, thin-walled, and rectangular or rounded. Their paper-making properties are inferior. The *prosenchymatous cells* consist of elongated, usually dead cells with typically narrowing ends, functioning as conducting and supporting tissues. Softwood tracheids and hardwood fibers and vessels belong to this group. The vessels are actually water pipes, the diameter and length of which increase from twigs to branches, down along the stem, to the long, ropelike roots. Wide and long vessels are more efficient conductors than narrow and short ones,

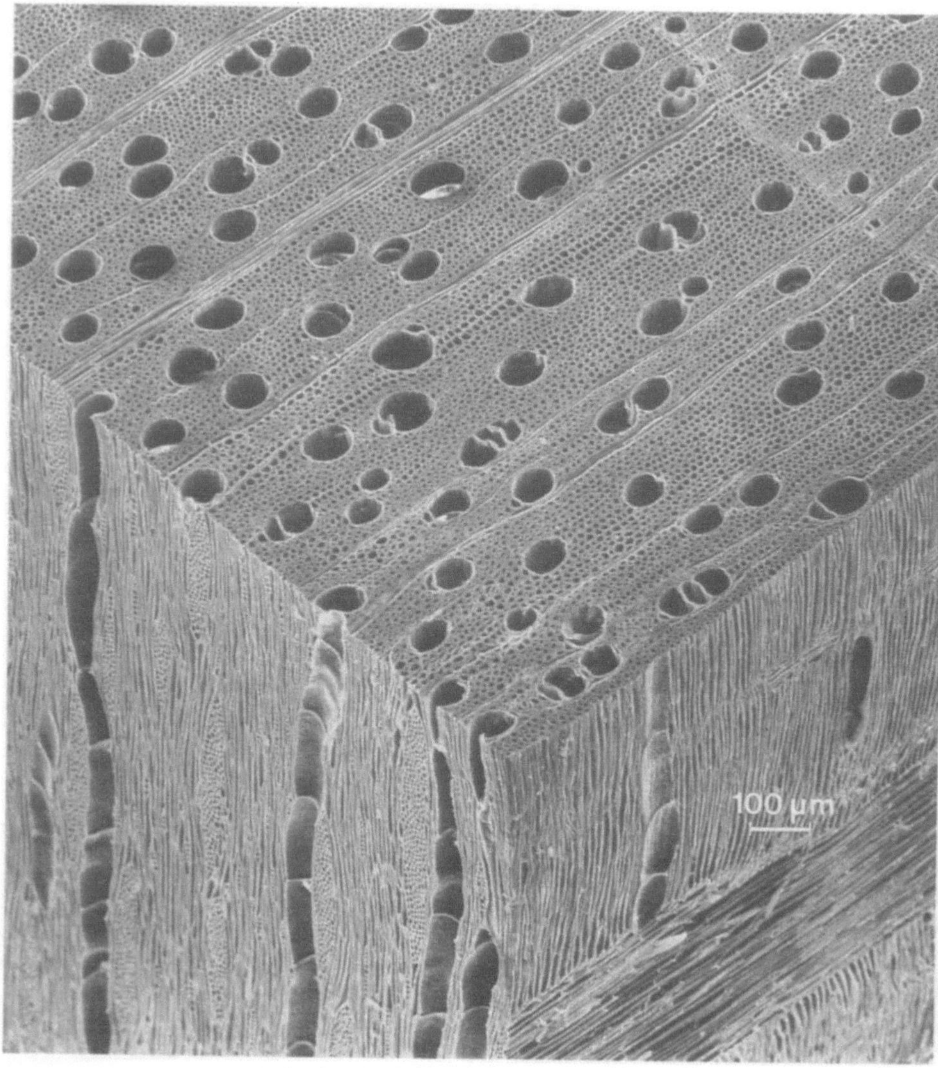

Fig. 3.3. Scanning electron micrograph of normal diffuse-porous hardwood. Stem wood of *Acer saccharum* (courtesy of W.A. Côté, N.C. Brown Center for Ultrastructure Studies)

but small units are safer. Since xylem pressures always decrease from bottom to top due to gravity and flow resistance, it is an advantage to the tree to have the conducting compartments in the top smaller, i.e., safer. Thus, in *Acer rubrum* the vessel diameters increased from 54 µm at the 10-m level to 72 µm at the base of the stem, and the longest vessel length was 20 cm at the top but as much as 32 cm both at midheight and at the base of the stem (Zimmermann and Potter 1982).

If wind, snow, or other factors force a stem to lean or bend, the tree tries to regain its correct orientation by growing specific wood tissue in order to exert pressure along the grain and restore the vertical growth pattern. In fact, only sup-

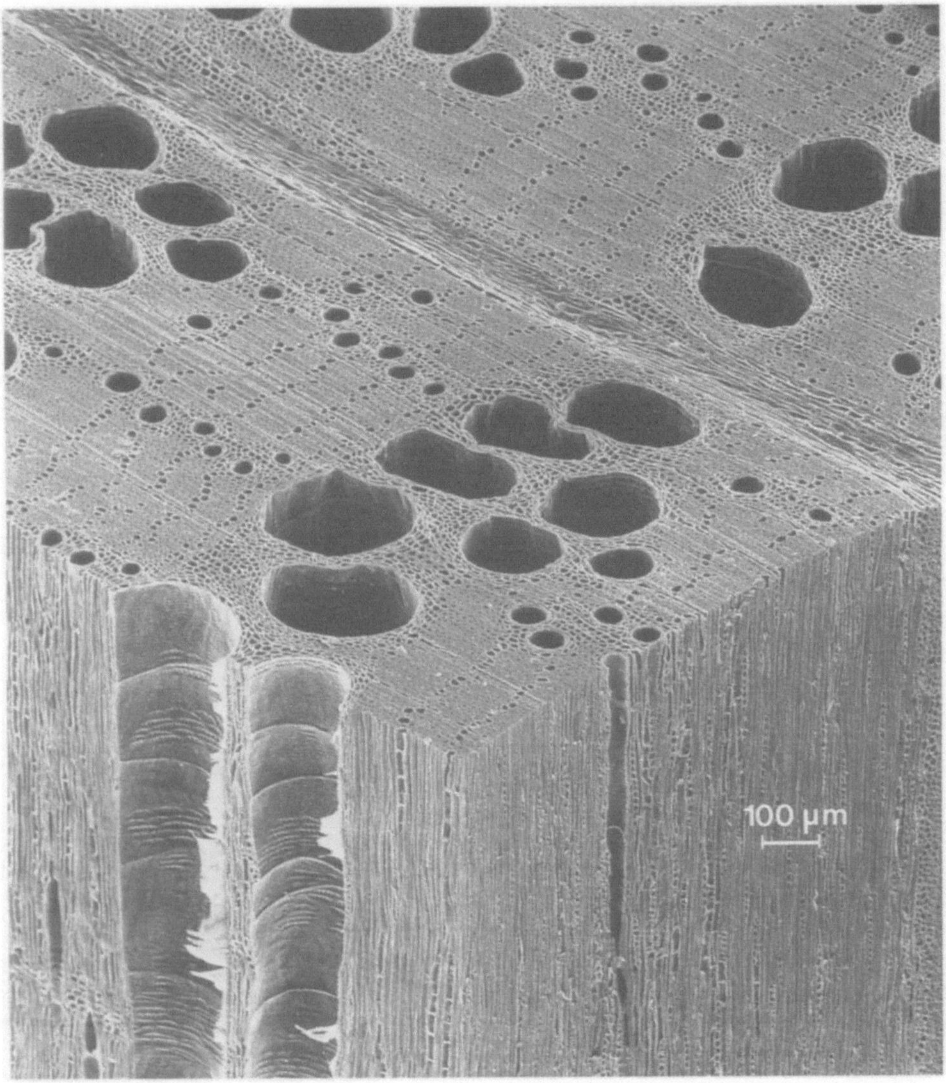

Fig. 3.4. Scanning electron micrograph of normal ring-porous hardwood. Stem wood of *Quercus rubra* (courtesy of W. A. Côté, N. C. Brown Center for Ultrastructure Studies)

porting cells, i.e., softwood tracheids and hardwood fibers, are affected. Since the development of this abnormal tissue is a result of the tree's reaction to an outer disturbance, it is referred to as *reaction wood.* In conifers it forms on the lower side and in hardwoods on the upper side of a leaning or bended stem. In conifers reaction wood is called *compression wood* (Sect. 3.1.1) and in hardwoods *tension wood* (Sect. 3.1.2). The tissue on the opposite side of the stem is called *opposite wood.*

A branch of a tree may be likened to a cantilever beam, i.e., a beam fixed at one end only. The mass of the beam acts downward and exerts a pull on the upper side of the branch which consequently tends to be stretched, i.e., to be in a state of tension. The lower side of the branch is simultaneously in a state of compression (Jane et al. 1970). To maintain proper branch angle, the tree forms compression wood on the lower side of coniferous branches and tension wood on the upper side of hardwood branches. As reaction wood behaves quite differently from unaffected, normal wood, and as its presence in branches is the rule rather than the exception, it decisively affects many important properties of branch wood. Branches also develop opposite wood. The normal tissue between reaction wood and opposite wood is called *side wood.*

The abnormal structure of reaction wood is similar in stem and branches. Although most studies referred to below actually concern reaction wood in stems, their results are believed to be applicable for reaction wood in branches as well. According to Timell (1986a, b), the evidence of possible differences between stem and branch compression woods is contradictory.

Many of the characteristic properties of reaction wood are undesirable from the utilization point of view. Particularly when pronounced, this abnormal woody tissue is classified as a serious defect. Thus, reaction wood has been a research subject in wood science and technology since the 19th century (Sanio 1860; Hartig

Table 3.1. Comparison of some properties of compression and tension woods (T.E. Timell, pers. commun.)

Characteristics	Compression wood	Tension wood
Gross properties	Radial growth promotion and compression wood on lower side. Dark, reddish color. Present in entire growth ring but modified in most first-formed earlywood.	Radial growth promotion usually on upper side but sometimes on lower. Tension wood on upper side. Bright, silvery sheen or, e.g., in eucalypts, dark. Absent in the last-formed latewood.
Anatomical properties	Tracheids with a rounded outline and intercellular spaces, almost all with a thick wall. Thick S_1 and S_2 layers with helical cavities and ribs in inner S_2. Large microfibril angle in S_2. Ray parenchyma cells unchanged. No S_3.	Gelatinous fibers present in most but lacking in a few species. The thick G-layer most often replacing S_3. Microfibrils in G-layer parallel with fiber axis. Compression failures common. Vessels reduced in size and frequency. Parenchyma cells unchanged.
Chemical properties	Less cellulose and galactoglucomannan but more lignin and galactan than in normal wood. Presence of a 1,3-β-glucan. Outer S_2 has a high concentration of lignin. The lignin has fewer methoxyl groups and is more highly condensed than in normal wood.	More cellulose and less lignin than in normal wood. Some species have a high galactan content. The G-layer usually consists entirely of cellulose. The lignin is the same as in normal wood. The xylan is also the same but is reduced in amount.
Physical and mechanical properties	Density much higher than that of normal wood. Hard and brash wood. Exceedingly high longitudinal shrinkage. Compressive and bending strengths high. Modulus of elasticity and tensile strengths low.	Density often somewhat higher than that of normal wood. High longitudinal shrinkage. Longitudinally sawn boards have a woolly surface. Compressive and bending strengths low. Tensile strength high.

1901). Extensive reviews have been published. The most recent and complete review is an interdisciplinary, encyclopedic book by Timell (1986a, b, c), titled *Compression Wood in Gymnosperms*. The massive, three-volume work with 8100 references also reviews the specific structure and properties of coniferous branch wood. Comparative features, properties, and structure of compression and tension woods are summarized in Table 3.1.

Roots, too, differ in structure from stem wood. Since roots presently have little utilization value, they have not received much attention in wood science. Structural variations in root wood in relation to distance from the stem, or actually from the synthesizing crown, and from the pith outward, probably have their origin in differences in function. The thin sections of the lateral roots are primarily organs of storage and food translocation in the living tree, and these functions affect their structure. Storage is served by abundant parenchyma, while food translocation requires abundant pitting. Exposure of roots to air and light causes variations in cellular structure. Exposed roots tend to produce wood with anatomical characteristics resembling stem wood (Tsoumis 1968). The anatomical structure of root wood is discussed in Section 3.1.4.

The within-tree variation in the anatomical structure and properties of wood occurs not only between the stem and other biomass components, but is also found within the stem. Variations in both the radial and longitudinal direction of the stem result in remarkable differences in cell dimensions and other technical properties between the conventional timber and the unmerchantable top of the tree. A fundamental cause of these differences is the presence of juvenile wood in a cylindrical column surrounding the pith (Sect. 3.1.3).

3.1.1 Compression Wood in Softwood Branches

The anatomical structure of compression wood, whether in the stem or a branch of a coniferous tree, has several distinctive features. These characteristics of compression wood influence the technical properties of branch wood. Although it is hard and dense, it is also brash and, mass for mass, weaker in strength than normal wood.

The presence of compression wood is usually indicated by *eccentric growth rings*. A cross-section of a softwood branch shows the pith nearer the upper side, compression wood appearing on the lower side of the branch, commonly as a dark arch or crescent-shaped patch (Fig. 3.5). It is characterized by a nonlustrous, lifeless appearance and a *gradual transition from earlywood to latewood* even in tree species which normally exhibit an abrupt transition. Compression wood appears to contain an exceptionally *large proportion of latewood* in the region of fastest growth (Fig. 3.6).

Contrary to tension wood in hardwoods, which is seldom found in the latewood region of a growth increment, compression wood has a definite tendency to be absent from or only partly developed in earlywood. However, Timell (1986a) pointed out that while the earlywood-latewood concept is of great theoretical and practical importance when applied to normal wood, it is of less significance in dealing with compression wood. In fact, its uncritical use has resulted in many

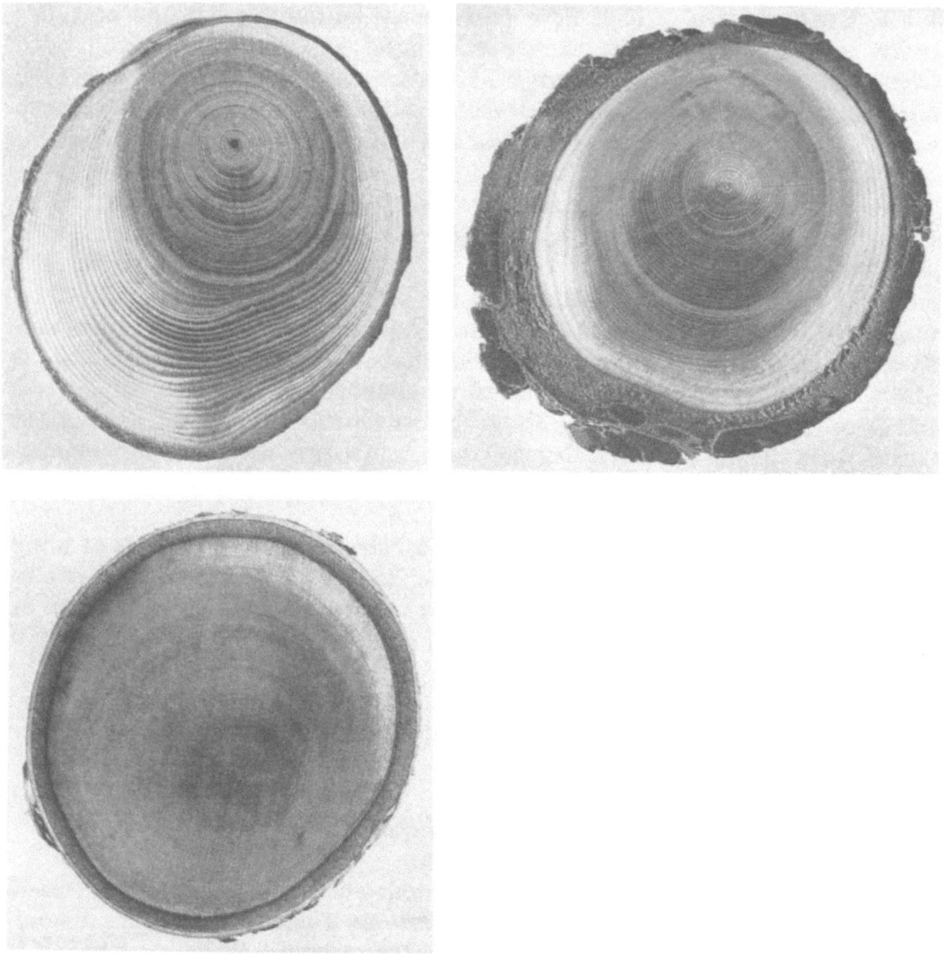

Fig. 3.5. Transverse sections of branches from *Pinus sylvestris* (*above left*), *Picea abies* (*above right*), and *Betula pubescens* (*below*) (courtesy of FFRI)

contradictory statements in the literature. Thus, according to Timell, the terms earlywood and latewood should be applied to compression wood only sparingly or not at all, since they do not have their usual meaning in this context. Compression wood is produced primarily under conditions of prolonged and very high cambial activity, possibly as a result of a high concentration of auxin. Conversely, in normal wood, high concentrations of auxin are associated with the formation of earlywood, whereas lack of auxin and the presence of growth inhibitors in con-

→

Fig. 3.6. Normal stem wood (*above*), branch compression wood (*middle*), and branch opposite wood (*below*) in *Pinus sylvestris*. Transition from earlywood to latewood is abrupt in normal wood but gradual in compression wood. Scanning electron micrographs of transverse sections (courtesy of Pekka Saranpää)

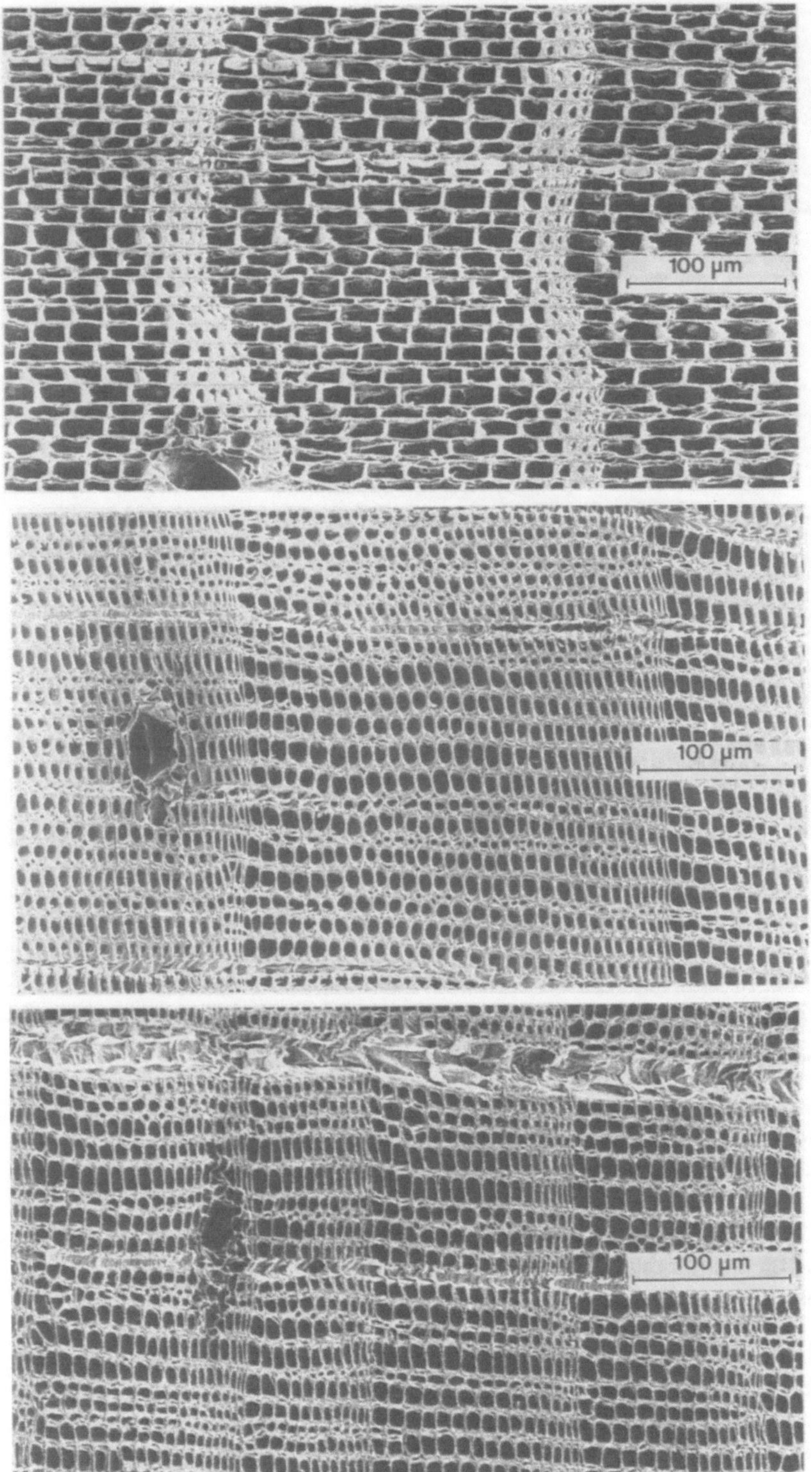
100 µm
100 µm
100 µm

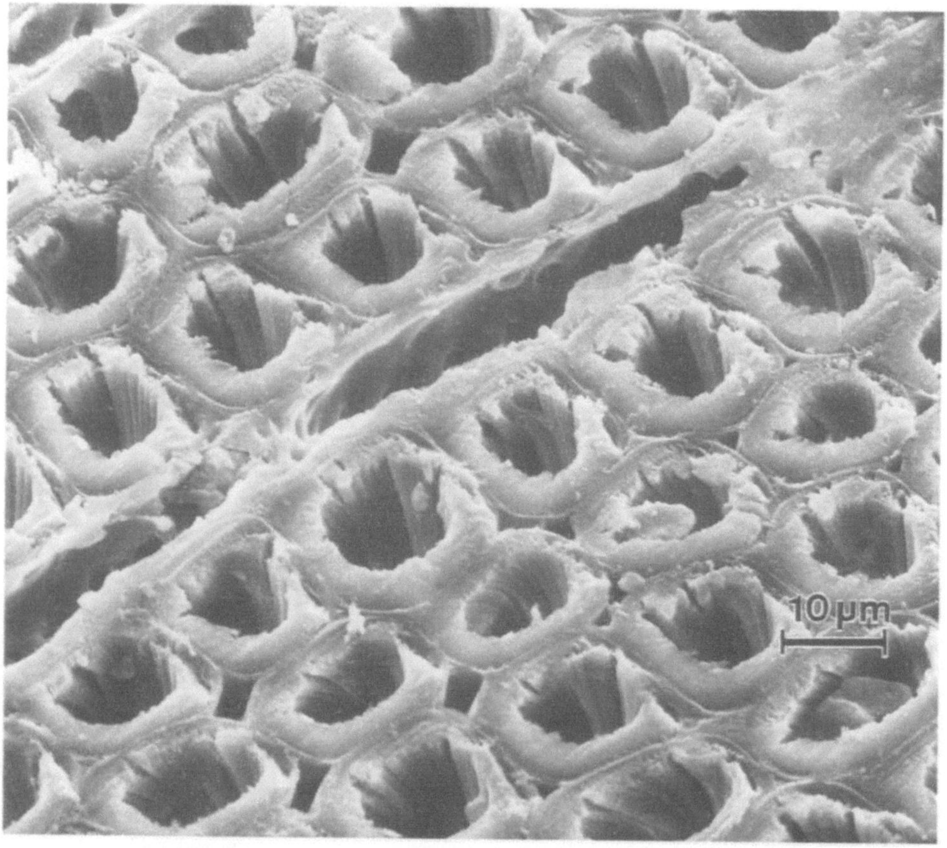

Fig. 3.7. Scanning electron micrograph of transverse section of stem compression wood in *Picea abies*. Conspicuous anatomical features: rounded shape of tracheids; intercellular spaces; helical cavities and ribs on the secondary cell wall (courtesy of T. E. Timell)

junction with an abundant supply of photosynthate, result in the formation of latewood.

Under microscope examination, compression wood has many structural peculiarities. Compression-wood tracheids are chiefly *rounded in cross-section*, in contrast to the angular outline of normal tracheids. As a result of the rounded form, the tracheids are not close-fitting but exhibit *intercellular spaces*. The nearly circular or, occasionally, oval tracheid cross-sections, together with the intercellular spaces at the junction of four tracheids and a somewhat distorted arrangement, represent some of the most conspicuous anatomical features of compression wood (Fig. 3.7).

Except for the narrow zones of the first-formed earlywood and last-formed latewood, tracheids in severe compression wood are of fairly uniform size and have a uniform wall thickness. They have a much *thicker wall* and a *narrower lumen* than the tracheids in normal earlywood. The difference between normal latewood and compression wood is in this respect, however, very slight. It would appear that the wall is generally somewhat thicker and the lumen smaller in normal latewood than in compression wood. Normal latewood is also harder and heavier than compression wood (Timell 1986a).

The tracheids are separated from one another by the middle lamella (M). In normal wood, the tracheid wall consists of a primary (P) and secondary (S) cell wall. Both the intercellular region of the middle lamella and the primary wall are encrusted by lignin. The secondary wall is composed of the outer (S_1), the middle (S_2), and the inner (S_3) layers, distinguishable from one another by the different orientation of their cellulose microfibrils. Following the S_3 layer is a warty layer, which terminates the secondary wall toward the lumen. Helical thickenings are sometimes part of the S_3 layer.

In compression wood tracheids, the secondary cell wall *lacks the S_3 layer.* The modified S_2 layer has a large, about *30–50° microfibrillar angle* to the cell axis, as compared with 10–30° in normal wood. Narrow, about 0.1-µm-wide *helical cavities* spiral around the tracheid in the same direction as the cellulose microfibrils, extending from the lumen toward, but never reaching, the S_1-S_2 boundary. Between the cavities are 1–2-µm-wide *helical ribs.* Warts often cover the helical cavities and ribs (Fig. 3.8). When subjected to stress, the compression tracheid wall will tend to crack where it is weakest, usually along the helical

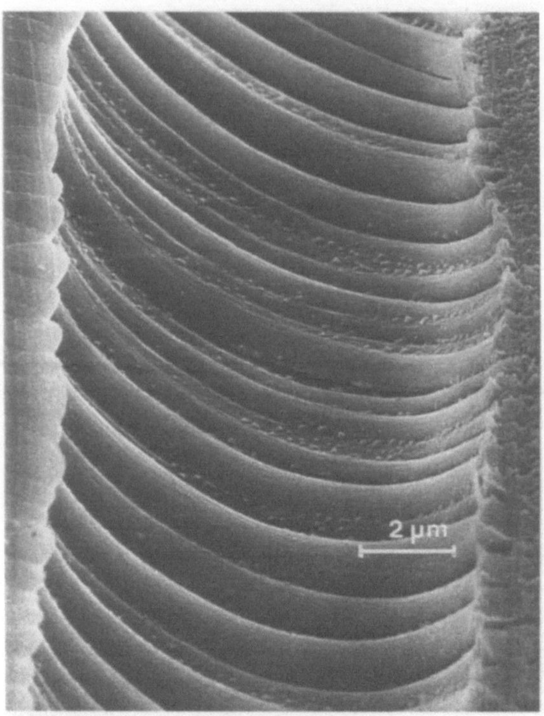

Fig. 3.8. Scanning electron micrograph of the cell wall of a compression-wood tracheid in a branch of *Pinus sylvestris*. Warts on helical cavities and ribs on the secondary cell wall (courtesy of Pekka Saranpää)

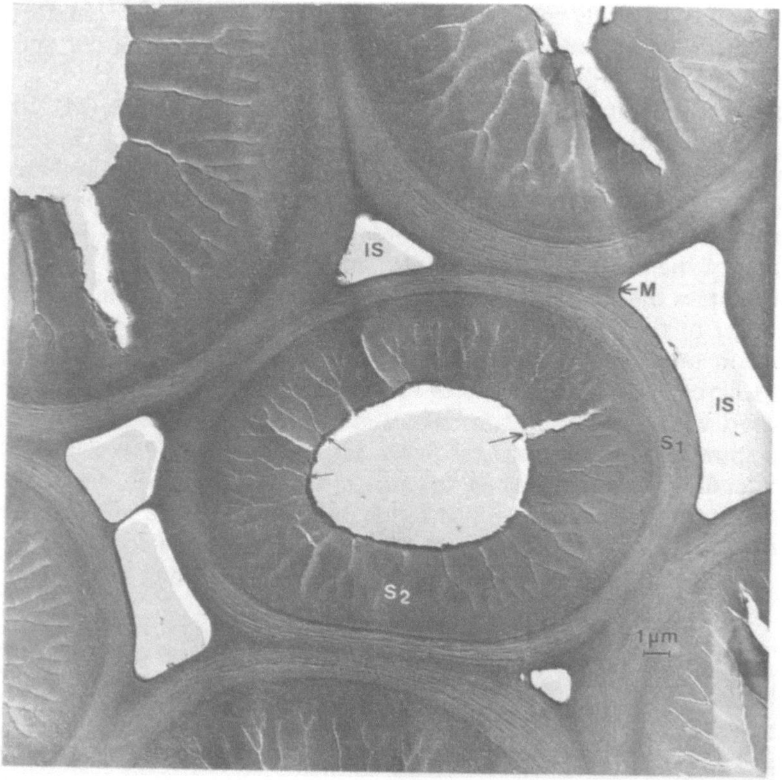

Fig. 3.9. Transmission electron micrograph of compression wood in a *Larix laricina* stem. Tracheids with rounded shape and intercellular spaces (*IS*), one of which is transversed by a band of lignin. Middle lamella between adjacent cells, S_1 and S_2 layers of the secondary cell wall, and helical cavities (*small arrows*) and helical drying checks (*large arrow*) on the S_2 la layer (courtesy of T.E. Timell)

cavities, thus forming oblique 0.5–1.5-µm-wide *helical checks* (Fig. 3.9). The significance of the cavities and ribs for the function of compression wood in the living tree is uncertain. Perhaps they contribute to the high resilience so characteristic of compression wood which makes it possible for displaced stems and branches to return rapidly to their original position. Perhaps they serve a function in making it possible for compression wood to expand in the living tree and thus cause the necessary movements of orientation (Timell 1986a).

Helical cavities are believed to be contributory to the *high longitudinal and low transverse shrinkage* of coniferous compression wood. A tenfold increase in longitudinal shrinkage, compared to that of normal wood, is not unusual. While longitudinal shrinkage in normal wood is only 0.1–0.2%, in extreme cases it has been known to reach 6–7% in compression wood. Consequently, where bands of compression wood occur adjacent to normal rings, there is a great tendency for bending, twisting, and splitting due to differential shrinkage (Trendelenburg 1939; Kollmann and Côté 1968).

Compression-wood *tracheids are shorter* than those of normal wood. Their tips are frequently distorted, while normal tracheids are usually quite simple in outline. Reports in the literature regarding the rays are contradictory, but the majority of investigators have found that *rays are more frequent* in compression wood than in normal wood and that both rays and ray cells are larger. It would appear that normal-type *resin canals are less frequent* in compression wood than in normal wood (Timell 1986a).

The cellulose content of compression wood is lower, and the lignin content higher than that of normal wood. There are also significant differences in the composition of pentosans between compression and normal wood. These are discussed in Section 3.3.1. Compression wood contains 10% of a galactan and 3–5% of a 1,3-glucan not present in normal wood.

The formation of opposite wood above the pith of a leaning softwood stem or branch is associated with a low cambial activity (Larson 1969). Although the mass of opposite wood in a branch is much smaller than that of compression wood, its presence should not be ignored when assessing the technical properties and utilization possibilities of branches. Figure 3.10 shows distinct differences in transverse sections of compression wood and opposite wood of *Pinus sylvestris*. Typical *structural features of opposite wood* in conifers are narrow growth rings, a low proportion of latewood, long tracheids, a thick S_3 layer, high tensile strength and low compression strength. Münch (1938) regarded normal softwood

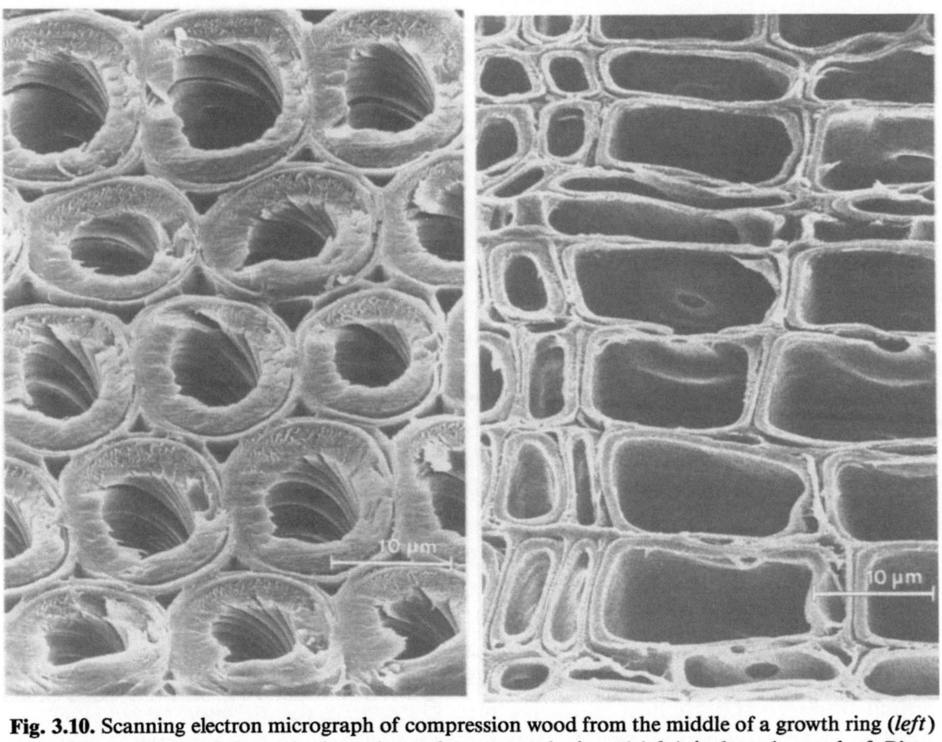

Fig. 3.10. Scanning electron micrograph of compression wood from the middle of a growth ring (*left*) and of opposite wood from the boundary of two growth rings (*right*) in branch wood of *Pinus sylvestris*. In contradistinction from compression wood, opposite-wood tracheids are characterized by an angular outline and thinner cell walls without helical cavities or ribs (courtesy of Pekka Saranpää)

Table 3.2. Comparison of some properties of compression, normal, and opposite woods in conifers (Timell 1973)

Property	Compression wood	Normal wood	Opposite wood
Growth rings	Wide	Intermediate	Narrow
Proportion of latewood	Very large	Intermediate	Small
Arrangement of tracheids	Irregular	Regular	Very regular in earlywood, often irregular in latewood
Tracheid length	Short	Intermediate	Long
Outline of tracheids	Round	Angular	Square, rectangular, or angular
Microfibril angle in S_2 layer	30–50°	10–30°	Variable
The S_3 layer	Absent	Thin	Thick
Helical thickenings	Rare	Occasional	Frequent in latewood
Tensile strength	Low	Intermediate	High
Compressive strength	High	Intermediate	Low

in many respects as a transition form between compression and opposite wood. According to Table 3.2 (Timell 1973), normal wood really occupies a transitional position between compression and opposite woods.

3.1.2 Tension Wood in Hardwood Branches

Tension wood occurs on the upper side of leaning stems and branches of hardwoods, usually associated with *eccentric growth rings.* In a cross-section, the pith is generally, but not always, nearer the lower side. Tension wood thus usually appears in a region of accelerated growth, while the region of suppressed growth below the pith produces opposite wood (Fig. 3.11).

Compared with compression wood, tension wood is more difficult to detect. The most characteristic feature is the appearance of a *woolly surface* on boards that are sawn green. In such boards, the tension-wood tissue tears loose in bundles of fibers which under extreme conditions can choke and overheat a saw. When dry, tension wood does not tear as readily in sawing and machining, although the fibers do tend to tear out to some extent. When cutting veneers from tension wood, the surfaces are rough and the veneers buckle. Some species of wood also give evidence of woolliness on the cross-section, and others may have a darker color in tension-wood zones on the cross-section. Unusual bowing and irrecoverable collapse may also indicate the presence of tension wood (Panshin and de Zeeuw 1980). Due to differences in chemical composition, tension wood can be easily detected under microscope examination by treatment with chlor-zinc-iodide, or with a combination of safranin and light green, etc.

When tension wood is formed in hardwood stems and branches, it is the fibers which undergo the greatest modification. Generally, the *percentage of fibers is greater* than in normal wood. Tension-wood fibers have somewhat *smaller diameters*, they usually have *greater lengths*, and they exhibit unusually *thick walls* with somewhat rounded outlines. On the other hand, the relative *frequency*

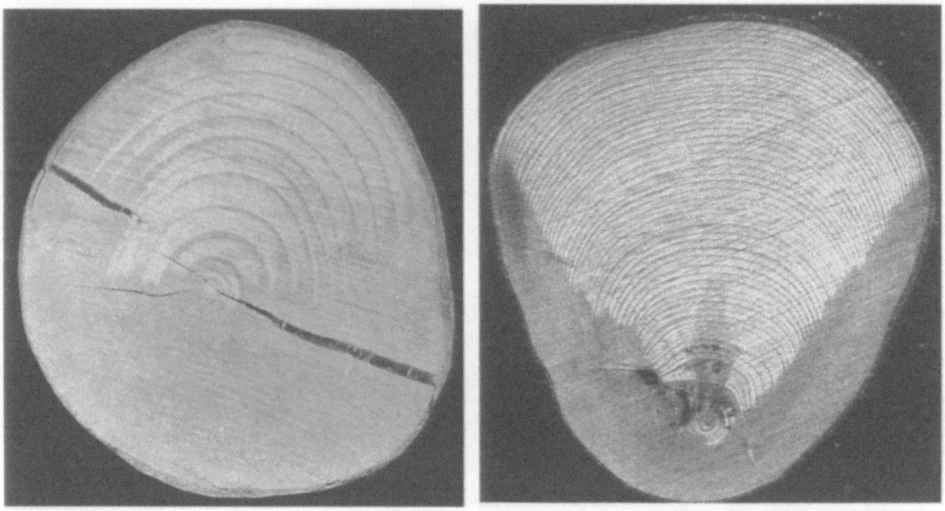

Fig. 3.11. Branch cross-sections from *Acer platanoides* (*left*) and *Fagus grandifolia* (*right*) showing tension wood on the upper side (courtesy of T.E. Timell)

and size of vessels may be reduced (Timell 1969). In a study on *Betula pendula* by Bhat and Kärkkäinen (1981b), the proportion of fibers decreased in normal stem wood from breast height to crown but, at a given stem height, increased from stem to branch. The trend along the branch was not clear. The change was opposite in the proportion of vessels. Koch (1985a) studied tissue proportions in stem and branch woods of 22 hardwood species grown on pine sites in the southern United States. In branch wood, the proportions of fibers and longitudinal parenchyma were higher and those of vessels and rays lower. However, the differences were not great (Sect. 3.1.4).

The characteristic anatomical modifications of tension wood are associated primarily with the fibers. A prevailing feature is the *gelatinous fiber*. The innermost, rather thick, unlignified part of the fiber wall has a gelatinous appearance and is, therefore, called the *gelatinous or G-layer*. This layer, formed almost exclusively of pure cellulose microfibrils, may sometimes nearly fill the fiber lumen. Frequently, it appears to be convoluted and partially separated from the cell-wall layers beneath (Fig. 3.12). This feature is now considered to be a preparation artifact.

According to Dadswell and Wardrop (1955), three variations of cell-wall layering exist in gelatinous fibers, and more than one type may be present simultaneously in the same specimen. In the first alternative the G-layer is deposited immediately within the S_1 layer of the secondary fiber wall, the S_2 and S_3 layers being absent. In the second alternative, which is most common, the G-layer is adjacent to the S_2 and only the S_3 is replaced (Fig. 3.13). The third, less common alternative contains the three normal, lignified layers of the secondary wall, the G-layer being inside the S_3. Thus, the secondary wall of gelatinous fibers may have any of the following arrangements:

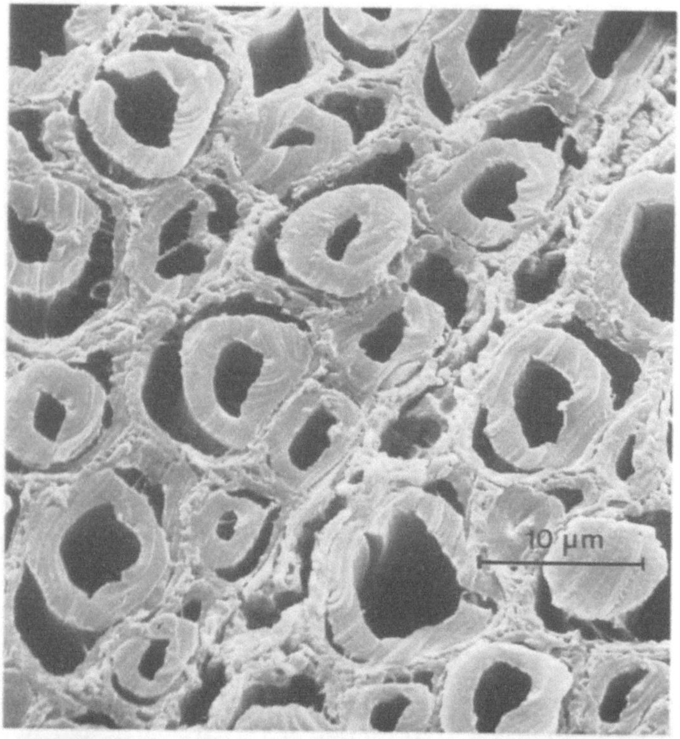
10 µm

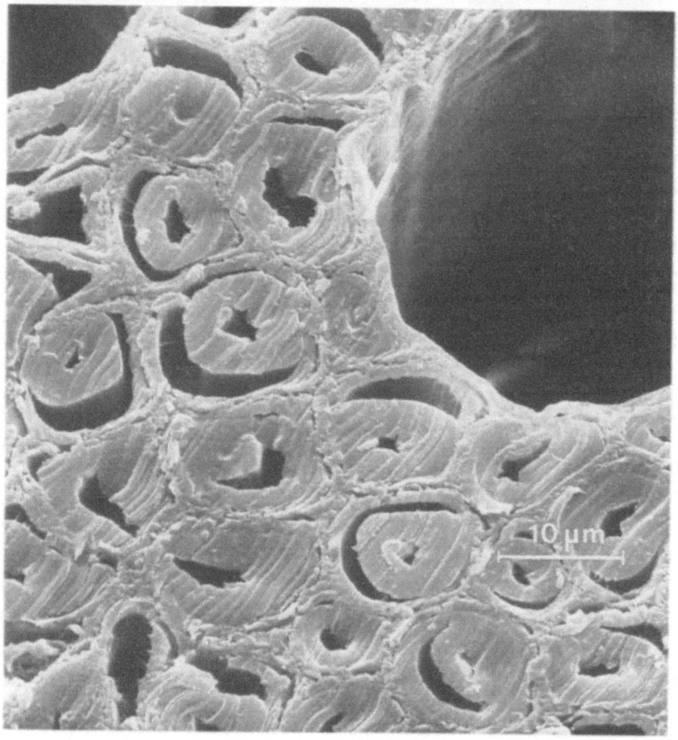
10 µm

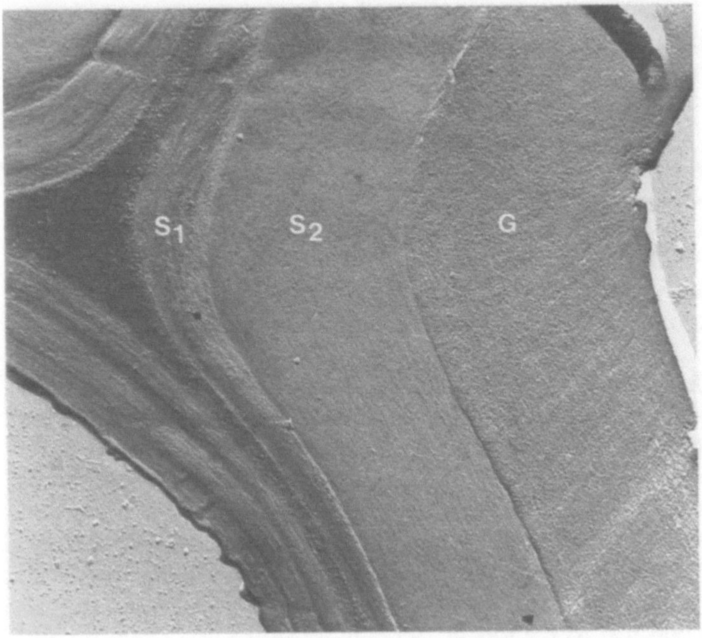

Fig. 3.13. Transmission electron micrograph of the cell wall of a gelatinous fiber in tension wood of a *Fagus grandifolia* stem, showing S_1, S_2 and G-layers of the secondary cell wall (courtesy of T.E. Timell)

$S_1 + G$
$S_1 + S_2 + G$
$S_1 + S_2 + S_3 + G$.

Gelatinous fibers characteristically have *slitlike pits* with their apertures parallel to the longitudinal axis of the cell, whereas the pit apertures in normal fibers lie obliquely. In longitudinal sections gelatinous fibers almost always show oblique, incipient *slip planes* and minute *compression failures*. Slip planes are deformations in the walls of individual fibers, whereas compression failures extend, more or less in straight lines, across a number of fibers (Jane et al. 1970).

The G-layer consists largely of cellulose, which is more highly crystalline than the cellulose of the other wall layers. With a large number of gelatinous fibers, tension wood is typified by a *high content of cellulose* and a *low content of lignin*. The composition of hemicelluloses differs from that of normal wood as well (Sect. 3.3.1). The cellulose microfibrils of the G-layer are almost parallel to the vertical axis of the cell. Despite this almost *axial orientation of microfibrils*, tension wood has exceptionally *high longitudinal shrinkage*. Even though it is seldom greater than 1%, an increase of one to two times from normal to tension wood

Fig. 3.12. Scanning electron micrographs of gelatinous fibers of stem tension wood in *Acer rubrum* (*above*) and *Fagus grandifolia* (*below*). The G-layer nearly fills the lumen in many fibers, appearing convoluted and partially separated from the S-layers of the cell wall (courtesy of T.E. Timell)

is fairly common. In extreme cases longitudinal shrinkage in tension wood may be fivefold, the degree of increase being directly related to the number of gelatinous fibers present. However, because of the nearly axial orientation of the microfibrils in the G-layer, and also because it is quite loosely attached to the adjacent cell wall, there is almost no longitudinal shrinkage in the G-layer itself. Instead, the reason for the increased longitudinal shrinkage in tension wood is usually attributed to the S_1 layer, where the microfibrils are oriented in a very flat helix in contrast to S_2 with its steep helix of microfibrils. Because of the presence of the G-layer, the S_2 is generally much reduced in width compared to the situation in normal wood, thus making it possible for the S_1 layer to exert a dominating influence on the longitudinal shrinkage behavior of the gelatinous fiber.

Tension wood is also characterized by a general *strength reduction* in comparison with normal wood except for tensile strength. The lack of strong bonding between the G-layer and the S_2 layer essentially eliminates a large volume of high-strength cellulose from any effective influence on the strength of the wood, except for toughness (Panshin and de Zeeuw 1980).

In hardwood branches and leaning stems, modifications of tissues occur in the opposite wood as well. However, only the fibers are affected. The narrower opposite-wood zone on the lower side of the pith has been found to possess shorter than normal fibers in contrast to the longer fibers in tension wood (Panshin and

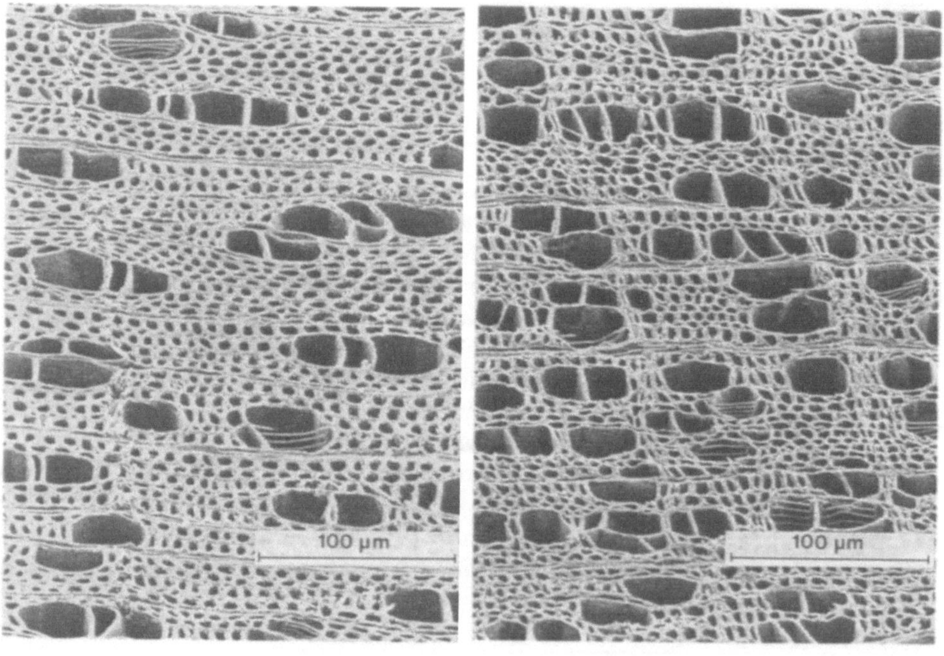

Fig. 3.14. Scanning electron micrograph of transverse sections of tension wood (*left*) and opposite wood (*right*) in a branch of *Betula pubescens*. Tension wood is characterized by wider growth rings, fewer vessels, and thicker fiber walls (courtesy of Pekka Saranpää)

de Zeeuw 1980). The difference in fiber wall thickness is illustrated in Fig. 3.14, opposite-wood fibers showing thinner walls.

3.1.3 Juvenile Wood in Stem Tops

The longitudinal or primary growth of a tree stem, branches, and roots is traceable to *apical meristems or growing points* consisting of cells capable of repeated division to perpetuate themselves and form new tissues. The radial or secondary growth of the tree is accomplished by cell division in the *vascular cambium*, which is a sheath of thin-walled meristematic cells, called *cambial initials*. This sheath of a single layer of living cells is situated between the secondary xylem and inner bark or secondary phloem, and extends from the apical growing points in the stem and branches into the corresponding regions in the roots. The cells of the *cambial zone* possess the ability to divide, thus producing new daughter cells which are capable of further division. It is through the activities of the cambial zone that the tree continues to increase in diameter, year after year, by addition of successive layers of secondary xylem and phloem.

In the region of the active crown, under the influence of the apical growing points, the cambium produces cells with abnormal structure. This wood, with generally inferior technical properties, is called *juvenile wood*, *core wood*, or *crown-formed wood*. As the crown moves further upward in a growing tree, the influence of the apical growing points on a given cambial region decreases. Gradually, wood cells with an anatomical structure considered normal for the species are formed. This wood is called *mature wood*, *adult wood*, or *stem-formed wood*. The stem may thus be divided into two regions on the basis of fundamental differences in the structure and properties of wood. Juvenile wood, formed as the result of the prolonged influence of the apical growing points, occurs in a cylindrical column surrounding the pith (Panshin and de Zeeuw 1980).

From the utilization point of view, in some tree species juvenile and mature woods may be considered as two different materials in the same stem. Although changes in the anatomical structure and chemical composition from the pith outward are gradual in most tree species and no distinct boundary can be recognized, juvenile wood is actually a core of rather uniform diameter along the length of a stem. The duration of the juvenile period, usually from 5 to 20 years, is quite variable among species. In a young tree, the stem may consist entirely of juvenile wood. In an older tree, above a certain height all wood may be juvenile. Consequently, depending on the tree species, growth rate, and minimum diameter of merchantable timber, the unutilized tree tops usually contain a high proportion of juvenile wood. Since the quantity of juvenile wood is thus a function of ring age, its occurrence is particularly problematic in fast-growing tropical and subtropical pine plantations with a short rotation and wide spacing of trees. However, although juvenile wood may be related to wide annual rings near the pith, it is not necessarily associated with fast growth.

The primary basis for defining juvenile wood is the structure and other properties of the cells. In a stem cross-section, juvenile wood is characterized by *wide growth rings*, a *low proportion of latewood*, and a *gradual transition from*

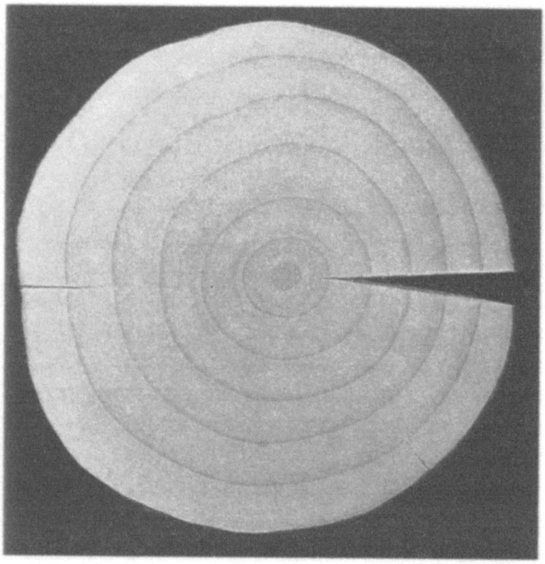

Fig. 3.15. Cross-section of a young stem of *Abies balsamea* with juvenile wood (courtesy of T.E. Timell)

earlywood to latewood (Fig. 3.15). Cell dimensions increase from pith to cambium until they reach a certain level. Although a few exceptions have been reported, the following trends are typical of many important tree species (Panshin and de Zeeuw 1980):

1. The length of softwood tracheids and hardwood fibers, both of which exhibit appreciable postcambial elongation, increases from the pith outward. The increase is rapid in the zone of juvenile wood near the pith and greatest in softwood tracheids. Vessel element lengths increase in successive growth rings from pith to bark as well, but to a much more limited extent, since vessel elements increase in size postcambially principally in a lateral direction.
2. The diameter of softwood tracheids increases systematically from the pith outward, but the rate of increase is considerably less than for length. According to a limited number of studies that have been reported, vessel elements in hardwoods also increase in diameter from the pith to the bark. Hardwood fibers, on the other hand, do not all conform to one pattern of diameter variation. However, in many tree species they also increase moderately toward the bark.
3. The wall thickness of softwood tracheids and hardwood fibers is relatively small near the pith (Figs. 3.16 and 3.17), but increases from the pith outward. For example, in latewood of *Picea sitchensis*, *Pinus echinata*, *P. pinaster*, *P. radiata*, *P. resinosa*, *Araucaria angustifolia*, *Chamaecyparis obtusa*, and *Cryptomeria japonica* the wall thickness increases progressively in successive increments from 15 to 70% at least up to 30 years of age. In the earlywood, the change in the cell wall thickness is less than that for latewood.

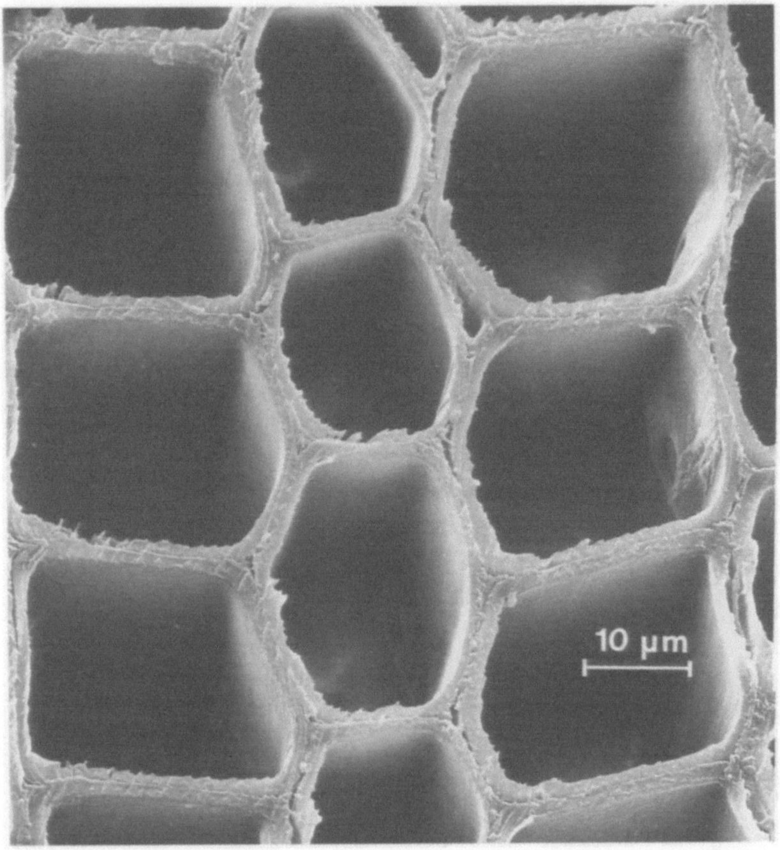

Fig. 3.16. Scanning electron micrograph of a transverse section of juvenile stem wood in *Abies balsamea* showing relatively thin cell walls (courtesy of T. E. Timell)

4. Microfibrillar angles in the walls of softwood tracheids and hardwood fibers are inversely related to the cell length. Because of this relationship the microfibrillar angle decreases from pith to cambium.

There is no region in softwood stems where compression wood is found more frequently than the first few juvenile growth rings near the pith. The *presence of compression wood* in the innermost growth rings of a stem was first observed by Sanio (1873). The reason for this frequent occurrence is that the stem of a very young tree and the upper top portion of a mature tree are lithe and slender, and thus easily displaced from their vertical, equilibrium position, for example, by wind, snow, or ice. Righting in young trees and tree tops is generally quite rapid, and the compression wood close to the pith is consequently usually limited in amount and restricted to a few growth rings. In stems which, for one reason or another, have remained inclined for several years, the compression wood at the center tends to occupy the entire increment and to be of a more severe type than that laid down later. In addition to its high content of compression wood, juvenile

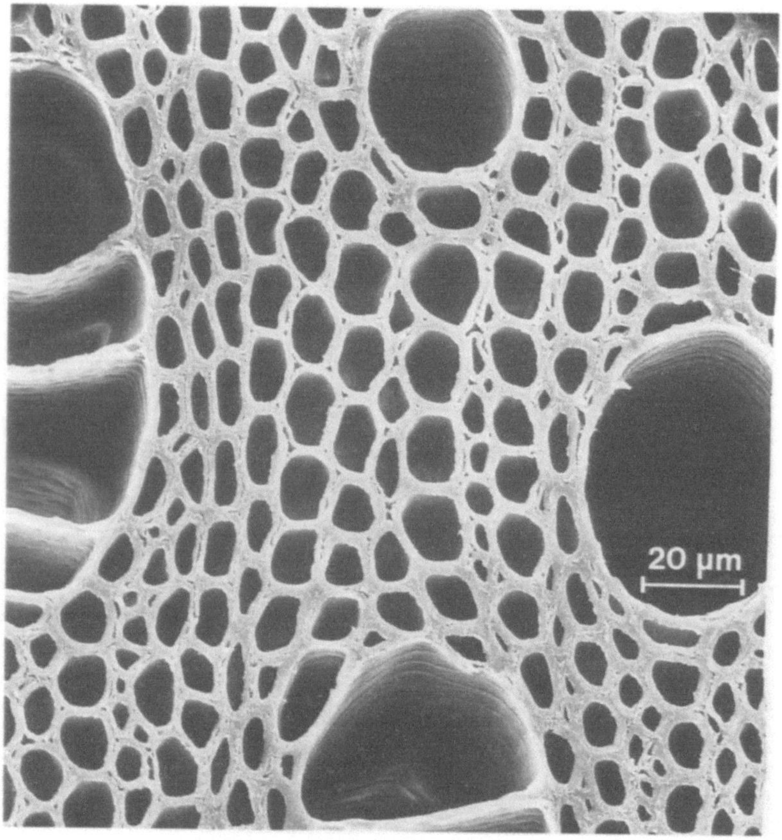

Fig. 3.17. Scanning electron micrograph of a transverse section of juvenile stem wood in *Acer rubrum* (courtesy of T.E. Timell)

wood tends to be associated with larger amounts of knots than does mature wood (Timell 1986b). Although little information is available, the same situation is believed to prevail in young stems and tops of hardwoods.

Accompanied by a low basic density (Sect. 3.4.2), these structural properties of juvenile wood contribute to the somewhat inferior raw-material quality of unmerchantable tops. Furthermore, the high content of lignin in the juvenile wood of conifers is undesirable from a pulping point of view. Conversely, hardwood tops have the advantage of a higher than average content of cellulose (Sect. 3.3.1).

3.1.4 Anatomical Characteristics of Root Wood

Stem, branches, and roots develop in different environments and perform different functions. This results in differences in their anatomical structure. Roots have three primary functions: absorbtion and conduction of water and minerals,

Fig. 3.18. Cross-sections of coarse roots of *Picea abies* showing eccentric growth pattern (courtesy of FFRI)

translocation and storage of food, and anchoring of the tree. Since most root studies have a physiological or botanical approach, they are primarily concerned with the smallest roots that take up the minerals and water and carry out most of the food-storage functions of a tree. These studies do not have much bearing on the possible use of stump and root mass, since from the utilization point of view, the main interest is in large roots and the central stump. Information available on the technical properties of the large sections of stump and root mass is meager and somewhat contradictory.

Characteristically, roots are twisted and only slightly tapered. The *cross-section is deformed* and may attain an ellipsoid, I-beam, or T-beam shape (Fig. 3.18). Such cross-section shapes are indicative of stresses which the roots derive from the mass of the tree, snow loads, winds, etc. In hardwood roots, more fibers and less vessels are formed where strength is required. Eccentricity is greatest in lateral roots near the root collar where stresses are high and in vertical roots wedged between stones, but decreases gradually with increasing distance from the stem. On the other hand, roots growing in sand are more or less round (Tsoumis 1968). According to Wilson (1964), the eccentricity results from depressed, often discontinuous cambial activity opposite the diarch protoxylem poles. Some growth rings are longitudinally discontinuous, disappearing toward the stem. Wilson suggested that branch roots emerging opposite the protoxylem poles divert materials moving

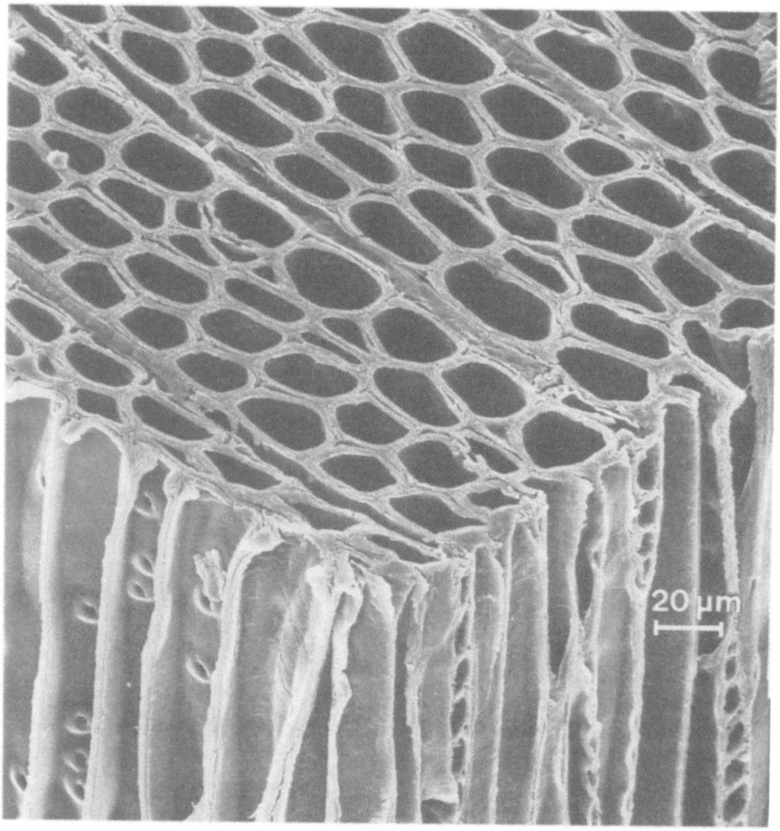

Fig. 3.19. Scanning electron micrograph of root wood in *Picea abies* (courtesy of T.E. Timell)

from the stem down the main roots, thus decreasing cambial activity further toward the root tip. This reduction could result in both discontinuous growth rings and the diameter decrease observed after branching. Twisting, on the other hand, apparently occurs in the root tips when they encounter obstacles while growing through the soil.

Heartwood forms in the roots of some species, for example, in *Picea abies, Larix leptolepis, Pinus sylvestris, Fagus sylvatica, Populus deltoides,* and *Quercus robur.* However, only the root region near the stem wood is capable of forming heartwood (Hillis 1987).

Unlike the stem and branches, *roots lack a pith*, the central core of parenchymatous tissue. Instead of the pith, the center of a root is marked by primary xylem and primary phloem. They occur in radial strands, as opposed to the collateral arrangement in the vascular bundles of stems, forming a star-shaped figure in cross-sections of many hardwood roots (Trendelenburg and Mayer-Wegelin 1955). At the root collar, this area of primary xylem is located near the lower root surface. Further out on the root it may be near the top. As the tree grows, this pattern progresses slightly along the root (Laitakari 1934; Fayle 1968).

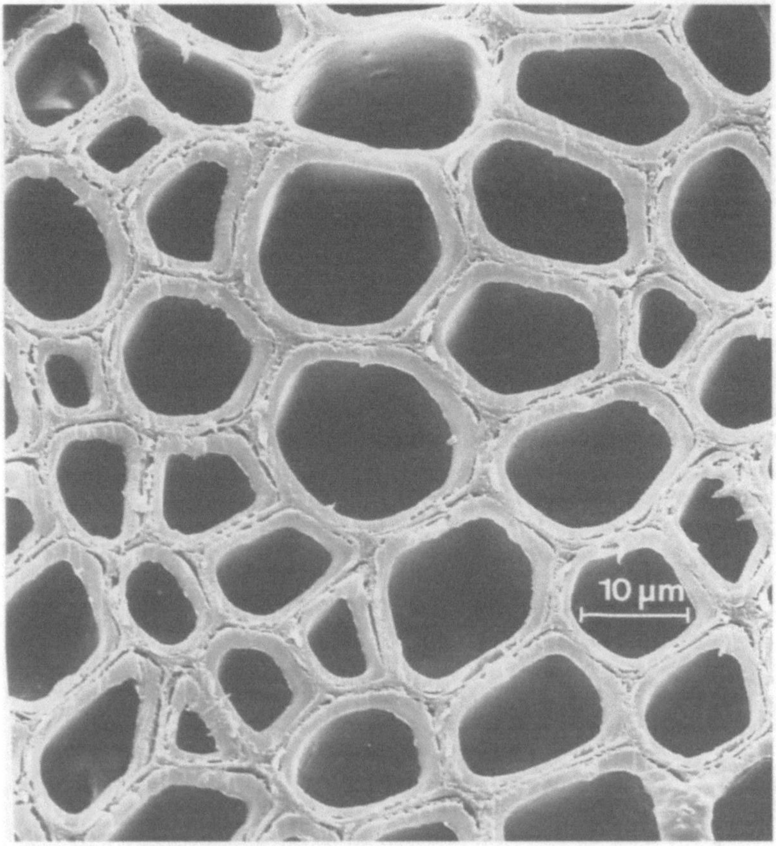

Fig. 3.20. Scanning electron micrograph of root wood in *Betula alleghaniensis* (courtesy of T. E. Timell)

The *ring width* varies, but tends to be narrow compared with stem wood. In swampy areas, unusually wide rings may appear in many tree species close to the root collar. Growth rings may or may not be well-defined. In conifer roots, the difference in wall thickness between earlywood and latewood tracheids decreases with increasing distance from the stem. On average, *tracheid walls are thinner* than in stem wood (Figs. 3.19 and 3.20). Both in softwoods and ring-porous hardwoods, latewood almost disappears with increasing distance from the stem (Tsoumis 1968). According to Koch (1972a), distinct bands of latewood are sometimes found in southern pine roots, but more often growth increments are distinguished only by several rows of radially flattened cells whose walls may not be thicker than those of the earlywood. In diffuse-porous hardwoods, root-wood vessels are larger in size but less in number (Fegel 1941; Bhat and Kärkkäinen 1981a). In ring-porous hardwoods the vessel arrangement is not retained and the root-wood structure may become diffuse-porous (Riedl 1937; Patel 1965). Roots may also possess false rings (Patel 1965).

Size and proportions of different types of cells vary in root wood depending on tree species and distance from the stem, but reported results are contradictory. Due to its storage function, root wood, at least in hardwoods, generally contains *more parenchymatous cells.* Although parenchyma with thickened, lignified walls occur frequently in stem wood of southern pine, virtually all parenchymatous cells in southern pine root wood are thin-walled (Koch 1972a). According to Knigge and Schulz (1966), the amount of parenchymatous cells may in some hardwood roots exceed the total amount of all other cells. Fegel (1941) compared the ray volume in stem, branch, and root woods in eight coniferous species (*Abies balsamea, Larix laricina, Picea mariana, P. rubens, Pinus resinosa, P. strobus, Thuja occidentalis,* and *Tsuga canadensis*), in eight diffuse-porous hardwoods (*Acer rubrum, A. saccharum, Betula alleghaniensis, Fagus grandifolia, Populus tremuloides, Prunus pensylvanica, P. serotina,* and *Tilia americana*), and in four ring-porous hardwoods (*Fraxinus americana, Fraxinus nigra, Quercus borealis* var. *maxima*, and *Ulmus americana*). In hardwood roots, the ray volume was considerably in excess of the ray volume in other components of the tree:

	Ray volume, percent of wood volume		
	Branch wood	Stem wood	Root wood
Conifers	5.1	5.2	4.9
Diffuse-porous hardwoods	12.5	11.1	15.6
Ring-porous hardwoods	15.3	14.1	18.7

Fegel also compared the *volume of vessels* in various wood components of hardwoods. In diffuse-porous hardwoods, the vessel volume in the stem was 27.2%, in the branches 22.4%, and in the roots 28.8% of total wood volume. In ring-porous hardwoods the corresponding figures were 26.5%, 18.2%, and 13.2%. Thus, the tendency of the root wood of ring-porous tree species to produce diffuse-porous wood in the roots resulted in an unusually low percentage of vessel volume.

Koch (1985a) studied the structure of wood tissues in stem, branches, and the central stump-root system including lateral roots up to 92-cm radius in 22 hardwood species in the southern United States. As shown below, the proportional differences among cell types were rather small. On average, stem wood had the largest proportion of vessels and vascular tracheids, wood from the central stump-root system had the largest proportion of fibers and rays, and branch wood had the largest proportion of other parenchyma and vasicentric tracheids. In thinner roots outside the 92-cm radius, the proportion of parenchymatous cells is probably higher than in the central stump and root components studied by Koch (see Scheme p. 127).

Coniferous species which normally possess *resin canals* in the stem wood contain these in branch and root woods as well. On the other hand, coniferous species lacking resin canals in the stem wood also lack them in the branch and root woods. In the study by Fegel (1941), the average number of resin canals in the stem, branch, and root woods in five North American conifers was 0.45, 1.09, and

Tissue	Proportion of volume (%)		
	Branch wood	Stem wood	Central stump and root wood
Vessels, vascular tracheids	19.7	20.5	17.9
Fibers	45.6	44.4	46.1
Rays	15.9	17.0	19.1
Longitudinal parenchyma	18.8	18.1	16.9
Total	100.0	100.0	100.0

0.26 per mm^2. The maximum tangential diameter of resin canals was 126, 85, and 124 μm, respectively. In branch wood, the number of resin canals was thus unusually large, and the size unusually small. In root wood, their number was small but the maximum tangential diameter normal in comparison with the stem wood.

Typically, the primary xylem in the root center of certain conifers has *primary resin canals.* Because of their translocation function, root-wood tracheids have *more pits* than stem-wood tracheids. In the root wood of 12 hardwood species analyzed by Fegel (1941), no tyloses were found, but gum deposits were present in the vessels of *Prunus pensylvanica.* In many softwoods and hardwoods, differences of other types occur between stem and root woods as well, but these anatomical characteristics have little effect on the utilization properties of root wood as industrial raw material. More important are differences in tracheid and fiber dimensions, which are discussed in Section 3.2.3.

Softwood roots may produce *reaction wood* but in far smaller amounts than stem and branches. With one exception, compression wood has been found only in exposed roots. In roots buried in the ground, formation of compression wood seems to be at best a marginal phenomenon of little or no importance. In hardwoods, tension wood can develop in both exposed and buried roots. In some studies, gelatinous fibers have been found to be absent, but on more numerous occasions they have been observed in such root wood (Timell 1986b).

3.2 Cell Dimensions in Residual Tree Components

The foundation for the study of dimensional variation in wood cells was laid by Sanio (1873) more than 100 years ago. His five main conclusions from within-tree variations of cell dimensions in *Pinus sylvestris* are today regarded as Sanio's Laws. These findings have continued to be studied and further amended by numerous wood scientists.

Differences between the merchantable stem and the residual components of the tree are in many cases remarkable. However, these differences are not always unequivocal, since they are dependent on factors such as tree age, tree size, the point of sampling in a tree component, and the laboratory technique used. The result of cell-length measurement is affected by the technique of cell separation, and both macerated samples and chemical pulps are used alternatively. Cell and lumen diameters, on the other hand, vary in the axial direction of the cell and are thus dependent on the point of measurement. If the cells are in the form of pulp, swelling and collapse influence the result considerably. Cell dimensions are sometimes reported as weighted and sometimes as arithmetic means. Because of the many variables, comparisons must be made with caution, especially since most fiber studies are based on a small number of specimens, and the exact point of sampling is not always reported.

Dimensional characteristics of cells have an important effect on the pulp and papermaking properties of raw material. Cell length, cell diameter, lumen diameter, cell-wall thickness, and their mutual ratios are essential wood quality indicators when residual tree components are used as a source of fiber. In other potential uses of residual biomass, however, cell dimensions are of minor significance.

Great importance was formerly attributed to cell length. Cell length influences many paper properties but seems to be decisive only in the case of tear resistance. Even here, however, the ratio of cell length to diameter might be a more important factor. On the other hand, the different morphological characteristics of earlywood and latewood have a striking effect on the properties of the derived paper (Timell 1986c). The dependence of tearing resistance on fiber properties changes with the degree of sheet consolidation. In a poorly bonded sheet, tearing resistance depends more on fiber length than on fiber strength, but the opposite is the case in a well-bonded sheet. For long, straight, and adequately bonded fibers, the tearing resistance is proportional to the square of fiber strength. This indicates that fiber strength is more important for tearing resistance than previously thought, particularly for softwood market chemical pulps. Among fibers of similar length and strength, coarser fibers make sheets with higher tearing resistance (Seth and Page 1988).

Thin-walled, large-diameter earlywood tracheids *collapse and become ribbon-like* in paper-making, producing dense, relatively nonporous sheets with high burst and tensile strength and high fold endurance but low tear strength and bulk. Thick-walled tracheids with a narrow lumen *retain the rounded cross-section form* on sheet formation, resulting in poor lateral conformation and a low interfiber bonding strength. The sheets produced are highly porous and opaque, have a low burst and tensile strength and a low folding endurance but a high tear strength and bulk (Runkel 1942; Watson and Dadswell 1962; Barefoot et al. 1964, 1972; Dinwoodie 1966; Hartler 1976; Timell 1986c). According to Page (1967), the percentage of collapsed cells depends mainly on the cell-wall thickness and the pulping process. For all pulps, complete collapse occurred in fibers with the thinnest walls. The percentage of collapsed fibers fell rapidly with increasing wall thickness. Collapse generally increased with decreasing yield due to higher delignification. The differences in resistance to collapse between sulfite and sulfate pulps were by no means small, a sulfate fiber being as rigid as a sulfite fiber when the wall thickness was less than half as much. It was suggested that this was because of a difference in the transverse shear modulus of the cell wall arising from a basically different wall structure.

Pulps from thin-walled, large-diameter cells are readily beaten, whereas thick-walled cells with narrow lumen are less responsive to beating (Clark 1969; Timell

1986c). When the proportion of thick-walled cells increases, the beating time requirement also increases, the sheet formation becomes more difficult, and more bleaching is required to reach a desired degree of whiteness (Edlin 1965). Thus, both earlywood and latewood cells have their advantages and disadvantages in paper-making and, consequently, the presence of both cell types in a particular ratio in pulp is desirable. For example, the wood from plantation-grown pines in Australia should contain 15–50% latewood if high-grade pulp is desired (Dadswell and Wardrop 1959).

3.2.1 Cell Dimensions in Branch Wood

Most of the lower, abaxial part of a conifer branch consists of compression wood, while some of the upper, adaxial region is composed of opposite wood. The remainder of the branch wood is normal. With some exceptions, the tracheid length is shorter on the lower side of conifer branches than on the upper side. Tracheids in normal branch wood are usually shorter than in normal stem wood, and the tracheid length of branch compression wood is evidently less than that of stem compression wood. Because of the high proportion of compression wood in most conifer branches, the net effect is that branch wood has, on average, shorter tracheids than stem wood. Tracheid length increases with increasing branch diameter, probably because small branches contain proportionally more juvenile wood, which outweighs the fact that older branches tend to have a higher content of compression wood (Timell 1986a).

At the basal end of the branch, the tracheids near the pith are comparable in length to those near the pith in the adjacent stem. However, the increase in tracheid length with successive growth increments from the pith is much smaller in branches. On average, the tracheids formed at later periods of branch growth are markedly shorter than those in the stem at the same age (Fig. 3.21). In the same growth increment, tracheid length usually first increases along the branch

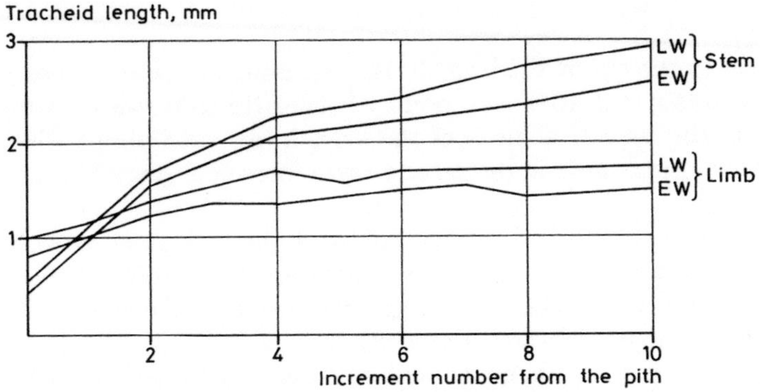

Fig. 3.21. Tracheid length in relation to the number of increments from the pith. Earlywood (*EW*) and latewood (*LW*) of the basal end of a branch and adjacent stem wood in a 35-year-old *Pinus resinosa* tree (redrawn from Panshin and de Zeeuw 1980)

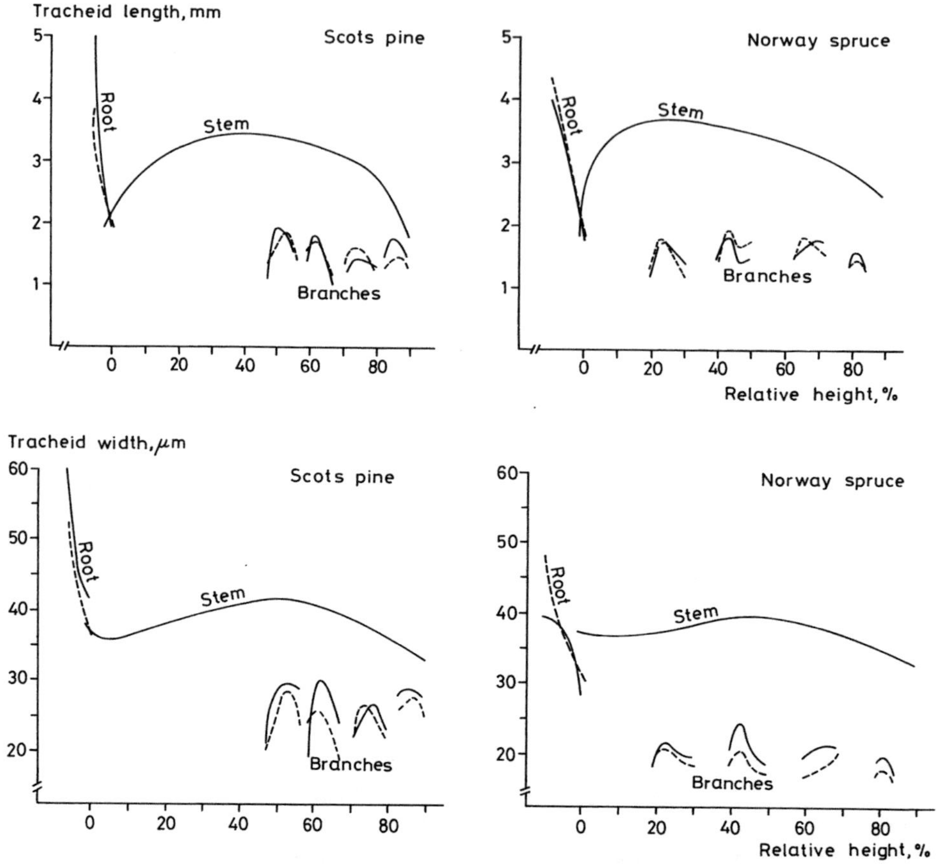

Fig. 3.22. The length and width of tracheids in the two outermost growth rings of stem, branch, and root woods in a 96-year-old *Pinus sylvestris* and 95-year-old *Picea abies*. In branch and root woods, the *solid lines* represent the upper side and the *broken lines* the lower side of a cross-section (redrawn from Atmer and Thörnqvist 1982)

axis and then decreases to the tip of the branch. For example, in *Pinus resinosa* the maximum increase was about 10% at a point of one-fifth to one-fourth of the branch length from the base. Tracheids at the branch tips were about 40% shorter than those at the branch base in the same growth increment (Panshin and de Zeeuw 1980).

Figure 3.22 shows the variation in tracheid length of the two outer growth rings in the stem, roots, and branches of a mature *Pinus sylvestris* and *Picea abies* tree as a function of relative tree height. On average, the tracheids are shortest in the branches. Their length increases first along the branch but then decreases toward the tip. In this case, branch tracheids are longest at a point of 25–50% of the branch length.

The range of variation in tracheid length is much narrower in branch wood than in stem wood. In Fig. 3.23 the main concentration of branch wood tracheids

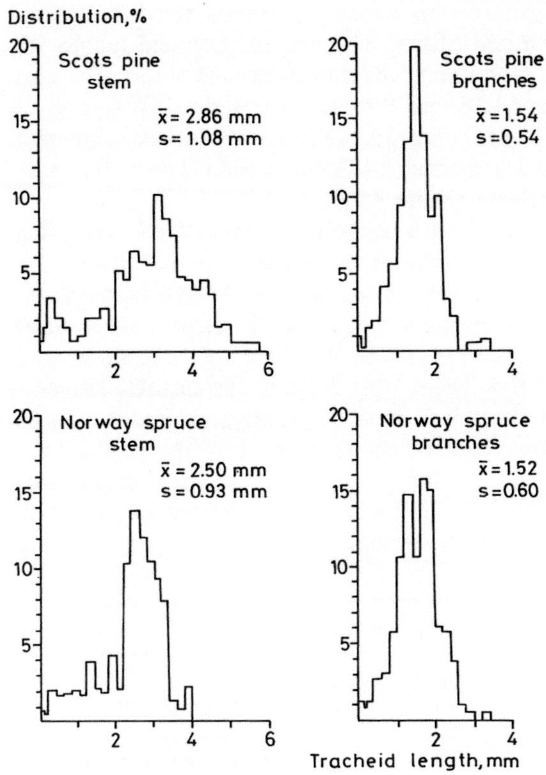

Fig. 3.23. Length distribution of tracheids in stem and branch woods of *Pinus sylvestris* and *Picea abies* (redrawn from Hosia et al. 1971)

of *Pinus sylvestris* and *Picea abies* occurs in length classes 0.5–2.5 mm. In the stem wood of the same trees, most tracheids in Scots pine belong to 2–4.5 and in Norway spruce to 2–3.25 mm length classes (Hosia et al. 1971).

In hardwoods also, cells are shorter in branch wood than in stem wood. In hardwood branches, the length of the fibers and vessel elements commonly increases from pith to bark and decreases from base to tip. The following average figures of *Betula pendula* and *B. pubescens* from a study by Bhat and Kärkkäinen (1981c) show an example of the variation of cell length along a hardwood branch:

Sampling point	Length (mm)	
	Fibers	Vessel elements
Branch base	0.89	0.43
25% Length of branch	0.83	0.43
50% Length of branch	0.80	0.38
75% Length of branch	0.72	0.35
Branch average	0.80	0.39
Stem average	1.14	0.62

Fegel (1941) determined cell length of stem, branch, and root woods in a number of tree species in the eastern United States. The average tracheid length for eight conifers was 3.44 mm in stem wood and 1.81 mm in branch wood. The corresponding fiber measurements were 1.16 and 0.83 mm for eight diffuse-porous hardwoods and 1.24 and 0.94 mm for four ring-porous hardwoods. Vessel element length was 0.54 and 0.45 mm for diffuse-porous hardwoods and 0.31 and 0.25 mm for ring-porous hardwoods, respectively (Table 3.7).

As mentioned earlier, studies of cell-length variation often differ in sampling procedure and techniques of analysis. Cell-length studies are thus not always unequivocally comparable. Nevertheless, results of average cell length in stem and branch woods from several independent studies are listed below. The average tracheid length in seven samples of softwoods is 2.83 mm in stem wood and 1.51 mm in branch wood, the latter thus being only 53% of the former. The corresponding average lengths in 43 recorded data of hardwoods are 1.20 and 0.90 mm, respectively, the latter being 75% of the former. However, tracheids in softwood branches are commonly longer than fibers in hardwood stems, and branch wood of some long-fibered hardwood species may have longer fibers than stem wood in short-fibered species such as maples.

Tree species	Stem wood	Branch wood	Reference
Softwoods:	Tracheid length (mm)		
Picea abies	2.50	1.52	Hosia et al. 1971
P. abies	3.51	1.50	Eskilsson 1972
P. abies	2.05	1.14	Young and Chase 1965
Pinus palustris	2.76	1.72	Gleaton and Saydah 1956
P. sylvestris	2.86	1.54	Hosia et al. 1971
Tsuga heterophylla	3.0	1.7	Worster and Vinje 1968
T. heterophylla	3.23	1.48	Keays and Hatton 1971a
Hardwoods:	Fiber length (mm)		
Acer rubrum	0.83	0.66	Manwiller 1974
A. rubrum	0.66	0.53	Young and Chase 1965
Alnus glutinosa	0.85	0.80	Vurdu and Bensend 1979
A. incana	0.98	0.86	Lönnberg 1975b
A. rugosa	0.64	0.48	Chase et al. 1973
Betula papyrifera	1.07	0.89	Young and Chase 1965
B. pendula	1.08	0.81	Bhat and Kärkkäinen 1981c
B. populifolia	0.73	0.61	Chase et al. 1973
B. pubescens	1.19	0.79	Bhat and Kärkkäinen 1981c
B. pubescens	0.98	0.86	Lönnberg 1975b
Carya sp.	1.29	0.98	Manwiller 1974
C. ovata	1.32	0.93	Taylor 1977
C. tomentosa	1.36	1.02	Taylor 1977
Celtis occidentalis	1.12	0.86	Manwiller 1974
Fraxinus americana	1.22	0.87	Manwiller 1974
F. pennsylvanica	1.16	0.84	Manwiller 1974
Liquidambar styraciflua	1.54	1.20	Manwiller 1974
L. styraciflua	1.78	1.12	Taylor 1977
Liriodendron tulipifera	1.39	0.97	Manwiller 1974
L. tulipifera	1.89	1.03	Taylor 1977
Magnolia virginiana	1.24	0.97	Manwiller 1974

Tree species	Stem wood	Branch wood	Reference
Nyssa sylvatica	1.76	1.40	Manwiller 1974
N. sylvatica	1.93	1.28	Taylor 1977
Populus tremula	0.68	0.75	Lönnberg 1975b
P. tremuloides	0.73	0.51	Chase et al. 1973
Prunus pensylvanica	0.64	0.50	Chase et al. 1973
Quercus alba	1.22	0.88	Manwiller 1974
Q. coccinea	1.30	0.98	Manwiller 1974
Q. falcata	1.30	0.94	Manwiller 1974
Q. falcata	1.58	1.06	Taylor 1977
Q. falcata var. *pagodaefolia*	1.30	0.94	Manwiller 1974
Q. laurifolia	1.34	1.01	Manwiller 1974
Q. marilandica	1.14	0.88	Manwiller 1974
Q. nigra	1.27	0.98	Manwiller 1974
Q. nigra	1.49	1.03	Taylor 1977
Q. rubra	1.27	0.90	Manwiller 1974
Q. schumardii	1.31	0.99	Manwiller 1974
Q. stellata	1.18	0.93	Manwiller 1974
Q. stellata	1.37	0.99	Taylor 1977
Q. velutina	1.30	0.98	Manwiller 1974
Salix caprea	0.81	0.85	Lönnberg 1975b
Ulmus alata	1.25	0.93	Manwiller 1974
U. americana	1.30	0.99	Manwiller 1974

To some extent, the variation in cell transverse dimensions appears to follow the same pattern in branch wood and in stem reaction wood. According to Timell (1986a), in conifer stems the tracheid diameter is slightly less in compression wood than in normal wood, when comparison is made with earlywood. In latewood, on the other hand, the tracheids in stem compression wood are either as wide as or slightly wider than in normal stem wood. The earlywood tracheids in severe stem compression wood have a much thicker wall and a narrower lumen than the earlywood tracheids in normal stem wood, while in latewood the difference in this respect is slight. In conifer stems the wall generally appears to be somewhat thicker and the lumen smaller in normal latewood than in compression wood. However, Timell emphasized that this comparison between normal and compression woods is of doubtful value for two reasons. First, tracheid diameter is very different in normal earlywood and latewood, and the earlywood-latewood concept is of limited usefulness when applied to compression wood. Second, compression wood tracheids have a circular or oval outline, whereas that of tracheids in normal wood is square or rectangular.

Correspondingly, branch wood cells are smaller in diameter than stem wood cells. In softwood branches, the diameter is smallest in the compression wood on the lower side of a branch, achieving its largest dimension at 25–50% distance from the base (Fig. 3.22). According to Panshin and de Zeeuw (1980), the tracheid diameter is about 25% smaller in branches than in stem wood.

Cell coarseness, i.e., the mass per unit length of tracheid, reflects the cross-section dimensions of the whole tracheid including the tapered ends. Eskilsson (1972) measured the tracheid coarseness in sulfate pulps made of branches and stem wood of 100- to 120-year-old *Picea abies* trees. With regard to all cell dimensions, a striking similarity was observed between tracheids from spruce branches and fibers from hardwood stems. However, in spite of dimensional similarities, the mechanical properties of cells differed considerably. Eskilsson concluded that due to the lower stiffness and strength of individual cells, sulfate pulp from spruce branches will form sheets with higher bulk and weaker strength:

	Branches 6–13 mm	Branches over 16 mm	Stem wood
Tracheid width (μm)	35.9	36.0	43.3
Wall thickness (μm)	1.6	2.1	2.5
Tracheid coarseness (μg/m)	70	85	205

In hardwoods as well, both cell diameter and lumen diameter of fibers are smaller in branch wood than in stem wood. Fegel (1941) found the fiber diameter to be 20% smaller in branch wood (Table 3.7). In Koch's (1985a) study of 22 hardwood species in the southern United States, the average radial diameter of branch wood fibers was 14.1 μm and that of branch wood fiber lumen 6.1 μm. Both diameters were 15% smaller than in stem wood. *Fiber flexibility*, i.e., the lumen diameter as a percent of the fiber diameter, was 43 in branch wood and 47 in stem wood. The tangential fiber wall thickness was 3.7 μm in branch wood and 4.4 μm in stem wood. In another study by Young and Chase (1965) in Maine, fiber flexibility in branch and stem woods was for *Acer rubrum* 45 and 53 and for *Betula papyrifera* 61 and 62, respectively.

Chase et al. (1973) studied fiber dimensions of six puckerbrush hardwood species in Maine. In both wood and bark, the fiber and lumen diameters were smallest in branches. In branch wood, the average diameter of the fiber was 18 μm and that of the lumen 13 μm. In branch-bark fibers the corresponding measurements were 13 and 3 μm. In all tree components of all species, bark fibers were longer and thinner than wood fibers and were characterized by thick walls and very small lumina (Table 3.3). These characteristics of bark fibers tend to produce paper with high bulk and low strength.

Table 3.3. Average dimensions of wood and bark fibers in various components of six puckerbrush hardwood species (*Acer rubrum*, *Alnus rugosa*, *Betula populifolia*, *Populus tremuloides*, *Prunus pensylvanica*, *Salix gracilis*) in Maine. Branch, stump, and root woods with bark intact, stem wood without bark. Fiber length of bark without *Salix gracilis* (data from Chase et al. 1973)

Biomass component	Fiber length (mm)	Fiber diameter (μm)	Lumen diameter (μm)	Wall thickness (μm)	Fiber flexibility
Stem wood	0.64	21.6	16.0	2.8	74
Stem bark	0.90	17.3	5.3	6.0	31
Branch wood	0.51	18.0	12.6	2.5	70
Branch bark	1.10	13.2	2.8	5.4	21
Stump and root wood	0.66	26.3	21.1	2.8	80
Stump and root bark	0.86	18.6	4.3	7.3	23

The vessel diameter is also smaller in branch wood than in stem wood. This is shown in Table 3.7 and by the average values of four Finnish diffuse-porous hardwood species (*Alnus incana, Betula pubescens, Populus tremula*, and *Salix caprea*) below. The samples were taken from 20- to 25-year-old trees (Lönnberg 1975b):

	Fiber width (μm)	Vessel width (μm)
Stem wood	24	87
Branch mass with bark	18	59

In pulp and paper-making, a high proportion of fibrous cells of great length, relatively thin walls, and wide lumen is desirable. The most valuable and abundant cells in stem, branch, and root woods are the tracheids in softwoods and the fibers and fiber tracheids in hardwoods. Since the percentage of fibrous cells is low in foliage, this tissue adds little or nothing to the fiber yield when using tree crowns for pulping. In a study of the needles of *Abies balsamea, Picea rubens, Pinus strobus*, and *Tsuga canadensis*, 70–90% of the tissue was composed of parenchymatous mesophyll cells which are of low value in paper. The proportion of fibrous cells was highest, 16%, in eastern white pine, but as low as 2% in eastern hemlock. In separate sulfate pulping of needles, the yield was 16.4–22.8% and the quality of pulp and paper poor because of the high percentage of nonfibrous tissue and short cell length (Hyland 1974). In another sample of defibrated *Picea abies* needles, cells of fibrous appearance constituted one-third of the total mass. Their arithmetic mean length was 1.56 mm, whereas that of stem wood tracheids was 2.12 mm. The fibrous cells from needles were narrower, less coarse, and considerably weaker (Eskilsson 1972).

3.2.2 Cell Dimensions in Stem Tops

The abnormal wood surrounding the pith of a tree stem is characterized by wide growth rings, a low proportion of latewood, gradual transition from earlywood to latewood, small cell dimensions, and a large microfibrillar angle in cell walls. This juvenile wood occurs in a cylindrical core along the length of a stem, consisting usually of 5–20 growth rings. All wood above a certain height in the stem is juvenile (Sect. 3.1.3). Therefore, depending on the tree species and rate of growth, a major part of wood in unmerchantable tree tops is juvenile. Consequently, cell dimensions in tree tops may differ significantly from those in conventional pulpwood.

Numerous studies on within-stem variation in cell length consistently show a significant increase from pith to bark. In the ring nearest the pith, the tracheid length in softwoods is 0.5–1.5 mm and the fiber length in hardwoods 0.1–1.0 mm, first increasing rapidly outward. After a few rings, the rate of increase declines until a maximum length is obtained, which is generally from three to five times the initial length. There is considerable disagreement whether this outward length increase is associated with an increasing ring number or with

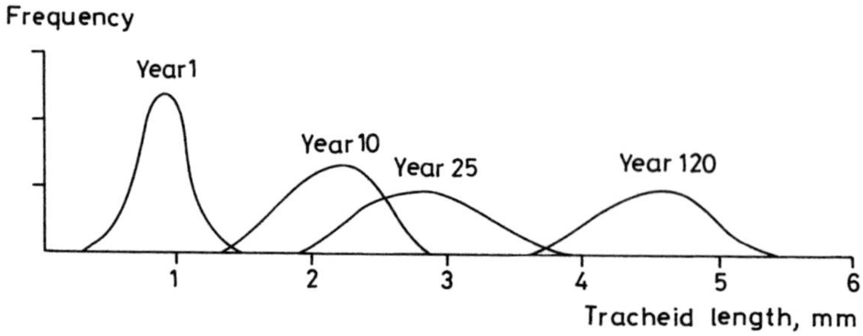

Fig. 3.24. Schematic drawing of the variation in tracheid length in subsequent growth rings (years 1, 10, 25, and 120) of a coniferous stem (Bailey 1957)

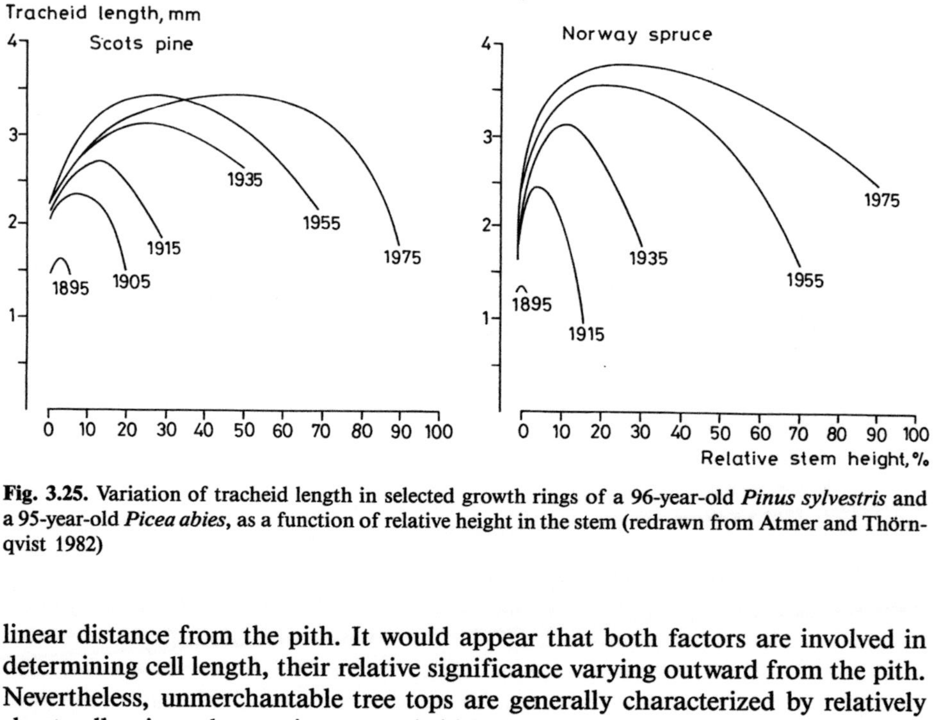

Fig. 3.25. Variation of tracheid length in selected growth rings of a 96-year-old *Pinus sylvestris* and a 95-year-old *Picea abies*, as a function of relative height in the stem (redrawn from Atmer and Thörnqvist 1982)

linear distance from the pith. It would appear that both factors are involved in determining cell length, their relative significance varying outward from the pith. Nevertheless, unmerchantable tree tops are generally characterized by relatively short cells, since the maximum tracheid length does not seem to occur until an advanced age after which there is a period of maximal length before a decrease occurs in extreme old age (Dinwoodie 1961). The increase in tracheid length over a cross-section of a coniferous stem is illustrated in a schematic drawing in Fig. 3.24.

The change in cell dimensions from pith to bark is reflected further in a longitudinal variation along the stem axis. This variation may be studied by following selected growth rings from the stem base to the tip. Within a growth increment, cell length first increases upward for a certain distance, but then decreases toward the tip of the tree. An example of this trend is shown in Fig. 3.25. As successive growth rings are formed, the maximum tracheid length within a ring occurs at progressively higher levels. However, even the last-formed rings of a mature tree contain relatively short tracheids at the top of the stem.

The position of the maximum cell length in a stem is dependent on tree species, age, and size. Sanio (1873) found the longest tracheids in that part of a *Pinus sylvestris* stem where the color, thickness, and texture of the outer bark changed. According to Helander (1933), the longest tracheids of *Pinus sylvestris* and *Picea abies* are to be found in the outer growth rings at 15–40% of the height of the stem. Dinwoodie's (1961) literature review on cell length variation includes several references showing maximum tracheid lengths at about 20 to 40% of the height of the stem and, thereafter, a decrease toward the top. It is thus quite evident that the wood cells in stem tops are shorter than those in conventional pulpwood. In the following examples, with one exception, the tracheids are shorter in tops than in merchantable stem wood.

	Tracheid length (mm)		Reference
	Merchantable stem wood	Unmerchantable top wood	
Picea abies	2.50	1.82	Hosia et al. 1971
P. rubens	2.05	2.08	Young and Chase 1965
Pinus palustris	2.76	2.52	Gleaton and Saydah 1956
P. sylvestris	2.86	1.84	Hosia et al. 1971
Tsuga heterophylla	3.0	2.4	Worster and Vinje 1968

In hardwoods also, the fiber length increases along any given stem radius, the most rapid increase occurring within a relatively short distance near the pith of the stem. Following this, the rate of increase is reduced markedly. Although the direct influence of height itself on fiber length is apparently slight, the variation in the radial direction of the stem results in a statistically significant variation in the longitudinal direction of the stem.

Figure 3.26 (Taylor 1979) shows the increase in fiber length from pith to bark in the stem wood of seven hardwood species in the midsouthern USA. In most species the fiber length reached a maximum in rings 30 to 40 and then decreased slightly in subsequent growth rings. Notable exceptions to this general pattern of variation were *Liquidambar styraciflua* and *Nyssa sylvatica*. These relatively long-fibered species continued to increase in fiber length from pith to bark at an almost constant rate. The same trends prevailed at all tree heights, albeit fiber length often increased more rapidly and attained maximum length after somewhat fewer growth increments at sampling points higher in the stem.

Figure 3.27 demonstrates how a *Quercus falcata* stem is characterized by a central core of uniform width containing fibers shorter than 1.1 mm. Immediately surrounding this area is a narrow shell with fibers ranging from 1.1 to 1.5 mm. Outside this shell, the fibers are longer than 1.5 but less than 1.9 mm (Hamilton 1961). In *Liriodendron tulipifera*, fiber length correspondingly increases rapidly through the first 10 to 15 growth rings and then continues to increase more slowly or levels off with increasing rings from the pith, resulting in a gradual decrease in fiber length from stem base to top. Simultaneously with a decrease in average

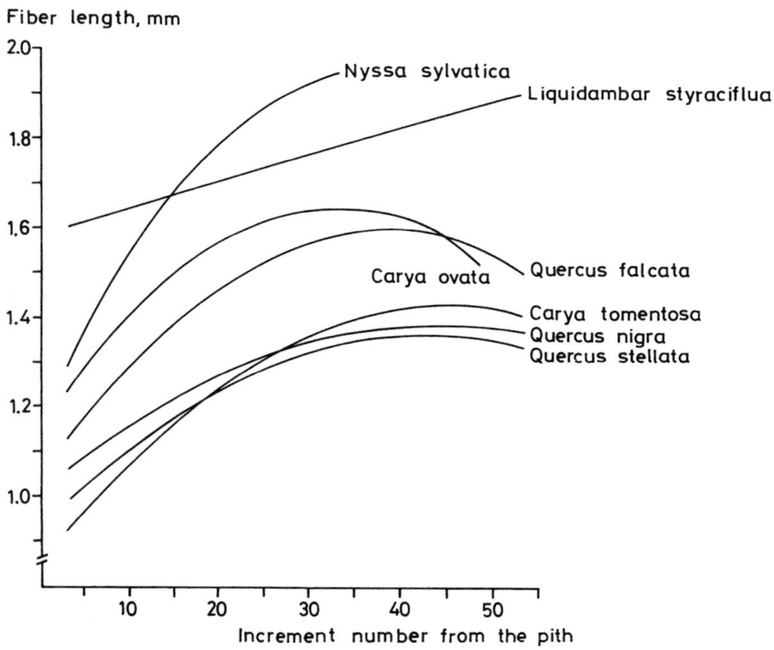

Fig. 3.26. Fiber length in relation to the number of increments from the pith at 1.5 m height in the stem wood of seven hardwood species in the US mid-south. All trees are about 28 cm at dbh (Taylor 1979)

fiber length, the proportion of fibers also decreases at the cost of an increasing volume of vessels (Taylor 1968):

	Height at the stem (m)			
	1.3	6.1	12.2	18.3
Fiber length (mm)	1.91	1.77	1.61	1.45
Vessel volume (%)	40.6	42.3	43.7	44.4
Ray volume (%)	12.5	11.2	11.5	12.0

Since tree tops contain only few growth rings and, thus, consist mainly of juvenile wood, their abnormal cells are characteristically not only short, but also thin-walled and small in diameter. However, measured numerical data on transverse cell dimensions in unmerchantable tree tops are scarce.

Significant differences in pulp and paper properties between juvenile and mature woods have been reported by many researchers. A typical example is a study by Jackson and Megraw (1986) of kraft pulping of juvenile and mature woods of *Pseudotsuga menziesii.* Juvenile wood gave two percentage points lower yields. Tear strength was 30% lower with juvenile wood, and this difference was maintained on beating. Burst strength was 60% higher with unbeaten pulp from juvenile wood, but this difference diminished on beating. Although it is not pos-

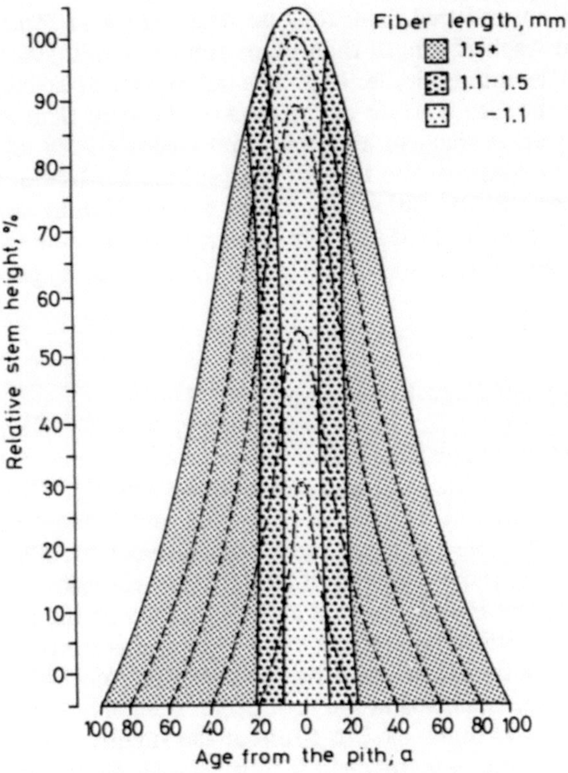

Fig. 3.27. Variation in fiber length of *Quercus falcata* as a function of number of growth rings from the pith and relative height in the stem (Hamilton 1961)

sible to present a single, unified picture for the yield and quality of pulps from unmerchantable tops, a few generalizations may be made. As summarized by Keays (1971a), for many tree species the tops, compared with the average values in stems, will show the following properties in pulping:

	Softwood tops	Hardwood tops
Wood characteristics:		
Wood density	Lower	Lower
Fiber length	Shorter	Shorter
Cell wall thickness	Thinner	Thinner
Proportion of knots	Higher	Higher
Proportion of reaction wood	Higher	Higher
Proportion of lignin	Higher	Lower
Proportion of cellulose	Lower	Higher
Pulp characteristics:		
Pulp yield	Lower	Higher
Tear factor	Lower	Lower
Burst factor	Higher	Higher
Tensile strength	Higher	Higher
Beating time	Faster	Faster
Proportion of rejects	Higher	Higher

According to Megraw (1985), chips reduced from tops or veneer cores of *Pinus taeda* are on average shorter in tracheid length than chips from saw mill slabs, edgings, or other outer material, assuming similar logs. The rate at which tracheid length increases with age from the pith is lower at the base of the stem than at heights of 4–6 m and above. Average tracheid length is approximately 30% less near the base and breast height regions of the tree for a given number of rings from the pith than it is at upper heights. Chips taken from young thinnings of *Pinus taeda,* where wood mass is concentrated near the base, can therefore be expected to have shorter fibers than chips from tops of older trees which contain the same number of rings.

3.2.3 Cell Dimensions in Stump and Root Woods

In the roots, cell dimensions tend to change from center to bark and from the root collar, i.e., the transition zone between the stem and roots, to the root tip. They seem to be affected also by the direction of orientation of the root. Failure to recognize this in sampling is responsible for many conflicting research results.

Figure 3.28 by Atmer and Thörnqvist (1982) illustrates the variation of tracheid length in the roots of mature *Pinus sylvestris* and *Picea abies* on the upper side of the root cross-section. Variation over the cross-section was irregular and without a clear trend, whereas a significant increase occurred from root collar to root tip along the root. While the tracheids at the root origin were only about 2 mm long, the average length was as much as 4 to 5 mm at the smaller end of the root. In thin root sections of Scots pine, tracheid lengths of over 6 mm were frequently recorded. A similar increasing trend along the root was found in

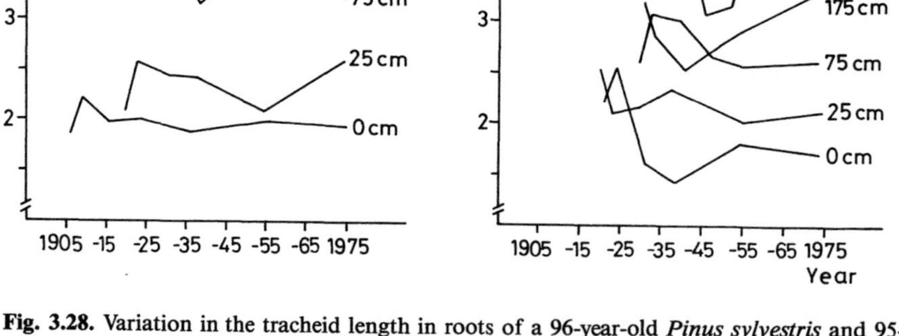

Fig. 3.28. Variation in the tracheid length in roots of a 96-year-old *Pinus sylvestris* and 95-year-old *Picea abies* in the upper side of a root cross-section by annual increments, from center to bark at 0–2.25 m distances from the root collar (redrawn from Atmer and Thörnqvist 1982)

tracheid diameter, but the rate of change was relatively smaller. The same pattern of variation was observed on the lower side of the root cross-section.

Very few data are available on the variation of cell dimensions within the root system in hardwoods. In a study by Vurdu and Bensend (1979) on 8-year-old *Alnus glutinosa*, the fiber length increased from root center to bark by 21%. From the third annual ring on, wood fibers had the same length in the roots as at the base of the stem. Bhat and Kärkkäinen (1981 c) found fibers and vessel elements of 65- to 95-year-old *Betula pendula* and *B. pubescens* to be shorter in the roots than in the stem. Fiber length increased slightly along the root from base to tip, and from root center to bark. The trend was not clear in vessel elements.

Compared with the average dimensions in stem wood, the tracheids tend to be longer and wider in the smaller parts of lateral roots but shorter and narrower in the central stump. In a sample of three *Picea abies* trees, 100 to 120 years old, the average tracheid length was 3.1 mm in stem wood, 2.2 mm in stump wood, 2.3 mm in thick parts of lateral roots, and 3.3 mm in 3–25 mm diameter root wood (Eskilsson 1969). According to Fig. 3.22 and the following table by Atmer and Thörnqvist (1982), the variation of tracheid dimensions appears to be greater in root wood than in other wood components of a tree:

	Root wood	Stump wood	Stem wood	Branch wood
	Tracheid length (mm)			
Pinus sylvestris	1.7–5.0	0.5–2.2	0.9–3.5	0.8–2.0
Picea abies	1.4–4.5	0.8–2.2	0.9–3.9	0.9–2.0
	Tracheid diameter (μm)			
Pinus sylvestris	33–61	18–41	23–42	18–31
Picea abies	27–51	18–40	21–42	15–26

Manwiller (1972) studied tracheid dimensions in stumps and roots of 12- to 89-year-old southern pine trees. The stump wood samples were taken from the above-ground portion of each stump, and the root wood samples from within 65 cm of the point of attachment of the major laterals. Within each root system, trends were observed along the taproot and side roots in three directions: horizontal, oblique, and vertical. In horizontal side roots, length, cell diameter, and lumen diameter of the tracheids increased but cell wall thickness decreased with distance from the stump. In vertical side roots, cell and lumen diameters increased steeply along the root, whereas length and cell wall thickness decreased. In oblique side roots, cell and lumen diameters decreased while length and wall thickness increased. Along the taproot, all dimensions increased to a maximum, and then decreased. On average, root wood had significantly larger but thinner-walled tracheids than stump wood (see Scheme p. 142).

Gleaton and Saydah (1956) studied fiber dimensions in pulp samples from a 46-year-old *Pinus palustris* tree. The largest tracheid dimensions were recorded in pulp made from stump wood. Surprisingly, stem wood tracheids were smaller in

Tracheid dimension	Stump wood	Root wood
Length (mm)	3.0	4.0
Cell diameter (μm)	38.8	52.5
Lumen diameter (μm)	25.7	41.7
Wall thickness (μm)	6.6	5.4
Length/width	77.3	79.4

Table 3.4. Tracheid dimensions in wood components of a 38-cm-diameter, 46-year-old *Pinus palustris* tree in Georgia (Gleaton and Saydah 1956)

Wood component	Tracheid length (mm)	Tracheid diameter (μm)	Length/width ratio
Root	2.50	32	78.3
Stump	2.93	38	77.1
Merchantable stem	2.76	26	106.1
Unmerchantable top	2.52	28	90.0
Branch	1.72	27	63.8

diameter than tracheids from any other biomass component, resulting in the highest length/width ratio (Table 3.4). Since a distinct, positive correlation occurred between the length/width ratio of the tracheids and the overall strength of paper, stump wood tracheids could not be expected to produce paper with the best physical strength properties, although they were longer than stem wood tracheids.

Patel (1971) compared tracheid dimensions in the stem and coarse roots of large *Pinus radiata* trees in New Zealand. The samples of stem wood were obtained at a height of 1.2 m above-ground level, and those of root wood from buried laterals at a distance of 30 cm from the root origin. The root wood tracheids were generally larger and thickerwalled in the growth rings close to the organic root center than in later growth rings. Although the roots were younger than the stems, their tracheids were wider and much longer (Table 3.5).

Figure 3.29 (Poller et al. 1973) illustrates the distribution of tracheids according to length by stump and root wood components in *Pinus sylvestris*. The shortest tracheids were concentrated in the root collar area in the uppermost portion of the stump. In the middle stump portion, taproots, and particularly in coarse lateral roots, the tracheids were markedly longer. Cell lengths of up to 7 mm were recorded in coarse roots.

Eskilsson (1972) found that the variation in cell dimensions from root center to bark was small in stump-root systems of 100- to 120-year-old *Picea abies*. Instead, a regular variation occurred along the roots, tracheid length correlating fairly well with the root diameter. The stump and roots larger than 25 mm in diameter had rather short and narrow tracheids with few pits. In the large roots, the tracheids were thick-walled resulting in a high basic wood density. In the smaller roots, the tracheids had a quite different appearance: long, broad, and extremely rich in bordered pits. The increase in tracheid length along the root was pronounced at a 20–40 mm root diameter. At the same time, a significant de-

Table 3.5. Tracheid dimensions in stem wood at 1.2-m height, and in root wood at 30-cm distance from root origin in 37-year-old, 37-m-high *Pinus radiata* trees in New Zealand (Patel 1971)

Tracheid dimension	Root wood		Stem wood	
	Range	Mean	Range	Mean
Length (mm)	2.1 – 9.6	4.5	0.7 – 4.8	2.7
Earlywood tangential diameter (μm)	20.2 – 88.9	49.7	21.8 – 56.6	38.4
Earlywood cell wall thickness (μm)	1.5 – 6.3	3.1	1.2 – 6.6	3.5

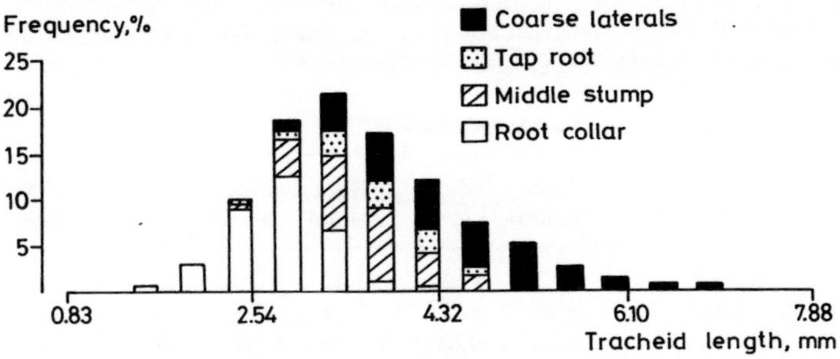

Fig. 3.29. Relative distribution of tracheids according to the length in various components of three stump-root systems of 48-year-old *Pinus sylvestris* trees (redrawn from Poller et al. 1973)

crease occurred in wood density. Average tracheid dimensions after sulfate pulping to a kappa number of about 30 are shown below. According to Eskilsson, the limit between large and small roots was set slightly too high, so that the pulp from small roots included a considerable proportion of large-root tracheids. More extreme figures would probably have been obtained if a smaller diameter had been chosen for the limit. Transverse dimensions of the tracheids in the stump and in large roots showed only minor differences from stem tracheids, with the exception of the thick wall in the tracheids in coarse roots. The tracheids in the small roots had the same mean dimensions as the stem tracheids, but a wider distribution:

Tracheid dimension	Root wood		Stump wood	Stem wood
	Under 25 mm	Over 25 mm		
Length (mm)	3.34	2.29	2.16	3.07
Diameter (μm)	56	50	51	56
Wall thickness (μm)	2.0	2.5	1.8	2.0
Coarseness (μg/m)	260	280	220	250

Koch (1985 a) studied fiber dimensions of wood and bark in the central stump-root system, including lateral roots within a 92-cm radius, of 22 hardwood species in the southern United States. *Nyssa sylvatica* had the longest wood fibers,

Table 3.6. Average fiber dimensions in wood and bark of the stems, branches, and central stump-root systems of 15-cm-diameter trees of 22 hardwood species in the southern United States (Koch 1985a)

Fiber dimension	Stump and root wood	Stem wood	Branch wood	Stump and root bark	Stem bark
Fiber length (mm)	1.27	1.27	0.96	1.22	0.98
Fiber radial diameter (μm)	20.99	16.72	14.12	20.68	17.86
Lumen radial diameter (μm)	10.27	7.84	6.71	5.08	2.45
Tangential wall thickness (μm)	5.32	4.44	3.71	7.80	7.69
Length/fiber diameter	60.51	75.96	67.99	58.99	54.87

Table 3.7. Average length and tangential diameter of stem, branch, and root wood cells in two pine species, six other conifers, eight diffuse-porous hardwoods, and four ring-porous hardwoods in the eastern United States; dbh of trees 30–40 cm. The species are indicated in Section 3.1.4 (Fegel 1941)

Species group	Cell type	Cell length (mm)			Cell diameter (μm)		
		Stem wood	Branch wood	Root wood	Stem wood	Branch wood	Root wood
Pines	Tracheid	3.17	1.93	4.44	28	23	31
Other conifers	Tracheid	3.53	1.77	2.11	26	20	30
Diffuse-porous hardwoods	Fiber	1.16	0.83	1.05	15	12	18
Ring-porous hardwoods	Fiber	1.24	0.94	1.01	14	11	16
Diffuse-porous hardwoods	Vessel elements	0.54	0.45	0.51	60	40	78
Ring-porous hardwoods	Vessel elements	0.31	0.25	0.31	100	71	90

1.90 mm. The shortest wood fibers, 0.79 mm, were found in stump and root wood of *Acer rubrum*. Bark fibers were longest, 1.65 mm, in *Ulmus americana* and shortest, 0.95 mm, in *Acer rubrum*. The radial diameter of the wood fibers was largest, 35.1 μm, in *Nyssa sylvatica* and smallest, 13.6 μm in *Celtis* sp. The diameter of bark fibers was largest, 30.0 μm, in *Nyssa sylvatica*, and smallest, 10.5 μm, in *Ulmus alata*. The average fiber dimensions in wood and bark of the stump and root mass of these 22 hardwood species are compared with those in the stem and branches in Table 3.6. Wood fibers in the stump-root system average about the same length, but are larger in radial dimensions and tangential wall thickness, than those in stem wood. Bark fibers in the stump-root system are longer and larger in diameter, but have about the same wall thickness as bark fibers in the stem.

Results of cell dimension comparisons by Fegel (1941), based on a comprehensive sample of 20 tree species, are summarized in Table 3.7. In both softwoods and hardwoods, the average cell dimensions were smaller in branch wood than in stem and root woods. Except for the vessels of ring-porous hardwoods, cell diameter appeared to be 10–20% larger in root wood than in stem wood. On the other hand, stem wood cells were longer than root wood cells in all other species except *Pinus resinosa* and *P. strobus*, the roots of which had very long fibers.

In summary, both softwoods and hardwoods generally possess short cells in the transition zone between the stem base and the roots. The cell length increases along the root, but the thin sections of the roots possessing the longest fibers are lost when stump and root wood is harvested. When the stump-root system is considered as a whole, the tracheids of pines seem to be longer but those in some other softwoods seem to be shorter than the average tracheids in stem wood. In hardwoods, fibers from stump-root systems are roughly of equal length with those from stem wood, or they may be slightly shorter. Root wood of softwoods and hardwoods is characterized also by large cell diameter together with wide lumina and thin cell walls. The longer the distance from the root base, the more distinct are these cell properties.

3.3 Chemical Composition of Residual Tree Components

Tree biomass is composed of three principal elements: carbon (C), oxygen (O), and hydrogen (H). About 50–52% of the dry mass of wood is carbon, usually more than 40% oxygen, and 6% hydrogen. Small and variable amounts of nitrogen (N) and mineral elements or ash are also present in forest biomass. The content of ash in temperate-zone trees is generally below 0.5% in wood but much larger in bark and foliage. The fuel value of forest biomass is largely determined by this proportional distribution of elements (Sect. 3.5).

Combinations of carbon, oxygen, and hydrogen form carbohydrates and lignin, which are the *primary cell-wall constituents* of wood. The carbohydrate portion of biomass comprises cellulose and hemicelluloses which are jointly referred to as *holocellulose.* Trees also contain smaller amounts of nonstructural carbohydrates, such as starch and sucrose, which function as energy reserves and are classified as extractives. Since the chemical structure of the primary constituents of the cellwall varies, the terms are generic in nature. All these substances are *polymeric*, i.e., composed of large molecules formed by the combination of small repeating structural units.

The gross structure and properties of woody cells arise from the nature and *organization of the chemical compounds.* With the initiation of daughter cells through cell division at the cambial zone, new walls are formed: the primary wall which encloses the new unit, and the middle lamella or intercellular layer which separates adjoining cells. Both of these regions are rich in pectic material. During the phase of cell-wall thickening that follows, cellulose and hemicelluloses are synthesized within the cell and deposited onto the primary wall and a new secondary wall. The formation of lignin begins before this phase is complete, starting at the cell corners and spreading along the primary wall and middle lamella. Finally, lignification proceeds to the secondary wall after which the cell dies. Cellulose, as the skeletal substance, contributes its high tensile strength to the complex of wood structure. The function of hemicelluloses is less clear, however, the possibility that they serve as a temporary matrix before lignification exists. The role of lignin is to provide rigidity to the tree, making upright growth possible. It also adds toxicity, thus making wood durable. Its concentration in the middle lamella serves the purpose of cementing individual cells together (Kollmann and Côté 1968).

In normal softwood tracheids, the thickness of the middle lamella, except at the cell corners, is 0.2–1.0 μm. The thickness of the primary wall is 0.1–0.2 μm.

The outer S_1 layer of the secondary wall contains three to four lamellae and has a thickness of about 0.2–0.3 μm. The middle S_2 layer is composed of 30–150 lamellae usually with a thickness of 1–5 μm, depending on the cell type. The inner S_3 layer is only 0.1 μm thick consisting of several lamellae. The warty W-layer is a thin amorphous membrane. The S_2 layer thus forms the main portion of the cell wall (Sjöström 1981).

Cellulose is the most abundant organic material on earth. It is a polysaccharide characterized by high resistance to chemicals, a property of great importance in the production of chemical pulp. It is distinguished analytically from the extractives by its insolubility in water and organic solvents, from the hemicelluloses by its insolubility in aqueous alkaline solutions, and from lignin by its relative resistance to oxidizing agents and susceptibility to hydrolysis by acids. Cellulose dissolves in strong mineral acids by hydrolysis to dextrins and oligosaccharides, and ultimately to D-glucose (Browning 1967b). Cellulose is present in all the higher plants, and it is the main constituent of wood, occupying roughly 40–50% of the dry wall mass in softwood and hardwood cells. It occurs predominantly in the secondary wall, usually in association with hemicelluloses and lignin.

Cellulose is composed of glucose molecules ($C_6H_{12}O_6$) produced by the tree through photosynthesis. These monomer units are joined through oxygen atoms end to end to form long-chain molecules or polymers $(C_6H_{10}O_5)_n$. The number of monomers per macromolecule shows the *degree of polymerization* in cellulose. According to Timell (1965b), for cellulose in the natural state the value of n lies between 8000 and 10000 in trees. This degree of polymerization would mean a linear chain length of approximately 5 μm in wood. In the cell wall, cellulose macromolecules are arranged parallel to each other, and are linked laterally by hydrogen bonds to form long strands called microfibrils which further expand into fibrils. Both pulp yield and fiber strength are positively correlated with the content of cellulose in wood. Besides the strength of the single fiber, the major physical factor influencing the properties of a pulp sheet is the degree of polymerization of the cellulose. In normal kraft and sulfite pulps, the degree of polymerization of cellulose falls within a range of 1000–2000 (Timell 1986c).

Hemicelluloses are low-molecular-weight, noncellulosic, heterogeneous polysaccharides. They are insoluble in water but soluble in aqueous alkali. This property can be used to separate them from the total carbohydrate fraction or holocellulose, leaving essentially pure or α-cellulose behind. Hemicelluloses also readily hydrolyze to their monomeric components in dilute acid. This process, forming sugars and sugar acids, is called *hydrolysis* (Sect. 8.4.4). Upon complete hydrolysis, the hemicelluloses of wood characteristically yield monosaccharides: hexoses (*D-glucose, D-mannose, D-galactose*) and pentoses (*D-xylose, L-arabinose*) together with a monomethyl *glucuronic acid* or oligosaccharides containing this acid. Other sugar units have been found in minor amounts in the hydrolysates of some species of wood. Corresponding to the respective monosaccharides, homopolymers which contain only identical sugar units are called *glucan, mannan, galactan, xylan,* and *arabinan*. The unsystematic terms *hexosans* and *pentosans* are often applied to those polymers which yield hexose and pentose sugars, respectively, on hydrolysis (Browning 1967b).

The distinction between the hemicelluloses and certain water-soluble extractives is not always clear. For example, some noncellulosic polysaccharides, such as arabinogalactans, are largely soluble in cold water and are, therefore, classed among the extractives. The term hemicellulose is thus neither systematic nor definitive.

Hemicelluloses constitute 20–35% of the dry wall mass of normal woody cells, the proportion being somewhat less in softwoods than in hardwoods. The degree of polymerization is rather low. The molecules have small side chains but are composed of only 150–200 monomers and thus relatively short compared with cellulose. While a cellulose chain molecule is linear and composed exclusively of glucose molecules, hemicellulose molecules are branched and include a large variety of monosaccharides. Two main classes of substances are separated: *xylans* and *galactoglucomannans*. The former class of chain molecules is formed by polymerization of the anhydro forms of five-carbon sugars such as D-xylose, L-arabinose, and 4-*O*-methyl-D-glucuronic acid. In the latter class, the chain molecules are formed by polymerization of six-carbon sugars such as D-glucose, D-mannose, and D-galactose. The properties of hemicelluloses have been discussed in detail by Timell (1964, 1965a, 1967).

The galactoglucomannans predominate in *softwood hemicelluloses*, accounting for 15–20% of the dry mass of wood. The glucose to mannose ratio is about 1:3, whereas the ratio of galactose to glucose can vary from 1:1 to 1:10. Softwoods also contain about 10% of a xylan. Arabinogalactan is a polysaccharide unique to members of the genus *Larix*, where it occurs in the lumina of heartwood tracheids in amounts ranging from 5 to 30%. The major *hardwood hemicellulose* is an acidic xylan, present in an amount of 25±5% by mass of extractive-free wood. In a few species, for example in some birches, the xylan content can reach 35%. Hardwoods also contain about 5% of a glucomannan. In papermaking, hemicelluloses serve as bonding, plasticizing, and swelling agents.

Lignin is a phenolic cell-wall component that characteristically differentiates wood from other cellulosic materials found in the plant kingdom. In its natural state, as it occurs in woody cells, lignin is an amorphous polymer known as native lignin or *protolignin*. Formation of the protolignin is believed to occur in the cambial zone by activation of lignin precursors which then polymerize in the cell wall as an indefinitely large, three-dimensional polymer. Its most important property is the rigidity and increased stiffness it imparts to cell walls. It is quite insoluble and, compared with cellulose, has a much lower hygroscopicity. It is present in the fine cavities within the cell wall where it acts as a bulking agent and reduces the dimensional changes with moisture changes because of its low hygroscopicity. Lignins are polymers, the basic structural unit of which consists of phenylpropane with a phenol ring that is substituted by zero, one, or two methoxyl groups. The addition of one methoxyl group to the phenol ring produces a *guaiacyl unit*, while the addition of two methoxyl groups results in a *syringyl unit*. Almost all softwoods contain a guaiacyl ligning, and all hardwoods have a guaiacyl-syringyl lignin, which is a copolymer of guaiacyl and syringyl residues (Sarkanen and Ludwig 1971).

The proportion of lignin in normal wood is 20–35%, depending on the tree species. It is considerably less in temperate hardwoods than in softwoods and tropical hardwoods. Each of the following figures shows average Klason lignin content and the standard deviation between species in normal stem wood for a large number of softwoods and hardwoods in various parts of the world (Pettersen 1984):

Country	Average content of Klason lignin (%)	
	Softwoods	Hardwoods
USA	28.8 ± 2.6	23.0 ± 3.0
USSR	29.0 ± 1.6	21.9 ± 3.2
Japan	29.6 ± 2.6	22.1 ± 3.0
Taiwan		25.0 ± 3.8
The Philippines		29.4 ± 5.6
Mozambique		27.3 ± 3.4

In chemical pulping, lignin is removed from the cell walls. The higher the content of lignin in the biomass, the lower is the yield of chemical pulp. A decrease in the lignin content of a pulp results in improved strength, because remaining lignin impairs the bonding ability of the cells. The two groups of lignin, guaiacyl and syringyl lignin, behave differently in pulping, thus complicating processing of softwood and hardwood as a mixture.

The three primary components introduced above by and large determine the chemical and physical properties of the tree components. In addition, a large variety of secondary or *extraneous components* are found in the woody cell walls and cell lumina. Ray cells include extraneous compounds such as terpenes, resins, fats, waxes, gums, starches, sugars, oils, tannins, alkaloids, etc. Many of the extraneous compounds can be separated, at least partly, from woody tissue with hot and cold water or neutral organic solvents, and they are therefore collectively referred to as *extractives* (Sect. 3.3.2).

Resin plays an important role in the extractive content of wood. Softwood resin is generated partly by the epithelial cells surrounding the resin canals, that are present in some genera, but in others it is produced mainly in the rays. Hardwood resin is located exclusively in ray parenchyma cells.

In extreme cases, especially in some tropical tree species, the proportion of the extractives may be as much as several tens of percentages of the dry forest biomass. However, in general they constitute less than 5–10% of the dry mass of wood. The largest concentration of extractives occurs generally in foliage and bark, and among wood tissues, in the heartwood. Although the extractives are not a structural part of the woody tissue, they contribute to many properties of wood: odor, color, durability, permeability, basic density, and strength. When abundant, the extractives cause problems in chemical and mechanical processing of wood. On the other hand, their presence presents a great opportunity and challenge to the chemical industries. Crown mass in particular possesses a large but very inadequately explored potential as a raw material reserve for chemical industries.

In all tree species, the biomass contains small amounts of *extraneous inorganic components* (Sect. 3.3.3), which the tree takes up from the soil through its roots and sap stream. In temperate-zone trees, the quantity of ash in stem wood rarely exceeds 0.5–0.6%, but in bark and foliage the proportion is much larger. The inorganic elements occur in wood either as components of the extractives or in the form of crystals. The alkaline earths, i.e., calcium, potassium, and magnesium, usually account for the majority of inorganic components. However, certain tree species may contain large amounts of silica, particularly at the base of the stem, causing chain saws, chipper knives, and other tools to become dull. Inorganic components also tend to cause wear and tear and chemical problems in pulping, especially in the chemical recovery systems. In composite board they are a serious defect, and ash formation must be taken into consideration in the energy use of wood. Furthermore, because intensive recovery of biomass accelerates the removal of nutrients from forests and may result in impoverishment of soil, the amount and composition of inorganic components in forest biomass are of significant ecological importance (Sect. 9.1).

The proportions and composition of chemical constituents vary greatly among species. Data in more than 350 references of the chemical composition of wood in 153 species of the temperate zones were compiled in tabular form by Fengel and Grosser (1975). On the average, stem wood in softwoods contains 40–45% cellulose, 25–35% ligning, and 25–30% hemicelluloses. Stem wood in temperate-zone hardwoods contains 40–50% cellulose, 20–25% lignin, and 25–35% hemicelluloses. However, prosenchymatous elements differ from parenchyma cells, earlywood differs from latewood, juvenile wood differs from mature wood, sapwood differs from heartwood, normal wood differs from reaction wood, etc. Even greater differences are found among wood, bark, and foliage components of a tree. Consequently, very significant variation may occur in the chemical composition of biomass from various sources of forest residue.

As long as the chemical composition is unknown, the utilization of residual biomass in an optimal manner is difficult. Detailed knowledge of the chemical nature of all tree parts would undoubtedly facilitate the utilization of residual forest biomass in traditional forest industries and open up new possibilities in chemical and biochemical industries. At present, however, applicable information on the chemical composition of crown, top, stump, and root mass in different species is scarce.

3.3.1 Primary Constituents in Residual Tree Components

Two major factors are responsible for the considerable variation in chemical composition among tree components. First, the proportion of wood varies among the biomass components of a tree, and second, marked differences may occur in the composition of wood proper. These differences result partly from quantitative and qualitative differences in the primary constituents: cellulose, hemicelluloses, and lignin, and partly from variation in the concentration of extractives and inorganic components.

Significant abnormalities are found in the chemical composition of branch wood. They result from the occurrence of compression wood on the lower side of softwood branches and tension wood on the upper side of hardwood branches. Since the opposite and side woods have an almost normal composition, the differences are largely dependent on the proportion of reaction wood in branches. Chemical differences between normal and reaction woods are actually known better than differences between stem and branch woods. Comprehensive summaries on the chemical properties of reaction wood have been published by Timell (1969, 1986a).

According to Timell (1982, 1986a), the differences in chemical composition between normal and *compression wood* in the majority of conifers are of a quantitative rather than qualitative nature. All polysaccharides so far isolated from compression wood are also present in normal wood, although in very different amounts. Characteristic chemical features of compression wood are its high content of lignin and galactan and low content of cellulose and galactoglucomannan. As shown by the following table by Timell (1982) of the average summative composition of 27 conifer species, the lignin content is almost 10 percentage points higher and cellulose content about 12 percentage points lower in compression wood. Laricinan or (1→3)-β-D-glucan and galactan are present in only trace amounts in normal softwood, whereas in compression wood the former is a minor and the latter a major constituent. The content of galactoglucomannan is twice as high in normal wood. The xylan content is similar, and the same probably applies to other polysaccharides which include pectin, arabinogalactan, arabinan, and xyloglucan. If the absolute amount of polysaccharides present in each tracheid is considered rather than the overall mass percentage, a compression wood tracheid contains as much cellulose as and slightly more xylan than does a tracheid in normal wood:

	Proportion in extractive-free wood (%)	
	Normal wood of average conifer	Compression wood of average conifer
Lignin	30	39
Cellulose	42	30
Laricinan	Trace	2
Galactoglucomannan	18	9
Galactan	Trace	10
Xylan	8	8
Other polysaccharides	2	2

Qualitatively, the chemical differences between normal and compression woods are less pronounced. Nevertheless, the lignin in compression wood differs in several respects in its structure from that present in normal softwood. It has a considerably higher proportion of p-hydroxyphenylpropane units and is more condensed than normal conifer lignin. It also contains fewer arylglycerol-β-arylether and probably more diphenyl and phenylcoumarin interunit bonds, and more carbon-carbon bonds. The proportions of different galactoglucomannans are probably the same in the two tissues, and their structure is identical, including the presence of acetyl groups. The arabino-4-*O*-methyl-glucurono-xylan in compression wood carries only half as many arabinose side chains as in normal wood. The exact frequency of the 4-*O*-methylglucuronic acid side chains is unknown. The

Table 3.8. Chemical composition of the upper and lower portions of a branch of *Pinus sylvestris* at different distances from the stem (Žerebov 1946)

Constituent	Branch zone	Distance from stem (cm)						Average
		8	100	200	300	400	500	
		Proportion (%)						
Cellulose	Upper	45.2	46.6	47.5	46.4	47.3	45.4	46.4
	Lower	40.0	41.1	42.3	40.7	40.9	39.7	40.8
Hemicelluloses	Upper	16.1	16.4	16.1	16.9	15.8	17.1	16.4
	Lower	16.3	15.7	15.2	16.2	16.0	15.5	15.8
Lignin	Upper	31.9	31.4	31.3	30.8	30.1	30.1	30.9
	Lower	38.5	37.0	34.5	34.0	35.7	34.6	35.7
Resin and fats	Upper	6.31	5.00	5.45	6.40	5.90	6.13	5.86
	Lower	6.32	4.90	5.34	6.50	5.72	8.57	6.23
Ash	Upper	0.25	0.31	0.55	0.54	0.42	0.22	0.36
	Lower	0.24	0.28	0.55	0.46	0.30	0.29	0.35

galactan and laricinan are both slightly branched, acidic polysaccharides. The galactan consists of (1→4)-linked *β*-D-galactose residues which have a few galacturonic acid side chains attached to C-6. Laricinan is a (1→3)-linked *β*-D-glucan which may contain both glucuronic acid and galacturonic acid side chains (Timell 1982, 1986a).

Table 3.8 shows the variation of the chemical composition in branch wood of *Pinus sylvestris.* Significant differences were found between the lower and upper parts of the branch cross-section, whereas no regular trend was observed along the branch. The average proportion of cellulose was 40.8% on the lower side but 46.4% on the upper side. The proportion of lignin was 35.7 and 30.9%, respectively (Žerebov 1946).

The specific chemical characteristics of hardwood branches result from the occurrence of *tension wood* on the upper side of the branch axis. It is the fibers, or actually only the earlywood fibers, that are chemically and anatomically modified in tension wood, whereas the latewood fibers, vessels, and rays remain unaffected. Thus, tension wood differs more from normal wood in its overall chemical composition in species with a high proportion of fibers than in species where the vessels and rays are more numerous.

Tension wood fibers differ typically from normal fibers in containing a rather thick gelatinous layer inside the S-layers of the cell wall (Sect. 3.1.2). The G-layer largely consists of highly crystalline cellulose microfibrils which are oriented parallel to the fiber axis. Hardly any lignin occurs in this additional cell-wall layer of tension wood. Nordberg and Meier (1966) found 98.5% cellulose and 1.5% xylose in the G-layer of tension wood fibers of *Populus tremula.*

Table 3.9 (Timell 1969) shows the chemical composition of normal and tension woods in ten hardwood species in the northeastern United States in terms of lignin, cellulose, and three hemicelluloses, which together represent 96–99% of the wood mass. Tension wood is characterized by an unusually high cellulose content, generally at least ten percentage points higher than in normal stem wood.

Table 3.9. Lignin and major polysaccharide composition of normal and tension wood in ten hardwood species in the northeastern United States (Timell 1969)

Tree species/ type of wood	Proportion (%)				
	Lignin	Cellulose	Glucomannan	Galactan	Xylan
Acer pennsylvanicum					
Normal wood	22.4	41.5	4.4	2.4	28.0
Tension wood	20.5	52.3	2.1	4.0	20.6
A. rubrum					
Normal wood	25.4	40.7	4.9	2.0	25.0
Tension wood	17.5	58.3	2.7	3.2	16.8
A. saccharum					
Normal wood	23.1	40.2	4.6	1.8	30.2
Tension wood	17.8	56.5	1.0	2.6	17.8
Betula alleghaniensis					
Normal wood	23.8	42.6	3.6	1.8	25.5
Tension wood	18.5	52.0	2.2	5.8	21.6
B. papyrifera					
Normal wood	22.0	38.3	4.0	2.6	33.0
Tension wood	16.1	50.0	2.6	10.8	24.0
Fagus grandifolia					
Normal wood	23.5	39.5	2.7	1.6	29.8
Tension wood	17.0	54.1	1.8	6.0	18.3
Populus tremuloides					
Normal wood	20.9	42.7	5.2	2.2	29.2
Tension wood	16.8	54.0	2.0	2.8	23.8
Prunus pensylvanica					
Normal wood	17.4	42.1	3.9	2.8	35.2
Tension wood	15.6	52.8	2.2	5.6	25.7
Robinia pseudoacacia					
Normal wood	30.1	40.9	3.3	1.6	23.0
Tension wood	25.2	50.6	1.8	2.8	19.5
Ulmus americana					
Normal wood	21.7	48.5	5.1	1.8	21.5
Tension wood	20.2	53.6	2.2	2.8	18.7

Although normal and tension wood fibers contain approximately the same absolute amount of lignin, the relative lignin content is several percentage points lower in tension wood fibers due to the presence of the extra cellulose in the unlignified G-layer. The xylan content in tension wood is low as well. According to Timell, this difference also is almost entirely a result of the incorporation of additional cellulose into tension wood fibers. In other words, the decrease in xylan content compared to normal wood is only apparent. In fact, each tension wood fiber contains only slightly less xylan than does the normal fiber or, in some cases, the same amount. According to Table 3.9, tension wood also has a low content of glucomannan but a high content of galactan, resembling in this respect the compression wood of conifers. In certain genera, such as *Betula*, *Eucalyptus*, and *Fagus*, tension wood is especially rich in galactan. The occurrence of an unlignified, or only slightly lignified, G-layer inside fiber walls thus largely determines the specific chemical characteristics of tension wood.

Table 3.10. Elementary carbohydrate composition of normal and tension wood and bark from five hardwoods in the northeastern United States. All values in percent of extractive-free wood or bark (Timell 1969)

Tree species/ type of tissue	Proportion (%)								
	Lignin	Ash	Acetyl	Uronic anhydride	Galactose	Glucose	Mannose	Arabinose	Xylose
Acer rubrum									
Normal wood	25.4	0.4	3.6	4.9	1.0	42.3	3.3	1.0	18.1
Tension wood	17.5	0.5	2.0	4.0	1.6	59.2	1.8	1.3	12.2
Normal bark	41.2	3.6	2.2	11.4	2.7	24.7	1.2	4.7	8.3
Tension bark	34.7	3.7	2.4	11.9	2.2	30.2	0.8	4.4	9.7
Fagus grandifolia									
Normal wood	23.5	0.6	4.3	5.9	0.8	40.4	1.8	0.9	21.7
Tension wood	17.0	0.6	3.3	6.1	3.0	54.7	1.2	0.8	13.3
Normal bark	35.0	9.4	2.3	9.1	2.5	25.4	1.3	4.7	10.3
Tension bark	36.7	7.7	2.5	10.2	2.2	25.3	1.0	4.0	10.4
Betula papyrifera									
Normal wood	22.0	0.4	3.9	5.7	1.3	40.3	2.0	0.5	23.9
Tension wood	16.1	0.4	2.8	4.7	5.4	51.3	1.3	0.5	17.5
Normal bark	43.0	4.3	1.9	8.8	2.0	19.6	1.1	4.5	14.8
Tension bark	42.4	3.6	1.8	6.7	1.9	21.9	0.8	3.5	17.4
Populus tremuloides									
Normal wood	20.9	0.4	3.9	3.7	1.1	44.4	3.5	0.9	21.2
Tension wood	16.8	0.4	3.1	4.6	1.4	54.5	1.3	0.8	17.1
Normal bark	28.3	6.8	2.8	11.0	2.0	30.3	0.8	6.7	13.3
Tension bark	20.5	6.9	2.9	9.9	2.1	34.3	0.9	6.2	16.3
Prunus pensylvanica									
Normal wood	17.4	0.4	4.5	4.7	1.4	43.4	2.6	0.7	24.9
Tension wood	15.6	0.4	3.1	4.2	2.8	53.5	1.5	0.8	18.1
Normal bark	35.3	3.6	1.3	4.6	3.0	35.0	1.7	5.7	9.8
Tension bark	33.9	4.1	1.2	5.0	2.7	36.5	1.7	4.9	10.0

Not only the xylem but also the phloem is modified when reaction tissue is formed. Table 3.10 (Timell 1969) shows the elementary carbohydrate composition of normal and tension woods and barks in five hardwood species. Unlike normal and compression barks in softwoods, normal and tension barks in hardwoods differ chemically, although not as much as do the two types of wood. It is evident that bark on the upper or tension side of a hardwood stem contains less polyphenolics, such as lignin and tannin, and has a higher cellulose content in comparison with bark on the lower or normal side. The presence of somewhat more cellulose and less lignin in tension bark is to be expected in view of the occurrence of gelatinous fibers in tension phloem. Surprisingly, tension bark appears to contain more xylan and less galactan than does normal bark.

As the chemical characteristics of branch wood are largely a result of the specific properties of reaction wood, so are those of *stem tops* considerably affected by the presence of *juvenile wood*, which occurs in a cylindrical column of 5–20 annual rings surrounding the pith. As mentioned in Sections 3.1.3 and

3.2.2, small cell dimensions, thin cell walls, a low proportion of latewood, and a high proportion of reaction wood are typical characteristics of juvenile wood. Because of their small number of annual rings, very little heartwood occurs in unmerchantable tree tops. Although the variation in cell wall composition in the longitudinal and radial direction of the stem has not been systematically investigated, certain trends with increasing age on a stem cross-section are evident. The anatomical abnormalities of juvenile wood result in differences in the chemical and physical properties between the merchantable stems and unmerchantable tree tops.

Larson (1966) analyzed microsamples of earlywood and latewood from three *Pinus resinosa* trees for lignin and five principal wood sugars. Figure 3.30 reveals definite differences between earlywood and latewood in most chemical components, and also shows changes with increasing age. The most obvious source of variation is a quantitative one arising from changes in the thickness of the tracheid walls. Another source of variation is a more subtle, qualitative one related to aging and/or position within the stem. Latewood of mature red pine, compared to earlywood, has a lower lignin content. The decrease in lignin appears to be primarily a dilution effect of the middle-lamella lignin, together with an actual decline across the secondary wall. The latewood hydrolysate yields more mannan but less arabinan and xylan. According to Larson, the increase in mannan yield apparently results from a progressive increase across the S_2 layer of the secondary wall which thickens most during latewood development. This mannan is presumably a glucomannan closely bound to, and at least partially extracted with, the alpha-cellulose. The decrease in arabinan and xylan yields may be considered dilution effects resulting from their primary localization in the outer part of the secondary wall and consequent decline inward. Although these changes are quantitative in the sense that they are directly related to wall thickening, they are also qualitative since definite compositional changes occur during latewood wall development.

Figure 3.30 shows a distinct variation in chemical composition of stem wood associated with aging. Rapid, but fluctuating, changes occur during the juvenile period, and relatively steady changes occur thereafter. Although a large share of the radial variation may be attributed to wall thickness, qualitative changes across the cell wall are, according to Larson, apparently inherent and represent developmental changes related to aging cambium combined with the declining influence of an upward receding crown. Lignin content declines gradually from pith to periphery. During the juvenile years, the lignin yields are somewhat erratic, but in mature wood the decrease is quite uniform. A very pronounced increase occurs in glucose with increasing age. The yield of mannose also increases, while those of galactose, arabinose, and xylose decrease. Both the glucose/xylose and mannose/xylose ratios increase but the glucose/mannose ratio remains constant or declines slightly. The data of Fig. 3.30, representing 1.8 m height in a red pine stem, did not differ significantly from corresponding measurements at 5.5 m height of the same tree. The inner annual rings of the stem thus supposedly reflect the proportional distribution of the primary chemical constituents in unmerchantable tree tops. However, the possible effect of heartwood extractives must be recognized. The differences between mature and juvenile woods, or merchant-

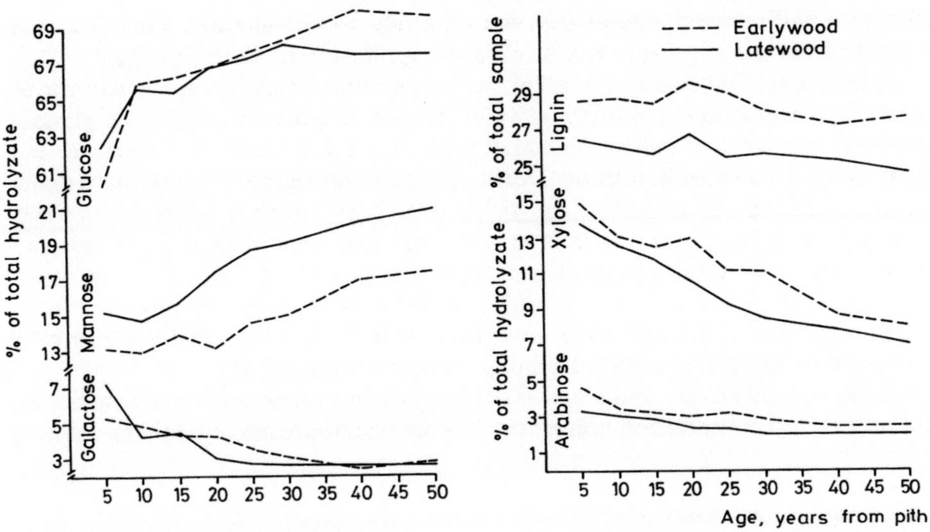

Fig. 3.30. Percentage of lignin (*right*) in total extractive-free wood and that of pentose (*right*) and hexose (*left*) sugars in the hydrolysate from microsamples of earlywood and latewood from 1.8 m height of a *Pinus resinosa* stem, as a function of age from the pith (Larson 1966)

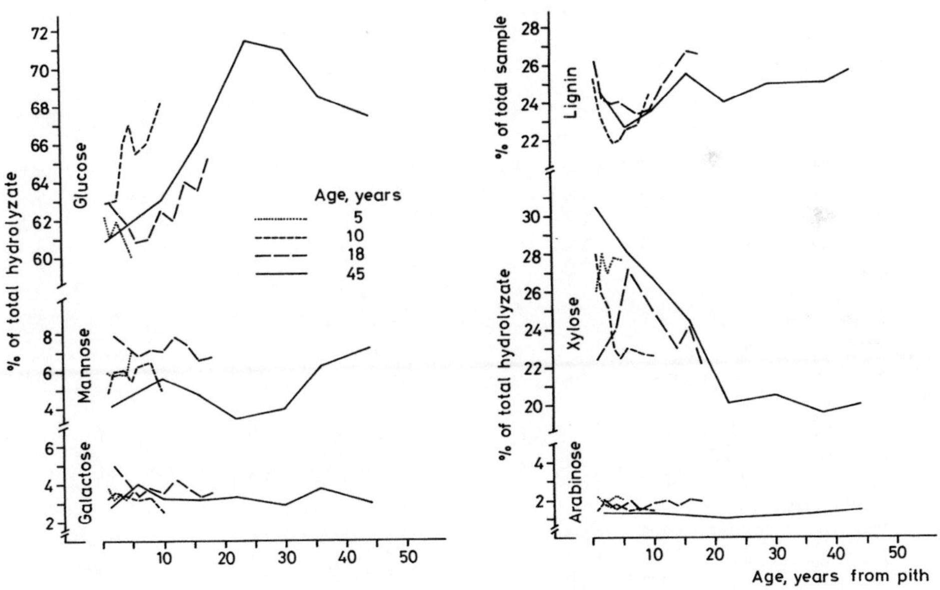

Fig. 3.31. Percentage of lignin (*right*) in total extractive-free wood and that of pentose (*right*) and hexose (*left*) sugars in the hydrolysate from wood from 1 m height at *Populus deltoides* stems of various ages, as a function of age from the pith. The 5- and 10-year-old trees from plantations, the 18- and 45-year-old trees from natural stands (redrawn from Dickson et al. 1976)

able stem and unmerchantable top, are naturally species-specific. Therefore, the example from red pine may not be directly applicable to other species.

Figure 3.31 (Dickson et al. 1976) shows an example of the aging pattern of sugars in a hardwood, *Populus deltoides*. In general, the proportion of glucose increased, and that of xylose decreased with age. There were no clear trends for mannose, galactose, or arabinose. During the juvenile period, no consistent aging patterns were apparent, and sugar yields were highly variable. On the average, however, small trees and tree tops may be expected to contain less glucose and more xylose than timber from larger trees. According to the authors, the sugar yield/age curves for the 45-year-old tree show what may be growth effects superimposed on the aging curve, since the annual rings after the 22nd year were very narrow, indicating either declining vigor or overmaturity.

Table 3.11 (Eskilsson and Hartler 1973) represents an example of the variation in the chemical composition among the biomass components of conifers. As may

Table 3.11. Chemical composition of various biomass components in 60- to 70-year-old *Pinus sylvestris* and 100- to 120-year-old *Picea abies* (Eskilsson and Hartler 1973)

Compound	Proportion (%)								
	Stem wood	Stump wood	Coarse root wood	Slender root wood	Coarse branch wood	Slender branch wood	Needles	Inner bark	Outer bark
Pinus sylvestris									
Undetermined	1	1	1	2	1		5	4	4
Extractives	3.0	4.0	3.0	3.4	3.3		12.6	4.5	3.4
Lignin	28.1	25.0	25.8	28.8	31.2		28.4	29.2	47.6
Carbohydrates									
Cellulose	40.8	43.6	43.6	37.2	32.1		29.1	36.2	24.5
Glucomannan	15.4	14.8	14.0	13.3	10.6		11.8	5.2	5.0
Galactan	1.1	1.7	1.8	2.1	5.1		3.7	4.2	3.5
Arabinan	0.8	1.2	1.6	1.5	1.5		3.4	9.2	4.5
Xylan	7.7	6.7	7.0	9.6	13.0		3.9	5.5	5.3
Uronic acid not in xylan	2.1	2.1	2.1	2.1	2.1		2.1	2.1	2.1
Total carbohydrates	67.9	70.0	70.1	65.8	64.5		54.0	62.4	45.0
Picea abies									
Undetermined	1	1	1	2		2	5	4	
Extractives	0.8	1.0	1.5	2.1		1.7	6.1	3.8	
Lignin	28.2	30.6	33.5	32.5		37.0	35.3	35.9	
Carbohydrates									
Cellulose	42.7	38.9	37.6	33.5		29.1	28.2	35.9	
Glucomannan	16.1	18.1	16.4	11.9		10.2	12.5	3.8	
Galactan	1.2	0.8	0.8	5.2		7.6	2.8	3.3	
Arabinan	0.5	0.6	0.5	1.1		0.5	3.4	6.1	
Xylan	7.4	6.9	6.7	9.5		9.9	4.6	5.0	
Uronic acid not in xylan	2.1	2.1	2.1	2.1		2.1	2.1	2.1	
Total carbohydrates	70.0	67.4	64.0	63.4		59.4	53.6	56.3	

Table 3.12. Relative elementary carbohydrate composition of biomass components of *Pinus sylvestris* and *Picea abies* according to (I) Eskilsson and Hartler (1973), and (II) Hosia et al. (1971)

Tree component	Percent neutral sugars					Undetermined
	Galactose	Glucose	Mannose	Arabinose	Xylose	
Pinus sylvestris (I)						
Stem wood	2.7	68.3	17.9	2.5	8.6	–
Stump wood	3.5	69.9	16.6	2.8	7.2	–
Coarse root wood	3.5	69.8	15.7	3.5	7.6	–
Slender root wood	4.3	64.4	16.1	4.1	11.1	–
Coarse branch wood	9.3	54.1	13.2	4.8	15.6	–
Slender branch wood	10.8	55.5	12.0	5.1	16.6	–
Needles	8.1	61.5	17.3	7.6	5.5	–
Inner stem bark	7.4	62.7	6.5	16.8	6.6	–
Outer stem bark	8.8	60.8	9.0	12.3	9.1	–
Pinus sylvestris (II)						
Stem wood	3.4	63.5	16.8	4.4	12.0	–
Stem top wood	7.7	61.3	15.8	4.2	11.1	–
Branch wood	9.8	57.0	12.2	6.0	15.0	–
Needles	5.2	64.5	14.6	7.2	5.1	3.4
Picea abies (I)						
Stem wood	2.7	68.6	18.6	1.9	8.2	–
Stump wood	2.4	66.0	21.7	2.0	7.9	–
Coarse root wood	2.3	67.0	20.7	1.9	8.1	–
Slender root wood	9.5	59.9	15.3	3.6	11.7	–
Slender branch wood	14.3	55.8	14.1	2.7	13.1	–
Needles	6.4	60.3	18.9	7.7	6.7	–
Stem bark	6.5	68.4	5.5	12.7	6.9	–
Picea abies (II)						
Stem wood	2.2	66.7	16.8	3.3	11.0	–
Stem top wood	3.3	63.9	18.0	3.3	11.4	–
Branch wood	8.5	62.3	13.0	3.6	12.6	–
Needles	3.6	65.5	15.3	7.2	6.0	2.5

be expected, lignin, galactan, and xylan contents are higher in branch wood than in stem wood at the expense of cellulose and glucomannan contents, typical characteristics of compression wood. The high lignin and low cellulose contents are serious drawbacks in the utilization of softwood branches for pulping. Table 3.12 shows that the proportions of galactose and xylose residues are, in fact, higher in branch wood than in any other biomass component of conifers. These general trends are supported by the findings of Žerebov (1946), Watanabe et al. (1962), Jarceva (1969), Hans (1970, 1971), Wood and Goring (1971), Howard (1973b), Virkola (1976), and Poller (1978), as well as those of Hägglund and Larsson (1937) and Wegelius (1939, 1946) regarding knot wood.

According to Tables 3.11 and 3.12, *stump wood*, and *root wood* sections more than 2.5 cm in diameter, seem to differ only slightly in their chemical composition from stem wood. Stump wood from older trees, however, contains more extractives. Root wood sections less than 2.5 cm in diameter have a low content of cellulose and glucomannan but a high content of lignin, galactan, and xylan. The table values are in agreement with the following results by Sjöholm (1977) on the lignin content of stump and root woods in conifers:

	Lignin content of unextracted wood (%)	
	Pinus sylvestris	*Picea abies*
Stem	28.0	27.7
Stump	26.5	28.9
Coarse roots (5 – 15 cm)	27.1	29.0
Slender roots (1 – 5 cm)	28.3	30.1

The highest percentage of glucose residues, reflecting a high concentration of cellulose, is found in stem wood. The chemical composition of stem tops is intermediate between those of conventional stem timber and branch wood. According to Hosia et al. (1971), the proportion of glucose residues in carbohydrates is two to three percentage points lower in the unmerchantable tops of *Pinus sylvestris* and *Picea abies* than in merchantable stem wood. This is obviously a result of the presence of juvenile wood.

The chemical composition of biomass recovered from logging sites is affected by harvesting decisions. For operational reasons, the most common method for harvesting residual forest biomass for the use of pulp and paper industries is whole-tree chipping. The raw material then essentially contains all above-ground tree components, although some small twigs and foliage is lost during logging. When stem tops and crown mass with bark and foliage are included, the chemical composition of the chips changes radically. The concentration of extractives, ash, and lignin increases at the expense of the holocellulose. For example, the holocellulose content of outer bark was found to be only 43.9% in *Pinus resinosa*, 40.3% in *P. strobus*, 42.2% in *Acer rubrum*, 45.3% in *Carya ovata*, and 47.1% in *Quercus rubra* (Harum and Labosky 1985). Stem wood of 8-year-old *Populus* hybrids in Pennsylvania contained on average 42.8% alpha-cellulose and 18.7% lignin, whereas stem bark contained only 29.4% alpha-cellulose but 28.8% lignin (Blankenhorn et al. 1985).

Table 3.13. Chemical analysis of whole-tree chips of *Pinus resinosa* and *Acer saccharum*, compared with conventional raw material from stem wood (data from Marton et al. 1976)

Tree species	Raw material of chips	Proportion (%)			
		Ash	Extractives	Lignin	Pentosans
Pinus resinosa	Stem wood	0.20	4.77	22.4	11.1
	Stem wood + bark	0.28	5.20	24.2	10.3
	Whole tree	0.65	9.59	25.4	10.5
Acer saccharum	Stem wood	0.28	2.73	20.6	19.9
	Stem wood + bark	0.43	2.80	21.2	19.8
	Whole tree	0.86	3.14	23.2	20.3

In a study by Marton et al. (1976), the lignin content of *whole-tree chips* in *Pinus resinosa* was 3.0 and in *Acer saccharum* 2.6 percentage points higher than in chips made from bark-free stem wood (Table 3.13). Powell et al. (1975) compared the chemical compositions of standard hardwood chips and whole-tree chips from mixed hardwoods in the eastern United States. Due to the presence of bark, whole-tree chips contained more ash, extractives, and lignin, but 2.3 percentage points less glucose residues than standard hardwood chips from bark-free stem wood. However, if the chips were screened and all bark was removed, the wood from both sources had the same chemical composition:

	Proportion (%)		
	Bark-free standard hardwood chips	Unbarked hardwood whole-tree chips	Wood from hardwood whole-tree chips
Ash (wet basis)	0.65	1.28	0.57
Alcohol-benzene extractives	1.86	2.88	1.97
Klason lignin	22.23	23.39	21.90
Acid-soluble lignin	5.20	4.41	4.57
Glucose	40.2	37.9	40.0
Other sugars	20.5	19.9	20.6

3.3.2 Extractives in Residual Tree Components

Extraneous components which are readily soluble in neutral organic solvents or water are called *extractives*. Some other extraneous components, such as proteins, salts of organic acids, and inorganic materials, are partly or wholly insoluble in solvents used to remove the extractives (Browning 1967a).

The extractives comprise an extraordinary diversity of compounds. Their amount varies considerably, ranging from only traces to 30–40% of the dry mass of wood. Important among extractives are terpenes, resin acids, starch, fats, waxes, tannins, and a large number of polyphenols, such as the lignans, and troponoles. The arabinogalactan present in larch heartwood could also be listed among the extractives, although it is more convenient to regard it as a hemicellulose (Timell 1986a).

The proportions of various extractives exhibit wide variation among tree species and biomass components. Some of them are found in significant quantities in only few species or genera. Compared to wood, bark and foliage usually contain considerably more extractives. Thus, biomass components of a tree are more definitely characterized by the presence and nature of extractives than by the proportions of the primary components of the cell wall. The extractives are of great importance in understanding the biochemistry and taxonomy of trees, and they also contribute to the raw material properties of the biomass, especially in pulping and biochemical industries. A comprehensive analysis of the significance of wood extractives to the pulp and paper industries was published by Hillis

(1962). Chemical composition and pharmaceutical use of extractives from conifers were reviewed by von Schantz and Hiltunen (1983).

Most extractives may be harmful in chemical and mechanical pulping. However, they are usually rapidly dissolved in the alkaline cooking liquor during delignification by the softwood pulping processes and retained to only a slight extent by the pulp produced. Accordingly, the *yield of pulp decreases* with increasing concentration of extractives, when calculated per unit mass of raw material. For example, larch wood is known to give low yields of kraft pulp on a dry mass basis due to its high content of water-soluble arabinogalactan.

Extractives are often responsible for *increased chemical consumption* since they react with cooking chemicals. They may also inhibit pulping reactions by preventing diffusion and penetration of the cooking chemicals into the raw material, reducing delignification, or decomposing cooking liquor. Furthermore, extractives may be detrimental to color, bleachability, and wettability of chemical pulps (Gardner and Hillis 1962). During preparation of mechanical pulp and newsprint and during storage of the latter, polyphenols and other extractives tend to cause color problems. Changes in color may be caused by the oxidation of chemical compounds when exposed to air and sunlight, an enzymatic catalytic effect of cell sap in rays and other tissues, and a combination of polyphenols or other compounds with metals (Hillis and Swain 1962).

In pulp and paper manufacture, free resin tends to coagulate and adhere to fibers and metal surfaces thus causing so-called *pitch troubles.* In papermaking, pitch tends to form deposits which may interfere with paper machine operation and cause spots or holes in the finished paper. Resin deposited on fiber surfaces prevents bonding between the fibers, and between a glue and the fiber material. The ability to form hydrogen bonds with one another, however, is a fundamental property of the fibers and a necessary requirement for paper strength. Pitch has its source in the resin acids that are incompletely removed in acidic and mechanical pulping processes or that are precipitated from the pulping liquor due to sudden changes in temperature, pH, or other conditions. Alkaline processes, on the other hand, can deal with resin so that pitch usually presents no problem in sulfate pulping of softwoods.

The resins and related materials comprising oleoresins from pines are examples of extractives of considerable importance. Commercial resin, fatty acids, and turpentine products are presently produced from living pines, extracted from chips made from mature pine stumps, or are recovered as by-products in sulfate pulping. On the other hand, the large variety of foliage extractives is almost completely outside commercial utilization.

The extractives are removed from tree biomass by extraction with neutral solvents such as hot and cold water, petroleum ether, ethyl ether, ethylene dichloride, dichloromethane (DCM), ethanol, or acetone, but no single solvent is capable of removing all the substances which are properly classed among the extractives. Thus, for reasonably complete removal of extractives, it is necessary to employ a sequence of two or more solvents. For example, the extractives may be removed by successive treatments with ether, alcohol, and water.

Extractives include a large number of compounds involved in the metabolic processes of a tree. The *primary metabolites* are the bioorganic intermediates common to essentially all organisms. Most of them can be interconverted in the organism into the other primary metabolites via the intermediary metabolic processes. Typical primary metabolites include simple sugars, amino acids, simple fats, and various carboxylic acids. These will always be found among the extractives of living trees, although the amounts will vary depending on the time in the growth cycle, the nutritional state, the tissue, and the season. *Secondary metabolites* are generally more complex compounds, and their formation within the organism is essentially irreversible. Glucose, the primary product of photosynthesis, is the starting material for producing both cell wall components and most secondary metabolites. Common secondary metabolites include starch, sitosterol, simple terpenoids, chlorophyll, phenylpropanoids, the common flavonoids, and simple tannins. In some species or tissues these intermediates accumulate in pools that can be isolated in large quantities, whereas in others they are metabolized further as fast as they are formed. These pools of intermediates may be limited to specific tissues or by specific conditions, which accounts for their variability. Many secondary metabolites have a restricted distribution because not all biogenetic pathways are operative in all species or tissues and they may terminate at different stages (Koch 1985a).

From the biochemical point of view, a majority of extractives are thus intermediate compounds from metabolic processes of a tree. This results in a wide qualitative and quantitative variation, depending on the characteristics of the producing tissue and the environment influencing the organism. Since photosynthesis takes place in the leaves, carbohydrates and other low-molecular-weight compounds are synthesized and stored primarily in the foliage. Consequently, foliage contains unusually large amounts of extractives with a complex composition. Sugars, starch, fats and fatty acids form a reserve food supply. Terpenes, resin acids, and phenols seem to be partly protective elements and partly toxic waste from metabolic processes. Sterols are assumed to function as growth hormones, and components of protein as enzymes in the synthetic processes (Hannus and Pensar 1970). Soluble carbohydrates and starch are in a dynamic equilibrium with each other, with photosynthesis, and with the demands of metabolic processes in various parts of the tree (Kozlowski and Keller 1966). Consequently, the concentrations change during the year, depending on the needs of the tree. Horntvedt (1983) found the proportion of soluble carbohydrates in the dry mass of *Picea abies* needles to be 10–17%. The concentration declined during the growing period, reflecting increasing export of reserve food to growth. It then increased sharply following the first autumn frost, which can be interpreted as a starch-to-sugar conversion. A decline occurred during winter until the concentration of soluble sugars started to rise again in the spring.

After a tree has been felled, the content of extractives decreases, and their composition changes because of various reactions. For example, the *resin reactions* involve both oxidation by air and enzymatic hydrolysis. These processes proceed simultaneously. Fats and waxes are mainly hydrolyzed enzymatically. Because of both hydrolysis and oxidation, the hydrophilicity of the resin constituents is increased. These reactions are largely influenced by conditions during storage. Initially, lingering life processes of ray cells are involved through respiration, changing gradually to degradation by microorganisms, as they invade the biomass. In acid sulfite pulping, these changes in resin are beneficial, and wood is usually stored for several months to minimize pitch problems and to lower the resin content of the pulp. In kraft pulping, which is the primary process when residual tree components are used as pulp raw material, storage of wood has detrimental consequences, since yields of both turpentine and tall oil are reduced (Sjöström 1981). As a consequence of microbiological degradation, polysaccharides are attacked during long storage, resulting in reduced yield and quality of pulp. If raw material is stored in the form of chips, the resin reactions are accelerated due to the high temperature in the chip pile (Sect. 8.1.4.2).

3.3.2.1 Extractives in Wood and Bark

Sapwood and inner bark generally are rich in simple monomers and nutrients such as fats, starch, sucrose, simple sugars, inositols, simple glycosides, free and esterified sterols, and phenylpropanoids and other simple phenolics. *Heartwood* and outer bark, by contrast, tend to be deficient in nutrients, glycosides, and metabolic intermediates, but are rich in compounds such as hydrolyzable and condensed tannins and many other phenolics, alkaloids, resins, essential oils, and a wide variety of specialized compounds. Heartwood and outer bark may also contain the various gums, kinos, and balsams that have evolved as part of the wound-response mechanisms, as well as the compounds capable of protecting these metabolically inactive tissues against biological attack (Koch 1985a, Hillis 1987).

In stem wood, the concentration of extractives tends to be highest in the heartwood. Within the heartwood zone, the content of many extractives, such as pinosylvin, is higher at the periphery than in the center. Usually there are only a few percent of extractives in the heartwood from species growing in temperate zones, but the amounts can be much larger in some species from tropical and subtropical regions. In cross-sections of a stem or branch the heartwood can often be detected visually by its darker color, which is caused by the larger amount of extractives present. Together with the increase in the amount of extractives as a group when heartwood is formed, new compounds also appear. As expected from observations of the intensity of color, the amount of extractives is usually higher in hardwoods than in conifers, although the amounts vary from species to species. Heartwood usually commences to form at about breast height, or within 1–3 m above-ground level. The heartwood tapers from the level of first initiation toward both the butt and crown. Consequently, wood from unmerchantable tops contains little heartwood and less extractives than conventional timber. However, the extent of taper in the heartwood zone toward the apex depends on the tree species (Hillis 1987).

No clear correlation seems to exist between formation of compression wood and extractives. Stem compression wood sometimes contains more and sometimes less extractives than corresponding normal wood. On the other hand, *branches* have almost consistently been reported to contain more extractives than stem wood, and knots can have a resin content as high as 30–40%. This high resin content of knots is typical also of conifer species that do not poccess resin canals. Most of this resin is formed by senescent branches in processes that are probably

Table 3.14. Proportion of extractives in dry stem, branch, stump, and root wood of a 25- and 100-year-old *Pinus sylvestris* (Poller 1978)

Tree component	Proportion of extractives (%)				
	Solvent				
	Methanol-benzene	Acetone	Ether	Hot water	Cold water
25-year-old tree					
Small branches	3.8	3.2		8.3	5.6
Large branches	4.1	3.3	2.6	4.5	2.7
Stem:					
3% relative height	2.1	1.8	1.2	2.8	
24% relative height	1.7	1.1	0.8	1.6	0.7
46% relative height	1.8	1.8	0.8	2.8	1.5
69% relative height	2.1	1.8	1.0	2.6	0.9
88% relative height	2.0	1.2	0.6	1.7	1.0
Average	1.9	1.5	0.9	2.3	1.0
Stump-root system:					
Root collar		1.7		2.4	1.6
Stump center	2.0	1.4	1.5	2.1	1.1
Taproot		1.3	1.0	1.4	0.8
Large laterals	1.9	1.6	1.4	2.0	1.0
Average	2.0	1.6	1.3	2.0	1.1
100-year-old tree					
Small branches	4.5	3.5	2.6	4.9	1.7
Large branches	3.1	2.7	1.6	3.7	1.6
Stem:					
1% relative height	3.2	3.0	1.7	3.4	3.0
12% relative height	3.5	3.0	2.0	3.7	2.7
23% relative height	3.3	2.7	1.7	3.7	2.4
34% relative height	3.1	2.1	1.8	3.3	2.0
45% relative height	3.5	3.1	2.1	3.6	1.6
56% relative height	3.1	2.4	1.7	3.1	1.6
68% relative height	4.8	4.2	3.1	3.4	1.7
78% relative height	3.8		2.6		
Average	3.5	2.9	2.1	3.4	2.1
Stump-root system:					
Root collar	6.6	5.9	4.9	3.6	1.5
Stump center	9.7	9.1	8.2	5.2	2.5
Taproot	6.9	6.6	4.7	2.7	0.9
Large laterals	2.8	2.0	1.6	2.2	1.2
Average	6.5	5.9	4.8	3.4	1.5

similar to those causing transformation of sapwood into heartwood (Timell 1986a). The purpose of the process is evidently to protect the stem when the branches die and are pruned naturally.

Since the composition of extractives varies, no solvent is equally applicable to all woods and other components of tree biomass. Table 3.14 represents an example of the yield of extractives when various solvents are applied. In a 25-year-old *Pinus sylvestris* tree, 2% of the dry mass in stem and root wood and 4–8% in

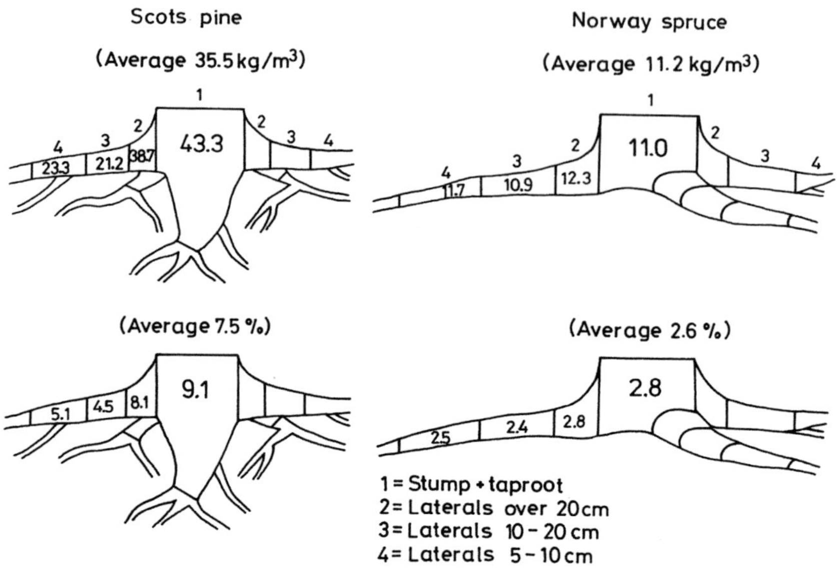

Fig. 3.32. Amount of acetone extractives in dry stump and root wood of large *Pinus sylvestris* and *Picea abies* trees (redrawn from Hakkila 1975)

branch wood was soluble in methanol-benzene and hot water. The amounts of acetone, ether, and cold-water extractives were significantly smaller. Samples from stem, branch, and root wood of a 100-year-old Scots pine showed larger concentrations of extractives, and the order of magnitude among the five extractives was not the same as in the younger tree. In the wood samples from the older tree, the largest concentrations of extractives, cold-water extractives excluded, were found in *stump wood.* Figure 3.32 and the following results by Sjöholm (1977) also show large amounts of extractives in the central stump but significantly less in roots of *Pinus sylvestris.* In *Picea abies*, the content of extractives is significantly lower:

Tree component	Proportion of extractives (%)			
	Pinus sylvestris		*Picea abies*	
	DCM	Acetone	DCM	Acetone
Stem	3.5	3.9	1.2	1.5
Central stump	7.1	7.6	1.5	2.8
Coarse, 5–15 cm roots	2.2	3.0	1.7	2.3
Slender, 1–5 cm roots	2.3	2.8	2.2	2.9

The content of extractives is generally much higher *in bark* than in wood, but information dealing with the chemical composition of bark in various tree components is sparse. The following figures represent average values from 22 hardwood species in the southern United States. The samples were extracted first with alcohol-benzene followed by a final extraction with ethanol. The highest contents of extractives, 21.8 and 20.9% of unextracted dry mass, were found in branch bark of *Fraxinus pennsylvanica* and *F. americana* (Koch 1985a):

	Proportion of extractives (%)	
	Wood	Bark
Stem	4.90	14.41
Branches	5.66	16.65

Lönnberg (1975b) analyzed the content of ethanol, dichloromethane, and hot-water extractives in stem wood and bark of 20- to 25-year-old *Alnus incana*, *Betula pubescens*, *Populus tremula*, and *Salix caprea* in Finland. Yields of extractives from wood samples were 1.3–3.4%. In bark samples, the concentrations of ethanol and hot-water extractives were six- to twelvefold, and those of dichloromethane extractives two- to sevenfold (Table 3.15).

Voipio (Table 3.16) studied the amount of hexane, acetone, ethanol, and hot-water extractives in the above-ground components of small-sized *Alnus glutinosa*, *A. incana*, *Betula pendula*, and *B. pubescens* in Finland. In four successive treatments, the average content of hexane and acetone extractives was highest in the outer bark and that of ethanol extractives in the inner bark, while the leaves were especially rich in hot water-soluble extractives. At the 20% height of the stem, the total content of extractives was 4.8% in the wood, 19.0% in the inner bark, and 26.7% in the outer bark. These proportions increased from stem base to top, and from stem to branch tip. On the average, the proportion of extractives was 6.7% at the 20% stem height, 8.8% at the 80% stem height, 10.4% in branches over 5 mm, and 13.4% in twigs under 5 mm. Leaves contained as much as 31.1% extractives. The same trends were clear in all four tree species. An exception was the outer bark of *Alnus glutinosa*, which contained less extractives than the inner bark of the same species.

Table 3.15. Yield of ethanol, dichloromethane, and hot-water extractives from stem wood and bark of 20- to 25-year-old *Alnus incana*, *Betula pubescens*, *Populus tremula*, and *Salix caprea* (Lönnberg 1975b)

Solvent	Tree component	Proportion of extractives (%)			
		Alnus incana	*Betula pubescens*	*Populus tremula*	*Salix caprea*
Ethanol	Stem wood	2.6	1.4	1.7	2.9
Ethanol	Stem bark	26.8	11.4	20.0	24.6
Dichloromethane	Stem wood	1.3	1.7	2.3	1.5
Dichloromethane	Stem bark	9.4	8.6	15.1	3.0
Hot water	Stem wood	2.6	2.1	2.6	3.4
Hot water	Stem bark	27.9	11.7	32.8	27.1

Table 3.16. The amount of extractives removed from the above-ground components of 10-cm-diameter hardwoods by successive treatments with hexane, acetone, ethanol, and hot water. Average values of *Alnus glutinosa*, *A. incana*, *Betula pendula*, and *B. pubescens* on the dry mass basis (unpublished data from Voipio, FFRI, 1988)

Tree component	Proportion of extractives (%)				
	Solvent				
	Hexane	Acetone	Ethanol	Hot water	Total
Stem at 20% height:					
Wood	0.8	1.6	1.4	0.9	4.8
Inner bark	1.3	5.4	6.7	5.6	19.0
Outer bark	10.6	9.0	3.9	3.2	26.7
All bark	4.0	6.5	5.8	4.9	21.2
All mass	1.2	2.2	1.9	1.4	6.7
Stem at 80% height:					
Wood	1.0	2.1	1.6	1.7	6.4
Inner bark	2.0	5.8	7.4	6.1	21.3
Outer bark	8.9	11.4	6.0	3.9	30.2
All bark	4.3	7.7	6.9	5.4	24.3
All mass	1.5	2.8	2.3	2.2	8.8
Branches, over 5 mm:					
Wood	1.0	2.4	1.8	1.4	6.6
Inner bark	1.3	5.7	9.5	6.4	22.9
Outer bark	6.5	7.4	6.0	4.8	24.7
All bark	3.2	6.4	8.2	5.8	23.6
All mass	1.5	3.3	3.2	2.4	10.4
Twigs, under 5 mm:					
Wood	1.1	2.9	2.5	2.4	8.9
All bark	3.1	5.6	6.5	6.4	21.6
All mass	1.8	3.9	3.9	3.8	13.4
Leaves	5.3	3.9	11.7	10.2	31.1

Betulinol is a naturally occurring triterpene compound typically forming the principal extractive substance of the bark in *Betula* spp. The white color of the outer bark of birch originates from betulinol. The content of betulinol is 10–30% in the outer bark but less than 2% in the inner bark of birch, while in the wood only traces can be detected (Jääskeläinen 1981).

Another characteristic compound in birch bark is suberin, although it is not considered an extractive. Suberin is a hydroxy acid complex found in the cork cells of bark consisting of esters of higher aliphatic hydroxy and phenolic acids. Birch bark may contain 20–39% and cork oak bark 35–40% suberin (Goldstein 1981), which occurs in other barks in much smaller amounts. Harun and Labosky (1985) found the suberin content of outer bark to be 5.5% in *Pinus resinosa*, 3.5% in *P. strobus*, 3.1% in *Acer rubrum*, 7.3% in *Carya ovata*, and 4.6% in *Quercus rubra*.

3.3.2.2 Extractives in Foliage

The largest proportions of extractives usually occur in foliage. In a sample of three 22-year-old *Pinus elliottii* trees the proportion of extractives was 9–14% in all other biomass components but as much as 26.2% in the needles (Koch 1972a). Even greater differences were found by Voipio (1988 unpubl.) among biomass components of four Finnish hardwood species (Table 3.16), and by Hosia et al. (1971) among biomass components of *Pinus sylvestris* and *Picea abies* (Table 3.17).

Results of comprehensive research on foliage utilization in the Soviet Union show that tree foliage contains essentially all the classes of organic compounds found in the kingdom of plants: proteins including irreplaceable amino acids, carbohydrates, lipids, waxes, sterols, steroids, essential oils, pigments, plastids, alcohols, fatty acids, phenolic compounds, fat-soluble vitamins, water-soluble vitamins, alkaloids, glycosides, organic acids and their salts, mineral substances, etc. The chemical composition of tree foliage differs greatly depending on species, season, geographic area, growth conditions, and other factors. Of this wide variety of compounds, vitamins, essential oils, and proteins found in common tree species such as *Pinus sylvestris*, *Picea abies*, *Larix sibirica*, and *Populus tremula* have been studied thoroughly, whereas the chemical composition of lipids, alkaloids, flavanoids, and phenolic compounds, and that of hardwood leaves in general, has been subject to less investigation. The amounts of separate substances such as essential oils, vitamins, and proteins are large enough to justify their commercial extraction (Ievin and Daugavietis 1986).

The volatile components of tree biomass consist chiefly of substances that are collectively called *volatile oils* or *essential oils*. The term "essential" is derived from a traditional belief that the fragrance of a substance contains its essence, or vitally important ingredient. They may be present in significant amounts in the stem wood of conifers but are usually negligible in hardwoods, occurring in larger amounts in roots, barks, and foliage. The exudates or oleoresins of many conifers are often rich in volatile materials. The components volatile with steam comprise a large variety of chemical compounds, including cyclic hydrocarbons (terpenes

Table 3.17. Proportion of dichloromethane, cold-water, and hot-water extractives in the dry mass of various tree components of *Pinus sylvestris* and *Picea abies* (Hosia et al. 1971)

Solvent	Tree species	Proportion of extractives (%)			
		Stem wood	Stem top wood	Branch wood	Needles
Dichloromethane	*Pinus sylvestris*	5.7	8.9	7.1	15.1
Cold water	*Pinus sylvestris*	2.2	2.0	3.5	19.0
Hot water	*Pinus sylvestris*	5.8	7.2	5.3	28.2
Dichloromethane	*Picea abies*	(2.1)	1.3	1.3	15.1
Cold water	*Picea abies*	1.1	2.1	3.8	19.0
Hot water	*Picea abies*	3.7	3.8	6.6	36.4

and terpenoids), alcohols, ethers, aldehydes, ketones, acids, lactones, aliphatic hydrocarbons, and phenols. They are usually liquids, although some components are solids at ordinary temperature. Their specific gravity varies from about 0.65 to 1.05 depending on the nature of the major constituents. Most of the components of the oils exhibit optical activity and rotate the plane of polarized light (Browning 1967a). Essential oils obtained by distilling were among the first foliage chemicals utilized. The main component of such essential oils consists of monoterpenoids which are recovered for pharmaceutical purposes. Since the essential oils can be separated by simple steam distillation methods, they are rather easily accessible and, therefore, of considerable interest in the medical and pharmaceutical industries. Products such as turpentine are of considerable commercial importance.

The essential oils usually have a fragrant odor, and some of them are highly prized as perfume ingredients. They belong to the terpene group of chemicals, which may be further divided into mono-, sesqui-, di-, and triterpenes. The most abundant of these compounds are the monoterpenes, which are subdivided into three groups: acyclic, monocyclic, and bicyclic. The yield varies depending on tree species, season, storage, etc. For example, the yield of essential oils from technical foliage of *Pinus sylvestris* in Finland was 1.0–1.2% of dry mass, containing 48% monoterpenes, 46% sesquiterpenes, 4% diterpenes, and 2% aliphatic acids. The oil constituted a valuable, marketable product. However, improper storage of raw material lowered its quality by increasing the amount of contaminants such as acidic and phenolic compounds (Hannus and Pensar 1973).

One of the most abundant terpenes is α-pinene, which is not only a major component of conifer foliage, but also of wood turpentine. Both α-pinene and its isomer, β-pinene, are characterized by easy conversion to other terpene structures by simple chemical techniques. This can be attractive commercially, since the market value of the converted derivative is likely to be many times that of the original precursor. α-Pinene can be treated with hydrochloric acid to give a mixture of products which are called collectively pine oil. This relatively inexpensive oil is used as a disinfectant, deodorant, and wetting agent. Other important uses for the terpenes are in flavoring agents, perfumes, and in medicinal preparations (Keays and Barton 1975).

Another group of interesting biochemicals in forest trees is formed by the pigment precursors of green foliage, which are essential to the photosynthetic process. The best known of these are the chemically similar *chlorophylls*, a and b, both contained in foliage in lipoprotein bodies called chloroplasts and characterized by the presence of magnesium. Chlorophyll a, the blue-green pigment, is three times as plentiful as chlorophyll b, the yellow-green pigment. They cannot be distilled by steam and they are not water-soluble, but they can be readily extracted from foliage tissue with alcohol, acetone, ether, benzene, or chloroform. Invariably associated with the chlorophylls in foliage are the yellow pigments: the *carotenes* and the *xanthophylls*. Collectively, together with certain other closely related compounds, these yellow-red pigments are called the *carotenoids*. They are not restricted in their occurrence to the chloroplasts but may also be present in the nongreen parts of the foliage. Carotenes are of considerable importance because they are the precursors of vitamin A, one molecule of β-carotene being split into two molecules of vitamin A by a simple hydrolytic reaction. Xanthophylls are important in animal feed supplements. For example, they increase the yellow color in eggs and butter. Other chlorophyll-related derivatives are pheophytin, chlorophyllin, phytol, etc. (Keays and Barton 1975). According to Dickson and Larson (1977), total chlorophyll constitutes about 0.5–1.0 and the carotenoids 0.05–0.10% of foliage dry mass in hardwoods.

In the Soviet Union, about 300 t of chlorophyll-carotene paste is produced annually from needles on extraction with organic solvents. One tonne of fresh needles yields 50–55 kg paste with 40% moisture content. The paste is applied in medicine for curing ulcer, skin, burns, frost damage, etc., in cattle feeding as a vitamin supplement, and as a bioactive component in cosmetic products such as skin lotions, creams, soaps, shampoos, deodorants, perfumes, toothpastes, etc. The paste can be further processed to products such as sodium chlorophyll, conifer wax, etc. (Ievin et al. 1981).

The vitamins in tree biomass are of interest in the medical and pharmaceutical industries and as fodder. They are the subject of much research in the Soviet Union. The vitamin content is greatest in foliage, including buds, but considerably smaller in bark and wood. For example, Levin and Repjah (1984) found the content of carotene in *Pinus sylvestris* and *Picea abies* in the winter to be 228 parts per million (ppm) in needles, 13 ppm in bark, and only 1 ppm in wood.

According to numerous studies, the content of *provitamin A or carotene* in tree foliage varies between 10–1000 ppm. Provitamin A is composed mainly of α- and β-carotenes, the proportion of β-carotene being 50–70% in softwood needles, and in hardwood leaves 75–95% of the total carotene supply (Voipio 1987). The amount of carotene varies with the season depending on temprature, humidity, and light climate. In softwood needles it is greatest during the winter (Levin and Repjah 1984; Fig. 3.33), and in hardwood leaves during the early sum-

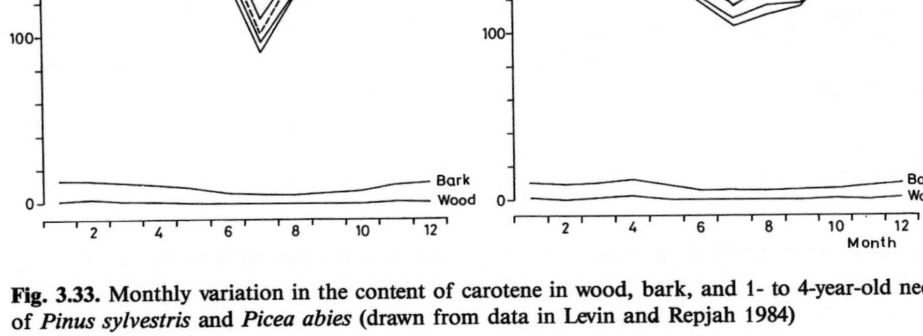

Fig. 3.33. Monthly variation in the content of carotene in wood, bark, and 1- to 4-year-old needles of *Pinus sylvestris* and *Picea abies* (drawn from data in Levin and Repjah 1984)

mer (Schertz 1929). Pine needles from trees growing in swamps were found to be richer in carotene than those from trees growing on dry mineral sites (Sirjaeva 1967). In *Abies sibirica* and *Picea abies* needles, the carotene content increased with altitude, and significant differences were found among needles of different age classes, young needles being low in carotene. For example, in a sample from *Picea abies* at a 900 m altitude, 1-, 2-, 3-, and 4-year-old needles contained 26, 65, 162, and 145 ppm carotene, respectively (Tomčuk and Ryzhilo 1968). The amount of carotene decreases radically during storage. In *Abies sibirica* the needles had an initial carotene content of 127 ppm, but the proportion decreased during 1-year inside-storage to 47 ppm. After the second year the carotene content was 17 ppm, and after the third year only 8 ppm (Nikiforov et al. 1973).

The most abundant of all the vitamins in tree biomass is *vitamin C or ascorbic acid.* In fact, the content of ascorbic acid is significantly higher in tree foliage than in many berries and fruits. In accordance with the carotene content, it is highest in softwood needles in the winter (Stancović and Senić 1954; Sabalitschka and Kleffner 1946; Fig. 3.34) and in hardwood leaves during the early summer (Wokes and Melville 1948). The concentration is greatest in the upper crown and in trees growing on fertile sites (Novikov and Gunjačenko 1964). During storage, reduction in the content of ascorbic acid proceeds faster in fresh than in dried foliage. On the other hand, some loss occurs in the drying process (Kaludin 1964). In an extensive study by Jones and Hughes (1984), the mean ascorbic acid concentration in foliage in the summer was 2928±270 ppm in 41 woody species compared to 1579±790 ppm in 192 nonwoody angiosperms. Some examples of the results are listed below:.

Tree species	Concentration of ascorbic acid in foliage (ppm)
Conifers:	
Cedrus libani	1437±275
Chamaecyparis lawsoniana	1428±283
Larix decidua	3045±463
Picea pungens	4047±254
Pinus sylvestris	2077±231
Hardwoods:	
Alnus glutinosa	2610±474
Betula pendula	4150±251
Corylus avellana	3178±321
Fagus sylvatica	2600±121
Fraxinus excelsior	1879±124
Juglans regia	9250±841
Populus nigra	3265±253
Quercus robur	2728±275
Salix caprea	3298±224
Sorbus aucuparia	2050±102

Vitamin E or tocopherol occurs in needles only in the form of α-tocopherol. During a 2-month storage experiment with pine needles, the content of tocopherol was reduced from about 300 to 210 ppm when stored dried, and to 160 ppm when

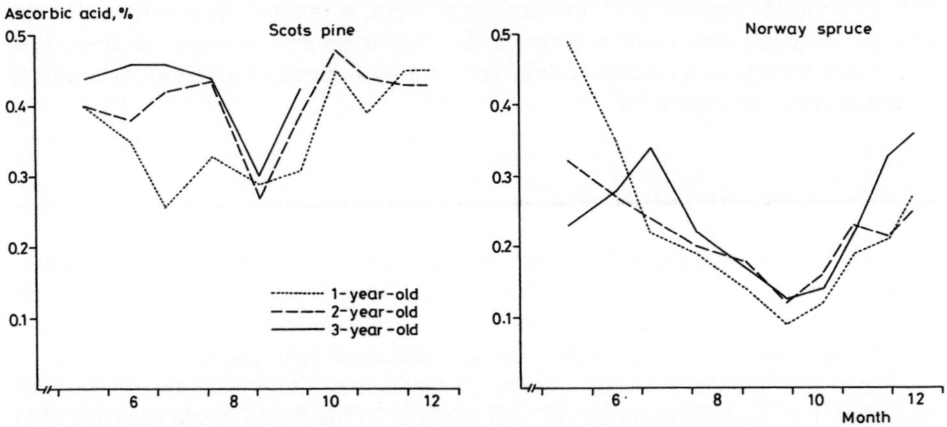

Fig. 3.34. Monthly variation in the content of ascorbic acid in dry needle mass of *Pinus sylvestris* and *Picea abies* in Germany (drawn from data in Sabalitschka and Kleffner 1946)

stored fresh (Kaludin 1964). As shown below, the contents are considerably greater than those of carotene (Kaludin and Kaludin 1967):

Tree species	Concentration in foliage (ppm)	
	α-Tocopherol	β-Carotene
Abies alba	455	91
Juniperus communis	395	79
Picea abies	342	105
Pinus maritima	323	70
P. montana	305	62
P. nigra	304	84
P. peuce	372	90
P. sylvestris	315	75

Other vitamins occur in tree foliage in smaller amounts. For example, Levin and Repjah (1984) found the concentration of vitamin K_1 in *Pinus sylvestris* needles to be 20 ppm and in *Picea abies* needles 12 ppm. Karepova et al. (1978) found that the concentration of compounds of the vitamin B group in *Pinus sylvestris* needles was 82 ppm and in *Picea abies* needles 89 ppm. Their distribution between individual vitamins was as follows:

Compound	Concentration in foliage (ppm)	
	Pinus sylvestris	*Picea abies*
Inositol	20.2	21.2
Thiamine (B_1)	10.5	19.4
Niacin (B_3)	17.3	19.8
Pantothenic acid (B_5)	19.5	16.8
Pyridoxine (B_6)	14.2	12.2
Biotin	0.0008	0.0008

Foliage contains considerable amounts of *protein*, which is only partly soluble in organic solvents. This protein forms a significant reserve of livestock feed. The use of tree foliage as a protein fodder and as a biologically active fodder additive is discussed in Section 8.5.2.

3.3.3 Inorganic Materials in Residual Tree Components

A tree takes minerals from the forest soil through its root system and transports them to the stem and crown by sap flow. In addition, foliage may take up minerals from rain-water and the atmosphere. The greatest concentration of minerals occurs in the parts of the tree where the life processes take place.

When the concentration of a chemical element in the liquid outside the root is greater than the concentration of that element in the liquid inside the root, that element will flow through semipermeable membranes into the cells of the root tips and root hairs. It is because of this basic physical fact that the amount of chemical elements in a tree biomass reflects not only the requirements of the tree but also the amount in the ground. Furthermore, the tree will also contain elements that it obviously does not need, such as lead, arsenic, etc. In a sense the tree is a sewer in that it has a lot of extraneous material in it. For example, there is a lot of magnesium in the soil in Maine and New Brunswick so that the spruce and fir needles may have 1500 ppm Mg, but in Pennsylvania where the amount of magnesium in the soil is normal, its concentration may be only 300 ppm in spruce and fir needles, respectively (Young 1987, pers. commun.).

As biomass is burned, the inorganic components form ash. This ash does not contain the nitrogen present in biomass because it is volatile and escapes in the burning process. The ash content of tree components varies according to site fertility, species, tree size and age, and season. Young trees tend to have a higher concentration of ash than mature trees. Hardwoods have a higher ash content than softwoods, and tropical hardwoods generally contain more ash than temperate hardwoods. However, the variation between tree species is large as shown by the following table. Each figure represents the average ash content of stem wood and the standard deviation between species for a large number of softwoods and hardwoods in several countries (Pettersen 1984):

Country	Average ash content in stem wood (%)	
	Softwoods	Hardwoods
USA	0.3 ± 0.1	0.5 ± 0.3
USSR	0.5 ± 0.4	0.6 ± 0.4
Japan	0.4 ± 0.4	0.5 ± 0.2
Taiwan		0.9 ± 0.4
The Philippines		1.2 ± 0.7
Mozambique		1.6 ± 1.1

In the stem wood of *Pinus taeda* in Louisiana the average ash content was 0.41%. Earlywood contained more ash than latewood, and the inner growth rings more than the outer ones. Ash content did not correlate with the rate of growth and basic density of wood (McMillin 1969). In a study by Timell (1986a), the ash content of all specimens from normal and compression stem woods from 27 conifers was 0.4±0.1%. In a comprehensive study in the southern United States, the average ash content in stem wood of 15-cm-diameter trees of 22 hardwood species was 0.75%. The percentage was highest, 1.33, in *Celtis* spp., whereas the lowest ash content, 0.43%, was found in *Liriodendron tulipifera* and *Quercus rubra* (Koch 1985a).

When biomass is used for the production of energy, all tree parts may be involved. Compared to stem wood, other tree components contain a much higher proportion of ash. In Koch's (1985a) study in the southern United States, branch wood averaged 0.94% ash and wood from the central stump-root system 1.23%. The largest ash concentration in stump and root wood, 2.54%, was found in *Quercus stellata*. The ash content of bark was very high in all tree components. The largest percentage, 16.42, was observed in stump and root bark of *Celtis* sp. Compared with respective wood components, the proportion of ash was tenfold in stem bark, sevenfold in branch bark, and eightfold in stump and root bark:

	Content of ash (%)	
	Wood	Bark
Stem	0.75	7.87
Branches	0.94	6.76
Central stump-root system	1.23	9.93

Under northern conditions, tree biomass appears to contain less ash. In a Finnish study, five tree species averaged 0.46% ash in stem wood, 2.97% in stem bark, 1.52% in unbarked branch mass, and 4.97% in foliage. With foliage included, the average ash content of whole-tree mass varied in the five tree species between 0.9−1.6%. By removing all foliage, for example, with flail treatment before chipping whole trees or with a sorter after chipping, the ash content could be decreased to 0.8−1.3% (Hakkila and Kalaja 1983). If, however, foliage is removed biologically by transpiration drying, a portion of its inorganic constituents is translocated into the branches and stem before abscission of the foliage. Decrease in the ash content of biomass is then less than if the foliage is removed fresh. For example, Wells and Metz (1963) found in dead needles of *Pinus taeda* only half as much nitrogen, phosphorus, and potassium but the same amount of calcium as in live needles.

Depending on tree species and degree of delimbing, a solid m^3 of small-tree fuel wood in Finland contains 3−5 kg pure ash. In practice, the total amount of ash residue from fuel wood is higher, since biomass collects sand and other inorganic contaminants during logging and transport (Sect. 9.4.2.1).

Table 3.18 summarizes results of mineral contents in tree components from three studies. The data for Maine softwoods represent averages for five species

Table 3.18. Average data on mineral concentration in dry biomass of softwoods and hardwoods in Maine (Young and Carpenter 1967), in Finland (Hakkila and Kalaja 1983), and in the southern United States (Koch 1985a)

Tree component	Concentration in biomass (%)					Concentration in biomass (ppm)								
	Primary elements					Trace elements								
	N	P	K	Ca	Mg	Mn	Fe	Zn	S	B	Cu	Mo	Na	Al
5 softwood species in Maine														
Stem	0.15	0.02	0.06	0.33	0.03	157	46	30			3	2		36
Branches	0.31	0.07	0.14	0.53	0.05	256	115	51			4	4		96
Foliage	0.93	0.15	0.42	0.69	0.10	660	121	46			3	5		132
Roots	0.21	0.05	0.10	0.18	0.14	138	176	31			3	2		131
2 softwood species in Finland														
Stem wood		0.01	0.06	0.12	0.02	147	41	13	116	3	2			
Stem bark		0.08	0.29	0.85	0.08	507	60	75	343	12	4			
Branches		0.04	0.18	0.34	0.05	261	101	44	203	7	4			
Foliage		0.16	0.60	0.50	0.09	748	94	75	673	9	6			
Whole tree		0.03	0.15	0.28	0.05	296	85	30	236	6	4			
3 hardwood species in Maine														
Stem	0.25	0.03	0.13	0.49	0.03	115	41	63			3	3		24
Branches	0.43	0.05	0.17	0.58	0.06	204	46	71			4	4		28
Foliage	1.35	0.19	0.75	0.78	0.19	725	148	77			5	6		72
Roots	0.29	0.09	0.21	0.48	0.05	94	252	61			3	3		71
3 hardwood species in Finland														
Stem wood		0.02	0.08	0.08	0.02	34	20	16	90	2	2			
Stem bark		0.09	0.37	0.85	0.07	190	191	131	341	17	13			
Branches		0.06	0.21	0.41	0.05	120	47	52	218	7	4			
Foliage		0.21	1.17	1.10	0.19	867	135	269	965	21	10			
Whole tree		0.05	0.21	0.25	0.04	83	27	39	212	6	5			
22 hardwood species in the southern US														
Stem wood		0.02	0.16	0.19	0.04	89	67	11			5		201	47
Stem bark		0.03	0.20	3.05	0.11	568	135	35			13		449	175
Branch wood		0.04	0.17	0.25	0.06	120	65	12			7		199	50
Branch bark		0.05	0.25	2.64	0.12	596	121	42			12		433	146
Stump wood		0.03	0.18	0.25	0.06	94	473	7			16		58	28
Stump bark		0.04	0.35	3.00	0.13	413	487	18			12		104	711

(*Abies balsamea*, *Picea rubens*, *Pinus strobus*, *Thuja occidentalis*, and *Tsuga canadensis*) and that for Maine hardwoods averages for three species (*Acer rubrum*, *Betula papyrifera*, and *Populus tremuloides*) from a study by Young and Carpenter (1967). The data on hardwoods in the southern United States represent an average of the following 22 species: *Acer rubrum*, *Carya* spp., *Celtis* spp., *Fraxinus americana*, *F. pennsylvanica*, *Liriodendron tulipifera*, *Liquidambar styraciflua*, *Magnolia virginiana*, *Nyssa sylvatica*, *Quercus alba*, *Q. coccinea*, *Q. falcata* var *falcata*, *Q. falcata* var. *pagodaefolia*, *Q. laurifolia*, *Q. marilandica*, *Q. nigra*, *Q. prinus*, *Q. rubra*, *Q. stellata*, *Q. velutina*, *Ulmus alata*, and *U. americana* (Koch 1985a). The study of Finnish trees represents two softwoods (*Picea abies* and *Pinus sylvestris*) and three hardwoods (*Alnus incana*, *Betula* spp., and *Populus tremula*) (Hakkila and Kalaja 1983). Generally, hardwoods contain somewhat more inorganic elements than softwoods but variation among species is large. In the southern United States hardwoods appear to contain more inorganic elements than hardwoods in Maine and Finland. Branch and root woods are richer in inorganic elements than stem wood. The concentrations are five- to tenfold in bark compared with wood, and even greater in foliage. According to Choong et al. (1976), the concentration of primary elements is in the inner bark higher than in the outer bark. As age of poplar hybrids in Pennsylvania increased from 1 to 8 years, ash content decreased in the wood but increased in the bark (Blankenhorn et al. 1985).

When leaf buds break and new foliage begins to grow, the leaf tissues often have high concentrations of N, P, and K. As the foliage matures and accumulates carbohydrates the initial concentrations of N and P are diluted. On the other hand, concentrations of Ca, Mg, and Fe usually increase with leaf age. The increase in calcium concentration results from calcium pectate deposition in cell walls and storage of calcium oxalate in cell vacuoles. After leaf maturation, the contents of N, P, and K remain relatively constant during the growing season until a strong reduction takes place in the autumn. Such losses represent active withdrawal of nutrients from foliage for reuse during the next year. Leaf nutrient contents are also affected by rainfall that leaches nutrients, particularly potassium, from the leaf surface. Leaching rates often increase as foliage undergoes senescence before abscission. The losses of nutrients in leaching follow the order K, P, N, and Ca (Waring and Schlesinger 1985).

Site fertility, possible seasonal fluctuation, differences in sampling and laboratory procedures, as well as small sample sizes result in random variation of ash content in biomass, making comparisons difficult. This problem is particularly true of trace elements, which occur in tree biomass at such a low level that their concentration is usually measured in parts per million (ppm) of dry matter. These difficulties notwithstanding, many similarities are found in the occurrence of mineral elements in various biomass components in the three studies of Table 3.18. The following general trends seem to prevail:

1. *Phosphorus* content in stem wood was of the order of 0.02%. It was two- to fourfold in unbarked branches and seven- to tenfold in foliage.
2. *Potassium* content in stem wood was of the order of 0.05–0.15% in Maine and Finland but 0.10–0.25% in the southern United States. Branches contained 0.15–0.20% and foliage 0.4–1.2% potassium.
3. *Calcium* content was 0.1–0.5% in stem wood, 0.5–0.9% in stem bark in Maine and Finland, and as high as 2–4% in stem bark in the southern United States. In foliage the concentration of calcium was in Maine and Finland 0.5–1.2%.
4. *Magnesium* content was 0.02–0.04% in stem wood, 0.05–0.15% in bark, about 0.1% in softwood foliage, and about 0.2% in hardwood foliage.
5. *Manganese* content was in stem wood 30–200 ppm, in bark 150–1000 ppm, and in foliage 500–1000 ppm. Variation was irregular.
6. *Iron* content in stem wood was 20–100 ppm. The highest concentrations were found in root wood and root bark, generally 150–500 ppm but in several southern hardwoods more than 1000 ppm.

7. *Zinc* content was 10–80 ppm in stem wood and significantly higher in other tree components, but the variation was irregular.
8. *Sulfur* content was 80–120 ppm in stem wood, 300–400 ppm in bark, and 500–1000 ppm in foliage. Measurements were made in Finland only.
9. *Aluminum* content was 20–60 ppm in stem wood and considerably higher in roots.
10. Contents of *boron*, *copper*, and *molybdenum* were in all tree components generally less than 20 ppm.

Tree components with excessive amounts of inorganic elements create technical problems in the forest industries. Consequently, the mineral constituents are a nuisance when biomass is pulped or burned. On the other hand, the primary and trace elements present in tree biomass are necessary for the growth of the trees, and therefore their recycling is of importance from a soil conservation and fertilization point of view (Sect. 9.4.2.3).

When wood is burned, the mineral elements remain in the ash residue in the form of oxides. The prevailing element in wood ash is calcium, which together with magnesium makes ash an effective neutralizing agent. Potassium and phosphorus are common fertilizers, and many trace elements are necessary for trees as well. Only elements such as natrium, silicon, aluminum, and some heavy metals are believed to be of no fertilizer value or even harmful. Table 3.19 shows the concentration of ten elements in pure ash from various components of newly felled, small-sized trees in Finland. Compared with ordinary power plant ash (Sect. 9.4.2.2), the concentrations in laboratory samples from fresh biomass are much higher because elements leach from wood, bark, and foliage during storage.

Table 3.19. Concentration of selected elements in pure ash from fresh biomass components of 10-cm-diameter softwoods (average for *Pinus sylvestris* and *Picea abies*) and hardwoods (average for *Alnus incana*, *Betula* spp., and *Populus tremula*) in Finland (Hakkila 1986)

Tree component	Concentration in pure ash (%)									
	Element									
	P	K	Ca	Mg	Mn	Fe	Zn	S	B	Cu
Ash from softwoods:										
Stem wood	2.4	12.4	22.4	4.3	2.9	0.8	0.3	2.3	0.05	0.04
Stem bark	2.8	9.8	28.5	2.8	1.7	0.2	0.3	1.2	0.04	0.02
Branches	3.0	12.2	23.6	3.5	1.8	0.7	0.3	1.4	0.05	0.03
Foliage	4.4	16.2	13.5	2.5	2.0	0.3	0.2	2.8	0.03	0.02
Whole tree	2.7	12.0	22.8	3.8	2.5	0.7	0.3	2.0	0.05	0.04
Ash from hardwoods:										
Stem wood	4.2	20.4	19.0	3.6	0.8	0.5	0.4	2.1	0.05	0.04
Stem bark	3.4	12.2	27.1	2.2	0.6	0.6	0.4	1.1	0.06	0.04
Branches	3.6	13.7	26.5	3.0	0.8	0.3	0.3	1.4	0.05	0.03
Foliage	3.7	20.2	19.1	3.3	1.5	0.2	0.5	1.7	0.04	0.02
Whole tree	3.9	18.0	21.4	3.3	0.7	0.5	0.3	1.8	0.05	0.04

Small amounts of some elements are also lost with combustion gases. Moreover, power plant ash may include a high proportion of sand and charcoal.

Ash from all softwood and hardwood components has a very high concentration of calcium. In the Finnish study under consideration, the calcium content of whole-tree ash was about 22% and that of bark ash 28%. Najdenow et al. (1982) reported even greater calcium contents for ash from bark-free stem wood of beech, birch, and oak in the Soviet Union. Another important mineral element in tree biomass is potassium, which appears in bark ash in lower concentrations than in wood and foliage ash. Pure whole-tree ash from fresh softwoods contained 12% and that from fresh hardwoods 18% potassium. Softwood ash had a lower concentration of phosphorus but a higher concentration of manganese. Hardwood ash thus seems to have a higher fertilization value.

3.4 Basic Density of Wood and Bark in Residual Tree Components

Density is probably the most commonly used quality indicator for wood raw material. It reflects many important use properties of the biomass, and it also offers a practical conversion factor for comparison of volume and mass measurements. *Density* denotes the mass per unit volume of a substance and is expressed in the metric system usually in kg/m^3 solid. If not otherwise mentioned, density values in this review refer to the oven dry mass per green volume, i.e., the *basic density* of a biomass component. Volume is thus measured in its natural, unshrunken condition when the moisture content is above the fiber saturation point.

The basic density must be distinguished from *the mass density*, which is calculated on the basis of both the mass and volume of the biomass measured at the same moisture content as received. The mass density per unit volume of biomass is calculated from the following formula:

$$\text{Mass density of biomass} = \frac{\text{Mass of biomass with moisture}}{\text{Volume of biomass with moisture}} .$$

Mass density of biomass, contrary to basic density, thus bears a direct relationship to the moisture content of the substance. The minimum values for mass density occur under oven-dry conditions, and the maximum when the wood is fully saturated. To predict the mass density of biomass at a moisture content below the fiber saturation point, it is necessary to know the volumetric shrinkage between green volume and the volume at the moisture content under consideration. The relationship between the mass density and moisture content is illustrated by the example below, where the basic density of biomass is 400 kg/m^3 and the total shrinkage 12% from the fiber saturation point at 23% moisture content (fresh mass basis) to the oven-dry condition:

$$\text{Basic density} = 400\ \text{kg/m}^3$$

$$\text{Mass density at 50\% M.C.} = 400 \times \frac{100}{100-50} = 800\ \text{kg/m}^3$$

$$\text{Mass density at 23\% M.C.} = 400 \times \frac{100}{100-23} = 519\ \text{kg/m}^3$$

$$\text{Mass density at 0\% M.C.} = \frac{400}{0.88} \times \frac{100}{100-0} = 455\ \text{kg/m}^3 \ .$$

When the basic density of a biomass component is divided by the water density at a temperature of +4 °C, the *specific gravity* of biomass is obtained. This is a unitless quantity and consequently independent of the measuring system. For example, if the basic density of biomass is 400 kg/m^3, then the specific gravity is 0.400.

The basic density of biomass is a result of three factors: the volume of cell walls in relation to the empty spaces occupied by cell lumina and intercellular spaces, the density of cell wall substance, and the amount of extraparietal substances present in the biomass. Consequently, density is affected by the proportions of tissue types, length and diameter of cells, thickness of cell walls, percentage of latewood, proportional distribution of the primary cell wall constituents, amount and type of extractives, and the amount of inorganic components.

The cell wall substance, as distinct from the cell wall which may contain voids, has been determined to have a density of about 1500 kg/m^3 in the oven-dry state (Kellogg and Wangaard 1969). Little variation exists between different kinds of woods, as long as the same experimental procedures are employed. This value agrees well with calculated densities for the entire cell wall determined as the sum of weighted percentages of the densities of the individual cell wall components (Sect. 3.4.1). However, there is little agreement with the computed densities for each cell wall layer determined separately, because the amounts of the components present in each layer are still incompletely known (Panshin and de Zeeuw 1980).

Basic density reflects the *mechanical properties of wood.* Tensile, compression, bending, and shearing strength and hardness are positively correlated with wood density. For example, wood hardness is directly proportional to density to the power of 2.25 (Markwardt and Wilson 1935). Because the use potential of residual biomass is restricted mainly to products such as pulp, chemicals, fodder, and fuels, the strength properties are not of great, immediate importance. However, the power requirement in chipping increases with increasing wood density, and hard wood contributes to rapid dulling of chipper knives.

Wood density affects the liquid absorption, swelling, and shrinkage of biomass. Low-density wood absorbs water and chemicals more readily than dense wood. Therefore, mixed cooking of woods of various densities causes problems in chemical pulping and tends to result in yield loss and excessive amounts of rejects.

Wood consumption in pulping, when measured by volume, depends on the basic density of the wood. Since latewood contains somewhat less lignin and more cellulose than earlywood, and since latewood cellulose is degraded less during pulping than earlywood cellulose, pulp yields increase with wood density not only on a volume but also on a mass basis (Thorbjörnson 1922; Gladstone et al. 1970).

Dense wood is composed of thick-walled cells. Wood density gives an indication of the cell-wall thickness compared with the lumen diameter, and it may be used to predict the *quality of paper* obtained from wood. Since extractives are a disturbing factor in this relationship, the density of extracted wood gives a better idea of the paper properties (Barefoot et al. 1964). As stated in Section 3.2, thick-walled cells, due to their stiffness, retain their original rounded cross-section form in paper making, whereas thin-walled cells collapse and become ribbonlike in form, which facilitates the bonding of the cells and formation of dense, opaque sheets (Runkel 1942). Tear strength and many other paper properties are affected more by variations in wood density than in cell length (Dinwoodie 1965, 1966; Barefoot et al. 1972).

Such behavior of the cells decisively affects the paper properties. The most suitable fibers for a given occasion depend on the type of paper and its intended use. In fact, one cannot speak of a pulp that is universally good for all paper qualities. However, according to Dadswell et al. (1959), it is desirable that the thickness of the cell wall should be less than half the lumen diameter.

If great tensile or burst strength is required of paper, thin-walled cells and a low wood density are advantageous. For great tear strength, which is an important property especially in kraft paper, thick-walled cells and dense wood are desirable (Hiett et al. 1960; Mitchell 1961; Farrington 1980; Kibblewhite 1980). The effect of wood density on tear strength is, however, considerably less and not the same for spruce sulfite and pine kraft pulps (Stockman 1962). If the smoothness and closeness of a sheet of paper are of primary consideration, it helps if a relatively high proportion of the cells are thin-walled (Curran 1938). In the manufacture of many types of papers, including newsprint and thin translucent papers with high resistance to fold and burst and with good tensile strength, low-density wood is often preferred (Edlin 1965). In the viscose process low-latewood pulps have been found to show better filterability of the viscose (Hedin et al. 1963). Some types of paper made of groundwood pulp need such thin-walled fibers that low-density wood may be desirable despite the smaller yield (Besley 1960, 1962). McMillin (1968) reported increasing strength in groundwood pulp from *Pinus taeda* with an increasing percentage of latewood, but decreasing wood density, on the other hand, was also associated with an increase in pulp strength. Proliferation of the number of thick-walled cells also lengthens the beating time necessary in pulp production, and makes sheet formation more difficult. Simultaneously, the paper surface becomes rougher and the fold endurance weakens (Dadswell and Wardrop 1959). Because of their pale color, thin-walled earlywood cells require less bleaching to reach a desirable degree of whiteness (Edlin 1965).

Both the fiber yield and the quality of paper thus depend on wood density. High-density wood improves the quality of some types, while low-density wood is good for other types, but this does not detract from the usefulness of density as an indicator of the quality of pulp raw material. It is important to know, on the one hand, what kind of wood is suitable for a certain use and, on the other hand, where wood of a certain kind is to be obtained.

The *quality of particle board* also depends on wood density. The suitability of different tree species and biomass components can be compared on the basis of wood density. For a constant density of particle board, its tensile strength perpendicular to the surface and bending strength decrease, and the moisture absorption, linear swelling, and thickness swelling increase when the wood density increases (Liiri 1964). According to Mitchell (1961), a wood density of 300 kg/m^3 is ideal for the strength and stiffness of the board, its dimensional stability, and weight. However, in manufacturing particle board of a fixed weight, the consumption of raw material is greater, the lower the density of the wood employed.

Wood density also indicates the *quality of a biomass fuel*, for the heating value is directly proportional to the wood density. The energy content of a unit volume of wood, bark, or foliage depends principally on its dry mass and moisture content and, to some extent, on the cell wall composition and content of extraneous components.

3.4.1 Basic Density of Wood and Bark in Branches

Branches possess a high content of reaction wood which is characterized by thick cell walls and narrow lumina. Since the proportion of cell wall substance in the total volume of wood is unusually large in compression-wood tracheids, compression wood is denser than normal wood. In his comprehensive review of compression wood in gymnosperms Timell (1986a) pointed out, however, that the *cell wall substance* itself is actually slightly lighter in compression wood than in normal wood, obviously because compression wood contains less cellulose and more lignin than does normal wood, and probably also because cellulose in compression wood is less crystalline than in normal wood. This difference in cell wall density is significant and cannot be neglected when attempts are made to calculate the absolute amount of material within a given volume of wood. Côté et al. (1968), using the flotation technique, found the cell wall density in normal and compression woods of *Abies balsamea* to be 1474 kg/m^3 and 1427 kg/m^3, respectively. Stamm and Sanders (1966) and Stamm (1969) computed the cell wall density in *Picea abies* on the basis of the following assumption: normal wood contains 42% cellulose, 28% hemicelluloses, and 30% lignin; compression wood contains 30% cellulose, 30% hemicelluloses, and 40% lignin; the density of wood cellulose is 1560 kg/m^3, the density of hemicelluloses 1500 kg/m^3, and the density of lignin 1330–1350 kg/m^3. With these assumptions, the hypothetical density of the cell wall substance was 1480 kg/m^3 in normal wood and 1457 kg/m^3 in compression wood. Stamm found the density of lignin in hardwoods about 2% higher than in softwoods.

Notwithstanding the presence of helical cavities and intercellular spaces and that it has about 2–3% lighter cell wall substance, compression wood, due to its thick-walled tracheids, is always heavier than normal wood. However, Timell's (1986a) review of *compression wood density* shows large differences between the reported values, since compression wood can occur in different forms, ranging from nearly normal via mild to more severe types, mild compression wood having a lower density than pronounced compression wood. Timell (1981, 1982) determined the density of normal and compression stem woods from the same growth rings in nine conifer species. The compression wood was of a pronounced type, and the few rows of angular, thin-walled tracheids formed at the beginning of the

Table 3.20. Basic density of pronounced compression wood and adjacent normal wood in stem wood of nine conifer species (Timell 1981, 1982)

Tree species	Basic density (kg/m^3)		Ratio
	Compression wood	Normal wood	
Abies balsamea	630	290	2.17
Picea abies, P. glauca, P. mariana, P. rubens	640	360	1.78
Pinus sylvestris	610	320	1.91
Pseudotsuga menziesii	610	360	1.69
Thuja occidentalis	590	280	2.11
Tsuga canadensis	580	350	1.66
Average	610	330	1.85

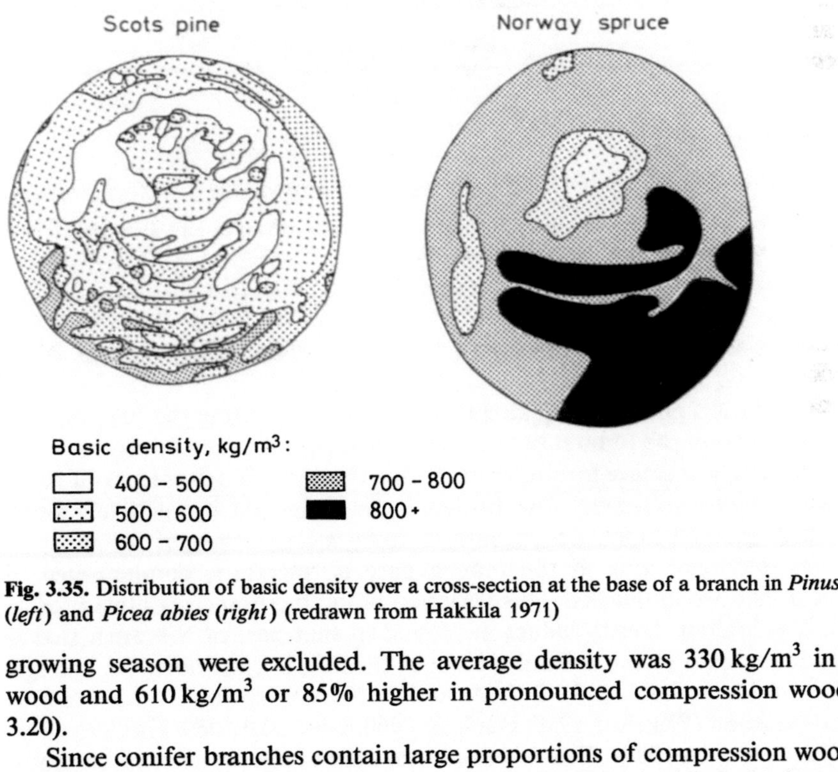

Fig. 3.35. Distribution of basic density over a cross-section at the base of a branch in *Pinus sylvestris* (*left*) and *Picea abies* (*right*) (redrawn from Hakkila 1971)

growing season were excluded. The average density was 330 kg/m^3 in normal wood and 610 kg/m^3 or 85% higher in pronounced compression wood (Table 3.20).

Since conifer branches contain large proportions of compression wood which occurs on the lower side of the cross-section, wood is significantly denser on the lower portion of the branch. This density difference between the lower and upper branch portions was found more than 100 years ago (Nördlinger 1878; Hartig 1896). Timell (1986a) reviewed several studies of *within-branch density variation* in conifers, the lower branch portion usually having 10–40% higher wood density than the upper one. Figure 3.35 also illustrates the density distribution over

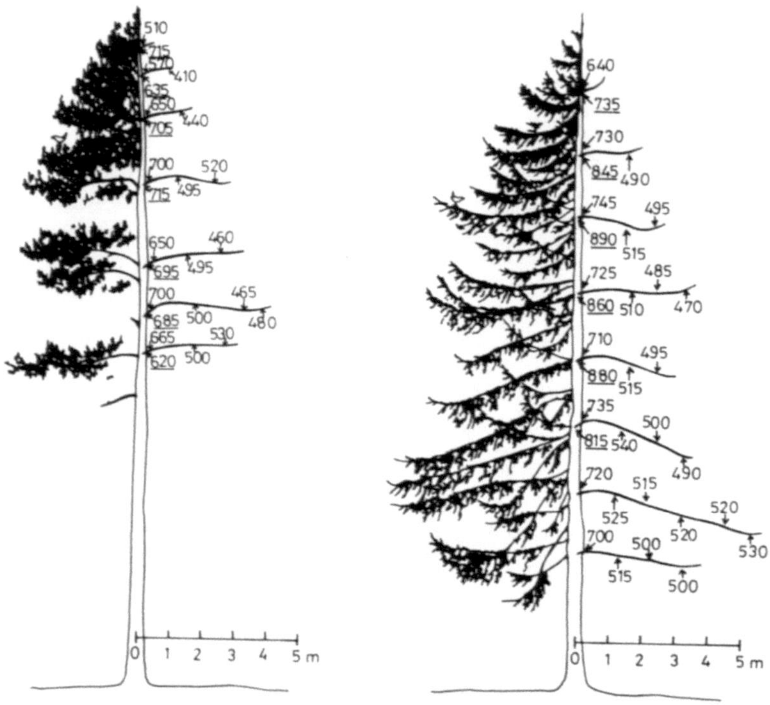

Fig. 3.36. Variation of basic density along branches in a *Pinus sylvestris* and *Picea abies* tree. *Underlined numbers* refer to knot wood (Hakkila 1971)

a branch cross-section at the base of a branch in *Pinus sylvestris* and *Picea abies*.

A clear variation pattern is apparent in wood density along the branch. The density declines from the branch base outward first rapidly, then levels, and may even turn to a slight increase toward the tip. This change is much steeper in *Picea abies* than in *Pinus sylvestris*. The former is characterized by relatively small branches with gradual taper, which have to carry a large mass of foliage. The smaller cross-sectional area at the branch base is, therefore, compensated in spruces by a high wood density. This variation along the branch is illustrated in Fig. 3.36. The highest density values are found in that part of a branch that is embedded in the stem by the natural growth of the tree, i.e., in knots. On average, knot wood in Scots pine is 1.5–2 times and in Norway spruce 2–2.5 times as heavy as stem wood (Wegelius 1939; Boutelje 1966; Levčenko 1969; Hakkila 1971; Enčev 1972; Lehtonen 1980). This is a result not only of the presence of compression wood but to a great extent also of an unusually high resin content.

Figure 3.37 (Atmer and Thörnqvist 1982) illustrates the variation of wood density along the branch at various tree heights in *Pinus sylvestris* and *Picea abies*. Compared with stem and root woods, the density was generally higher in the branches. From branch base to tip, wood density showed first rapid decrease and then slight increase. Taylor (1977) found only a slight difference between the

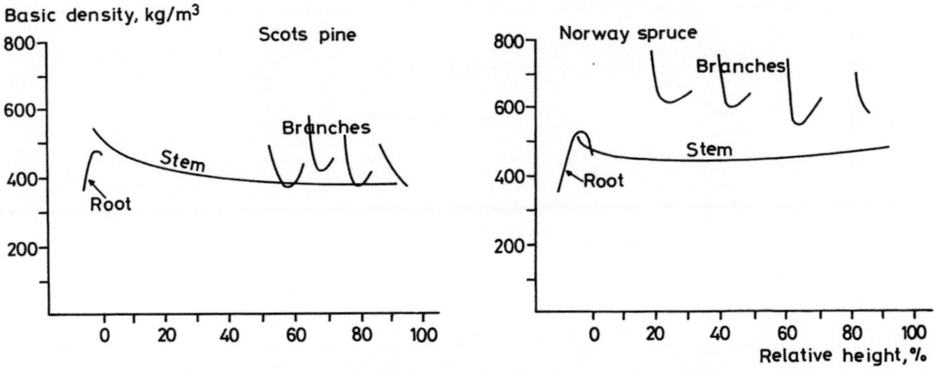

Fig. 3.37. Variation of basic density in stem, branch, and root woods of a 96-year-old *Pinus sylvestris* and 95-year-old *Picea abies* tree (redrawn from Atmer and Thörnqvist 1982)

average densities at one-fourth and one-half length of the branches from eight hardwoods in the southern United States. The basic density was 584 kg/m^3 at the former sampling point and 577 kg/m^3 at the latter. There was no difference in the basic density between top branches and bottom branches.

In southern pine, the density of wood and bark increases with branch diameter. Despite its high content of extraneous components, branch bark is considerably lower in density than branch wood for all species of southern pine, the average difference being about 20% (Phillips et al. 1976):

Diameter at branch base (mm)	Basic density (kg/m^3)	
	Branch wood	Branch bark
−13	428	347
14−49	449	365
50−	470	384
Average	456	366

If branches are cut into sections according to diameter, a certain diameter class may consist of butt parts from smaller branches and center parts from larger branches. This mixing levels the effects of diameter and longitudinal density variation along the branch. Although the average wood density apparently remains highest in the category of the thick branch sections, the differences between the diameter classes are small for *Pinus sylvestris* and *Betula* spp. but still significant for *Picea abies* (Hakkila 1971; Gislerud 1974; Kärkkäinen 1976a).

Due to increasing branch age and a corresponding increase in branch size, the average branch-wood density increases gradually with the breast height diameter of the tree. In Fig. 3.38 this trend is clear in all major species of southern pine except *Pinus elliottii* in which the wood density appears to remain relatively constant. Branch-bark density of the same species fluctuates irregularly without showing a constant trend over the breast height diameter (Phillips et al. 1976).

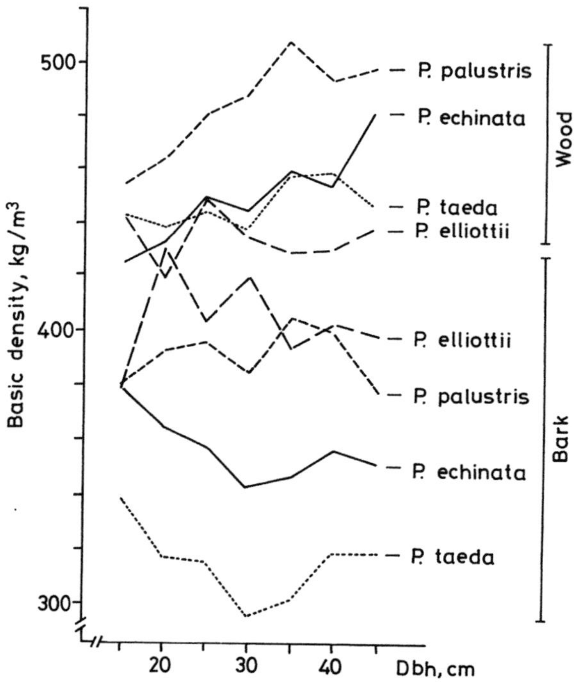

Fig. 3.38. Variation in average basic density of wood and bark in branches of the four major species of southern pine as a function of dbh of the tree (redrawn from Phillips et al. 1976)

Fegel (1941) studied the wood density *variation among tree components* of conifers, diffuse-porous hardwoods, and ring-porous hardwoods. In all groups, the average density was highest in branch wood. Attention was paid to the differences in the rate of growth in the stem, branch, and root woods and their possible effect on the basic density of wood. In conifers, the denser latewood remained nearly constant in width while the lighter earlywood varied with the rate of growth, resulting usually in a density increase as the rate of growth decreased. This relationship held to some degree in comparing the densities in stem and branch woods. The root wood, however, did not increase in density in direct proportion to the decrease in the rate of growth, since the tracheids were thinner-walled and larger in diameter than those of the stem and branches. In hardwoods, particularly in ring-porous species, the dense latewood increased while the lighter earlywood remained relatively constant as the growth rate increased, resulting in denser wood. This relationship appeared to hold in comparing roots and stems of the hardwoods, but the slow growth of branches was more than offset by a decrease in the size of vessels and fibers, and an increase in cell wall thickness. Although the major differences in wood density were traceable to differences in cell wall structure, the infiltration products present in wood were also thought to have a marked influence on the density of stem, branch, and root woods (Table 3.21). Many hardwoods thus have denser wood in their branches, but the difference between stem wood and branch wood densities appears to vary among species rather unpredictably (Fegel 1941; Taylor 1977; Phillips 1977; Manwiller 1979).

Table 3.21. Average ring width and basic density in stem, branch, and root woods of eight conifers, eight diffuse-porous hardwoods, and four ring-porous hardwoods in New York State; species indicated in Section 3.1.4 (Fegel 1941)

Species group	Ring width (mm)			Basic density (kg/m³)		
	Stem wood	Branch wood	Root wood	Stem wood	Branch wood	Root wood
Conifers	1.7	0.7	1.3	363	488	375
Diffuse-porous hardwoods	2.2	1.0	1.3	489	536	464
Ring-porous hardwoods	2.0	1.0	1.1	540	573	455

Table 3.22. Basic density of wood and bark in stem and branches of 11 oaks, 6 other ring-porous hardwoods, and 5 diffuse-porous hardwoods in the southern United States (data from Manwiller 1979)

Species group	Basic density (kg/m³)				
	Stem wood	Branch wood	Stem bark	Branch bark	Whole tree
Oaks	621	637	607	514	618
Other ring-porous trees	578	568	445	441	551
Diffuse-porous trees	456	455	434	409	451

The differences between wood and *bark densities* show a somewhat more regular pattern, with branch wood usually denser than branch bark (Fig. 3.38). Kärkkäinen (1976a) found the average basic density in branch wood of *Pinus sylvestris*, *Picea abies*, and *Betula* spp. to be 424, 547, and 514 kg/m^3, whereas the corresponding basic density values for bark were 311, 348, and 493 kg/m^3, respectively. In Manwiller's (1979) extensive study of 22 southern hardwoods in the United States, stem and branch woods differed only slightly in average density, whereas remarkable differences were found between stem and branch barks. *Ulmus alata* showed an unusually large difference between wood and bark; branch wood density was 610 kg/m^3 but branch bark density only 369 kg/m^3. In *Acer rubrum*, *Celtis* spp., and in some of the oaks, stem wood was unexpectedly lighter than stem bark, but in all 22 hardwood species branch wood was always denser than branch bark. Table 3.22 summarizes wood and bark density data from 11 oaks, 6 other ring-porous hardwoods, and 5 diffuse-porous hardwoods.

Little information is available on the *density of foliage*. In a small sample from *Pinus taeda* the density was 450 kg/m^3 in unextracted needles and 340 kg/m^3 in extracted needles (Koch 1972a). Gislerud (1974) reported 373 kg/m^3 for *Pinus sylvestris* and 353 kg/m^3 for *Picea abies* needles.

3.4.2 Basic Density of Wood and Bark in Stem Tops

The average wood density of a stem is affected by a large number of variables such as tree species, geographic location, site quality and other environmental factors, position in a stand, tree age and size, growth rate, stem taper, and heredity. The relationships are complex, differences among species are great, and research results are often contradictory. The average wood density of tropical species is considerably higher than that of temperate species.

Unmerchantable small trees that are left behind as residue in forest operations may be fast-grown trees felled in precommercial and early commercial thinnings of young plantations, or they may be slow-grown understory trees from unmanaged older stands. Wood quality in these undersized trees is strongly influenced by stand history, since both the tree age and growth rate affect properties such as the proportion of juvenile wood, proportion of latewood, and basic density. At least in conifers, age is probably a more decisive factor for wood density than growth rate (Spurr and Hsiung 1954), but the effect of a variable is species-specific. The following general conclusions can be drawn as to the effect of ring width on wood density (Kollmann and Côté 1968):

1. Among the coniferous woods the density of spruces increases with decreasing width of growth rings, whereas the density of pines and larches at first increases and then decreases again, with the maximum density at 1–1.5 mm ring widths. The pattern of firs and Douglas-fir apparently lies between the two.
2. In ring-porous hardwoods, wide annual rings imply wood of high density. Exceptions are mostly due to variations in the amount and structure of latewood.
3. In diffuse-porous hardwoods, ring width is not a significant criterion of density.

Since latewood cells have thicker walls than earlywood cells, wood density strongly correlates with the *percentage of latewood*. The contrast in density between earlywood and latewood is more pronounced in conifers and ring-porous hardwoods than in diffuse-porous hardwoods. The density ratio of latewood to earlywood in conifers is usually 2.2–2.8 (Kollmann and Côté 1968). According to Müller-Stoll (1948), this ratio was only 1.2–1.8 for three diffuse-porous hardwoods, while ring-porous hardwoods were intermediate between conifers and diffuse-porous woods. Figure 3.39 (Paul 1939) illustrates the density variation in earlywood and latewood samples from southern pines which are characterized by a pronounced contrast between these two tissues.

In many tree species, the percentage of latewood is unusually low in the juvenile wood near the pith of the stem, resulting in a low initial wood density. In pines, the percentage of latewood and the wood density increase from the pith toward the bark until they reach a maximum. An overmature pine may start producing less dense wood again. In spruces, the density first decreases but then increases from the pith to the bark. These changes are usually smaller in hardwoods. The variation pattern may be disturbed by infiltration of extractives in conjunction with heartwood formation.

The *radial density variation* from pith to bark is of great significance in wood utilization. For example, since sawmill slabs are usually composed of dense wood

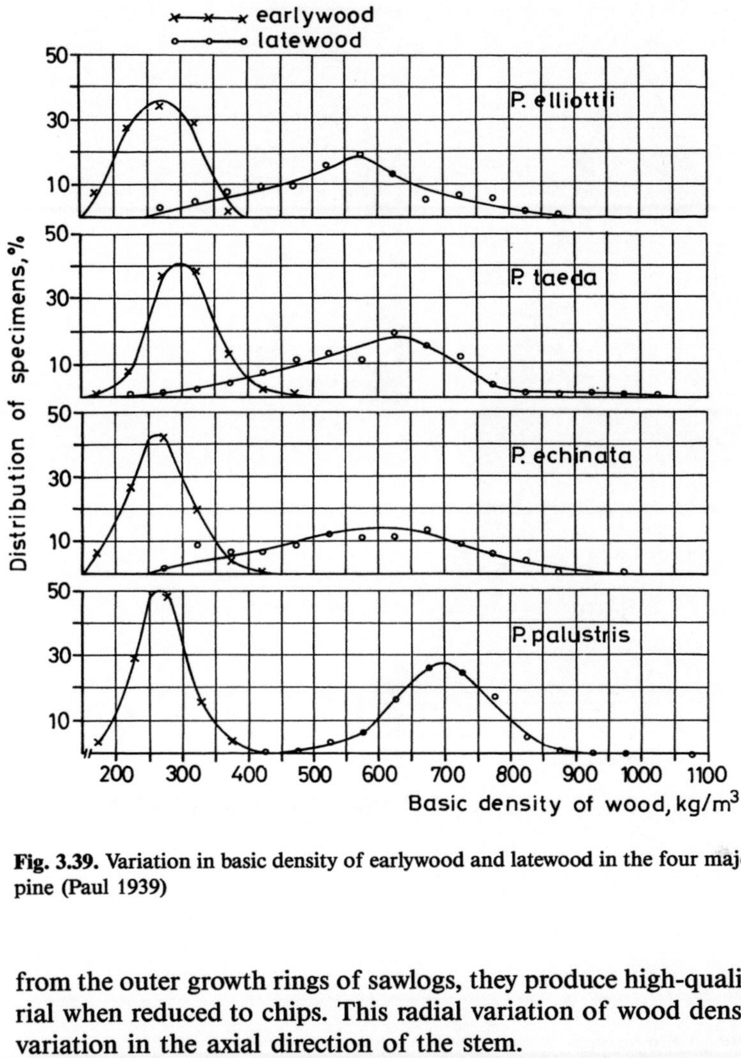

Fig. 3.39. Variation in basic density of earlywood and latewood in the four major species of southern pine (Paul 1939)

from the outer growth rings of sawlogs, they produce high-quality pulp raw material when reduced to chips. This radial variation of wood density causes further variation in the axial direction of the stem.

Since the unmerchantable stem tops are an essential component of forest residue, information on *longitudinal density variation* is important for effective utilization and value estimation of top wood. However, the variation pattern is species-specific, and a general comparison between conventional stem wood and top wood is therefore difficult. According to Knigge and Schulz (1966), in most conifers wood density decreases from butt to top. This decrease is rapid at the stem base but less conspicuous at the upper portion of the stem. It is affected by changes in ring width and, above all, the presence of juvenile wood. Some conifers, especially spruces, show only a slight longitudinal variation in wood density.

The differences among species are larger in hardwoods. Among diffuse-porous hardwoods, density either first decreases notably and then increases toward the top, or increases or decreases steadily from butt to top. In various ring-porous hardwoods such as oaks, the longitudinal density variation is often

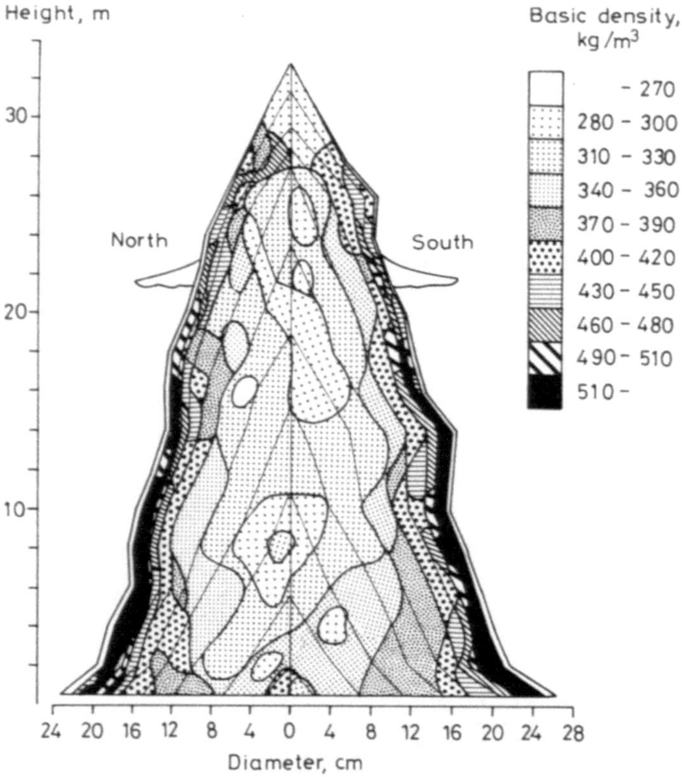

Fig. 3.40. Distribution of basic density in stem wood of *Abies grandis* (Knigge 1960; Knigge and Schulz 1966)

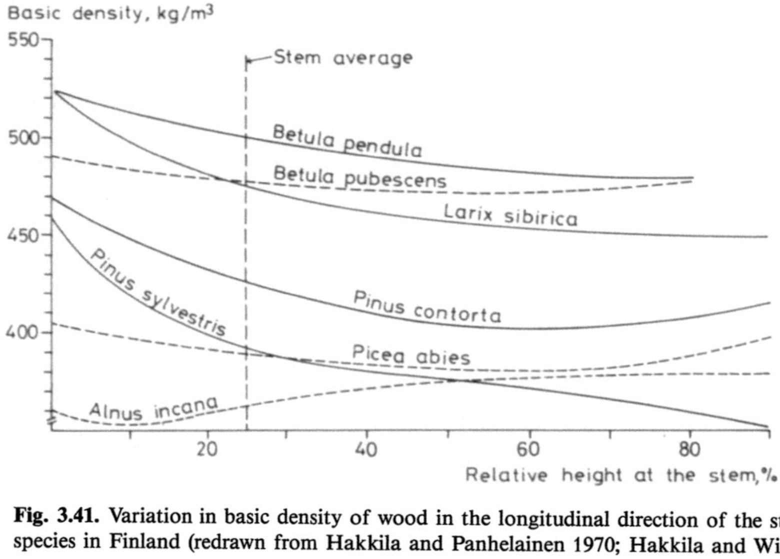

Fig. 3.41. Variation in basic density of wood in the longitudinal direction of the stem in seven tree species in Finland (redrawn from Hakkila and Panhelainen 1970; Hakkila and Winter 1973)

negligible. A stem analysis of *Abies grandis* in Fig. 3.40 (Knigge 1960) is an example of density variation within a stem, showing the relatively low density near the pith and, consequently, in the unmerchantable top. In northern European tree species, the average wood density of a stem is approximately the same as the density at 25% stem height (Nylinder 1961; Hakkila 1966, 1979). Figure 3.41 illustrates patterns of longitudinal variation in wood density in seven tree species in Finland. *Pinus sylvestris*, *P. contorta*, and *Larix sibirica* represent species with a distinct difference between the butt and the top, whereas in *Picea abies* the differences are minor.

The *longitudinal variation in bark density* does not necessarily follow the same pattern as the wood density in the same species. For example, data from Tamminen (1962, 1964) show for *Pinus sylvestris* bark first a moderate decrease and then a slight increase, and for *Picea abies* bark a constant decrease in density from butt to top. These trends are distinctly dissimilar from those in wood density:

	Relative stem height (%)				
	1	10	30	60	90
	Basic density of bark (kg/m^3)				
Pinus sylvestris	304	301	280	277	289
Picea abies	419	353	333	318	295

Figures 3.42 and 3.43 (Manwiller 1979) illustrate the longitudinal density variation in wood and bark of 22 hardwood species in the southern United States. In most species the densest wood is found at the stem base but a density increase in the top wood is not uncommon. The density range is considerably wider in bark than in wood. In some species the variation patterns are similar in wood and bark, while in others they differ remarkably. The densest bark is often found at the top of the tree and the difference in the basic density between butt and top barks may exceed 100 kg/m^3.

Kärkkäinen (1976b) studied the bark density in sawlogs of *Pinus sylvestris* and *Picea abies*. In Scots pine, the density of rough butt bark was 309 kg/m^3 and that of thinner bark 301 kg/m^3. In Norway spruce, the corresponding values were 409 and 365 kg/m^3, the top bark thus showing considerably lower density. Erickson (1972b) pointed out that a natural or induced difference in density and moisture content of bark and wood can provide a basis for segregating the bark particles from the wood chips using processes such as air flotation or liquid medium segregation.

As the density of wood and bark thus commonly varies in the longitudinal direction of a stem, the residual wood from unmerchantable tops may be expected to differ in density from conventional timber. This difference is species-specific and is affected by tree age and environmental factors and naturally also by the minimum top diameter of merchantable timber. Compared to the merchantable portion of the stem, wood and bark from stem tops may thus be either higher or lower in density, depending mainly on tree species (Table 3.23).

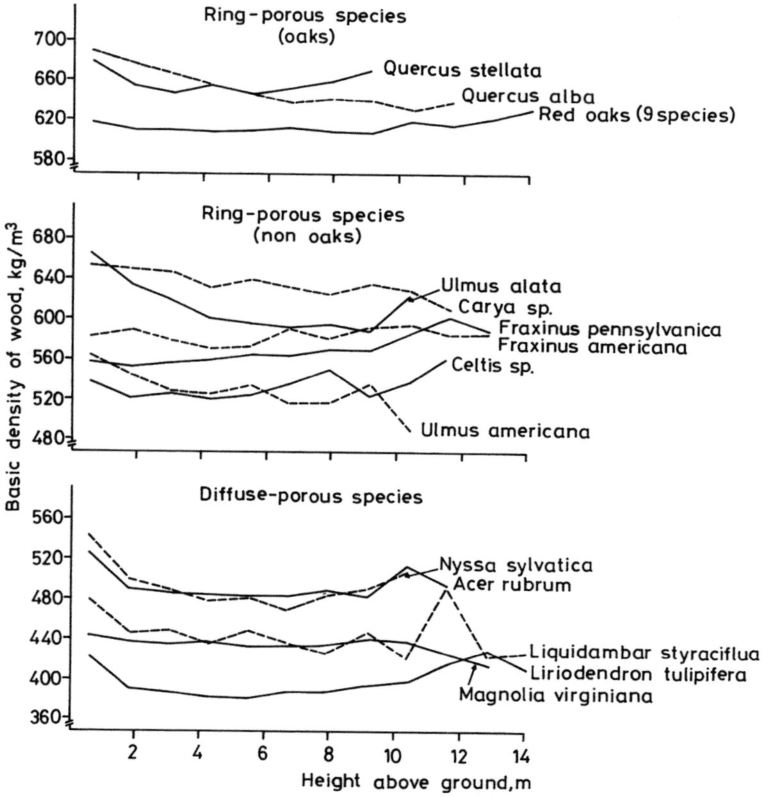

Fig. 3.42. Variation in basic density of wood in the longitudinal direction of the stem in 22 hardwood species in the southern United States (redrawn from Manwiller 1979)

3.4.3 Basic Density of Wood and Bark in Stumps and Roots

The cell structure and basic density of wood are determined largely by the functions of the tissue. In roots the primary functions are absorption and conduction of water and minerals, translocation and storage of food, and anchoring of the tree. The anchoring function is most important in the taproot and base sections of laterals near the root collar, where the strength is increased by means of thick-walled cells and eccentricity of the root cross-section. The wall thickness decreases from root base to tip, and latewood almost disappears in conifer and hardwood roots with increasing distance from the stem. Compression wood, which is characterized by thick walls, is almost nonexistent in conifer roots (Sect. 3.1.4).

Most tree species produce dense wood at the stem base. Accordingly, the central stump-root system with a taproot, as the natural below-ground continuation of the stem, might be expected to have a high average basic wood density. However, the wood from the wide growth rings of the butt swell is often less dense, and in many tree species the average basic density is actually lower in the central stump-root system than in the stem. Koch (1985 a) found that the average wood

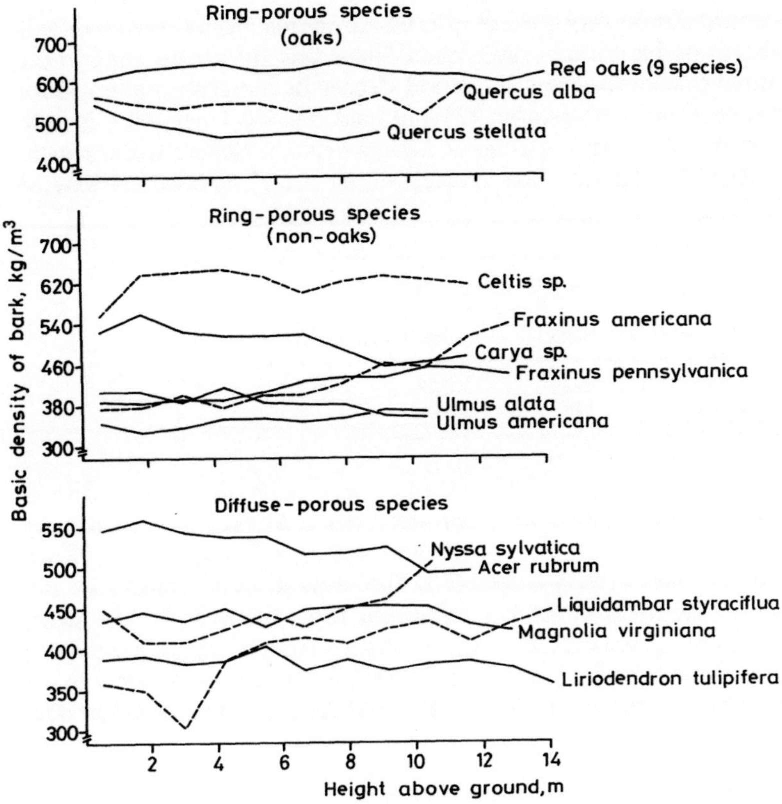

Fig. 3.43. Variation of basic density of bark in the longitudinal direction of the stem in 22 hardwood species in the southern United States (redrawn from Manwiller 1979)

Table 3.23. Examples of differences in wood and bark densities between merchantable stem and unmerchantable stem top in various tree species

Tree species	Basic density (kg/m^3)					Reference
	D_{top} (cm)	Merch. wood	Top wood	Merch. bark	Top bark	
Conifers						
Abies balsamea		317	331	366	326	Erickson 1972b
Picea abies	Under 5	392	393			Hakkila 1971
P. glauca		336	340	290	267	Erickson 1972b
Pinus banksiana		371	361	322	259	Erickson 1972b
P. elliottii	Under 10	470	380			Koch 1972a
P. resinosa		376	333	256	268	Erickson 1972b
P. sylvestris	Under 5	417	372			Hakkila 1971
Tsuga heterophylla	Over 5	417	471			Worster and Vinje 1968
Hardwoods						
Acer sp.		562	555	548	531	Erickson 1972b
Populus tremula		367	385	505	501	Erickson 1972b

density in the central stump-root system at a 92-cm radius around the stem axis in 22 southern hardwood species in the United States was 10% lower than in the stem. In only three tree species was the wood denser in the central stump-root system. The difference was even greater, 22%, in bark density. Unexpectedly light bark was found in the stump-root system of *Liriodendron tulipifera* with an average density as low as 193 kg/m^3. The average values for 22 hardwoods were as follows:

	Basic density (kg/m^3)	
	Wood	Bark
Stem	572	523
Central stump-root system	513	408

In the central stump-root systems of 22-year-old *Pinus taeda* trees, the density in unextracted wood was 380 kg/m^3 and in extracted wood 340 kg/m^3. The density was significantly higher in stem wood, 470 and 430 kg/m^3, respectively (Koch 1972a). Young and Chase (1965) determined the average basic density of unbarked stump and root mass at 2.5-cm minimum root thickness in 15–40 cm diameter trees in Maine. In accordance with the results from southern hardwoods above, all three hardwood species analyzed in the study had a lower density in the stump-root system than in the stem. Conversely, four conifers showed higher density in the stump-root systems:

	Basic density of unbarked wood (kg/m^3)	
	Stem	Stump-root system
Conifers:		
Abies balsamea	297	317
Picea rubens	435	461
Pinus strobus	387	428
Tsuga canadensis	408	438
Hardwoods:		
Acer rubrum	581	572
Betula papyrifera	525	507
Populus sp.	342	336

Differences in the basic density between stem and root woods are greatest when the roots are small and branched and the samples are taken far from the stem axis (Knigge and Schulz 1966). According to Atmer and Thörnqvist (1982), the root wood of *Pinus sylvestris* is densest at the base of the root; the density decreases rapidly toward the root tip. In *Picea abies*, density first increases along the root, reaches a maximum at about 0.5 m distance from the stem, and then decreases toward the tip (Fig. 3.37). Eskilsson (1969, 1972) reported the same trend in Norway spruce roots, although the maximum density occurred further from the stem.

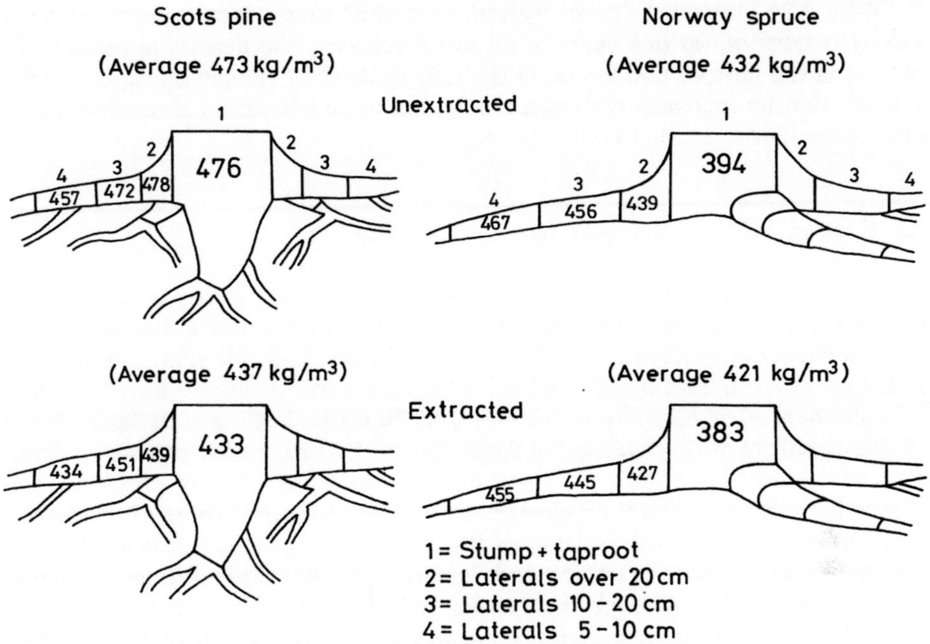

Fig. 3.44. Basic density of stump and root wood in large *Pinus sylvestris* and *Picea abies* trees before and after extraction with acetone (drawn from data in Hakkila 1975)

Table 3.24. Basic density of wood in various components of stump-root systems of 23- to 110-year-old *Pinus sylvestris* (Poller et al. 1973)

Biomass component	Basic density (kg/m^3)		
	Young trees	Middle-aged trees	Old trees
Root collar	470	510	540
Stump center	470	500	510
Taproot	500	510	600
Coarse sections of laterals	490	470	530
Thin sections of laterals	450	440	460
Average	480	490	530

The reduction was remarkable at a root diameter of 20–40 mm. Similar results were obtained by Hakkila (1975), but the sample did not include root sections under 5 cm. When the acetone-soluble extractives were removed, density was reduced in Scots pine by 21–43 kg/m^3 and in Norway spruce by 11–12 kg/m^3, and in Scots pine the point of maximum density moved slightly toward the root tip (Fig. 3.44).

Table 3.24 illustrates density variations within stump-root systems of *Pinus sylvestris* trees of various ages. In all age categories, the density appears to be highest in the taproot and lowest in the thin sections of the lateral roots. On the average, density increases with age, partly due to an increasing concentration of extractives (Poller et al. 1973).

3.5 Heating Value of Residual Tree Components

Green plants produce biomass by converting carbon dioxide and water in photosynthesis to simple carbohydrates that serve as basic raw materials for further biochemical synthesis. In this process the plants absorb energy from solar radiation through chlorophyll and other pigments in the chloroplasts and store it in photochemical reactions in the biomass. Photosynthesis uses radiation in the visible spectrum with wavelengths from 400 to 700 nm, which represents about 50% of the global radiation. On a foliage mass basis, the leaves of deciduous tree species have the potential for higher photosynthetic rates than those of evergreens. In terms of carbon uptake per unit of leaf surface area, the difference is less but sometimes still apparent (Waring and Schlesinger 1985). This difference is compensated by conifers through their larger foliage mass.

The *photosynthetic conversion efficiency* of a forest is obtained by dividing the energy content of the net production of biomass by the incident solar energy in a given time. The figures below show the conversion efficiency of temperate and tropical forests according to Jordan (1971), showing the net production of biomass in relation to the energy of visible solar radiation during the growing season. Long-term storage of energy refers to woody tissue and short-term storage to foliage, fruit, and other short-lived tree components. However, Young (1987, pers. commun.) pointed out that the tonnage of root tips and hairs that form, live, function, and die may very well be significantly higher than heretofore published.

	Conversion efficiency of forests during the growing season (%)
Temperate climate:	
Long-term storage (woody tissue)	0.57 – 3.14
Short-term storage (foliage, etc.)	0.37 – 0.84
Tropical climate:	
Long-term storage (woody tissue)	0.28 – 1.25
Short-term storage (foliage, etc.)	0.15 – 1.15

The energy fixed in photosynthesis is released through an oxidation process in natural decomposition or burning of biomass. Thus, forest biomass forms a reserve of renewable energy which can be used as fuel in various forms, presently primarily by burning it for heat.

The heating value of a fuel is a measure of the maximum amount of energy that can be released on burning a given quantity of the fuel under given conditions. It is determined by combusting biomass in an adiabatic or isothermic oxygen bomb calorimeter at a constant volume. Part of the thermal energy released in the calorimeter is consumed by vaporizing the water originally present in the biomass and that formed in the combustion process, but since the calorimeter is closed, the water vapor cannot escape from the system. In the analyses, the combustion products are cooled to the original temperature of the fuel, usually 20 °C, resulting in condensation of water vapor and subsequent release of the heat of vaporization. The total heat thus released from the biomass is referred to as the *calorimetric or higher heating value.* It is independent of the moisture content of the biomass. In older literature, heating values were commonly given in calories per gram of wood (cal/g). Presently, the heating values are expressed in the metric system in MJ/kg or kWh/kg, and in the American system in British thermal units per pound (Btu/lb). One MJ is equivalent to 238.9 kcal, 0.2778 kWh, and 947.8 Btu.

The thermal energy contained by a fuel depends on the chemical composition of the substance. In the complete combustion of biomass, oxygen combines with carbon, producing carbon dioxide, and with hydrogen, producing water. It is important to note that the carbon dioxide released in the combustion process was originally taken by plants through photosynthesis from the atmosphere. Therefore, unlike the use of fossil fuels, combustion of forest biomass does not result in an increase in the carbon dioxide content of the atmosphere. At the same time, thermal energy which was absorbed by plants from solar radiation through photochemical reactions, is released according to the following formulas:

$$C + O_2 \rightarrow CO_2 + 32.8\ \text{MJ/kg of C} \ ;$$

$$2H_2 + O_2 \rightarrow 2H_2O + 142.2\ \text{MJ/kg of } H_2 \ .$$

Figure 3.45 shows a schematic drawing of the composition of freshly felled tree biomass. The major combustable elements in biomass are carbon and hydrogen. A high carbon and hydrogen content thus means a high heating value, whereas oxygen, nitrogen, and inorganic elements have a contrary effect. On the

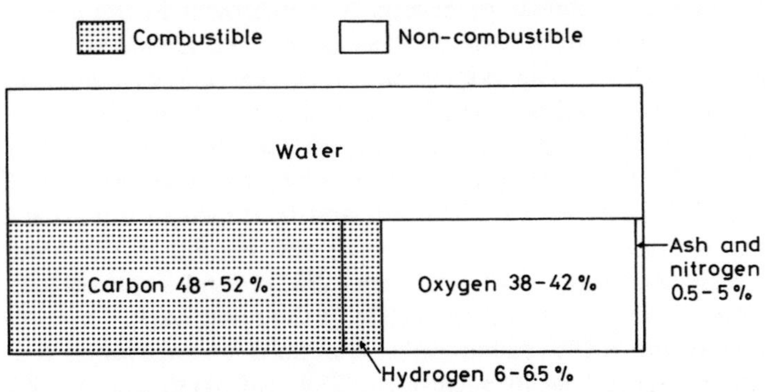

Fig. 3.45. Chemical composition of bark-free wood by elements

Table 3.25. The content of carbon and hydrogen in the dry mass of various above-ground components of *Pinus sylvestris*, *Picea abies*, and *Betula* sp. trees removed in early thinning, and of logging slash from newly felled trees (Thörnqvist 1985a)

Tree species	Content of carbon (%)			
	Stem wood	Stem bark	Small branches	Foliage
Pinus sylvestris	49.9	51.8	52.2	51.5
Picea abies	50.2	49.8	51.5	49.8
Betula sp.	48.5	52.3	53.2	
Fresh slash	50.5	50.4	52.2	51.6
	Content of hydrogen (%)			
Pinus sylvestris	6.3	5.9	6.4	6.4
Picea abies	6.2	6.0	6.3	6.2
Betula sp.	6.3	6.3	6.5	
Fresh slash	6.5	6.0	6.0	6.6

average, softwoods contain 50.7% carbon and 6.2% hydrogen. The corresponding figures are 49.0 and 6.0% for hardwoods (Kollmann 1951).

Ultimate analysis shows the proportion of various elements in a fuel. The following table by Arola (1976) presents an example of an ultimate analysis of woody biomass in typical hardwood species in the northern United States:

Element	Proportion (%)	
	Wood	Bark
Carbon	50.8	51.2
Oxygen	41.8	37.9
Hydrogen	6.4	6.0
Nitrogen	0.4	0.4
Ash	0.9	5.2

Thörnqvist (1985a) studied the content of carbon and hydrogen in the above-ground components of young *Pinus sylvestris*, *Picea abies*, and *Betula* sp. trees from early thinnings in Sweden. The concentration of carbon was highest, 52–53%, in the stem bark and small branches of birch, and lowest, 48.5%, in the stem wood of birch. The content of hydrogen varied from 5.9% in Scots pine bark to 6.5% in birch bark and 6.6% in needles sampled from newly-felled logging slash (Table 3.25). The high percentages of carbon and hydrogen in birch bark appear to be the result of a high content of suberin and betulin in the outer bark. In an ultimate analysis of birch bark, Virtanen (1963) found a carbon content of 52.0% in the inner bark but as high as 65.8% in the outer bark. The content of hydrogen was 6.0 and 8.4%, respectively. It should be mentioned that a typical composition of fuel oil is 86% carbon, 12% hydrogen, and 1–2% sulfur.

Compared to other compounds of biomass, cellulose and hemicelluloses are richer in oxygen but poorer in carbon and hydrogen. Consequently, they contain less thermal energy. Lignin, resin, terpenes, and waxes have considerably higher heating values. According to Kollmann (1951), the effective heating value is 17.4–18.2 MJ/kg in cellulose, 25.5 MJ/kg in lignin, and 35.6–38.1 MJ/kg in resins. Howard (1973a) determined the calorimetric heating value of liquid resin collected from mechanically injured southern pine trees as 34.0–37.8 MJ/kg. Since conifers usually contain more lignin and resin than hardwoods, they tend to have slightly higher heating values per unit mass of stem wood. For the same reason, in conifers the heating values are higher in foliage and branches than in stem wood, whereas in hardwoods the lignin content and heating value are slightly lower in branch wood (Harris 1984). For example, in a study by Koch (1985a), showing surprisingly low calorimetric heating values, average results of 22 southern hardwoods for wood and bark were as follows:

	Calorimetric heating value (MJ/kg)	
	Wood	Bark
Stem	18.21	17.66
Branches	18.11	17.75
Central stump-root system	19.38	17.81

Howard (1973a) analyzed the *calorimetric heating value of biomass components* of southern pine. The average heat of combustion for samples from *Pinus taeda* stem wood was 20.0 MJ/kg, and the range of variation from 19.3 to 21.7 MJ/kg. When calorimetric heating values were regressed against data on wood properties, growth rate, wood density, and percentage of latewood proved nonsignificant. Extractive content was positively correlated with the heating value and accounted for 54% of the variation, whereas variation in the proportions of the primary chemical constituents had only minor effects. A negative correlation with the number of rings from the pith reflected the distribution of extractives in the stem. Unusually high heating values were measured in resinous wood from mature stumps of old-growth *Pinus palustris* with 27–35% alcohol-benzene extractives, and in resinous, decay-resistant knots with 29–40% alcohol-benzene extractives, selected from the forest floor in a stand of the same species. The heating value of mixed stem bark samples was negatively correlated with the height above ground, obviously because the proportion of inner bark was largest at greater heights, and because it probably had a lower heating value than the outer bark. Needles in the four major species of southern pine had a higher heating value than either stem wood or bark. Variation in the calorimetric heating value of samples from different components of southern pine is seen below (see Scheme p. 198).

The content of heat energy is higher in the outer bark than in the inner bark. The outer bark of birch wood has an unusually high calorimetric heating value which significantly exceeds, for example, that of pine needles. This is the result of its high content of suberin (Table 3.26).

Fuel from southern pine	Calorimetric heating value in southern pine biomass (MJ/kg)
Liquid resin	34.0 – 37.8
Commercial charcoal from wood and bark	26.1 – 29.6
Resinous wood from mature stumps (*P. palustris* only)	23.8 – 25.2
Decay-resistant knots from the forest floor (*P. palustris* only)	23.6 – 26.7
Needles	20.8 – 21.2
Stem bark at 0.3-m height	20.6 – 21.3
Stem bark at 50% height	20.1 – 20.7
Stem bark at 10-cm top	19.9 – 20.5
Stem wood (*P. taeda* only)	19.3 – 21.7
Earlywood	19.7 – 20.4
Latewood	19.5 – 20.4
Root wood	19.9 – 20.2
Tops (wood and bark at 2.5-cm top)	18.6 – 20.3
Old cones	18.8 – 19.0
Dried kraft black liquor after removal of tall oil	13.5 – 14.3

The calorimetric heating value per unit of dry mass of fuel may *change during storage*, if dry matter loss caused by decay is directed to only certain constituents of biomass (Sect. 8.1.4.2). However, in practice the changes are usually small. Thörnqvist (1986a) stored chips from fresh coniferous logging slash and from oak stems in large piles in Sweden from May-June to January-February. Although dry matter losses ranged from 1.4 to 12.0%, the calorimetric heating value per dry mass of chips remained unchanged. The only exception was chips from logging residue when the pile was compacted with a wheeled loader and covered. This treatment resulted in an increase in the calorimetric heating value from 20.77 to 21.15 MJ/kg or 1.0%. Salmi (1969) found the calorimetric heating value of small birch stems to be unaffected by the degree of decay after 1- and 2-year storage except in specimens with an unusually light yellowish color, reflecting an advanced stage of decay. In these samples the calorimetric heating value was 0.9% lower than in sound wood. Heating values may be reduced also by water storage of timber due to loss of extractives. For example, Zhitkov (1958) found a reduction of 6.5% in the calorimetric heating value of bark from spruce timber when floating was used instead of land transport.

Three consecutive stages can be distinguished in the combustion of biomass. First, the moisture is evaporated to dry the wood. Second, the volatile substances are driven off and combusted. Third, the fixed carbon is combusted. The volatiles burn in the gaseous phase, whereas the fixed carbon burns in the solid phase.

Proximate analysis of solid fuel is used mainly for classification by rank of coal and determination of certain combustion characteristics, but it may be used for biomass fuels as well. The result of the analysis is a material balance as percentages by mass of moisture, ash, volatile matter, and fixed carbon in the fuel.

Table 3.26. Calorimetric heating value of above-ground components of 10-cm-diameter *Pinus sylvestris*, *Picea abies*, and *Betula pubescens* (unpublished data from Nurmi, FFRI, 1988)

Tree component	Heating value (MJ/kg)		
	Pinus sylvestris	*Picea abies*	*Betula pubescens*
Stem:			
Wood	20.7	20.3	19.9
Inner bark	20.3	19.2	20.2
Outer bark	21.8	21.9	33.3
All bark	20.8	20.1	23.8
All mass	20.7	20.3	20.5
Branches, over 5 mm:			
Wood	21.1	20.9	20.0
Inner bark	20.9	19.7	20.3
Outer bark	22.5	22.0	27.5
All bark	21.5	20.3	23.0
All mass	21.3	20.8	20.7
Twigs, under 5 mm:			
Wood	21.7	20.9	20.0
All bark	22.7	21.8	23.4
All mass	22.2	21.3	21.2
Leaves	22.7	20.7	21.1

Table 3.27. The proportion of volatile matter, fixed carbon, and ash in dry wood and bark of six coniferous species in the western United States at 950 °C furnace temperature (Mingle and Boubel 1968)

Tree species	Proportion, percent dry mass					
	Wood			Bark		
	Volatile matter	Fixed carbon	Ash	Volatile matter	Fixed carbon	Ash
Abies concolor	84.4	15.1	0.5	73.4	24.0	2.6
Pinus ponderosa	87.0	12.8	0.2	73.4	25.9	0.7
Pseudotsuga menziesii	86.2	13.7	0.1	70.6	27.2	2.2
Sequoia sempervirens	83.5	16.1	0.4	71.3	27.9	0.8
Tsuga heterophylla	84.8	15.0	0.2	74.3	24.0	1.7
Alnus rubra	84.9	14.5	0.6	74.3	23.3	2.4

From a combustion standpoint, the moisture content will predict the heat required to evaporate and superheat the water, the ash content predicts the residue handling requirements, and the volatile matter and fixed carbon contents predict the requirements for the division of combustion airflow between over-fire and under-fire air. This division of airflow is useful in smoke abatement and air pollution control. However, for the heating value and complete combustion calculations, an ultimate analysis of fuel is required (Mingle and Boubel 1968).

Table 3.27 shows the results of proximate analysis of wood and bark from six tree species from the western United States. The volatile component is about 85% and fixed carbon about 15% for wood of each species. Bark appears to have 70–75% volatiles and about 25% fixed carbon. Bark also has a higher concentration of ash (Sect. 3.3.3). The proportion of volatiles is much higher in biomass than in coal, where the volatile component is usually less than 40% (Mingle and Boubel 1968).

When biomass is burned in a furnace, part of the released heat must be consumed for vaporizing water. This water originates from two sources: water present in the fuel and water generated when hydrogen and oxygen combine to form water. The heat consumed in vaporization is lost to the atmosphere. Consequently, even in complete combustion it is impossible to exploit fully the calorimetric heating value of biomass. Hence, it is feasible to deduct from the calorimetric heating value the energy spent on vaporization, assuming that the water cools down to its initial temperature but retains the vapor form. This unit, referred to as the *effective or lower heating value*, is thus effected not only by the calorimetric heating value but also by the moisture content and hydrogen content of biomass.

Presuming that the biomass does not contain water, the formula below can be used for conversions between the calorimetric and effective heating values. In the formula, 2.45 MJ/kg is the heat energy required to vaporize water at 20 °C, and the factor 0.09 is because one part of releasing hydrogen and eight parts of oxygen combine to form nine parts of water. For example, if the hydrogen content of dry forest biomass is 6%, then the effective heating value is 1.32 MJ/kg less than the calorimetric heating value:

$$W_{ea} = W_c - 2.45 \times 0.09 \times H_2 = W_c - 0.22\, H_2 \ ;$$

W_{ea} = Effective heating value of dry biomass, MJ/kg dry mass ;

W_c = Calorimetric heating value of dry biomass, MJ/kg dry mass ;

H_2 = Hydrogen content of dry biomass (%) .

In practice, forest biomass always contains water which must also be taken into consideration in the formula of effective heating value. Due to the latent heat of condensation, a further deduction is required from the calorimetric heating value, depending on the moisture content of biomass. The formula is then as follows:

$$W_{em} = W_{ea} - 2.45 \times \frac{MC}{100 - MC} \ ;$$

W_{em} = Effective heating value of biomass with moisture, MJ/kg dry mass ;

MC = Moisture content in the biomass on a fresh mass basis (%) .

In extractive-free wood, the heating value varies little among tree species and tree components. The calorimetric heating value is typically about 20.5 MJ/kg and the effective heating value about 19.2 MJ/kg of dry wood. Figure 3.46 shows the influence of moisture content on the effective heating value, given in MJ/kg of fuel with moisture as delivered, or in MJ/kg of dry mass of the fuel.

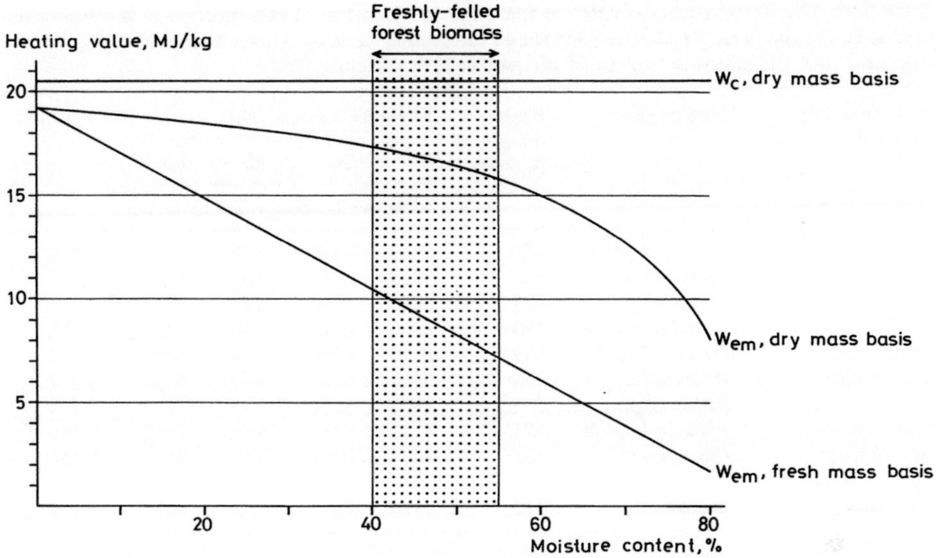

Fig. 3.46. The calorimetric and effective heating value of bark-free wood in MJ/kg dry and fresh mass as a function of moisture content on a fresh mass basis

Table 3.28. Effective heating value of selected assortments of fuel chips prepared from residual forest biomass in Finland, as a function of moisture content on fresh mass basis (Hakkila 1978)

Raw material of fuel chips	Tree species	Moisture content (%)						
		0	20	40	60	20	40	60
		MJ/kg of dry mass				MJ/kg of fresh mass		
Small-sized stems	*Pinus sylvestris*	19.3	18.7	17.6	15.6	15.0	10.6	6.2
	Picea abies	19.1	18.5	17.5	15.5	14.8	10.5	6.2
	Betula spp.	19.5	18.9	17.9	15.8	15.1	10.7	6.3
Whole trees from sapling stands	*Pinus sylvestris*	20.5	19.8	18.8	16.8	15.8	11.3	6.7
	Picea abies	19.6	19.0	18.0	16.0	15.2	10.8	6.4
	Betula spp.	19.6	19.0	18.0	15.9	15.2	10.8	6.4
Whole trees from comm. thinnings	*Pinus sylvestris*	19.6	19.0	18.0	15.9	15.2	10.8	6.4
	Picea abies	19.2	18.6	17.6	15.5	14.9	10.6	6.2
	Betula spp.	19.0	18.4	17.4	15.4	14.7	10.4	6.2
Branch mass without foliage	*Pinus sylvestris*	20.4	19.8	18.7	16.6	15.8	11.2	6.6
	Picea abies	19.7	19.1	18.1	16.0	15.3	10.9	6.4
	Betula spp.	19.7	19.1	18.1	16.0	15.3	10.9	6.4
Crown mass with foliage	*Pinus sylvestris*	20.5	19.9	18.9	16.8	15.9	11.3	6.7
	Picea abies	19.8	19.2	18.2	16.2	15.4	10.9	6.5
Stumps and roots	*Pinus sylvestris*	19.5	18.9	17.9	15.9	15.1	10.7	6.4
	Picea abies	19.1	18.5	17.4	15.4	14.8	10.4	6.2

Table 3.29. The effective heating value of fuel chips in MJ/m^3 solid as a function of moisture content on fresh mass basis. The last column shows the heating value of chips in tonnes of oil equivalent (toe/m^3) when the moisture content of biomass is 40% (Hakkila 1984)

Raw material of chips	Tree species	Basic density (kg/m^3)	Moisture content (%) 0	20	40	60	Toe/m^3
			Effective heating value (MJ/m^3)				
Whole trees	*Pinus sylvestris*	385	7450	7310	6920	6130	0.170
	Picea abies	400	7430	7430	7020	6210	0.173
	Betula spp.	475	9040	8750	8270	7300	0.203
	Alnus incana	370	6810	6810	6340	5680	0.158
Branch mass excl. foliage	*Pinus sylvestris*	405	8250	8000	7590	6720	0.187
	Picea abies	465	9160	8880	8400	7450	0.207
	Betula spp.	500	9850	9550	9040	8020	0.222
Crown mass incl. foliage	*Pinus sylvestris*	395	8113	7872	7470	6628	0.184
	Picea abies	425	8424	8164	7731	6824	0.190
Unbarked saw-mill slabs	Softwood	415	7970	7720	7300	6470	0.179

Since the extractives increase and inorganic materials decrease the heating value of biomass, the range of variation is wider in bark than in wood. Loss of vigor due to suppression or aging appears to be associated with the reduction of the heating value in foliage, and seasonal variation occurs as well (Madgwick 1970). Crown components show higher values than stem components.

The thermal energy actually recoverable from forest biomass depends on both the effective heating value and the moisture content of the fuel, high moisture content being detrimental in two ways. First, moisture reduces the *available heat* of the fuel because there is simply less combustable matter per unit mass. Second, moisture also reduces the effective heating value, because the temperature of water must be raised from ambient to the boiling point, and energy is further required to vaporize water to steam at 100 °C.

The effective heating value of a fuel thus decreases with increasing moisture content. It is also slightly decreased by an increasing proportion of wood in dry mass (Nielson et al. 1985). Table 3.28 shows examples of the effective heating value of various types of fuel chips produced from residual biomass of *Pinus sylvestris, Picea abies*, and *Betula* spp. in Finland, as affected by the moisture content. The table is based on heating value measurements by Olofsson (1975) for wood, bark, and foliage, and on the actual distribution of dry matter by biomass components (Hakkila 1978). The results are presented on both a dry and fresh mass basis of biomass. The heating value per kilogram of fuel is highest in coniferous logging slash from clear cuts as slash consists mainly of branch and foliage mass, both of which are rich in lignin and resin.

As biomass is frequently measured and bought by volume, it is important to know *the effective heating value per cubic meter*. The basic density of biomass serves as a conversion factor from mass to solid volume. Since variation in basic density among tree species and biomass components (Sect. 3.4) is considerably larger than that in the effective heating value of dry mass, differences among biomass assortments are largest when the calculations are made on a volume basis.

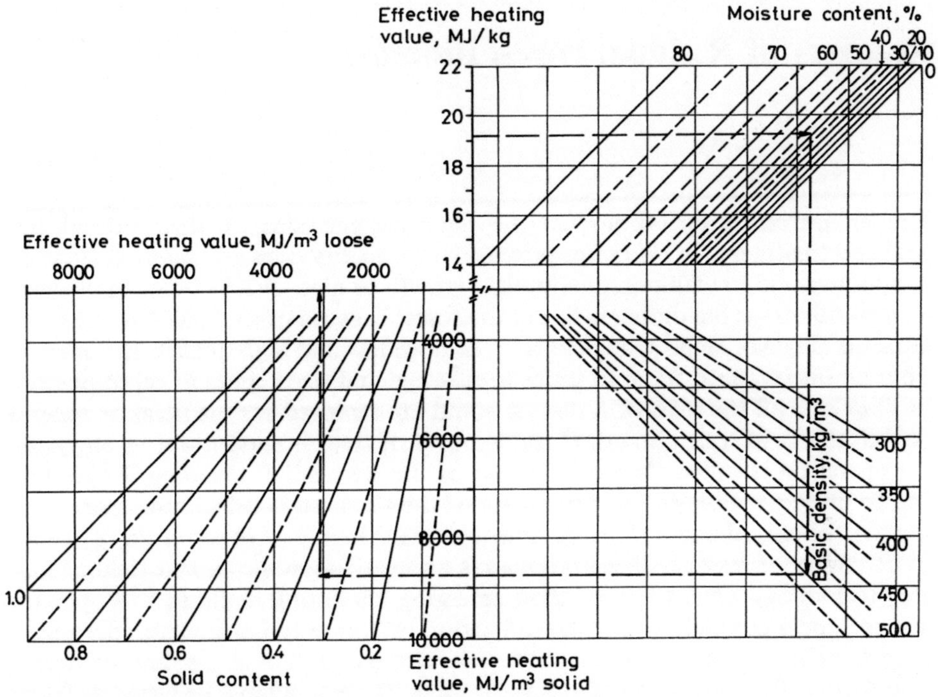

Fig. 3.47. Nomogram for determining the effective heating value per m^3 solid and m^3 loose on the basis of effective heating value per kg, moisture content, basic density, and solid content of biomass (redrawn from Nylinder 1979)

A high heating value per cubic meter of biomass is thus always a result of a high basic density. Foliage, although it has a high heating value per kilogram dry mass due to plentiful extractives, actually reduces the average heating value of biomass fuels per m^3 solid, because its basic density is low. Thus, in Table 3.29 fuel chips made from crown material show a higher heating value per m^3 solid when the foliage is excluded. The last vertical column of the same table compares the effective heating values of 1 m^3 of biomass and 1 tonne of fuel oil, when the moisture content of wood is 40% on a green mass basis. For instance, a solid cubic meter of whole-tree chips corresponds to 0.203 t of oil equivalent if *Betula* spp. is used as raw material, but only 0.158 t if *Alnus incana* is used. Southern woods with a high basic density have remarkably higher heating values per unit volume.

If the volume of biomass is expressed in loose measure, variation in the solid content of biomass must also be taken into consideration (Sect. 6.1). Figure 3.47 may be used to determine the effective fuel value of a solid or loose m^3 of biomass, when the effective heating value per kilogram of dry mass, moisture content, basic density, and solid content of biomass are known. The example of birch chips in Fig. 3.47 assumes an effective heating value of 19.2 MJ/kg, a moisture content of 40% on a fresh mass basis, a basic density of 500 kg/m^3, and a solid content of 0.35. With these assumptions, the effective heating value of the chips is 8800 MJ/m^3 solid and 3100 MJ/m^3 loose (Nylinder 1979).

Chapter 4

Recovery of Residual Forest Biomass

Timber harvesting operations may be classified according to their silvicultural goal and treatment. In *thinnings* primarily low-quality trees are removed from an immature stand to maintain or stimulate growth of high-quality trees. In precommercial thinnings the timber is left in the forest because of its small size. In *selection cuts* single or small groups of trees of both low and high quality are removed from an uneven-aged stand in order to mix age and size classes of reproduction. In *final cuts* all the trees of a mature stand are removed for artificial or natural establishment of a new stand. If the removal of trees is carried out in one pass, the operation is called *clear-cut.* If part of the trees is left standing for seeding and to protect seedlings, the first phase of the final cut is called *seed-tree cut* or *shelter-tree cut.* Biomass recovery may take place in any of these cutting systems.

Recovery of residual forest biomass is an insufficiently explored field of logging technology. Over the long term emerging harvesting methods will open up new technical possibilities for the utilization of formerly unmerchantable small-sized trees and logging residue as industrial raw material or as a source of renewable energy. The logging schedules for residue harvesting must be based as far as possible on standard equipment and only when necessary on special machines. Depending on conditions, residue recovery may be integrated with the harvesting of conventional timber in a *one-pass system*, or it may be carried out as a separate independent operation in a *two-pass system.*

Low costs, high productivity of labor, a high degree of recovery, ergonomically improved work methods, and purity of raw material from soil contaminants are concurrent targets. A great number of machines and methods of residue harvesting have been developed for large-scale industrial operations since the beginning of the 1970s, especially after the worldwide energy crisis in 1973. Because these special machines require high capital expenditure, a prerequisite for profitable operation is high production and full employment of the equipment. On the other hand, light equipment has been developed for the temporary needs of self-employed private forest-owners, especially in northern Europe.

Compared with the merchantable stem, residual biomass involves technical drawbacks such as small unit size, low areal yield, and low bulk density of the material that do not favor profitable recovery. In spite of mechanization, the productivity of work tends to be inadequate and the cost level higher than in conventional logging. Hence, the relatively low value of the product cannot always support the capitalization.

Where prices of conventional timber are high, the relative competitiveness of residual biomass is improved by its low stumpage price. Because unmerchantable biomass is often available in profusion, it is possible to concentrate activities on sites where the recovery is economically and silviculturally the most feasible.

The recovery of residual biomass and ultimately complete-tree utilization signify an epochal change in forestry. Special attention must therefore be paid to the maintenance of the nutrient balance and production capacity of the ecosystem. In certain habitats it will be necessary to limit recovery or to embark upon countermeasures in conjunction with harvesting (Sect. 9.1.4). Hence, the degree of biomass recovery and the choice of sites and logging techniques are affected by ecological aspects more than in conventional logging of only stems.

4.1 Harvesting Whole Trees

The concept of handling, transport, and processing of a *whole tree* applies here to a tree with intact branches even if it has been bucked into two or more parts or *tree sections* (Fig. 4.1), as they are commonly called in Sweden and other North European countries. The concept thus refers to the source and composition of the uncomminuted biomass and does not necessarily presuppose that the aboveground portion of the tree is kept in one piece and transported whole.

The technical and economic advantages of whole-tree logging are by far the most attractive when harvesting small-sized trees. Consequently, the majority of the logging methods and schedules introduced in this review are aimed primarily at small-tree operations, where the need for rationalization is the most urgent. However, whole-tree logging may as well be used for large-tree operations.

4.1.1 The Basis of Whole-Tree Harvesting

A small-sized tree is defined differently in different countries, depending mainly on the demand for low-quality wood. In North America trees under 20 cm at breast height are usually considered small, while in western Europe the small tree limit is often only 10 cm dbh. However, excluding natural tropical forests, the problems of harvesting small-sized trees are almost universal.

Typical objects of small-tree harvesting are early thinning of young softwood plantations and clear-cutting of unproductive hardwood stands of inferior spe-

Fig. 4.1. Whole tree in one piece (*above*) and bucked with the naked eye to three tree sections (*below*)

cies. In the *early thinnings* of pine plantations in various parts of the world, such as the southern United States, Brazil, New Zealand, Australia, the Republic of South Africa, and Europe, the technical logging characteristics of the stands are surprisingly similar.

Although the importance of timely thinnings in forest management is recognized all over the world, proper execution of the thinning regimes meets growing difficulties everywhere due to the shortage of labor and excessive logging costs. For an optimal response, thinning treatments should be selective from below, moderate, and repeated frequently. Economic and technical factors force a compromise, however. Technical considerations increasingly influence the intensity and scheduling of thinnings. Even if systematic thinning methods are less desirable from the silvicultural and environmental points of view and may result in yield loss, they are accepted in some countries, because opening corridors by removing one or more entire rows of trees provides convenient access for forest machines and elimination of marking costs.

The basic problems in harvesting small-sized trees are generally the same: preparing the traditional bolt-form timber is labor-intensive, slow and costly, and a high proportion of the biomass is left unutilized. Compared with clear-cuttings, technical logging conditions are always less favorable in thinnings. The following distinctive problems are aggravated in the first commercial thinning of a young stand:

1. Danger of damage to standing trees and soil restricts the moving of machines.
2. The trees to be cut are small, often partly unmerchantable.
3. Timber yield per unit area is low, which often leads to injurious postponement of the operation and finally reduction of plantation growth.

Because of these problems, the productivity of conventional, motor-manual logging methods is low in early thinnings. The machines, on the other hand, are often too large and uneconomical to operate in immature stands, since the output of mechanized methods is strongly dependent on tree size. As long as logging machines process small-sized trees singly, their output remains low and the economy of the operation weak. In view of this background and the expected scarcity of the raw material, forestry is faced with the necessity of seeking new solutions to the rationalization of small-tree harvesting.

Small-tree harvesting was developed earlier within the bounds of a handling and processing technology based on delimbed timber. Long-distance transport, scaling, and debarking above all presuppose the removal of branches. But if small-sized trees have to be handled and delimbed singly, rationalization is hardly possible. Instead, multiple-tree handling and abandoning delimbing provide greater scope for intensive biomass recovery and increased productivity of work. This leads to the principle of whole-tree harvesting, the application of which first started on a practical scale in the United States in the early 1970s. The essential *targets of whole-tree harvesting* can be outlined as follows:

1. As large a proportion of the above-ground biomass of the tree as ecological and technical limitations permit is recovered.
2. In-woods delimbing is avoided and bucking minimized. Processing of raw material is moved to the mill or special terminals.

3. Piece-by-piece handling of timber is abandoned at the earliest possible stage of the logging schedule and the raw material transformed into a mass article.
4. The mill-yard and industrial processes are adapted to handle the new type of raw material.
5. Integrated use for all the components of forest biomass delivered to the mill is set as a target.
6. The long-term sustained production capacity of the forest soil must not be sacrificed for increased biomass recovery, and the harvesting methods are to be adjusted accordingly.

4.1.2 Felling for Whole-Tree Harvesting

For all harvesting operations, either manual or mechanized, *directional felling* is a basic requirement. Directional felling improves work-site organization and safety, and reduces the time consumption for bunching and off-road haulage. The right felling pattern will also reduce the damage to the remaining trees.

A *felling angle* of 30° to 45° as seen from the transport direction is recommended for skidding trees by animals, winch, or tractor. The shallowness of the angle reduces the skidding damage to the standing trees. Because of visibility problems of the driver and a lower utilization of the reach of the crane, this shallow angle is less suited for forwarding with load-carrying tractors. Instead, a felling angle of 90° is preferred in forwarding of unlimbed tree sections and in-stand chipping

Fig. 4.2. Directional felling in selective thinning from below for off-road haulage of whole trees with a skidder (*left*) and tree sections with a forwarder equipped with a grapple saw (*right*)

of whole trees in selective thinnings (Fig. 4.2). This felling angle is not suitable for skidding, since swinging whole trees around into the strip-road system would cause widespread damage (Kofman 1985).

In whole-tree logging the trees are usually felled with the *butt to the strip road* for easier grasp with a forwarder crane or faster setting of the skidder chokers. Felling with all the *tops to the strip road* is rarely used, because the roads are then covered by the trees. However, a combined *butt-top felling* system can be used to widen the distance between strip roads. In this method the trees within crane reach at each side of the strip road are felled with the butt to the road, while the more distant trees are felled with the top toward the road. The theoretical maximum distance (L) in meters between the center lines of two adjoining strip roads is then determined by the crane reach (R) and the average height (H) of the trees as follows. The two meters in the formula are deducted from the sum of crane reach and tree height in order to give the crane a sufficient hold of the trees (Kofman 1985):

$$L = 2(R + H - 2) \ .$$

Both in motor-manual and mechanized whole-tree logging methods felling is integrated with bunching of trees. Correct planning and implementation of felling and bunching is extremely important for cost-effective and damage-free off-road haulage.

4.1.2.1 Motor-Manual Felling

When timber is harvested by conventional, motor-manual stem-only methods, a large part of the productive time of the cutter is used for delimbing (Fig. 4.3). In whole-tree harvesting, felling alone accounts for at least half of the time consumption of the cutter, because delimbing is eliminated and some other work phases possibly essentially reduced.

In whole-tree logging the cutter becomes a felling specialist. His primary task is to fell the trees in such a way that the tractor operator can pick up the load as smoothly as possible. A central goal is to speed up haulage by selecting a suitable *felling pattern*, which determines the location and orientation of the trees, thus affecting the operation of the subsequent machine (Olsson 1984). Collection into bunches by the cutter increases haulage productivity.

If the tree is large, the cutter can hardly do more than direct the falling tree correctly. But if the breast height diameter is less than 10–12 cm, especially in thinnings, trees can and should be bunched together. The *bunching* requirements are governed by the operation and reach of the tractor used for haulage.

A circular blade clearing saw is fast for mere felling of small-sized trees, but attempts to adapt it for combined, directed felling and bunching have not been successful. Hence, if the trees are to be recovered, felling must be done with a chain saw. But when a standard chain saw is used in whole-tree logging of small-sized trees, felling and bunching are ergonomically difficult. The following drawbacks are evident:

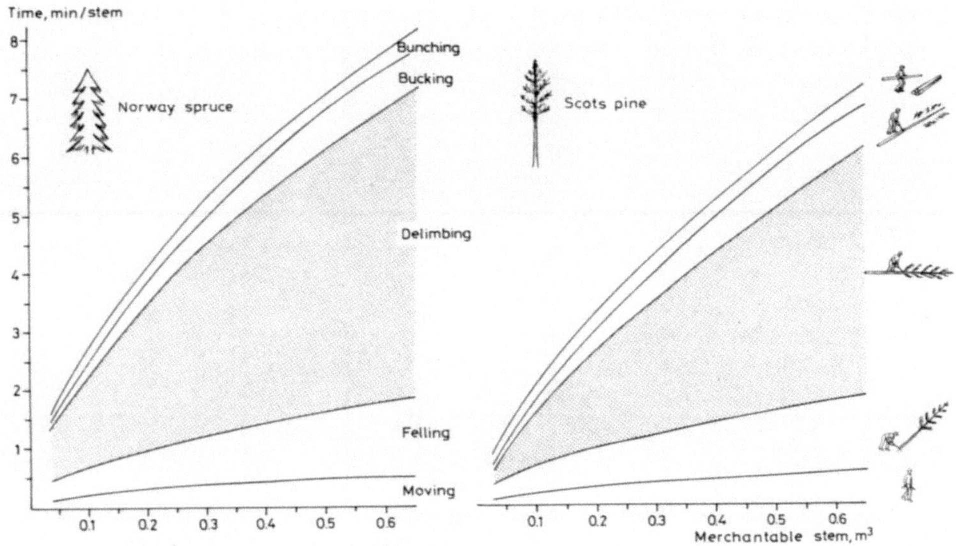

Fig. 4.3. Time consumption by work phases for motor-manual preparation of 3-m *Pinus sylvestris* and *Picea abies* pulpwood under average stand conditions in Finland. Rough delimbing, bucking with the naked eye, and light bunching for a long-reach crane (drawn from data in Kahala 1980)

1. The cutter has to do the felling in a bent position. The smaller the trees, the more frequently he must assume this awkward posture.
2. The cutter must push and guide the falling tree with one hand while sawing in a bent position. The risk of accident is increased.
3. Moving a fallen, unlimbed tree manually to a bunch is cumbersome.

In the Soviet Union, chain saws used for felling are equipped in conventional logging operations with a fixed frame, which permits the cutter to stand upright when felling the tree. The same principle is applied to small-tree operations in Finland, where a special work technique has been developed for *combined felling and bunching* of trees smaller than 12 cm in diameter, using an easily attachable felling frame as an auxiliary device of light chain saws. The gas trigger is placed in the upper part of the frame, but the original trigger of the saw also allows the saw to be used for bucking when necessary. The 1-kg frame must be adjusted separately for each chain saw model.

The working position of the cutter is improved by the felling frame. Rather than holding the saw in the air, the cutter lets it rest in contact with the ground. As a result of greater control over the tree due to his upright position, he is better able to guide the tree. The erect position also provides the cutter with good visibility, allowing him to plan ahead while working, which is imperative for feasible operation. The risk of accidents also diminishes. In ergonomic tests carried out in Czechoslovakia, the felling frame considerably reduced the pulse rate and energy consumption of the chain saw operator, and the noise level affecting the operator's ear (Petříček 1986).

Fig. 4.4. Use of a felling frame to synchronize motor-manual felling and manual bunching in an early thinning of a pine plantation

An essential feature of the work technique is the smooth synchronization of felling and bunching. As soon as the stem is severed from the stump, the cutter lays the saw down quickly and catches the falling tree with both hands. By guiding, pushing, or pulling the tree, and using its kinetic energy while it is still in the air, the cutter can move it a few meters to a bunch, generally placing the butt toward the strip road or winching trail (Fig. 4.4). The technique is easy to learn, but its adoption requires training.

Under favorable conditions a cutter's performance can be improved with the felling-bunching method by 20 to 30%. If the breast height diameter of the trees is 5–12 cm, the average productivity under Finnish conditions during the bare ground season is 2–4 m^3/h. In a North American study on very small-sized alder with more than 80% of trees less than 5 cm in diameter at breast height, the output was only 1.0 green short ton per productive hour (Mattson and Winsauer 1985). In Finland the method is well suited to small-sized pine and hardwoods but does not lend itself as well to spruce with its long and heavy crown (Hakkila et al. 1979). In an experiment in Maine the method worked well with softwood species with narrow cylindrical crowns. In a dense spruce-fir sapling-pole stand the production rate for 5-cm diameter trees was 175 and for 10-cm trees 109 pieces per effective hour. Multiple-stem hardwoods were more difficult to cut, because the saw had to be held above the ground and one stem might interfere with the cutting of the others (Hoffman and Mills 1983; Hoffman 1986). In Dutch experiments in young coniferous stands with an average height of 4 m, performance with the felling frame did not differ significantly from other equipment but because of the ergonomic advantages the felling frame was recommended. In thin-

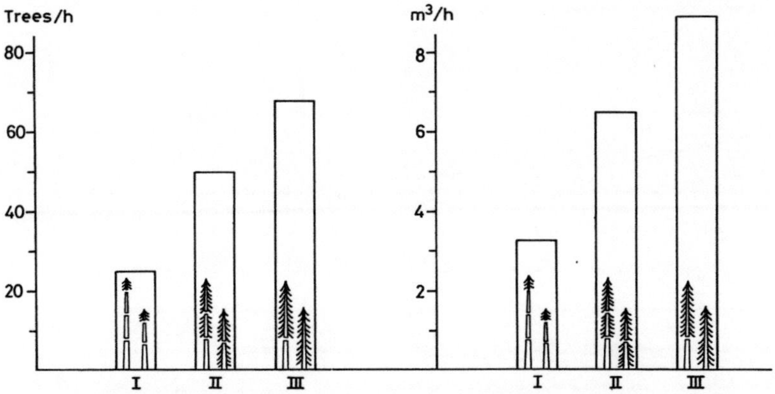

Fig. 4.5. Productivity of a chain saw operator in clear-cutting of a *Picea abies* stand by the conventional log-length method (*I*) and in two tree-section method (*II* and *III*) in Sweden. Average dbh is 14 cm and stem volume 0.13 m^3 (redrawn from Ahlgren et al. 1982a)

nings felled for machine processing, the conventional chain saw was more economical (Schaafsma and Hoksbergen 1980).

The felling frame is applicable only to small-tree operations when the majority of the trees are thinner than 12 cm at breast height. In most whole-tree harvesting operations the trees are too heavy to be bunched manually while branches are intact. Then the cutter uses an ordinary chain saw for directed felling. Productivity varies with the tree size and felling pattern employed. When the breast height diameter of the tree is between 7 and 16 cm, the average output per productive hour is between 100 and 50 trees (Olsson 1985).

Figure 4.5 represents an example of the productivity of a chain saw operator in clear-cutting of a Norway spruce stand with an average dbh of 14 cm and stem volume of 0.13 m^3. Three motor-manual cutting methods are compared. Alternative I represents the conventional log-length method, in which all timber assortments are delimbed and bucked. In alternative II the cutter delimbs and bucks the sawtimber logs in the traditional way, but only bucks the thinner pulpwood parts of the trees to unlimbed tree sections of a maximum length of 5.5 m. Alternative III is the same as the second one except that the unlimbed pulpwood is not bucked by the chain-saw operator. When the delimbing of pulpwood is abolished, the productivity of the cutter in alternatives II and III is 2.0 and 2.7 times as high as in the conventional method. In another Swedish study the differences were even greater under deep snow conditions (Ahlgren et al. 1982a, b).

Figure 4.6 compares the conventional, motor-manual log-length method with a motor-manual tree-section method, in which sawlogs are delimbed and bucked in the traditional way and pulpwood left unlimbed but bucked to about 5 m lengths. In the smallest diameter classes, the chain saw is equipped with a felling frame. The smaller the trees, the greater is the relative saving in the cutting time. The difference is most conspicuous for spruce, which has a much larger crown than either pine or birch. The figure also shows that time-saving in cutting is achieved partly at the expense of forwarding output (Sect. 4.1.3).

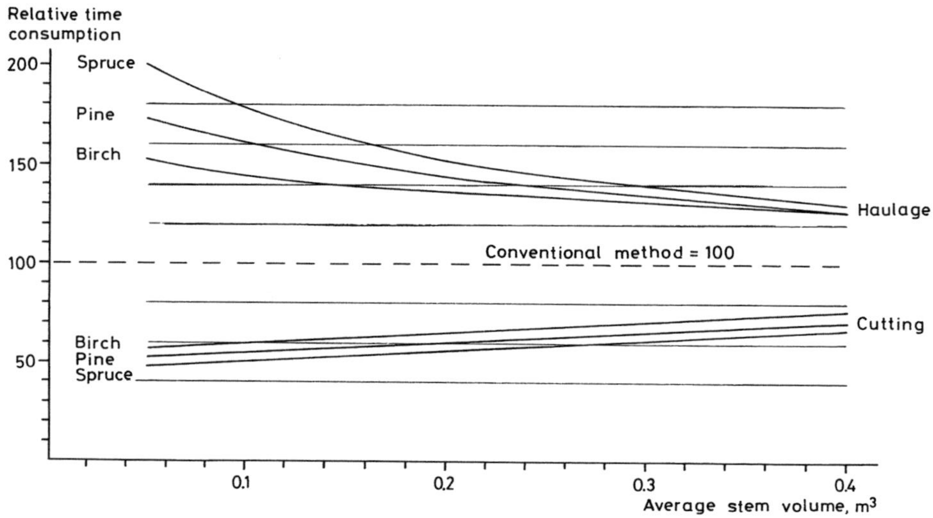

Fig. 4.6. Relative time consumption in preparation and forwarding of unlimbed tree sections of *Pinus sylvestris, Picea abies,* and *Betula* spp. in Finland as a function of stem volume, when time consumption of conventional log-length method is 100 (redrawn from Eskelinen et al. 1984)

4.1.2.2 Mechanized Felling and Bunching

Mechanization of felling generally solves the problems of bunching, which are critical to overall productivity and total costs of the entire logging schedule. Because the machines are able to move taller trees over longer distances than a chain-saw operator, bunches can be made larger and positioned more suitably for flexible loading. Even sorting the trees according to size or species becomes technically feasible. From the mechanization point of view, systematic thinnings, such as row or corridor thinnings, are advantageous. The output of subsequent off-road haulage improves as well, but the increase in system performance may be at the expense of the future timber yield.

A large variety of feller-bunchers is available. They have operational characteristics that suit different stand and terrain conditions and forest management patterns. *Machine-to-tree method* is based on chassis-mounted feller-bunchers which approach the trees individually, accurately position the felling head, sever the tree, and then travel with one or more trees to a bunch. *Boom-to-tree method* uses feller-bunchers with the felling head mounted on a crane which can reach, sever, and move several trees by swinging the boom from tree to tree while the vehicle itself remains stationary.

Since a felling-bunching machine may not perform more than about 120 running cycles per hour, productivity remains low in small-tree operations if trees are handled one by one (Ievin and Andreev 1985). The travel or swing time, respectively, is decreased by using an *accumulating felling head* allowing multitree handling. The accumulator reduces travel time and dump activities and speeds up placement and activation of the cutting mechanism. The following theoretical simulation results show the effect of a large accumulator, capable of collecting up to eight small trees, on the time distribution of a wheeled loader equipped with a chassis-mounted shear, in thinning of a young softwood plantation (Winsauer 1984):

	Travel per tree (m)	Travel time (%)	Shear time (%)
No accumulator	30.8	51.9	23.4
Large accumulator	4.5	20.6	70.8

The *severing principal* may be based on a single- or double-action shear, chain saw, circular saw, cone saw, or auger bar. Under all conditions it is necessary to use directional felling and have good control over the tree. Some feller-bunchers are capable of severing trees up to 50–60 cm in diameter at stump level, but the largest machines are uneconomical for small trees and unacceptable in thinnings.

A felling head may be mounted on various kinds of *prime movers*. They may be wheeled or tracked vehicles, and the frame construction may be rigid or articulated. If a felling head is mounted on the crane of a modified forest tractor, the machine can be used as a feller-skidder or feller-forwarder.

Excluding steep terrain, numerous reliable and effective feller-buncher combinations are available for clear-cutting operations, with tracked excavators and wheeled loaders as the most common prime movers. Figure 4.7 shows a heavy-duty feller-buncher designed to handle large trees under clear-cut conditions in North America. In thinning stands, especially in early thinnings, mechanization is much more difficult for economical and environmental reasons. Due to the small size of the trees, the productivity of machines tends to be lower in thinnings. The machine must be able to remove trees selectively without causing excessive damage to the soil or to the stems and roots of the remaining trees. Ecological factors limit the allowable width, length, and mass of the machine.

In the *strip road-based operation pattern*, or boom-to-tree method, the felling head is mounted on the end of the boom of a tractor crane. The machines do not cause soil compaction or root damage outside the strip roads. Even if the operation principle is applied increasingly in grapple processors and harvesters, it is used only rarely in felling-bunching for whole-tree harvesting. Some new boom-to-tree feller-bunchers, equipped with a tiltable pillar and cabin, can be used in thinning stands on steep slopes.

The reach of the crane determines the density of the strip-road network. To lengthen the reach of a strip road-based feller-buncher to more than 10 m, wheel could possibly be used to support the boom carrying the felling head. The technique might make it possible to accumulate several trees in the felling head and to convey them to the strip road in an upright position for a distance of 15–20 m (Knutell and Sundin 1982).

Stand-operating feller-bunchers, using a machine-to-tree operation pattern, place no technical limitations on the strip-road spacing. However, the road density naturally affects the performance and the extent of damage to the soil and trees. When this principle is applied, the strip roads are generally 25–40 m apart. In early commercial thinnings in northern Europe the machines have to move in

Fig. 4.7. The heavy Drott 40 feller-buncher clear-cutting a hardwood stand; boom-to-tree method (courtesy of Forestry Sciences Laboratory, Houghton Mich., U.S. Forest Service)

a stand where the number of remaining trees is commonly 1000–1200 per hectare. In central and eastern Europe the number of trees may be much higher but in tropical and subtropical plantations usually lower.

The stand-operating machines must be small. Consequently, they are affected adversely by rough terrain. The risk of damage to soil and trees is high in poor and soft terrain and with tree species with a superficial root system. The method is therefore used mainly in easy, flatland conditions, as in pine plantations on dry, sandy sites in Europe and the southern United States.

After accumulating a grapple-full of one to six small trees, a stand-operating light feller-buncher conveys the trees in an upright position placing them along the strip road and leaving them according to the requirements of the subsequent machine of the logging schedule. *Performance* is affected by stand characteristics such as the size and branchiness of the trees, undergrowth, areal yield, number

Fig. 4.8. The light Makeri 33 T feller-buncher in selective thinning of a young pine stand; machine-to-tree method (courtesy of FFRI)

of remaining trees per hectare, strip-road spacing, and terrain. Travel speed, shear speed, accumulator head size, thinning pattern, back-up distance, and required bunch size are the most important machine and work organization factors affecting the output (Winsauer 1984). In second thinnings of southern pine, average tree diameter and the number of trees per accumulation are the most important factors affecting feller-buncher productivity. Less important, but still significant, are bunch size, strip-road spacing, and the average distance between trees (Greene et al. 1987).

For example, a Finnish 3.7-t, rigid-framed, tracked feller-buncher (Fig. 4.8) produced 105–160 trees or 3–17 m^3 and a Swedish 5.2-t, articulated, tandem-bogie feller-buncher 115–180 trees or 3–20 m^3 of stem volume per effective hour in selective thinning of young *Pinus sylvestris* stands (Hakkila and Wojcik 1980; Kahala 1984). An American 7.9-t, rigid-framed, four-wheel feller-buncher had an output of 15–20 m^3 of stem volume per productive hour in row thinning of a 20-year-old *Pinus echinata* plantation, when the undergrowth was light. Dense undergrowth of small, unmerchantable hardwoods, vines, and other vegetation reduced visibility for the machine operator, resulting in a 20% lower output (Granskog and Anderson 1981).

On steep slopes the cost of logging is usually too high for separate recovery of low-quality and small-sized wood. To be profitable, recovery must be integrated with the harvesting of ordinary sawlogs and pulpwood. For example, whole trees

Fig. 4.9. Menzi Muck 3000 EHA feller-buncher prototype for steep terrain; boom-to-tree method (courtesy of Forestry Sciences Laboratory, Houghton Mich., U. S. Forest Service)

are deposited on the landing, and high-grade logs are removed by chain saw before chipping the rest of the biomass. In steep-terrain hardwood stands in the eastern United States, biomass recovery was increased 30–60% over conventional hardwood operations by combining a running skyline with a whole-tree chipper (Castaños and Arola 1982).

Physical and economical constraints make the use of ordinary feller-bunchers impractical and often impossible on slopes steeper than 20–30%. Timbco Hydro-Buncher and Tree Power are examples of special feller-bunchers for steep terrain. Due to characteristics such as a self-leveling cab and boom platform and a low center of gravity, they can be operated on slopes of up to 60% to improve the efficiency of cable yarders. An unusual Swiss machine, called the Menzi Muck, originally designed for construction and mining industries, was equipped in the United States with a 30-cm felling-bunching head for harvesting whole trees from steep terrain on slopes up to 100% (Fig. 4.9). The semiwalking machine has a pair of nonpowered wheels equipped with a ratchet-type locking mechanism which, when activated, will allow the wheels to roll in only one direction. Other major operational components are a pair of extendable stabilizer legs with pads and a knuckle-boom with a hydraulic shear without an accumulator. The 5-t machine propels itself up and down slopes and over obstacles in an inchworm fashion by pulling or pushing with the boom. A winch and cable secured to an upslope tree provides added safety against downslope slippage on very steep slopes. In two hardwood clear-cutting sites on slopes too steep to harvest by a conventional feller-buncher, the production rate was 85 and 50 trees or 6.4 and 14.9 green tonnes per productive hour for trees with an average breast height diameter of 11 and 18 cm, respectively (Arola et al. 1981; Arola 1983b). Another steep-terrain feller-buncher, mounted on a Lichtenstein-made prime mover of a similar type, has been tested successfully in the Pacific Northwest on slopes ranging up to 60% (Schiess et al. 1983).

As a rule, bunching should be integrated closely with felling, since a work organization based on separate machines for felling and bunching is usually less cost-effective. However, under certain conditions the separation of these work phases is feasible, especially if the felling cannot be mechanized. This is often the case in the seasonal operations of small, private forest-holdings using low-investment systems.

The most common method of separate bunching, or rather preskidding, to the side of the strip road, is *winching* over short, 10–50 m distances. The winching goal is to improve the haulage productivity of larger machines and reduce site impact. Winches are used either separately and anchored to a tree or stump, or mounted on a tractor, commonly on the three-point hitch of a farm tractor. The winch is controlled mechanically or by radio. Blocks are used to change the direction of skid when required.

In flat terrain, winching represents light mechanization suitable for timber harvesting in farm forests by a self-employed owner himself, and often for conditions prevailing in developing countries. Winching makes it possible to widen the strip-road spacing in first commercial thinnings to 50 m and in later thinnings even more. However, pulling whole-tree bunches to the side of the strip road may be technically difficult in a dense stand. Because the cable must be dragged manually to each tree, widening the strip-road spacing increases the physical strain of the work, particularly in steep and rough terrain and under snow conditions. The risk of damage to the root collars of standing trees grows with the winching distance.

The output of winch-bunching increases with tree size and decreases with widening strip-road spacing. The more closely spaced the strip roads are, the greater is the proportion of trees that are left within reach of the tractor by the cutter. Including the trees already within reach of the 5.5-m knuckle-boom of a tractor working on strip roads spaced at 30-m intervals, the output of a farm tractor-mounted winch in a first thinning under bare-ground conditions is $3-8\ m^3/h$ of whole trees. If a winch is combined with a mechanical or hydraulic boom loader, or with a skidding grapple, the farm tractor can combine winch-bunching, loading, and haulage to the landing.

Another solution to *separate bunching* in thinnings is a 4-t South African tricycle vehicle featuring two hydrostatically driven wheels, a smaller trailing caster wheel, and a simple telescopic grapple boom loader for bunching, preskidding, and loading of trees. In level terrain this compact machine, which does not fit into the traditional categories of skidders, forwarders, or loading cranes, is easy to integrate into established harvesting operations so as to improve the efficiency of other machines (Bell 1985). For example, the machine can operate feasibly on minor 3-m wide corridors opened by a feller-buncher in a herringbone fashion, in order to collect and consolidate small bunches into larger formations alongside the strip roads for further haulage with heavier equipment (Stokes and Lanford 1985). In pine plantations in flat terrain this machine is highly maneuverable, but ergonomic and traction problems occur on relatively gentle slopes and in uneven terrain (Wells 1980). The three-wheeled prime mover also can be equipped with a feller-buncher head. Instead of carrying trees upright, the machine directs the falling tree toward the pile and then quickly positions it on the pile with the grapple (Stephenson 1986).

4.1.3 Off-Road Haulage of Whole-Tree Material

Two main types of equipment are used for off-road transport of whole-tree material in terrain where the slopes are not steeper than 25–30%. *Skidders* drag the whole trees on or partially off the ground, and there is thus no technical upper limit to the tree length. *Forwarders* carry the whole load completely off the ground, and hence the length of the material is generally limited to 5–8 m.

Steeper slopes are inaccessible for most ground vehicles, and the tractors must be replaced by *cable yarders*. The cost of harvesting is then essentially higher. Consequently, the method is seldom used for the recovery of small trees and biomass residue. Another reason limiting the recovery of crown mass in mountainous terrain is the danger of erosion on steep, bare slopes.

4.1.3.1 Off-Road Haulage with Skidders

Skidders are generally simpler and cheaper than the forwarders. Load volume is strictly dependent on the tree size, so that the use of large skidders is not feasible for small trees or short tree sections. Except with very tall trees, the load is smaller than in the forwarders. Consequently, despite a faster traveling speed, the competitiveness of the skidders is best in the haulage of large trees over short distances. Three types of skidders are recognized.

The simplest of these is the *cable skidder*, equipped with a single or double drum winch, cable, and cable or chain chokers. Cable skidding is a rather labor-intensive operation because both manual pulling of the cable and coupling and removal of chokers around the stems are time-consuming, especially in whole-tree haulage. The average production of a cable skidder in off-road haulage of 10–25-cm-thick hardwoods is about 5 m^3 per man-hour, while that of a grapple skidder is 10 m^3 per man-hour (Koch 1985b). A cable skidder can cross muddy areas and steep slopes by disconnecting and then rewinching the load. In small-tree operations the size of the load is limited by the number of chokers rather than the load capacity of the skidder.

Fig. 4.10. Hauling whole trees with a grapple skidder

The *grapple skidder* has a special kind of loader-mounted grapple, i.e., bottom-opening hydraulic jaws, which are designed for collection of scattered trees or bunches of trees primarily by reciprocating motions along the vehicle. During haulage the trees are held by the grapple (Fig. 4.10).

The *clam-bunk skidder* has a bunk with top-opening hydraulic jaws mounted on the rear of the vehicle. The trees are held by means of sharp, inside edges and chains or cables built into the jaws to suspend the load. Loading is carried out with a regular hydraulic crane mounted on the tractor (Fig. 4.11). Load is larger and its mass distribution better in comparison with other types of skidders, because the center of gravity of the load is moved over the rear axle which reduces the friction of the load and improves the traction of the wheels. The clam bunk skidders are most competitive when the hauling distance is long and trees large.

Compared with large logging organizations, the economic conditions for mechanization are less desirable in the small-scale operations of self-employed, private forest owners. Off-road haulage takes place naturally with farm tractors, which can be equipped for skidding as well as for forwarding. An example of the special equipment for light mechanization of *preskidding* is the Swedish 2.2-t Cutting Ant feller-skidder (Myhrman 1983; Fig. 4.12). The driver operates the small, articulated, four-wheel drive vehicle while standing on the ground. The trees are felled and placed in the mini-skidder's clam bunk with a crane-mounted hydraulic felling shear. The load is then preskidded to alongside the strip road for further haulage with a larger tractor. The method allows the use of wide strip-road spacing.

A great deal of *biomass breaks off* in the skidding of whole trees. In a Swedish experiment the average loss of crown and top material from unlimbed sawtimber

Fig. 4.11. Hauling whole trees with a clam-bunk skidder

Fig. 4.12. The Cutting Ant feller-skidder for light mechanization of small-tree harvesting. The driver operates the vehicle from outside on the ground (courtesy of Hydrovåg)

trees was 56% for Scots pine and 52% for Norway spruce in clam-bunk skidding under bare-ground conditions over a distance of 600 m (Rheborg 1972). Under winter conditions, when the temperature varied between −4° and −17 °C and the skidding distance was 470 m, the loss was 63 and 58%, respectively (Gunnarfelt 1973). In a Russian experiment, 48% of crown mass was lost when large, whole trees were skidded the top foremost. Only 37% was lost when the butt was foremost (Gugeleb and Emeljanov 1970). In a Norwegian experiment the biomass loss was only 4% in mild winter weather when small-sized trees were loaded manually to a farm tractor and hauled over a distance of 400 m (Gislerud 1973).

4.1.3.2 Off-Road Haulage with Forwarders

In most countries the off-road movement of unlimbed trees is accomplished by skidding. Canada and the United States use mainly wheeled skidders, and the Soviet Union and China track skidders. In some other countries, such as Sweden and Finland, whole-tree material is hauled to the landing site with forwarders, whereupon the trees must be bucked to 5–8 m sections (Fig. 4.13). The following reasons may be mentioned:

1. In northern Europe, conventional timber assortments are transported almost entirely by forwarders. For operational reasons, whole-tree raw material should preferably suit the same equipment.

Fig. 4.13. Hauling unlimbed tree sections with a forwarder. Bucking with a grapple saw in conjunction with loading

2. Whole-tree raw material picks up grit and other contaminants when dragged along the ground. Grit blunts chipper knives, wears out mill equipment, lowers the quality of composite boards, and creates ash and slag problems in the combustion of forest biomass. A load-carrying forwarder keeps the biomass much cleaner.
3. Whole-tree harvesting is applied in northern Europe mainly in small-tree operations where skidders are not cost-efficient because of the small load size. In forwarder haulage the tree size affects the load volume much less.
4. Skidders require larger landing areas. Forwarders stack the trees or tree sections at the landing, or at the road side, into 3–4-m-high piles, thus saving landing space. Furthermore, forwarders can, if necessary, do some sorting while unloading the whole-tree material.
5. Due to a better mass distribution, lower driving speed, and more sophisticated construction, the ergonomic properties of modern forwarders are superior to those of most skidders. Besides this, the larger load size of the forwarder results in a shorter daily driving distance and thus reduces the whole-body vibration to which the operator is exposed.
6. Because of shorter load, forwarders cause considerably less damage to standing trees in thinnings.

The skidder characteristics referred to above are typical mainly for cable and grapple skidders, while modern clam-bunk skidders have more resemblance to forwarders in some respects. A weak point of standard forwarders in whole-tree harvesting is that the short load space is not adequate for long, whole trees. It can easily be lengthened by 1–2 m, but trees longer than 6–8 m generally have to be bucked with a chain saw motor-manually by the cutter or mechanically in conjunction with the haulage operation. In the latter case the tractor crane is furnished with a hydraulic grapple saw.

A forwarder is preferably equipped for whole-tree transport with a *long-reach slide- or telescope-boom crane*. The reach of the boom used for whole-tree operations in thinnings is 8–11 m. The trees are first skidded with the extending boom to the strip roadside by sliding them along the ground, and lifted only when they are close to the forwarder. When the strip-road spacing is not more than 30 m, the forwarder can collect all single and bunched trees aligned toward the strip road. However, when the forwarder has to pick up part of the trees from far off, its transport output is reduced by 10–15%. A further decrease is caused by grapple sawing if motor-manual bucking is not used. Since the plates between adjacent arms in a standard grapple may hamper the handling of tree sections and catch stones and other impurities, they are often removed (Bengtsson and Myhrman 1982).

A *grapple saw* is a special chain saw with its hydraulic motor mounted outboard on an ordinary timber grapple. The operator grasps a single tree or a bunch of trees at a desired point, lifts them clear of the ground, bucks them by activating the saw, and loads the material onto the forwarder. Since the grapple saw is somewhat wider than the original grapple, problems may occur in reaching trees in dense forests. Grapple sawing reduces forwarder productivity in haulage but, given certain conditions, improves the overall productivity of the whole-tree harvesting schedule.

Regardless of the size of the trees, a *forwarder load* of bulky, unlimbed timber is roughly 40–50% lighter than a load of conventional cordwood of the same length. The low bulk density increases the cost of haulage particularly over longer distances. The following figures show the average volume of a 5-m pulpwood load in a medium-sized forwarder in Finland (Kahala 1984):

	Biomass (m^3 solid/load)	
	Delimbed	Unlimbed
Pine pulpwood	9–11	5.5–6.5
Spruce pulpwood	9–11	4.5–5.5

The forwarding output of whole-tree material suffers from the underutilization of the mass capacity and extra time consumed for grapple sawing. According to Kuitto (1987), combined forwarding and bucking of tree sections requires 1.4–2.5 times longer as conventional forwarding of log lengths in thinning stands. The following figures illustrate the *performance* of a medium-sized forwarder in hauling unlimbed, small-sized material from the first thinning of Scots pine in connection with three various logging methods (Hakkila et al. 1979). In the first method, felling, bucking, and bunching are carried out motor-manually. The forwarder crane has a reach of 10 m. In the second method, felling takes place motor-manually as well. Bunching to the side of the strip road is carried out with a separate boom-type buncher. The forwarder is equipped with a 5.5-m knuckle boom crane and grapple saw for bucking in conjunction with loading. In the third alternative, felling and bunching are carried out by a light feller-buncher, resulting in a larger bunch size. The forwarder is equipped again with a 5.5-m knuckle boom crane and grapple saw. Mechanized bunching to the side of the strip road clearly improves the output of forwarding:

Method of bunching	Output in forwarding, m^3 solid per productive hour	
	Haulage distance (m)	
	100	500
Manual bunching	6.9	5.4
Strip road-operating boom-type buncher	8.2	6.4
Stand-operating feller-buncher	9.4	7.0

In later thinnings of a softwood plantation the haulage output is higher due to larger trees and a higher biomass yield per unit area. In clear-cutting areas the output for the same tree sizes is an additional 10–20% above the average level in the thinnings.

Extraction of tree sections with conventional forwarders presents problems. The load capacity is poorly utilized, the long load swings out on cornering, and the machine can easily become tail-heavy. According to Hallonborg (1984), an *optimum tree-section forwarder* for thinning operations should have the following characteristics:

1. Damage to the stand and area occupied by strip roads must be kept to a minimum in thinnings. In conventional operations the free-space requirement of forwarders is kept down with single tracking, i.e., with the rear wheels following precisely the tracks of the front wheels. Because 5 – 8 m timber swings out when cornering, the free-space requirement of a tree-section forwarder is actually most narrow if the rear wheels are designed to cut in somewhat on cornering. Then the tail end of the load will extend only slightly outside the arc followed by the outer front wheel.
2. The ground pressure of a tractor is influenced by the distribution of mass between the front and rear wheels and also between the bogie wheels. Conventional forwarders tend to become tail-heavy in tree-section haulage because of greater overhang. The problem can be solved by lengthening the rear section of the tractor. It may also be justifiable to modify the design of the bogies to prevent excessive load on the rearmost wheels.
3. A load of tree sections contains a large proportion of voids. Compaction equipment could provide advantages.
4. To reduce the free-space requirement when driving over obstacles and on lateral slopes, it is customary in conventional forwarders to incline the side stakes of the load space slightly inward. However, especially when handling tree sections, this makes the loading work more difficult. A feasible solution might be a combination of folding stakes and compaction.
5. A forwarder is equipped with a long-reach crane and a grapple saw for the tree-section method. For smooth operation, the crane should be mounted on a tilting base plate. This enables the boom pillar to be kept vertical when the machine is either on an uphill or downhill slope, thereby reducing the damage to standing trees. To make this possible, the cab guard on many machines has been moved back, which again increases the tendency for the machine to become tail-heavy. But since the tree sections do not slide in the load easily, it should be possible to cut a larger recess in the middle of the cab guard to accommodate the boom post.
6. When equipped with a long-reach crane and a grapple saw, a forwarder needs stability. An effective locking mechanism for the articulated joint is therefore necessary, since the front section of the forwarder provides a large part of the overall stability. When the machine is positioned with the front and rear sections at an angle to each other, additional stability can be gained by locking the bogies.

As long as tree sections are hauled only some of the time, it would not be feasible to reconstruct a forwarder specifically for unlimbed timber. However, in many countries the need to increase thinnings of young plantations is urgent. Activation of silvicultural management programs together with the labor shortage will probably open up possibilities for expanded whole-tree logging, improving the economic conditions for the use of special equipment.

A special *feller-forwarder* concept has been developed for large clear-cutting operations in Canadian boreal forests and in the northeastern United States. The heavy 26-t, articulated, four-wheel drive vehicle, equipped with a crane-mounted, accumulating felling head and a rear-dumping load space, fells the trees onto the

Fig. 4.14. The heavy-duty Koehring K2FF feller-forwarder for large-scale clear-cutting operations. Load capacity is 15 t (courtesy of Koehring Canada)

machine and carries the up to 15-t whole-tree load off the ground to the landing (Fig. 4.14). Where applicable, the method offers many advantages: simple work organization, over 100 m^3/man-day productivity, low harvesting cost, high biomass recovery, and clean raw material assuring long runs between knife changes in chipping (Kurelek 1984). The method is especially suitable for whole-tree chipping schedules, but an absolute prerequisite for cost-efficient use of the machine is continuous, large-scale clear-cutting activity.

4.2 Relogging Slash from Cutovers

Logging residue in a cutover area consists of two parts: logging slash and living residue (Sect. 2.5). *Logging slash* is defined as all above-ground residue left on ground in timber harvesting, consisting of crown material, unmerchantable tops, undersized trees, culls, defective stem sections, unmerchantable tree species, breakage, and possibly even some merchantable timber. The relative amount of slash depends on the tree species, size and age, history and characteristics of the stand, minimum dimensions and quality requirements of various timber assortments, demand for small-sized wood, and logging practices. Under good market conditions and in intensively managed forests the main components of slash are crown and top, while in remote, unmanaged virgin forests the mass of culls, defective stem parts, and dead trees often exceeds that of the crowns. In natural tropical forests much of the logging slash consists of unmerchantable, inferior species.

For several reasons, relogging of slash is restricted mainly to clear-cuts. First, the amount of slash is generally highest on clear-cut areas, and hence the cost of recovery is lower than from thinnings. Secondly, the nutrients released from decomposing slash in clear-cut areas are largely lost through leaching, while in thinning stands the residual trees can take them up immediately. Thirdly, the removal of slash from clear-cuts facilitates soil preparation, planting, and direct seeding for the establishment of a new stand.

4.2.1 Collection of Slash

The *location of slash* on the cutover is important from the viewpoint of recovery. It is largely a result of the felling pattern and delimbing method. Manual and motor-manual stem-only logging leaves the slash fairly evenly distributed around the logging site, although directed felling toward strip roads causes some concentration. Self-propelled logging machines, on the other hand, leave slash in various forms of heaps or windrows depending on their operating principle (Fig. 4.15). The reach of the crane, or the ability of the machine to process several trees at every setup, naturally affects the concentration. The size of the piles depends also on whether the tree is driven through the delimbing device parallel to or across the strip road, whether the unmerchantable top is also fed through the delimbing device, and whether the machine continues delimbing when moving. In the case of crane-mounted one-grip processors and harvesters the result depends more on the driver's decisions.

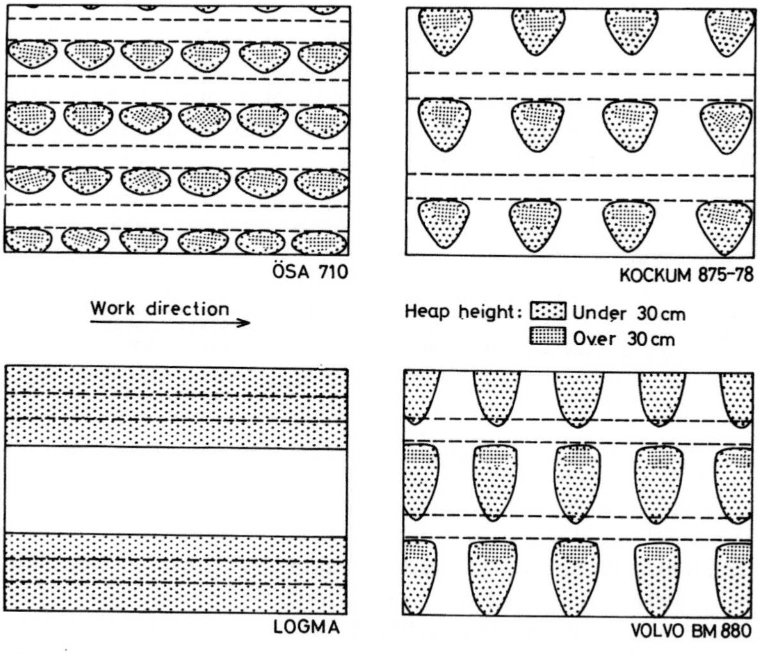

Fig. 4.15. Location and height of slash heaps in a clear-cut area after motor-manual felling and mechanized delimbing with various chassis-mounted processors in Sweden. Removal of conventional timber 150 m^3/ha. Strip roads shown by *dashed lines* (redrawn from Simonsson 1974)

For feasible forest haulage, the slash has to be in large enough formations. After motor-manual delimbing the first phase of residue recovery is collection of slash into heaps, and then the use of special equipment is essential. If the timber is prepared by logging machines, additional bunching of slash is seldom needed. A simple method is *bulldozing* the slash in windrows with a forest or crawler tractor. However, so much sand, stones, and other impurities become mixed with the slash that the raw material is unsuitable for industrial processing and possibly unsuitable even for combustion.

With regard to the purity of slash and the carefulness of the recovery, a satisfactory result is achieved with minitractor-type *front-end loaders* equipped with a light *slash jaw*. Because of their low ground clearance, however, these machines do not perform well under a variety of terrain conditions. They cannot be used on soft or stony sites and in deep snow, and the whole-body vibration to which the driver is exposed may be excessive for continuous work in forest terrain. In a Finnish study, the size of piles gathered with minitractors was 7–18 m^3 loose or 0.3–1.0 t dry matter. The hourly output was 3–4 t dry matter, or 0.1–0.3 ha (Hakkila and Kalaja 1974).

A special slash jaw, mounted on a heavy, four-wheel log loader, has been developed for large-scale operations in the United States. The device has eight lower teeth opposed by three upper, incisorlike arms. The lower teeth are all hydraulically operated to pop back automatically when they hit a boulder or a stump. The teeth reset after clearing the obstacle. The machine is capable of working in rougher terrain than the light minitractors, but due to the high investment cost it requires large work sites and basically year-round employment. The unit can pile slash in windrows at a rate of 1 ha/h.

The South African *tricycle buncher* principle, which is described in Section 4.1.2.2, shows good potential for collecting pieces of residual stem mass for subsequent extraction by a forwarder or skidder. Bunching with the standard equipment of the machine increased the skidding output of residual stem sections of large *Pinus contorta* in a cutover area in New Zealand by 55–65% (Wells 1980). The limiting factors are insufficient terrain capability on slopes and on uneven or soft ground, and the small grapple size.

Under northern European logging conditions, a workable solution for slash collection after motor-manual cutting is to replace the timber grapple of a long-reach boom with a 1.5–2-m-wide *slash rake*. The forwarder can rake slash in windrows from both sides up to a distance of about 10 m. In deep snow the work is difficult, but on a bare-ground experiment the bunching output was 9 t dry matter or 0.4 ha per effective hour (Melkko 1976). Excluding the rake, no additional investment cost is involved when a forwarder with a long-reach crane is available.

Hardwood residue is a major potential source of additional biomass in the eastern United States. Selective logging of hardwood sawtimber leaves about 40% of the above-ground mass behind in the form of tops and branches (Castaños and Arola 1982). Salvaging these heavy-branched tops of large hardwood trees poses technical problems. For example, in a survey in the Upper Peninsula of Michigan the unmerchantable tops from sawtimber trees had an average length of 11.0 m and width of 7.6 m (Christopherson 1983b). Skidding the bulky tops causes exces-

Fig. 4.16. The top-wood processor prototype developed by the U. S. Forest Service (courtesy of Forestry Sciences Laboratory, Houghton Mich., U. S. Forest Service)

sive damage to the residual stand especially if the work is performed during the loose-bark season. Moreover, large tops reduce skidder payloads and may need trimming before feeding to a chipper. Reducing the top by chain sawing is costly and unsafe.

A hydraulic shear specially mounted on the knuckle boom of a carrier vehicle has been designed by the U.S. Forest Service to sever large branches and align them with the top for more effective skidding. The principal feature of the *top-wood processor* prototype is the mounting mechanism between the shear and boom, permitting the operator to rotate the shear assembly in practically any direction within the limits of a 90° wrist action relative to the final boom axis (Mattson et al. 1978). In the second prototype of the topwood harvester the shear has been replaced with an auger cutter, allowing limbs to be severed much closer to the stem. The prime mover is a clam-bunk skidder which also performs haulage (Fig. 4.16). In a work study the output of processing and haulage of 2-year-old tops was 1.7 t dry matter per operating hour. About 50–60% of the productive time was used for processing and loading the tops onto the clam bunk (Christopherson and Barnett 1983; Christopherson 1983a).

4.2.2 Off-Road Haulage of Slash

If slash has been bunched with special machines, the pile size is larger than directly after logging machines, resulting in higher haulage output. However, the total

Fig. 4.17. Off-road haulage of slash with Kockums 84-31 forwarder (courtesy of Hannu Kalaja)

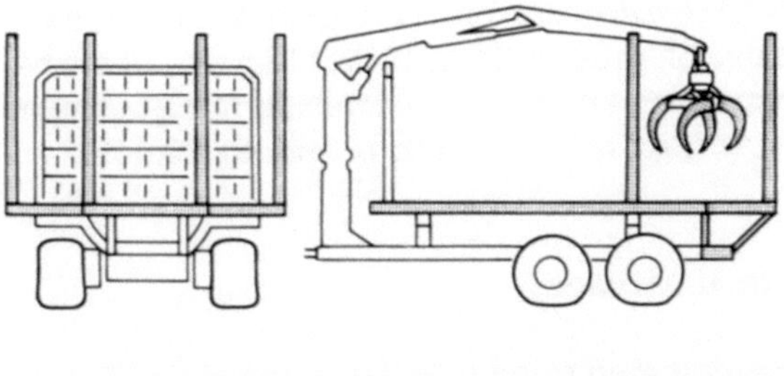

Fig. 4.18. Enlargened load space of a forwarder for off-road haulage of slash (Mellström and Thörlind 1981)

economy of slash harvesting is best in conjunction with mechanized logging systems since the cost of bunching is avoided.

The most feasible solution for off-road haulage of branch and top residue from softwoods is the load carrying farm or forest tractor (Fig. 4.17). Because the bulk density of slash is only 75–150 kg/m^3 piled, the mass of a load stays considerably below the *capacity* of a forwarder. Although, due to the load crown, the frame measure of a slash load may be up to twice the nominal load space of the tractor, the efficiency of the transport is still unsatisfactory. The load space is therefore often lengthened with a deck or tailgate and widened with special side stakes (Fig. 4.18) to give a load volume of up to 50 m^3 piled. The bulk density may be compacted slightly by pressing the slash with the grapple in conjunction with loading, but the gain is not large because of the elasticity of the branches. With enlargement of the load space and careful loading, the net load of green slash in a medium-sized forwarder may be increased to 8–10 t and the costs of haulage lowered roughly by 10%.

A conventional timber *grapple* is not ideal for slash handling. The grapple may be modified by widening, increasing the number of arms, and removing the plates from between the tips of adjacent arms. A horizontal grapple cylinder allows a higher slash load than a vertical one.

The *productivity* of slash haulage improves with increasing pile size, which again depends on the total amount of slash per unit area and the preceding bunching method. If the slash is hauled during the snow season immediately after the preparation of timber, the output is higher than under bare-ground conditions, except at very low temperatures when the load size has to be reduced due to weak stability. On the average, a production rate of 7 m^3/h is achieved over a transport distance of 300 m in practice (Mellström and Thörlind 1981). This is as much as the productivity in forwarding of unlimbed tree sections, but considerably below the haulage output of conventional timber from clear-cuts under corresponding conditions.

A part of the biomass is lost in the bunching and forwarding phases, especially if logging machines have moved over the slash earlier and added soil and stones to the material. The *efficiency of recovery* is affected by several factors. The more slash available per unit area, the higher is the relative level of recovery. If slash is not harvested until several months after the clear-cutting, all foliage is left on the site and thinner twigs start breaking when handled with a grapple. In very cold weather branches tend to break and fall off the load. The driver's experience and skill also affect the result.

After mechanized logging, the level of slash recovery under bare-ground conditions in Sweden is between 50–60% (Mellström and Thörlind 1981) or even 66–75% (Sondell 1984) of the available on-ground slash. If the efficiency of recovery is raised very high, the cost of harvesting and the risk of mixing foreign matter with the biomass will increase. A very high recovery ratio is seldom feasible.

4.3 Continuously Progressing Swath Harvesters

In the northeastern United States it is common to refer to dense young coniferous stands as *thickets* and dense young deciduous stands as *puckerbrush*. The number of successional deciduous trees and woody shrubs of various species is very large in a puckerbrush stand, and the annual dry matter production is comparable to the climax deciduous species. In many countries there is an enormous tonnage of puckerbrush that cannot be economically utilized until special harvesting and transport equipment for this small-sized material is developed and commercially available (Young 1980a).

All the whole-tree harvesting methods described in Section 4.1 presuppose single-tree felling. Even with an accumulating shear it is difficult to reach a felling output of 3 trees/min or 180 trees/h. If the tree volume is very small, e.g., 0.01 m^3, techniques based on single-tree felling are not sufficient. The costs of felling and subsequent bunching tend to become too high to allow large-scale utilization of this low-value raw material.

Swath harvesters offer an alternative for the highly mechanized recovery of forest biomass from brush stands composed of a large number of undersized trees. Without stopping to sever single trees and process the biomass, these continuously progressing, multipurpose machines fell all trees growing in the swath that the advancing machine covers. The trees are handled in all phases of harvesting as a *mass article*. As a consequence of the operational pattern, the principle of swath harvesting is applicable only in clear-cutting and corridor thinning.

Although large areas of land throughout the world are covered by unmerchantable puckerbrush, no commercial swath harvesters are available for puckerbrush operations, probably due to lack of demand for low-valued, small-tree biomass. However, several prototypes have been built and tested, and some of them have proven the feasibility of recovering above-ground biomass from brush stands. Accordingly, Section 4.3 is based solely on experiences from prototypes of swath harvesters reflecting the potential of the concept rather than proven, practical technology.

4.3.1 The Basis of Swath Harvesting

Tree size is probably the most important single factor affecting the productivity of a recovery system and the cost of harvesting forest biomass. In single-tree systems, regardless of the degree of mechanization, the productivity of work is bound to remain low when the tree volume is below 0.05 m^3. Compared with motor-manual methods, many machines are more sensitive to varying tree size.

Due to excessive logging costs, the minimum breast height diameter of a merchantable tree is, under the most favorable market circumstances, seldom less than 7–10 cm. Consequently, brushy stands are commonly excluded from commercial utilization. Yet their total quantity of biomass per hectare may be considerable. Furthermore, removal of inferior brush vegetation would facilitate regeneration of the site to more desirable tree species.

The productivity problems of small-tree operations can be solved within certain limits by whole-tree logging methods in which delimbing is abandoned and the mass handling of the crop is started at an early phase of harvesting. But because in present-day whole-tree logging methods trees are still felled singly, the tree size strongly affects the cost.

An essential factor limiting the performance of existing felling concepts, including the accumulating multitree feller-bunchers, is the number of decisions and actions an operator must make per tree. The concept of a continuously progressing harvester provides a highly significant reduction in the demands made on the operator in controlling the movement of the machine, the positioning of the felling head, and the handling of the tree. In a hypothetical, simplified assessment, 34 decisions were required per tree from the operator of an excavator-mounted feller-buncher in harvesting young, plantation-grown eucalyptus, while only four decisions per tree were needed to operate a continuously moving machine. The operation and output of a continuously moving harvester is governed not by an individual tree but rather by the area of the swath that the advancing machine covers, or by the basal area of the growing stock confronting the machine. Continuous movement reduces the power requirement of the machine. For example, a 13-t tree-to-tree harvester needs 18 kW merely to accelerate its mass on flat terrain. For a continuously moving machine the need is zero (Kerruish 1975).

From beginning to end the trees are treated as a mass product. If the operation is restricted to moving the trees in a swath, the machine is called a *swath feller.* If the machine also performs other operations in addition to felling, it is called a *swath harvester.* In each of the following alternatives the operation must be organized to permit flexible and undisturbed haulage to the landing:

1. Felling and windrowing;
2. Felling and bunching;
3. Felling and baling;
4. Felling and chunking;
5. Felling and chipping.

The largest application potential for swath harvesters is in *naturally regenerated brush stands.* The biomass of these forests is concentrated largely in densely spaced, low-quality small trees which originate from seed or sprouts. However, the diameter range is large, and part of the trees is probably oversized for continuously progressing machines and must be circumvented. Irregularly shaped and multistem trees are not uncommon. The terrain may be uneven and rocky, requiring rugged equipment.

Another typical situation where the swath harvesting principal could be applied is the *short-rotation hardwood plantations* producing biomass mainly for energy purposes. Here the stand characteristics, such as age and spacing, are controlled and even. The terrain is generally level and free of stones, but often moist and soft. The conditions for mechanization are much easier than in natural brushwood stands, except that disturbing the stool must be avoided. To prevent damage to the roots, the lateral pressure against the stumps must be kept low, and a clean-cut stump surface is required. Requirement of vigorous coppicing limits the harvesting to the dormant season. Development of swath harvesters for short-rotation hardwood plantations has been most active in the Netherlands, Sweden, Canada, Ireland, and the United States. Golob (1986) analyzed in more detail the basic functions of short-rotation harvesters: in-feeding mechanisms, severing devices, tree-gathering systems, tree accumulators, and unloading mechanisms.

The existing short-rotation harvester prototypes are designed for very young hardwoods planted in straight rows and regenerated by coppicing. They all have a cutting mechanism based on a single or twin circular saw. The only fundamental difference between these rather light, mobile machines lies in the scale of operation for which they are intended. Their operating principles are as follows (Mitchell and Säll 1983):

1. Travel readily through the crop without causing undue damage;
2. Sever stems as near to the ground as possible;
3. Lift the cut material;
4. Comminute the cut material or process it into bundles;
5. Transfer the processed material from the site.

4.3.2 Operation of the Swath Harvesters

The smaller the trees, the more competitive is the swath harvester concept compared with single-tree felling. When the tree diameter increases, the feasibility of continuously moving swath harvesters vis-à-vis other alternatives weakens, and the operation becomes technically more difficult and rugged. The maximum stump diameter for the existing prototype machines varies from 10 to 50 cm.

The *swath felling head* is either front- or side-mounted. Severing can be done with one or several circular saws, a chain saw, shear, chopping knife, or auger bar with a straight or helical flute configuration. For short-rotation plantations, where vigorous sprouting is essential for regeneration, an auger bar might offer important advantages as a felling solution. It is able to cut anywhere along the length of the bar, it does not need an anvil to cut against, and it leaves a clean cut on the residual stump. Moreover, because auger cutting edges are more massive than the teeth of conventional circular or chain saws, they are less vulnerable to damage and require less sharpening (Mattson 1982, 1983).

As the machine travels over irregularities in the forest terrain, the swaying of the chassis produces kinetic forces which increase the strain on the cutting head, especially when wide swaths are cut. To prevent the cutting head from coming into contact with the soil and breaking, chassis movement should be minimized with constructions such as bogie wheels.

The most critical phase of the continuous harvesting operation is probably controlling the tree after severing from the stump and moving it forward for bunching, bundling, or comminution. A pusher bar above and in front of a felling head is used to facilitate cutting and prevent trees from falling back on to the advancing machine. If a tree or its top falls to the ground in front or to the side of the machine, it is difficult to recover. In a crosswise position the tree is not only lost but also readily blocks the controlled movement of other trees.

Another source of *biomass loss* is the stubble, i.e., overlong stumps (Fig. 4.19). The stubble height depends mainly on the terrain and the swaying properties of the prime mover. Stumps of larger trees, boulders, and other creators of unevenness increase the stump height, which tends to be more than in single-tree felling.

The *operation pattern* of the harvester must be designed to fit smoothly with the subsequent off-road transport of the product. The use of the harvester itself for haulage is technically possible, but the unit is generally too expensive and slow as a transport vehicle. If the product of the swath harvester is an unbound bunch, a bound bundle, or bale, it can be dropped on the site singly or in larger formations, to be picked up easily later on with the grapple loader of a farm or forest tractor. The bunches can be left on site for seasoning, and haulage to the landing can be timed independently of harvester operations. If the haulage tractor is equipped with a crane-mounted grapple saw, the brush mass can be bucked to desired lengths in conjunction with loading.

All existing, continuously progressing prototype swath harvesters designed for brush stands reduce the biomass to chips. The trees seldom touch the ground and, therefore, there is very little dirt and grit in the product. But on-site comminution causes a *hot logging schedule* where the organization and timing of short- and long-distance transport must be carefully synchronized with the operations of the

Fig. 4.19. High stubble may be a significant cause of biomass loss in swath harvesting of puckerbrush (courtesy of FFRI)

harvester. At least the following options are available for the organization of off-road haulage:

1. The chips or chunks are blown into a separate vehicle, moving alongside or behind the harvester. Two chip transport vehicles are required to enable the harvester to work without interruption (Fig. 4.20).
2. The harvester pulls its own trailer. After filling the attached trailer, it is replaced with another. A single tractor suffices for hauling the chips to the landing unless the transport distance is exceptionally long.
3. The harvester blows the chips into its dumping bin. The bin is unloaded on site onto a tractor or on the landing onto a roadside pile, chip pallet, trailer, or truck.
4. The continuously moving harvester sacks the chips and drops the bags on the logging site. A forwarder moves the chip bags to the landing (Fig. 4.21).

Chip bags have certain theoretical advantages over the other chip haulage alternatives. The method gives the harvester greater maneuverability around obstacles and enables it to reverse more freely. These are important considerations on small work sites. It also permits the formation of buffer storage on the site and at the landing, easing the scheduling problems and the vulnerability of the hot operation. However, broken bags and moving the bags within and between work sites are additional cost factors, and sacking itself may require an extra hand for the harvester. No completely satisfactory chip bags are yet available.

Fig. 4.20. Residue recovery schedule based on Nicholson-Koch felling-bar harvester and two companion chip forwarders. The chips are deposited in roadside inventory piles (Koch and Nicholson 1978)

Fig. 4.21. Puckerbrush harvesting with the Pallari swath harvester prototype. Chips are stored temporarily in bags which are forwarded to the roadside and discharged into an exchangeable trailer (Hakkila and Kalaja 1980)

4.3.3 Swath Harvester Prototypes

Several swath harvester prototypes have been built and tested. Partly because of technical problems, and partly due to the inadequate demand for and the low price of the brush biomass, none of them are in commercial production, and some may no longer exist. Nevertheless, because the concept of swath harvesting opens up interesting future solutions for biomass recovery, examples of swath harvester prototypes are presented below. These prototypes, although not ready for commercial operations, show that many alternative solutions to the problem are evolving.

The Finnish 10-t *Pallari swath harvester* fells and chips the brush in a 230-cm swath along its path. The prime mover is a six-wheel, 100-kW farm tractor (Fig. 4.22). As the harvester advances, a 2.3-m-high pusher frame bends the trees forward and prevents them from falling to the side of the machine. Two horizontal, 110-cm cutting plates, each with three 20-mm-thick, sicklelike shear knives, counterrotate in front of the machine at a speed of 60 rpm. The rotating sickle knives shear the tree and throw its butt between two vertical feed rollers. With the help of the forward movement of the machine, the tree glides to a 440-mm-wide self-feeding drum chipper. A conveyor belt takes the chips to 1.5-m^3 bags which are eventually dropped to the ground and hauled later with a conventional forwarder to the landing (Fig. 4.23). Because of their low speed, the knives do not hurl pieces of wood around, thus increasing work safety.

The output of harvesting small-sized *Alnus incana* and *Betula pubescens* brush was in a field test 5 t green matter per effective hour. Major limiting technical factors were insufficient engine power, trees thicker than 10 cm at the butt, plugging of the chipper, and feeding crossed, leaning, and too densely growing trees to the chipper. The sacking system required one extra hand for the harvester

Fig. 4.22. The Pallari swath harvester prototype from behind (courtesy of Hannu Kalaja)

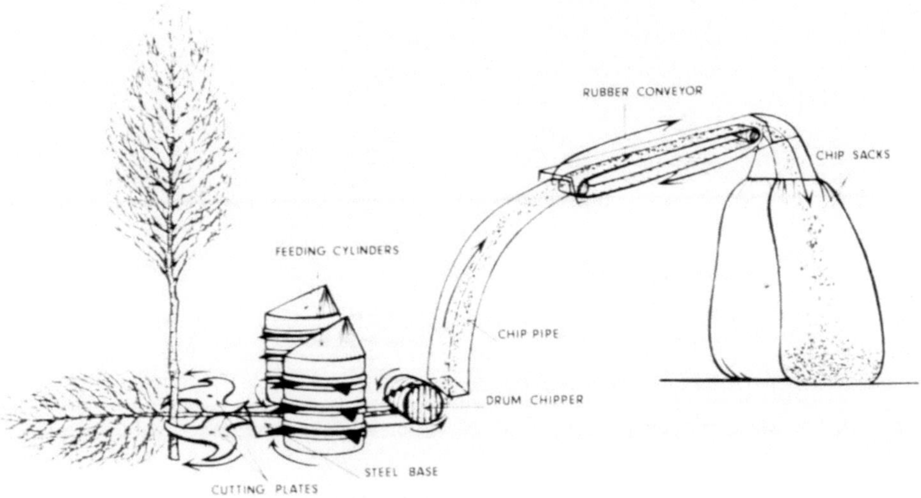

Fig. 4.23. Operation principle of the Pallari swath harvester prototype (Hakkila and Kalaja 1980)

and another for unloading sacks at the landing (Hakkila and Mäkelä 1975; Hakkila and Kalaja 1980).

The *Crabe Combine* is a redesigned Canadian prototype version of the Pallari swath harvester. New features are a rigid-framed tracked carrier, an additional 200-kW rear-mounted power pack to operate the two rotating shears and the chipper, an extra powered feed roll ahead of the chipper, and modification of the sickle knives and the chipper. The concept of sacking the chips has been abandoned as unsuitable for large-scale operations. Instead, the chips are blown onto a tractor following the harvester.

In the first tests in stands covered by small-sized dense brush, the machine demonstrated the ability to function as a continuously progressing unit and to harvest biomass, although a stop and go operation prevailed during the test program. Over 80% of the biomass was recovered, but the machine availability and utilization as well as the output of 0.2 ha per effective hour were unsatisfactory. Increasing the machine performance to an operational level requires further design improvements (Sutherland 1985).

The *Nicholson-Koch felling-bar harvester* is a 30-t, 385-kW commercial prototype for large-scale postlogging residue recovery from the stone-free sandy soils of the southern United States. The central feature of the tracked machine is a four-knife, 100–600-rpm felling bar, extending 2.4 m across the front of the machine near ground level. The upward-rotating felling bar mills through standing trees and with the help of bottom and side feed rollers delivers them along with ground residue into the 1.2×1.2 m in-feed opening of a 550-rpm, one-knife chipper drum (Fig. 4.24). The chips are blown to the rear of the advancing machine and discharged directly into one of a pair of tracked chip forwarders, each carrying a quick-dump chip bin with a holding capacity of 9 t (Fig. 4.20).

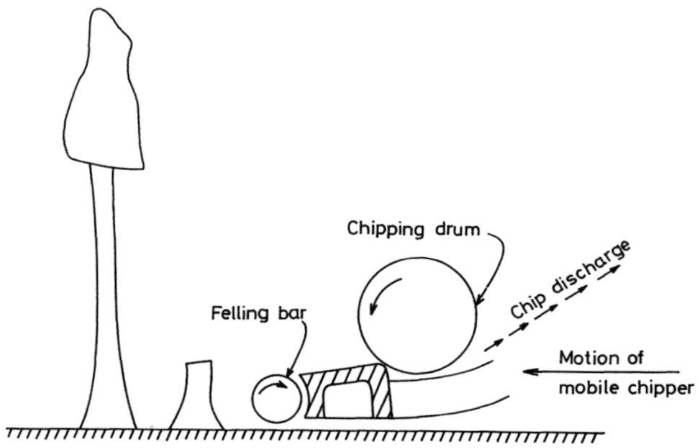

Fig. 4.24. The felling-chipping principle of the Nicholson-Koch harvester (Koch and Savage 1980)

The operational performance of the machine has been quite satisfactory. On a test site containing residual southern pine the output was 0.4 ha or 18 t green biomass per effective hour. The efficiency of recovery was 76%. Problems were caused by trees lying crosswise to the machine's route of travel and excessive soil compaction which prevents the use of the harvester on soft soils (Koch and Nicholson 1978; Koch and Savage 1980; Sirois 1981 a, b).

The *Georgia-Pacific Jaws III* prototype is mounted on an 18-t, four-tracked articulated vehicle. The felling apparatus, chipper, operator's cab, and chip bin are mounted on the front section, and the 320-kW power plant is located in the rear section. The 2.7-m felling head, fitted with a push bar and horizontal in-feed rollers, consists of two counterrotating, toothed cutter disks. The cutters lift and direct the biomass into the feed rollers from which it moves to the disk chipper. The chip bin has a capacity of 2.5 t green matter. It is dumped to the side directly into a chip forwarder, which carries a bucket loader for transferring the chips further to open-topped vans for on-road transport.

The harvester is intended for stone-free sites in the southern United States, partly for precommercial strip thinning in dense, naturally regenerated pine stands, where the diameter of the trees is generally less than 13 cm. However, it is capable of harvesting standing trees up to 30 cm in diameter. The design operation speed is about 1.6 km/h. Production rates have averaged 5.5 t green biomass per hour in the forest and somewhat higher on powerline rights-of-way.

The French 10-t *Scorpion* prototype is designed to fell and chip Mediterranean hardwood brushes. The prime mover is a six-wheel drive special tractor designed to work on slopes (Fig. 4.25). The felling head consists of two counterrotating toothed cutting plates and two feeder rolls mounted around the same vertical axles with the plates. The butt of the cut tree is forced into a chipper behind the felling head and rolls. The chips are blown into a 4–7 m^3 tipping bin. The dumping height is 4.2 m, which makes it possible to discharge the bin directly onto a truck. The width of the swath is 2.1 m, and the maximum tree diameter 15 cm. The daily output is about 1 ha or 30 t (Anonym 1983).

Fig. 4.25. The Scorpion swath harvester prototype (courtesy of Roland Bourquet, AFME)

The *Recufor M* is a heavy, 34-t, 240-kW wheeled prototype machine designed for collecting, reduction, and haulage of on-ground residue from cutovers under eastern Canadian conditions. On the front of the harvester is a pivotally mounted gathering drum rotating counter to the direction of travel. The drum has 15 curved teeth with beveled edges, placed in three rows of five. As the vehicle moves forward, the teeth on the drum rotor skim the surface of the ground, collect residue, and force it through a series of fixed knives located at the back of the rotor, thus comminuting the material to chunks. A conveyor system transfers the product to a bin which is discharged by dumping (Bérard 1981).

In field tests the harvester was able to gather slash at a speed of 0.6 km/h. The output of fresh biomass was 4 green t/h, but a production rate of 10 t/h was considered possible with higher rotor power, design adjustments, and rationalizing the operational layout. The most serious problem was the mineral soil picked up with the residue. The amount of grit, 2% of the gross load mass, was considered unacceptably high even for energy use (Hamilton 1984).

Fully mechanized recovery of very small trees from precommercial selective thinnings was tested in Czechoslovakia with an unconventional, *boom-mounted chipping head* principle. Although the machine recovers the sapling trees singly and selectively, and does not progress continuously, it is otherwise comparable with the prototype machines presented in this section. The harvester performs felling, chipping, and off-road haulage in an integrated operation. The chipping head, mounted on the crane of a tractor, is guided to the tree top and moved downward to chip the standing tree from above. The reduced chips are moved pneumatically to a trailer hauled by the tractor. The method respects the biological requirements for selection with minimal damage to the remaining trees, but due to limited reach of the crane the distance between the machine passes is only 15 m. In a naturally regenerated 2-m-high *Picea abies* sapling stand the biomass recovered, due to its high foliage content, was suitable for animal fodder. However, the hourly output of fresh biomass was not more than 1.3 t (Petříček 1985).

4.4 Harvesting Stump and Root Mass

Stump removal is not a new art. To alleviate the shortage of fuel wood in England, J. Evelyn (1729) published a sketch of a stump uprooting machine as early as 1670. There were about 500 kinds of stump pullers on the American market in the 1800s, and stump-pulling became a common occupation. Two men with a yoke of oxen and a stump-puller could travel indefinitely across the country, pulling 20–50 stumps a day at $ 0.25 per stump, the standard price in 1850 (Sloan 1958; Koch 1985b). Stumps were earlier a common source of fuel wood, and in Finland the utilization of pine stumps for tar production reached a high level as late as during World War II (Kärkkäinen 1975).

Stump extraction is still carried out to some extent. Stumps are removed mainly to clear the site for farming or construction purposes or in order to facilitate mechanized reforestation operations. For the industrial utilization of stump and root wood, the recovery activity is at present restricted primarily to mature pine stumps that are used in the chemical industries. Utilization of immature new stump and root wood takes place only in few countries and only on a restricted pilot scale. An exception since 1977 is a commercial operation in Mackmyra, Sweden, annually producing in the mid-1980s some 150000 m^3 of sulfate pulp chips from fresh *Pinus sylvestris* and *Picea abies* stumps from cutover areas (Sect. 8.6.4).

4.4.1 Logging Characteristics of Stump-Root Systems

The technical logging characteristics of stump and root mass differ radically from those of the above-ground biomass. Special machines and working methods are required, and the harvesting schedule must be rearranged. Great forces, and consequently heavy machines, are needed for the extraction work. Uprooting is restricted to the season when the soil is free from snow and neither frozen nor very wet. If the special equipment has to stand idle for a large part of the year, capital costs per unit produced increase.

The mass of a stump-root system is roughly only one-fifth of the stem mass. Consequently, the biomass yield per hectare and work site is bound to be low. Yet the harvesting is performed with heavy machines, which often have to be carried from site to site on trailers.

The bulk density of whole stump-root systems is so low that their long-distance transport as such is precluded. The solid content has to be raised by splitting the raw material into smaller pieces before trucking, but splitting is hampered by impurities in the stump-root systems. Stones and loose earth are additional

Fig. 4.26. An entrapped stone inside a stump (courtesy of Kemi Company)

weight, and they preclude the use of sharp knives for comminuting. Stones are also found inside the stumps (Fig. 4.26).

Technical stump data are important for machine design, work organization, recovery estimation, and cost calculations. Stump dimensions and power requirement in extraction are discussed in Sections 4.4.1.1 and 4.4.1.2, and dry mass of stump-root systems in Section 2.1.3.3. Generalization must be avoided because the characteristics of the stump-root systems are specific to species, soil, and climate.

4.4.1.1 Dimensions of Stump-Root Systems

Planning of conventional logging operations often includes preharvest estimation of the total amount of timber by measuring the breast height diameter of the standing trees. If *the regression between dbh and stump diameter* is known, the size of the stump-root systems left in a cutting area can be predicted.

In boreal forests, the breast height diameter is approximately 75% of the stump diameter. No significant differences were found in this ratio between the southern and northern parts of Sweden, although the taper at the top of the stem varied geographically (Ager et al. 1964). Figure 4.27 gives examples of this relationship in eastern Canada and Finland. Bylin (1982) presented regression equations for predicting dbh from the stump diameter of inside and outside bark for 15 southern tree species in the United States. Dbh was predicted better for the conifers than for hardwoods.

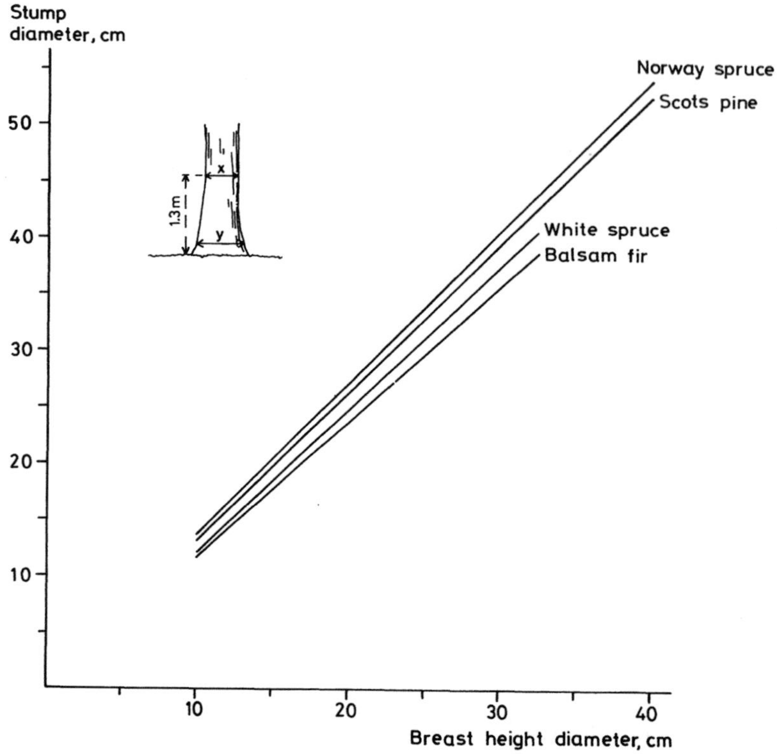

Fig. 4.27. Stump diameter as a function of dbh in *Picea glauca* and *Abies balsamea* in eastern Canada (Golob and Silversides 1978) and in *Pinus sylvestris* and *Picea abies* in Finland (Hakkila 1972)

Stump height, in managed forests commonly 10–25 cm, refers to the distance of the stump cross-section from ground level. The theoretical stump height is the lowest height at which a tree can be cut by chain saw without serious disturbance to the work or the log form by the uppermost root collar. Stump height is affected by the butt swell of the tree, which depends on species, soil, and site topography. In motor-manual winter felling under snow conditions, the stumps tend to exceed the theoretical height. In mechanized bare-ground felling, especially with a hydraulic shear, the tree is often cut below the theoretical stump cross-section. Stump height data are needed in feller design and for estimating the mass of a stump-root system.

The depth of the stump-root system is the vertical length of the stem projection underground, disregarding the parts thinner than 5 cm of the tip of the taproot. The depth affects the force and techniques which have to be applied in the extraction process. On it also depends the size of the holes after uprooting.

A well-developed taproot is characteristic of many pines in well-drained sandy soils in a warm climate. For example, the average taproot depth is about 150 cm in 17-cm-thick *Pinus taeda* trees in Georgia (Coughran and Koch 1974). A 250-year-old *Pinus palustris* in Florida had a 440-cm-deep taproot with a diame-

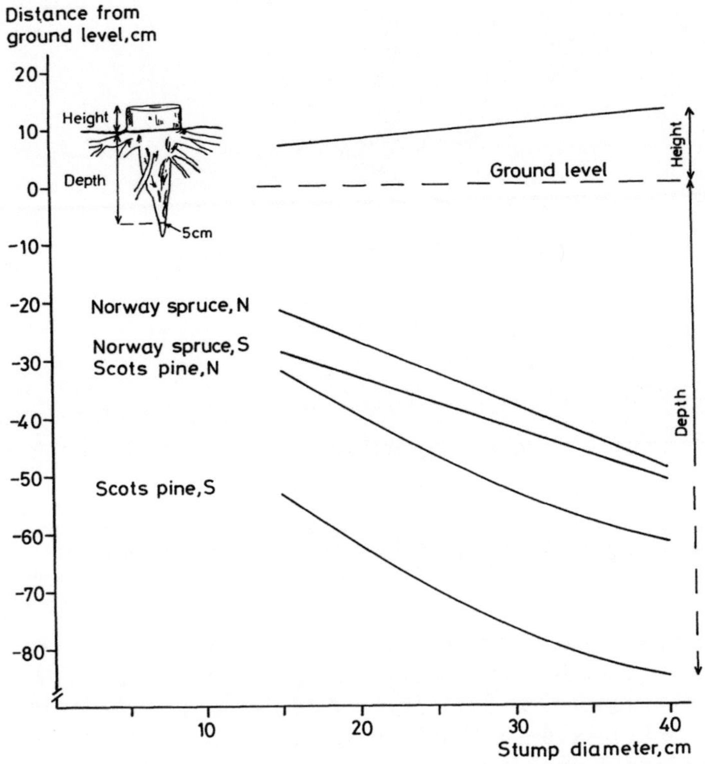

Fig. 4.28. The above-ground height and the below-ground depth of the stump-root systems of *Pinus sylvestris* and *Picea abies* in mineral soils in southern (*S*) and northern (*N*) Finland as a function of stump diameter (Hakkila 1972)

ter of 43 cm at a depth of 160 cm, which was the same as the breast height diameter of the tree (Heyward 1933). On the other hand, in cold soils near the northern tree limit the taproot of pine develops poorly or is totally missing. At the arctic circle in northern Finland the average taproot depth of a 30-cm-thick Scots pine stump is only 50 cm, and in waterlogged peatlands many pines do not grow a taproot at all. Norway spruce is a typical tree species that has a superficial root system with frequent long laterals but never a taproot. An increase of 1 cm in the stump diameter is accompanied with a 0.9 cm average increase in the depth of the central root system of Scots pine and Norway spruce in Finland (Fig. 4.28). However, the correlation is not strong.

The *horizontal dimensions of the stump-root system* affect the off-road haulage and design of the splitting equipment. They also determine the area of the mineral soil spots that are exposed by stump extraction. The root system of a tree normally occupies a more or less circular area of topsoil, depending on the competition of the neighboring trees. When the stump diameter is 40 cm, the area covered by root parts thicker than 5 cm in Finland is 1.6 m^2 for Scots pine and 12.0 m^2 for Norway spruce.

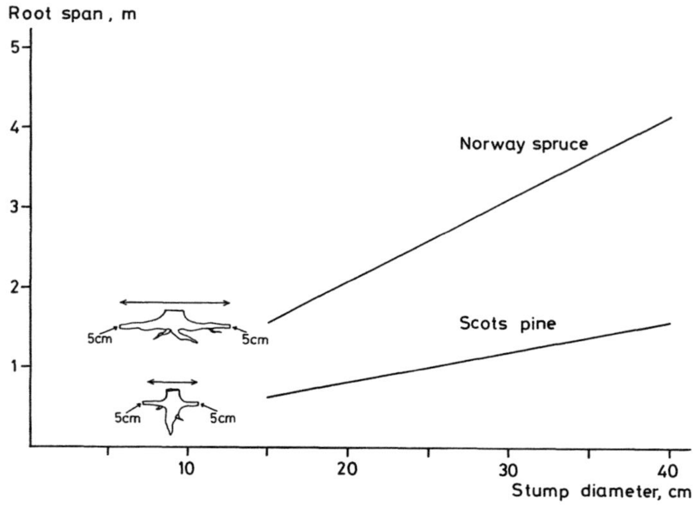

Fig. 4.29. The 5-to-5 cm central root span in *Pinus sylvestris* growing in mineral soil in northern Finland and in *Picea abies* growing in drained peat land in southern Finland (Hakkila 1972)

The horizontal distance is best described by the *central root span*, which is the maximum distance between the 5-cm-thick tips of the opposite lateral roots, thus excluding the root parts thinner than 5 cm. Figure 4.29 shows the width of the root span of Scots pine in northern Finland mineral soils and that of Norway spruce in southern Finland drained peatlands.

The force required in the root-shearing method for severing the lateral roots from the central stump is related to their *total cross-sectional area.* This variable is closely correlated with the diameter of the stump. In *Picea glauca* and *Abies balsamea* in Ontario, the total cross-sectional area of laterals was 7 dm^2 in 20-cm stumps and 15 dm^2 in 30-cm stumps (Golob and Silversides 1978).

4.4.1.2 Power Requirement of Stump Extraction

The force required to extract a stump-root system depends on the diameter and age of the stump, the form and depth of the roots, the composition and moisture content of the soil, and the weather conditions. It is also greatly affected by the techniques used in uprooting as well as by possible combined work phases such as splitting and root shearing. According to the direction of the force acting on the stump, four main alternatives can be identified (Czereyski et al. 1965):

1. The force acts principally in the horizontal direction;
2. The force acts principally in the vertical direction;
3. The force acts simultaneously in both directions;
4. The force acts from inside the stump, i.e., the blasting.

Figure 4.30 shows the *horizontal force* required to pull out stumps of some eastern Canadian conifers. When a stump is extracted by pulling horizontally, the dis-

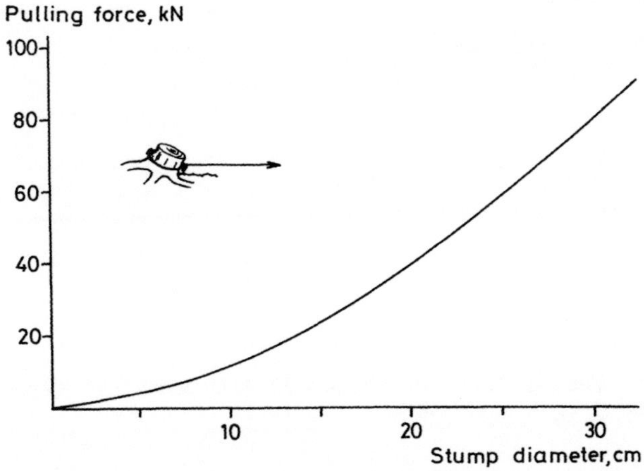

Fig. 4.30. Forces required to pull out horizontal conifer stumps in Ontario (Golob et al. 1976)

tance traveled by a winch or tractor in the process varies in length from 1 to 1.5 stump diameters. The pulling force varies irregularly because of the erratic distribution of the resisting lateral roots (Golob et al. 1976). On an average, 40 kN is needed for a 20 cm and 90 kN for a 30 cm softwood stump. Experiments with *Betula papyrifera* and *Populus tremuloides* gave almost equal results (Golob et al. 1976). The force requirement for vertical lifting is much higher. For example, pulling out Scots pine stumps 20 and 30 cm in diameter required a *vertical force* as high as 480 and 600 kN, respectively (Czereyski et al. 1965).

Because decay in sapwood and bark loosens the cohesion between the roots and the soil, the resistance decreases with the maturation of the stump. The horizontal force requirement decreased to one-half of the original in 3–5 years under Canadian conditions (Golob et al. 1976). For the vertical extraction of 20–35 cm mature Scots pine stumps a force of 15–150 kN was sufficient (Fiolek 1961).

The problem of high removal resistance can be overcome by better application of power and reduction of resistance. Stump-root systems can be uprooted by pushing or pulling the complete tree and using the stem and crown mass as leverage. Removal resistance can also be reduced by breaking the stump in advance into more easily extractable pieces, or by shearing lateral roots (Walker 1976).

If, instead of separate stump-root systems, complete trees are uprooted, the pulling or pushing force can be applied to the tree stem several meters above ground level. Due to a favorable *torque*, the power requirement is reduced radically. The force is initially applied to the standing tree to throw it off balance. Next, the gravity of the falling tree is used to wrench the stump-root system from the ground. An outer force is still required in the final stage of uprooting to extract the roots remaining anchored after toppling the tree (Fig. 4.31). The torque intended to off-balance the tree increases as the force is applied at higher points along the stem. According to Kiščenko et al. (1962), a 70-kN force is sufficient to loosen 26–30 cm Scots pine stumps by overthrowing, and for Norway spruce the requirement is 40% less. The torque requirement for overthrowing complete

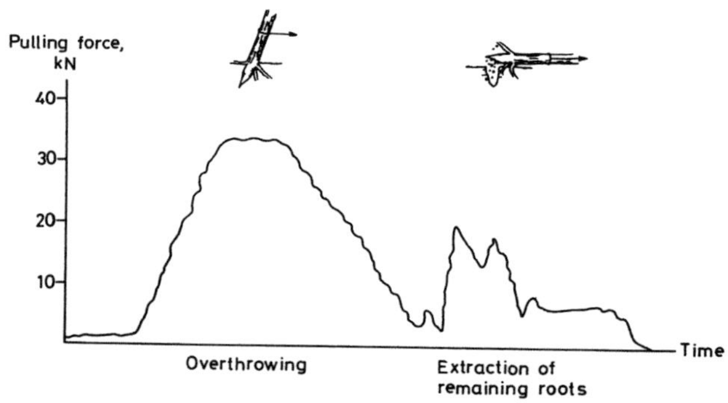

Fig. 4.31. The resistance pattern of a *Pinus sylvestris* tree when overthrown by pulling with a rope in Poland; dbh 28 cm (redrawn from Czereyski et al. 1965)

Scots pine trees on sandy soil in Poland is seen in the following table (Czereyski et al. 1965):

Breast height diameter (cm)	Torque requirement (kNm)	
	Pinus sylvestris	*Betula pendula*
−25	70	70
26−30	100	110
31−40	160	170
41−50	200	...

Another alternative to reduce the force requirement is to shear the lateral roots in advance. The laterals of a standing tree can be severed with downward-penetrating, curved blades which encircle the tree. The tree is then lifted out of the ground, recovering what would be the normal stump and its below-ground projection along with the stem.

The force required *to pull the tree out* of the ground without the laterals depends mainly on the resistance of the taproot, but also partly on the friction between the soil and the shear blades. Figure 4.32 shows the extraction force needed to pull out *Pinus taeda* trees with taproots on loamy-clay soil in Georgia and *Pinus strobus* trees in Ontario. The maximum extraction force requirement for a 20-cm loblolly pine on moist soil was 130 kN and on extremely dry soil 175 kN, and for white pine 290 kN. For a 30-cm-diameter stump the corresponding requirements were 160, 250 and 355 kN. All these values are considerably less than the power requirement of a complete stump-root system in vertical lifting.

However, before the tree can be lifted with its central stump-root system, force is required to shear the lateral roots, and simultaneously to push the shear blades through the soil. There is only an indirect relationship between the tree diameter and the maximum *root shearing force*, which is dependent mainly on the cross-sectional area of the lateral roots, basic density of the wood, soil properties, and the structure and form of the shear blades. When the maximum force required with the FMI shear prototype for root shearing of *Picea glauca* in Ontario was

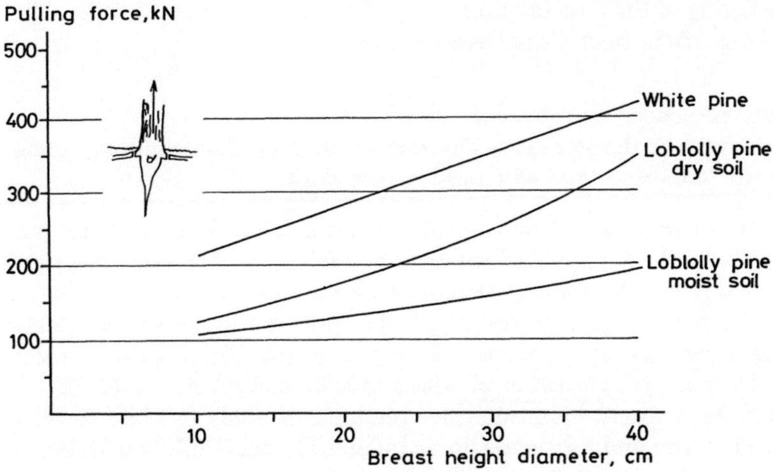

Fig. 4.32. Maximum force required for vertical lift of *Pinus taeda* on dry and moist loamy-clay soil in Georgia (Coughran and Koch 1974) and *Pinus strobus* in Ontario (Golob and Silversides 1978), after severing the laterals with soil-penetrating, curved shear blades

355 kN for a 20-cm stump and 465 kN for a 30-cm stump, about 95 kN of the total was actually used to shear the soil. The total force was higher than the force required for pulling up the tree without the laterals (Golob and Silversides 1978). The maximum forces required for root-shearing and then lifting the tree were almost mutually equal in tests carried out with the Rome root shear on moist and dry loamy-clay soilds in Georgia (Coughran and Koch 1974).

4.4.2 Extraction of Complete Trees

The recovery of stump and root mass is usually carried out as an independent operation after conventional stem-only harvesting. In complete-tree logging the entire work schedule is reorganized, and these two operations are integrated by extracting the stump-root system together with the stem. The stump-root system or part of it is separated from the stem immediately after the uprooting or in a later phase of the logging schedule. The system has the following advantages and disadvantages.

Advantages:

1. When the stump-root system is recovered together with the stem, the requirement of extraction force can be reduced by a favorable torque which is produced by using the stem as a lever.
2. When the stump is kept as part of the stem, some of the additional raw material may be used for sawtimber instead of pulp. Under Swedish conditions an average of 20 cm of stem below the stump cross-section still has a vertical fiber direction. This wood is a potential raw material for high-quality lumber (Wiklund 1970, 1971).

3. Combining felling of the tree and uprooting of the stump-root system in a single work phase offers operational advantages.

Disadvantages:
1. Damage may be caused to the stem.
2. Soiling of the butt of the stem may cause extra wear on the delimbing, bucking, debarking, chipping, and saw milling machines.

Complete trees are sometimes removed *by pulling* to prepare the area for reforestation, cultivation, or land clearing. The work can be performed manually by a hand winch, but manual removal requires severing of lateral roots and even digging of ditches around the tree. With a tractor winch the roots are loosened completely from the ground provided that the tree is thrown onto a sulkylike support, a wooden stand, or a short log (Czereyski et al. 1965; Débroussaillement ... 1968).

The productivity of these labor-intensive methods is much too low for economic recovery of stump and root mass on an industrial scale. Pulling with heavier machines requires an essentially smaller labor input, but even mechanized removal by pulling requires a helper to handle the cable. Cable breaks cause interruptions and additional expense, and are a safety hazard. A crawler tractor or a wheeled forest tractor can remove 250 trees/day by pulling. Most of the work time is used for manual transfer, rigging, and disengaging the cable (Koivulehto 1969).

Pushing is a more effective method for removing complete trees than pulling (Kiščenko et al. 1962), but it requires heavy machines. The daily achievement of complete-tree felling by pushing with a crawler tractor is 300 trees. In pushing the output is greater, the procedure includes preliminary cleaning, no helper is necessary, and the cable costs are avoided. Various kinds of auxiliary devices for tractors have been developed for removing trees by pushing (Débroussaillement ... 1968).

When the stump-root system comes out with the falling tree, much loose earth and humus remains in the roots. The working machine can do preliminary cleaning by raising the butt of the tree, shaking it, and letting it fall to the ground from a height of 2–4 m. This work phase can be combined more easily with pushing than with pulling. On the other hand, a drawback of pushing is damage to the stem. A bulldozer blade may cut several centimeters into the tree. If the wheels or tracks of the pushing machine come to rest partly on the lateral roots of the tree, the stem may break in two. The value of the stump and root mass will then not offset the loss in stem wood quality.

Removal by pulling or pushing is a simple way to uproot complete trees, but it leaves the other phases of stump mass recovery unsolved. Off-road haulage and handling at the landing of complete trees with the roots attached is difficult with existing machines. The payload decreases in skidding, contaminants hamper bucking of the butt by a chain saw, stem suffers damage, and the landing area requirement grows. In addition, after a stump-root system has been severed from the stem, it still has to be split to improve the bulk density before on-road transport. In fact, it is splitting rather than uprooting that is the true bottleneck in the harvesting of stumps and roots.

A more sophisticated solution for complete-tree harvesting is offered by the Swedish *Tree Max* principle. The wheeled loader-mounted prototype machine first lifts the complete tree from the ground with a hydraulically controlled clamp, using a combination of lifting and levering forces. To brace the machine during the extraction, a support pad is placed on the ground as the tree is clamped. After the uprooting, an adjustable-diameter cylindrical shear is run down the stem to shave off the lateral roots. Using a clam bunk located at the top of the cab, the machine skids the trees with the taproot to the landing. The laterals can be salvaged separately with a forwarder. Another alternative is to sever the whole stump-root system from the stem with a hydraulic chain saw. The method has not found practical application.

The *Rome Tree Extractor* head is designed to harvest southern pine with a deep central stump-root system, leaving the lateral roots in the soil. The harvester head can be mounted with a quick-hitch mechanism on a number of prime movers, such as wheeled loaders (Fig. 4.33). Two elements are central to the design. The first is a scissors-type grip achieved with a pair of stout horizontal knife blades that close at the ground line and bite into the stem from opposite sides. With this grip as an anchor, the second element in the design, a 55-cm-diameter

Fig. 4.33. A wheeled loader, equipped with Rome tree extractor, pulling and bunching southern pine with the central stump-root system intact (courtesy of P. Koch)

clamshell-hinged tubular shear, comes into action. The shear, sharpened on the lower edge, is forced vertically into the ground to a depth of 28 cm, severing the lateral roots around the tree. At this point, broad steps on opposite sides of the shear strike the ground and prevent further penetration. An additional stroke of the hydraulic cylinder raises the grip 23 cm while the steps remain pressed against the soil surface. The effect is to jack the stem and break it free of the ground. Finally, the complete tree, excluding the laterals, is lifted into the air and bunched for skidding. A considerable proportion of the soil clinging to the stumps and taproots can be shaken free when the trees are bunched (Koch and Coughran 1975; Koch 1976). Adhering dirt is more common in clay soils than in sandy soils.

This method has been developed for harvesting southern pine pulpwood trees with a breast height diameter of about 15–30 cm. Pilot-scale industrial operations in Florida have proven the method both technically and economically feasible for recovering additional fiber raw material from the stone-free soils of the southern United States. The output of pulling-bunching was 1–2 trees/min or 7–11 m^3/h, depending on the tree size (Koch 1976). Subject to certain soil types, terrains, and tree-size limits, the machine gave a satisfactory output of 10 m^3/h in pulling and bunching 18-cm *Pinus radiata* trees on sandy soils in Australia. In heavy clays, however, many trees simply broke away from their stumps during the jacking (Shanks et al. 1979).

Machines of this design have been tested in regular commercial operation in Florida, pulling and bunching 45–90 pines/h with a maximum breast height diameter of up to 30 cm. The pullers have been employed in Mississippi to harvest pines with central stump-root resin enriched through injection at the root collar with paraquat. The resin-rich wood is used in the production of naval stores, and the upper sections are pulped (Koch 1985b).

The root-shearing principle is applied also in the Canadian *FMI harvester head.* The prototype is mounted on a prime mover suitable for ordinary felling-bunching work. The root shear itself consists of a frame supporting a grapple which has a number of curved, shallow blades biting into the tree. Fastened to the frame as an integral part of the assembly are three curved, vertical blades, which encircle the tree and sever its lateral roots when pressed down through the soil. The tree is then pulled up using extraction pads (Golob and Silversides 1978). Another root-shearing complete-tree harvester has been built in the Soviet Union. This prototype machine severs the stump from the stem after lifting the tree from the ground (Tauber 1979).

4.4.3 Separate Extraction of Stump-Root Systems

Extraction of stump-root systems was performed previously by simple methods with a high input of human labor. A great variety of manual tools and machines were evolved for land clearing and small-scale local utilization of stumps for fuel or production of tar. These machines and methods have been described in detail in the literature (Moltesen Nielsen 1940; Czereyski et al. 1965).

Large-scale recovery of fresh stump and root mass for industrial purposes presupposes higher productivity of work, lower costs, and cleaner wood. In addition to uprooting, the essential phases of the work schedule are precleaning, improving the bulk density by splitting, and off-road transport to the landing.

4.4.3.1 Stump Extraction by Blasting

A simple and proven method of stump extraction is blasting, which combines uprooting and splitting. It can also be used to split extracted stump-root systems. Blasting is done practically by a team of three to four workers, with one man drilling the holes and the others doing the actual blasting. The productivity of work is fairly low. Exclusive of the gathering of the wood pieces, the output in Finnish experiments was 70–80 stump-root systems or 5–15 m^3/man-day, depending on the size of the stumps (Hakkila 1972).

In deep blasting, the full cartridge is placed under the stump, while in shallow blasting several smaller cartridges are placed under the roots. Shallow blasting is more labor consuming and requires more blasting materials, but it causes less soil compaction and leaves smaller holes in the ground. Large craters sometimes have to be filled after blasting.

The material cost of explosives, detonators, and fuses is high, depending on the composition and moisture content of the soil and on the form, size, and age of the stump-root system. A suitable charge for the extraction of a medium-sized Scots pine stump consists, for example, of 200–300 g ammonium nitrate, trotyl, and aluminum. An oak stump requires an even stronger charge, while about 150–200 g is enough for a Norway spruce stump.

Working with blasting materials is dangerous. Only specially trained, authorized personnel are permitted to transport, store, and use the explosives. The stumps disintegrate into small parts, which must be collected by hand. Some of the roots may remain in the ground, and the wood loss is relatively great. Yet the splitting result of blasting alone is not always enough.

Blasting is thus not a feasible method for industrial stump-harvesting operations. However, it has local significance on small work sites and when removing large isolated stumps, especially in land-clearing operations. In harvesting stump mass for resin processing in Central America blasting is still a large-scale practice (Weldwood 1979b).

4.4.3.2 Stump Extraction with Chassis-Mounted Equipment

The most common mechanized methods of stump extraction are based on the use of heavy *crawler tractors*. Special stump tools are attached to the forward frame or rear tool bar of the tractor. The crawler tractors are used for stump removal mainly in land-clearing operations but, in the southern United States, also for harvesting mature pine stumps.

Crawler tractors usually extract the stump-root system intact, but the power requirements can be reduced by presplitting the stump into two or more parts. A stump can be split in two prior to extraction with a *wedge or knife blade* fitted to the rear of the machine by driving over the stump with the blade lowered. Large stumps may be split in four by driving in two directions.

A more common presplitting solution is to mount a long splitting blade on a *stump rake* in front of the crawler tractor. This principle is commonly used in the Soviet Union and in the southern United States. Crawler tractors hoist a whole or split stump-root system with a rake consisting of a variable number of strong teeth on hydraulically or mechanically controlled framework in front of the tractor in place of the ordinary dozer blade. Smaller stumps are extracted with one push, while larger stumps may require several pushes depending on the machine power. Simultaneous pushing and lifting may be used. A stump rake can also be mounted at the rear of the crawler tractor which first runs over the stump, drops the rake down, and then lifts the stump up by moving forward.

The output of crawler tractors is 100–400 stump-root systems per day, depending on the stump age and size, number of stumps per hectare, soil characteristics, and the equipment. Performance is often reduced by inadequate visibility from the driver's cab. Compared with the output, the cost of a heavy crawler tractor is high. One drawback is the need to use carriage trailers for transport between work sites, which increases the minimum size of an economic operation.

Presplitting in conjunction with stump removal is not enough to achieve an adequate bulk density. To improve the efficiency of transport, it is therefore necessary to shear off the lateral roots in a separate work phase. Cleaning presents additional difficulties, for the crawler tractors leave a great deal of loose soil in the stump-root systems. However, stump removal is taxing work, and crawler tractors are in this respect more durable than most other machines.

The minimum size of an operation can be reduced with machines that do not necessarily require a separate trailor for transport from site to site. One alternative is extraction with *wheeled loaders*. They move faster than crawler tractors within the work site and their average per-hour costs are lower. They are capable of versatile work and can be used not only for extraction but also for preliminary cleaning and bunching of the stump-root systems. Further advantages are good visibility and a center of gravity that enables the machine to be used as a lever in the extraction process. However, wheeled loaders require bearing and relatively even terrain. Their mobility and strength are inadequate under difficult conditions. They are not the equal of crawler tractors in durability and, consequently, may require structural reinforcement for continuous use in stump removal.

A 10-t wheeled loader can uproot stumps under light soil conditions with a conventional *log grapple*. The machine grips the stump from above, simply lifts it up, and collects the stumps in small piles. As it moves, the wheeled loader cleans the stump-root system by shaking the grapple arms. Another working technique is based on a *stump cub*, developed from a conventional gravel cub by adding two lifting teeth in its front and sharpening the cub edge between the teeth for shearing of laterals. In addition, gaps on the sides and bottom of the cub help to reduce the amount of loose soil from the stump-root systems. The actual uprooting is done partly by pushing and partly by hydraulic lifting of the cub. Preliminary cleaning is done by shaking the stump-root system in the stump cub. Ahonen and Mäkelä (1972) studied both methods on a sandy pine site in Finland. Where the ground was free of stones, the output was 290–400 stumps/day, but stones slowed down the work.

Extraction device prototypes of special construction have been designed for wheeled loaders, but the cost efficiency of the recovery schedules based on these is poor. In general, the use of wheeled loaders is restricted to rather small stumps, easy terrain, and temporary operations. If the raw material is to be recovered and transported further, a separate machine is needed for splitting.

4.4.3.3 Stump Extraction with Crane-Mounted Equipment

The difficulties in harvesting stump and root mass are not confined to uprooting. Cleaning, splitting, and transport may present more problems. For productivity and operational reasons, it is often profitable to combine some of these work phases with uprooting. Machines performing uprooting and one or more other work phases are called stump harvesters. The most versatile solutions are based on *crane-mounted stump-harvester heads.* They offer important advantages over chassis-mounted equipment. It is easier to combine work phases, and several stump-root systems may be recovered and collected into one pile from one place. This improves the output, increases the size of the piles, and reduces soil compaction.

The prime mover of a crane-mounted stump harvester may be a farm tractor, a forest tractor, or an excavator. Despite their low travel speed, *excavators* are competitive in stump harvesting. They are constructed for just this kind of work and are durable in extracting tough stumps. The great mass and stability of the excavators reduces machine sway and thus improves the ergonomic environment of the driver. However, in uneven, rocky terrain the driver is subject to machine sway. The relatively high working pressures make it possible to use auxiliary hydraulic apparatus requiring great power, and the strong boom can be used to produce a favorable torque in uprooting. The excavators can move in difficult terrain, and when equipped with wide tracks they can also work on soft sites. Moreover, large production series make excavators cheaper than forest tractors. On the other hand, they must be moved from site to site on trailers, and the travel speed from stump to stump is low.

Excavators have been tested and used in stump extraction in a number of countries. In a work study in New Zealand a 70-kW tracked excavator was equipped for stump uprooting by replacing the bucket with a *stumping hook*. Later, the hook was modified by fitting a large plate to it to increase the leverage on the stumps. In a 40-year-old *Pinus ponderosa* plantation the uprooting output was originally 69 and after the modification 95 stumps/h. When the stump diameter exceeded 40 cm, the performance dropped sharply. The output was lower for *Pinus radiata* stump-root systems which have large surface laterals and a central core consisting of a mass of small diameter sinker roots (Page 1975).

When stump-root systems are recovered for utilization as industrial raw material or fuelwood, they must be converted to a suitable form for off- and on-road transport. This requires fragmentation into smaller pieces to increase the bulk density, at least rough precleaning to reduce the amount of soil, and collection of the pieces into piles to allow effective loading of off-road vehicles. If the stumping hook is replaced by a multipurpose harvesting head, these work phases can be integrated.

Fig. 4.34. Uprooting and splitting stump-root systems with the Pallari Stumpharvester head mounted on an RH 4 L excavator (courtesy of FFRI)

The first multipurpose machine for harvesting stump and root mass, the *Pallari Stumpharvester*, was developed in Finland at the beginning of the 1970s. The preferred prime mover is a tracked excavator with the bucket replaced by an uprooting-splitting harvester head (Fig. 4.34). The simple, original KH-10 version consists of a shear blade moved by a hydraulic shovel arm cylinder, a forked anvil, and two uprooting teeth fixed to the under side of the device. The uprooting leverage is produced with the jib and arm cylinders of the crane. The shovel hood cylinder activates the shear blade against the two-pronged anvil for grabbing and splitting. In a more effective modified version, the Pallari Stumpharvester KH-120, the positions of the shear blade and the cutting anvil have been reversed, both sections have been made independently movable with an additional hydraulic cylinder on the shovel arm, and the uprooting teeth have been mounted linearly opposite the anvil prongs instead of perpendicular to the anvil (Fig. 4.35).

The work techniques of the Pallari Stumpharvester depend on the form, depth, and size of the stump-root system and the soil conditions. Stumps smaller than 25–30 cm in diameter are extracted intact and then split with the shear. The most suitable work technique for superficial Norway spruce stumps over 25–30 cm in diameter is to tear off first the thickest laterals singly by twisting crosswise. Extraction and splitting are thus done concurrently. The taproot of Scots pine stumps over 30–35 cm in diameter is so firmly lodged in the ground that it is often difficult to extract intact stumps. To expedite the work, the stump

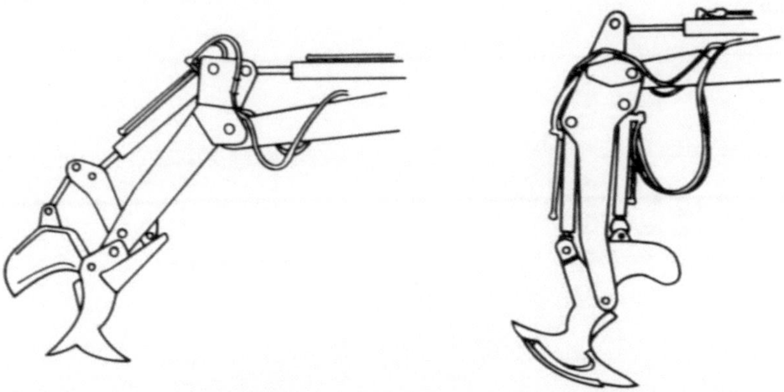

Fig. 4.35. The original KH-10 (*left*) and the modified KH-120 (*right*) version of the Pallari Stumpharvester head for combined uprooting and splitting of stump-root systems

portion is split by the shear blade into two or more parts before extraction. The final phase may include some additional splitting to remove unusable thin roots and to improve the bulk density, precleaning by shaking the crane, and bunching the stump and root mass pieces into piles (Hakkila and Mäkelä 1973, 1974).

The time consumed in the uprooting phase is affected primarily by tree species, stump size, and soil. Splitting is affected by the tree species and stump size as well, and a very important factor is the required piece size. Shaking time for cleaning depends on the soil, tree species, and piece size. The thinner the humus layer around the stump and the coarser the soil structure, the shorter is the time needed for precleaning. The following figures show the average distribution of the effective work time of the modified Pallari Stumpharvester in a large study in Sweden (Nylinder 1977):

	Time consumption (cmin/stump)	
	Pinus sylvestris	*Picea abies*
Moving from stump to stump	15	15
Uprooting	118	49
Splitting	65	33
Cleaning by shaking	60	35
Collecting into heaps	37	28
Other	14	14
Total effective time	309	174
Number of stumps/h	19	34

The stump procurement organization of the Mackmyra Cellulosaflis AB in Sweden is based on the use of about 30 Pallari Stumpharvesters. Figure 4.36 shows the number of stumps harvested per effective hour as a function of stump diameter. In continuous work, the average output of excavator-mounted modified Pallari Stumpharvesters has been about 3 m^3 solid per operating hour (Y. Jonsson 1985). Several other stump-harvester head prototypes, more or less resembling

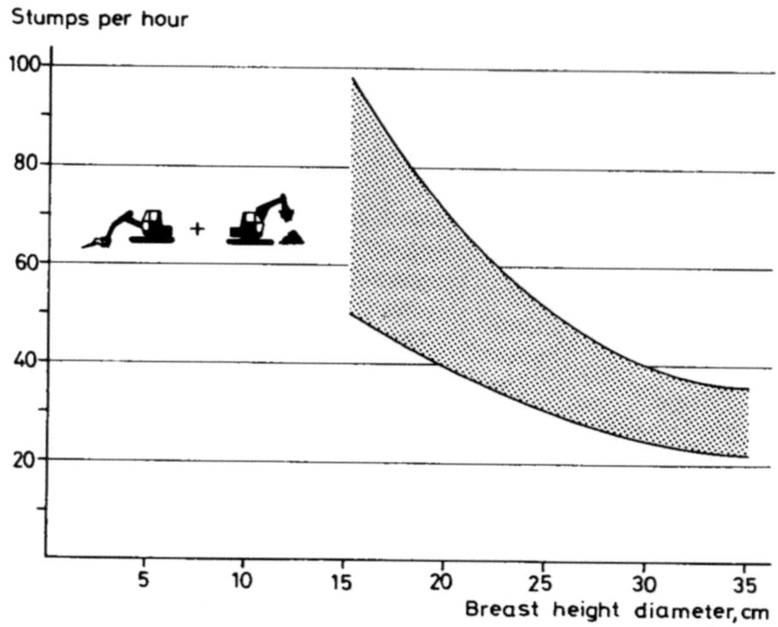

Fig. 4.36. Number of stump-root systems uprooted and split per effective hour by the KH-120 Pallari Stumpharvester, as a function of stump diameter (redrawn from Jonsson 1985)

the Pallari Stumpharvester, have been built more recently for the same purpose. With regard to output, reliability, and applicability under variable ground and stand conditions, they have not introduced basic improvements.

An attractive solution is the *use of vibration* in conjunction with uprooting. Vibration at a given amplitude and frequency is transmitted to the stump-root system (Fig. 4.37). The idea was first tested in the Soviet Union at the end of the 1950s. According to Skalskij (1960), a crane-mounted vibrator grapple had the following advantages:

1. Power requirement in pulling up Scots pine stumps was reduced by more than 50%.
2. All root mass was recovered.
3. The soil layers were neither broken nor mixed up.
4. Practically all loose soil was removed from the stumps.
5. The stump holes were small.

In the Soviet Union a vibrator stump grapple is mounted on the 8–10 m crane of a tracked skidder. During the extraction the stability of the machine is improved with support pads. The vibrator grapple is lowered in an open position over a stump, and the grapple prongs are forced under the lateral roots and closed together. The stump-root system is then extracted with the vibration and vertical lifting power of the crane. The output of combined extraction, cleaning, and bunching is about 30 stumps/h. The machine is used to extract mature pine stumps in boreal forests (Demin et al. 1970).

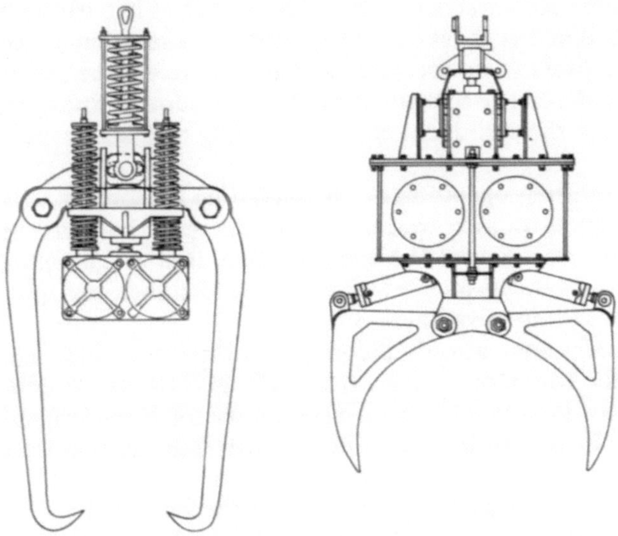

Fig. 4.37. Crane-mounted vibrator grapples for combined uprooting and precleaning of stump-root systems. Russian AKP-1 (*left*), American Vibro Stump Puller (*right*)

In the United States the Vibro Stump Puller head is used widely for harvesting mature stumps of southern pine for the use by chemical industries. An elastomeric suspension isolates the vibration from the crane of the prime mover, usually an excavator. To vibrate and uproot a stump, the operator lowers the open grapple over the stump, with prongs rotated to avoid visible outstretched roots. Hydraulic pistons close the grapple prongs under the root system, and the vibrator is started. The loader crane raises the grapple and extracts the stump-root system. Excess dirt is shaken off as the operator swings the vibrating stump-root system onto a pile. This cycle takes 90 s or less (Koch 1985 b). The amount of soil can be reduced to less than 1% of the total mass of the raw material. Contrary to the observation of Skalskij (1960), this technique was reported to mix up the ground some.

4.4.3.4 Splitting and Cleaning as Part of the Stump Harvesting Schedule

Effective methods are readily available for the extraction of stump-root systems from the ground, but as such the raw material is not yet in a transportable form. Before transport to the mill, the stump-root systems must be reduced to smaller pieces. The object of *size reduction* may be presplitting the stump proper into pieces of desired size, severing large lateral roots from the stump, cutting laterals into shorter pieces, or removing the thin parts of the roots from the raw material. The following advantages can be gained through size reduction:

1. Improved efficiency of haulage through a higher bulk density;
2. Simultaneous decontamination of raw material;
3. Easier handling in the transport phase and at the mill yard.

In-woods presplitting of stump-root systems with mobile machines is a difficult technical problem. In fact, it is in this phase of the harvesting schedule that a substantial proportion of the recovery costs rise. Sand and stones prevent the use of sharp blades, thus restricting the choice of equipment. The reduction of transport costs must be weighed against the additional costs of splitting and the possible loss of fiber.

The position of the presplitting operation in the harvesting schedule is very important. As the primary aim is to lower transport costs, the reduction must be done at the forest end, either on site or at the upper landing. It is usually more feasible to split the stump-root systems on site before off-road haulage, but splitting at the landing also has certain advantages.

Presplitting can be partially done already before the extraction by shearing, cleaving, or cone screwing the stump or tearing off laterals. In this case presplitting and uprooting are usually done with the same machine, the force requirement of uprooting being reduced by the work order. Another alternative is to extract the whole stump-root system and do the presplitting and decontamination afterwards with the same or a separate machine on the site or at the landing. In separate splitting the stump-root systems are handled either individually or as a mass.

Decontamination is usually combined with uprooting, splitting, or haulage. It can be done by shaking, vibrating, rolling in a drum, beating in a hammer mill, or with water (Y. Jonsson 1985). The more thorough the cleaning, the higher are the costs and the loss of wood. The most economical solution for an industrial enterprise is determined not merely by the transport costs but also by the capacity of the stump mass processing equipment of the mill. The intensity of decontamination is thus an optimization problem. Under Swedish conditions, decontaminated stump mass can be delivered to the mill at the same cost as stump mass without any decontamination at all (Jonsson 1978).

From the operational point of view, it is advantageous to integrate uprooting, splitting, and decontamination. Examples of combined operations are blasting (Sect. 4.4.3.1) and crane-mounted harvester heads (Sect. 4.4.3.3). However, in some cases a separate splitting procedure is technically simpler to implement.

Presplitting can be done even with a *chain saw*, but the work is hazardous and contaminants soon blunt the chain. The useful life of the chain is reduced to only a few hours. Intervals between sharpening can be lengthened by using chains with cutting parts of special metal, but this does not prolong the life of the chain as sand continues to wear the other parts. An advantage of chain-saw splitting is that the method is suitable for small work sites. Moreover, with a chain saw a stump-root system can be split in a way most feasible for bulk density and decontamination of the product. When green stump-root systems were reduced by severing off the laterals and reducing them to desired lengths, but leaving the central stump intact, the output of a chain-saw worker was only 3.5 m^3 stacked of split pieces per 8-h day. The output was considerably higher in removing laterals from mature pine stumps, but still not more than 13.4 m^3 stacked/man-day (Vasil'ev et al. 1970).

In *farm tractor-based splitting machines*, the stump-root systems are placed singly into the device. Rotary wedges and guillotine-type blades are commonly applied principles. For example, in a Russian splitter mature pine stumps are pressed against a fixed frame consisting of an outward-widening conical ring blade and eight radial straight blades. The productivity of such machines is only adequate for small-scale operations.

To reduce the amount of impurities in truck loads of split stump mass, various kinds of mobile decontamination machine prototypes have been constructed in Sweden. One of the machines is a *decontaminater-forwarder,* the load space of which is equipped with rotating elliptical disks in the bottom. The disks shake the stump pieces free of most of the contaminants which fall on the ground through openings. The machine is capable of cleaning about one-third of the load space at a time. The cleaned wood is dumped on the ground, and the load space is refilled with uncleaned wood until a loadful of decontaminated stump pieces is available. When this machine is used for off-road haulage, the stump harvester does not have to do decontamination in conjunction with extraction and splitting. This increases the productivity of the harvester. The quantity of contaminants is reduced from approximately 50 to about 30% of the total mass of the load (Y. Jonsson 1984, 1985).

More thorough comminution and decontamination can be carried out at the upper landing with a forwarder- or trailer-mounted stump-mass processor. This alternative is discussed in Section 5.2.3.

4.4.4 Off-Road Haulage of Stump and Root Mass

If trees are pulled up with the taproot intact after severing the laterals, the off-road haulage can be performed with ordinary skidders. The taproot does not cause major technical or organizational problems, and the additional haulage cost is low. On the contrary, haulage of separate stump-root systems or split stump and root mass to the landing is much more difficult to mechanize.

The *solid content* of a pile of whole stump-root systems is very low, 0.05 – 0.10. Consequently, efficient haulage requires size reduction and decontamination on site to increase the solid content of the load. An exception is mature pine stumps free of laterals. In the southern United States they are hauled to the landing usually unsplit on 10 – 15 m^3 pallets.

The solid content of green stump-root systems can be increased to 0.15 – 0.30 by presplitting, depending on the raw material, piece size, removal of thin roots, and the degree of decontamination. Compared with the conventional raw material of the pulp industry, the solid content of the load still remains low irrespective of the treatment. Impurities may raise the total load mass per solid cubic meter of split stump and root material to 1200 – 1500 kg, thus further increasing the cost of haulage. Moreover, as the irregular form of the pieces hampers loading, haulage represents a considerable proportion of the total cost of stump and root mass at the landing.

A modified forwarder can be used for off-road haulage of split stump and root mass. Due to the low bulk density, it helps if the tractor has a large bin with a solid bottom and sides. A heavy forwarder with a 30-m^3 bin takes an average load of 7 t, but 50% of the total mass may consist of contaminants. On average, a forwarder of the same size takes 12 – 14 t conventional timber (Y. Jonsson 1985).

The grapple of the forwarder crane must be widened and reshaped in the manner presented in Section 4.2.2. The crane may be used for both loading und unloading, but if a forwarder is used continuously for haulage of stump and root mass, equipping it with a dumping load space may be feasible. However, if a landing site is crowded and the unloading must be done onto high piles, the use of a crane is necessary.

The *performance* of a forwarder depends on the size of the piles in the clear-cut area, total yield of stump mass per hectare, amount of contaminants, haulage distance, load space, etc. Over a haulage distance of 300 m, the average output is 7 m^3/h (Y. Jonsson 1985).

Chapter 5

Comminution of Residual Forest Biomass

Residual forest biomass is by nature heterogeneous in composition, size, and shape. It is often too bulky to be hauled and handled economically by conventional methods. Frequently the mass capacity of the transport vehicles is not fully utilized.

Whether whole small trees, tree sections, branches, stumps, or roots, the biomass must eventually be reduced to smaller particles of relatively uniform size before it is used for industrial processing, fuel, or fodder. The diverse range of forest residue is converted either on site, at an upper landing, at a central terminal, or at the point of utilization into a more homogeneous material which can be handled in automatic systems and processed into various industrial products or biomass-derived energy. The residue is processed into merchantable mass articles, suitable for the use by large-scale industry, by reducing and uniforming the size of the particles and changing their form. This reduction of forest biomass by mechanical means into a higher valued and relatively uniform bulk material is called *comminution*.

Wood particles may be classified according to their size as follows (Förslag... 1986). Reduction to fine particles such as sawdust or powder is principally a secondary conversion operation and is not included in the present discussion.

	Size fraction (mm)
Powder	– 1.0
Sawdust	1.0– 5.0
Chips:	
Fine	5.0– 15.0
Coarse	15.0– 50.0
Small wood:	
Fine	50.0–150.0
Coarse	150.0–500.0
Full wood	500.0–

Traditionally, conventional pulpwood is converted to a suitable form for chemical pulping by reducing it with sharp knives into chips, i.e., particles with a fairly regular shape and limited size variation. The bulk of the material has a length of 5–30 mm. As a natural consequence of this common practice, the basic solution of forest residue comminution is usually chipping, regardless of the end use. However, depending on the type of the biomass to be comminuted and its end use, actions such as hammering or shredding with impacting blunt tools and chunking may be feasible alternatives to chipping.

5.1 The Purpose of Comminution

The primary purpose of comminution is to reduce the particle size, homogenize the composition, and improve the bulk density of forest biomass in order to facilitate transport, handling, storage, sorting, and further processing. The principal quality parameters of comminuted biomass are its physical and chemical properties, the amount of organic and inorganic contaminants, and the geometry of the particles. The relative importance and requirements of these variables depend on the end processing technology and the final use of the raw material.

The *geometry of the particles* is determined by the cutting techniques. The major dimensional particle characteristics are the length, width, thickness, ratio of thickness to the diagonal across the face of the particle, and particle size distribution in screening (Fig. 5.1). Other important quality parameters are compression damage, solid content, and bulk density of the comminuted material. To a limited extent, the amount of contaminants may be affected by the comminution techniques.

In an ideal case for cross-grain cutting during chipping, adjacent cuts should be equidistant and parallel, resulting in chips of constant length in the grain direction. In practice, however, many deviations occur, resulting in a chip length distribution of an approximately Gaussian nature. When the length distribution is as narrow as desired, the quantity of oversized chips in the upper tail and that of undersized chips in the lower tail of the distribution is small. In disk-type chippers with all knives moving at an oblique angle to the grain, chip formation does not allow separate control of either thickness or width. The corresponding fractures are random, which means that the distribution of thickness and width is relatively wider than that for the length. However, a straight-line relationship exists between average values for length and thickness. In good-quality pulp chips the coefficient of variation is 20% for length and 40% for thickness (Hartler and Stade 1977, 1979).

Due to the heterogeneous composition, small size, and irregular shape of raw materials, such as small whole trees, logging slash, or stumps, the characteristics of chips made from residual biomass differ greatly from those of conventional mill chips. Since the raw material properties vary with stand conditions and logging practice, and because the alignment of residues in the comminution process is extremely difficult, the geometry, purity, and composition of the particles obtained from residual biomass are never constant. The mills have less direct control over chip quality than in the case of roundwood supplies. Consequently, responsibility for the quality of the chips must rest with the logging organizations. Indi-

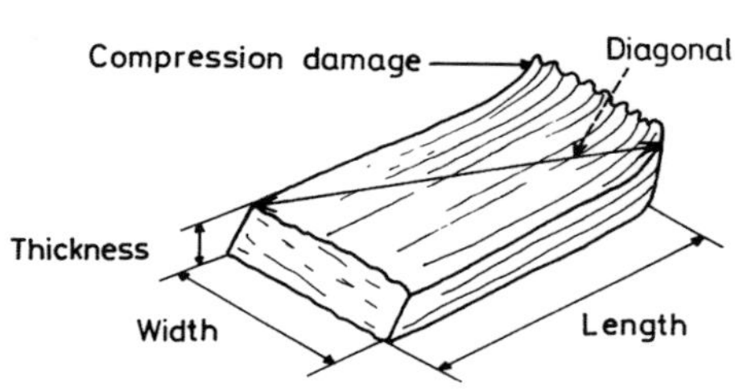

Fig. 5.1. The dimensions of a chip particle

vidual specifications relate to the process used, type of equipment, and the product requirements (Mitchell and Fischer 1985).

In chemical pulping the *impregnation of chips* takes place through diffusion and, especially in partly dry wood, penetration. The rate of diffusion, which is the predominant means of chemical transport in kraft pulping, is roughly the same in all directions of the wood. Consequently, the smallest dimension, i.e., the thickness of the particle, determines the diffusion rate. On the other hand, the penetration in the grain direction is 50–100 times as great as across the grain, and is thus determined by chip length. Slow and incomplete impregnation results in an excessive amount of screening rejects in pulping. Although the rejects are refined and recooked, poorly impregnated, overlarge and overthick particles will definitely yield inferior pulp.

In addition to the penetration rate, *chip length* also affects the final fiber length (Fig. 5.2) and the occurrence of compression damage, which has a deteriorating affect particularly on the sulfite process. Because short chips thus facilitate faster impregnation of chemicals but produce short and damaged fibers, the optimal length is a compromise. According to Hartler and Stade (1979), the preferred chip length is in the range of 15–25 mm in the kraft process and 25–35 mm in the sulfite process. The chip length is affected mainly by the distance of the knife edge from the disk or drum, and the position of the log during reduction so that the grain direction relative to cut is proper and constant.

Chip thickness, which is the determining factor in chemical diffusion, is roughly one-fifth to one-tenth of the chip length. Cross-grained wood around knots and poorly aligned branches yield chips that are overthick in relation to length. Increasing thickness results in less cooking uniformity and, therefore, inadequately delignified, undigested thick chips have to be rejected in the screening of pulp. The acceptable upper thickness limit, which must be kept under control,

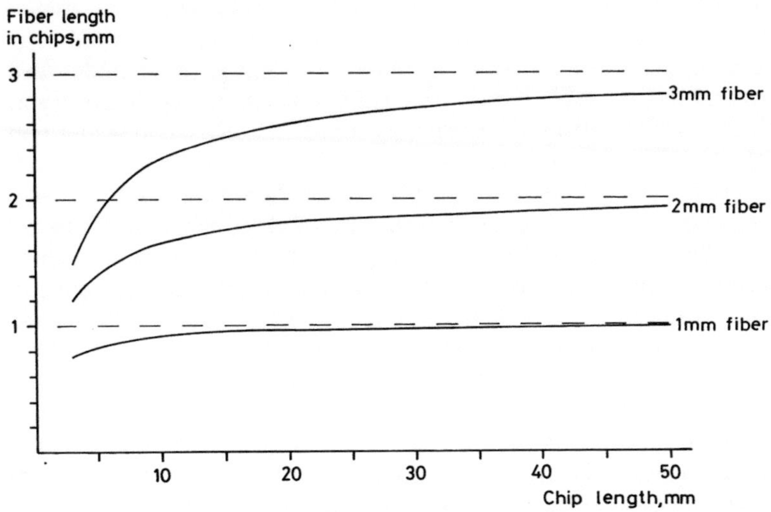

Fig. 5.2. The theoretical average fiber length in comminuted wood as a function of the particle length, when the original fiber length is 1, 2, or 3 mm

is determined largely by the cooking process, cracking inside the chips, and surface lamellations. Mill-cut chips with cracks and lamellations do not produce more screen rejects in kraft pulping than do thinner laboratory-cut chips without internal cracks (Colombo et al. 1964).

According to Hartler and Stade (1979), chip thickness is highly critical in kraft pulping, the preferred range being 1.5–4 mm, but in the sulfite process this dimension is of minor importance. According to Wahlman (1967), maximum-strength sulfate pulps are produced from 2–5-mm-thick chips, and screenings become excessive if the chips are thicker than 5 mm. A slot width of 8 mm is recommended for thickness screening of softwood and 6 mm for hardwood chips (Hatton 1977b). Overthick chips segregated in screening may be reduced with slicers, in which they are first aligned against the inner wall of a spinning knife rotor by centrifugal force and then propelled into the knives by the paddles of another, faster spinning rotor inside the knife rotor (Artiano 1986).

Chip width in the tangential direction of wood may still be reduced after the chip formation by impact between the wood particles and the chipper casing and discharge spout. In pulping, the chip width is of secondary importance.

If the width is no more than the thickness, the small-diameter particles are called *pin chips*. In a digester they tend to move more rapidly in relation to the bulk movement of the chips, causing plugging and circulation disturbances when abundant. Pin chips are produced especially when comminuting brittle, frozen, or very dry wood, or when the cutting velocity of a chipper is very high. Pin chips may also be produced by built-in screens around drum chippers or hogs or, for example, when bark is separated from chips by ball milling. The tolerance limit for pin-chip content in regular chips is 10–12% by mass (Hartler and Stade 1979).

Due to their short fibers, *undersized chips or fines* give inferior pulp properties. Chips smaller than 6 mm make it difficult to maintain a uniform temperature and liquor circulation throughout the digester, resulting in blockage in the circulation system and uneven cooking. In many respects pulps from undersized softwood chips resemble short-fiber hardwood pulps. The presence of undersized chips is more critical in the sulfite than in the kraft process. The tolerance limit, which is usually set by circulation difficulties, is approximately 1% (Hartler and Stade 1979).

The ideal chips for pulp production are *free of contaminants* such as bark, foliage, and inorganic materials. This goal cannot be achieved when the chips are made of unsorted forest residue. For sulfite pulping, the bark content of chips made in the woods remains unacceptably high even if the chips are upgraded at the mill, while many sulfate pulp mills are able to use forest residue chips mixed in a specified ratio with the ordinary raw material. Consequently, since sulfate mills are the main industrial user of forest chips, the most important dimensional variable is usually thickness.

If the size distribution of chips is wide, excessive impregnation in undersized particles may deplete chemical supplies, resulting in incomplete delignification of larger chips. A rough estimation of the *length distribution* of the particles is achieved with round-hole screens. A fairly accurate *thickness distribution* is provided by slotted screens made up of parallel rods. Combinations of various sizes of successive round-hole and slotted screens are used as particle classifiers to separate oversized, overthick, pin-chip, and undersized fractions from acceptable chips (Hartler and Stade 1977; Hatton 1977a, b).

Chip quality is thus determined by measuring a suitable combination of properties. In view of the different processes and machinery used in production lines, it is not easy to reach a well-balanced and complete combination. The following list represents a compromise for the pulp industries (Hartler 1986):

Property	Measured by	Target
Length	Hand-sorting	Uniformity
Thickness	Slot screening	No tail in the distribution above 8 mm
Oversized	Hole screening	Small proportion
Undersized	Special hole screening	Small proportion
Bulk density	Volume after standard filling conditions, and dry mass	Small variation
Mechanical stability	Increase in small-sized fractions	High stability
Chip damage	Sulfite cooking and evaluation through paper strength	Low for acid sulfite, less important for alkaline processes

Figure 5.3, based on an extensive quality evaluation of chips from small-sized whole trees for industrial processing and combustion in Finland, illustrates the thickness distribution of particles produced with large portable disk chippers. The proportion of 2–8-mm acceptable chips is considerably higher from delimbed stems than from whole trees. The average chip thickness decreases and the amount of fine material increases in winter chipping when the wood is frozen and brittle.

In the particleboard industry *chips are reprocessed* either by flaker knives, producing smaller engineered particles of determined dimension, or by hammermills, generating particles of indistinct geometry. The size and shape of the chips affect

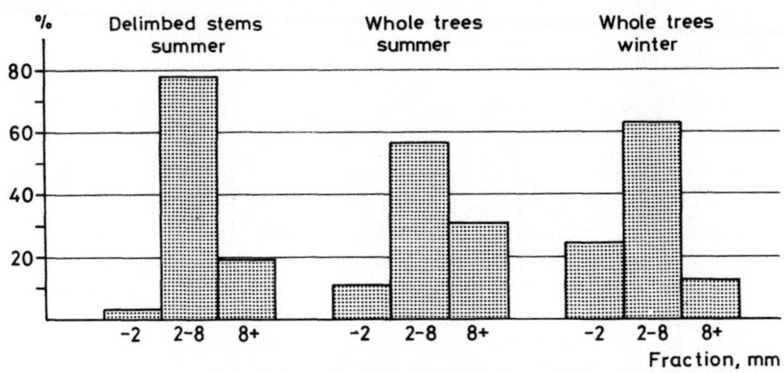

Fig. 5.3. Thickness distribution of particles reduced from 9–12-cm-thick, fresh *Pinus sylvestris, Picea abies*, and *Betula* spp. trees with a large disk chipper in summer and winter in Finland. Average values are from extensive material (drawn from data in Uusvaara and Verkasalo 1987)

the final geometry of the reprocessed particles, which in turn affects the mechanical strength, water absorption, surface smoothness, appearance, and machining properties of the final product. The common chip specifications of the pulp mills do not necessarily represent the optimal geometry for particleboard manufacturing. Conventional chips are so short in length in relation to width and thickness that flaking these along the grain with commercial ring flakers is virtually impossible. Since the conventional chips do not yield the desired long, thin flakes, somewhat larger pieces of wood would be desirable for the intermediate furnish of particleboard.

Larger particle size also appears to have advantages over conventional chips in some solid-fuel energy converters. Chunky fuel particles may offer the following advantages over conventional pulp chips (Arola 1983a, 1985):

1. They take less energy to produce.
2. They provide better air flow for natural or artificial drying and are thus less susceptible to the growth of fungi and spontaneous combustion during storage.
3. They give a better flow of underfire combustion air.
4. They have fewer particles per unit volume, which results in less solid particulates and easier control of emissions in large-scale combustion systems.
5. They are more suitable for blending with stoker-sized coal and constitute a better feedstock for producing a metallurgical grade of industrial charcoal.
6. Due to the larger particle size, fines can be segregated more easily.

The size of the reduced particles can be adjusted in most comminution devices within a limited range. A Finnish conescrew chipper-chunker with six interchangeable conescrew knives makes it possible to alter the particle length within an exceptionally wide range (Fig. 5.11). The particle size is controlled by the pitch of the thread and the number of parallel threads on the conescrew. Figure 5.4 illus-

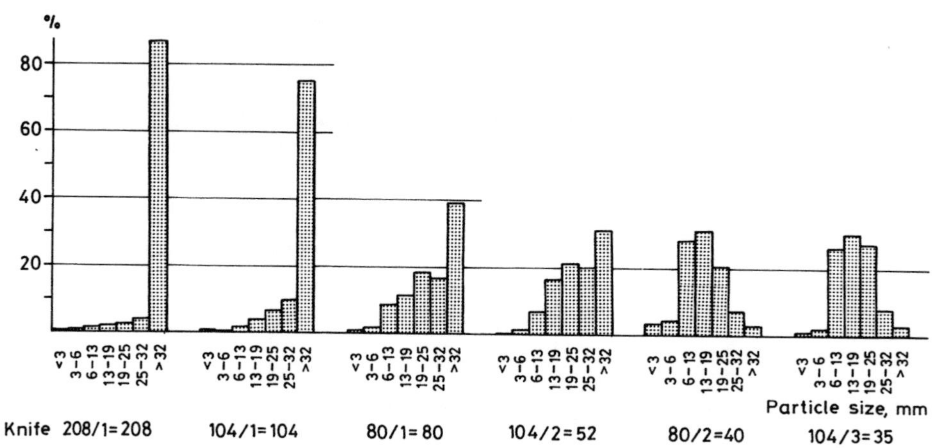

Fig. 5.4. Length distribution of particles reduced from small-sized, delimbed *Betula* spp. stems with a conescrew chipper-chunker equipped with a set of six exchangeable conescrew knives. The quotient of the pitch length in mm and the number of threads is indicated below the diagrams (drawn from data in Nurmi 1986)

trates the effect of knife setting on the size distribution in the round-hole screening of particles reduced from small-sized, delimbed birch stems with a conescrew chipper-chunker. For the uniformity of particle size and achievement of optimal chip quality, the following points are of particular importance for chipper design and maintenance (Hartler and Stade 1977).

1. Proper direction and speed of the log when it enters and passes through the chipper;
2. Exact positioning of all knives in relation to bed plate and disk;
3. As low a sharpness angle as possible for maintaining knife sharpness;
4. Limitation of the cutting speed, especially at low wood temperature.

5.2 The Techniques of Comminution

To separate chip from the workpiece during any wood-machining process it is first necessary to cause a structural failure at the juncture of the chip and workpiece. Chippers cut wood with high-speed revolving knives against stationary anvils. Configuration of the resulting particle, cutting power requirement, and the surface quality of the wood are all strongly affected by the direction of the cut in relation to the grain, as well as by the knife geometry. The orthogonal machining situation may be described with a two-number notation such as that used in Fig. 5.5 (Koch 1985b).

Comminution of forest biomass is performed by producing either engineered particles by *cleaving and shearing with sharp knives*, or particles of indistinct geometry by *crushing or shredding with blunt, impacting tools.* Particles generated with sharp knives have a more uniform size distribution, smoother surface, and less compression damage. They are therefore preferred in industrial processing.

Due to wear and tear caused by stones and sand, knives are a considerable cost factor in comminution with portable equipment. If the biomass is extremely dirty, it may be necessary to change or sharpen the knives twice a day, while one change a week or even less may be sufficient for very clean material (Pottie and Guimier

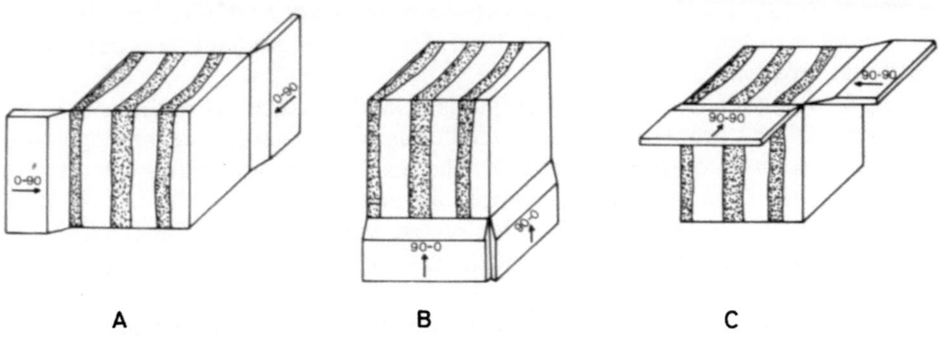

Fig. 5.5. Designation of three major machining directions of wood. The *first number* is the angle of the cutting edge to the grain, and the second the angle between cutter movement and grain (redrawn from Koch 1985b)

1985). Some chippers are equipped with a dirt separator to remove loose grit from the biomass before it reaches the knives (Fig. 8.29).

In Finland, when forwarder-hauled whole-tree material is reduced by various disk chippers, the interval between sharpenings is 40–70 m^3 and the life of a knife set is 900–1100 m^3 solid. On average, the direct cost of knives is 8% of the total cost of chipping (Hakkila 1984). Stones, sand, and other inorganic contaminants may completely prevent the use of sharp knives in comminution of slash and stump and root mass. Regardless of poorer product quality and higher consumption of energy, comminution has to be performed then by crushing.

Exclusive of light, hand-fed chippers, the biomass is taken to the *in-feed* system of a portable comminution machine by its self-contained or a separate crane. The feeding of long material usually takes place from the side or behind, but in a Danish mobile terrain chipper from the front. Many slash and stump hogs are fed from above.

The chipped or hogged product is usually discharged pneumatically through the *out-feed* spout, assisted by a separate or a flywheel-mounted fan. Throwing and blowing of such material with a powerful fan may damage the particles. To avoid damage from impact with the walls of the discharge spout, the chips should pass the chute as straight as possible. Gravity or bottom discharge to a conveyor belt, which causes less particle impact and damage and requires less power, is applied mainly in portable slash hogs.

If the comminution is carried out with a mobile unit in terrain, the product is usually discharged into the dumpable bin of the tractor. In landing comminution the discharge takes place into a truck, trailer, exchangeable container or pallet, dump bin, sack, or pile. The in-feed and discharge systems together determine the work organization and the positioning of the piles and machines on the landing.

The theoretical *output* of a self-feeding chipper is the result of the log diameter, particle length, and the number of cuts per minute. The following formula may be used to calculate the output:

$$V = RPM \times N \times L \times A \times 10^{-6},$$

where V = output, m^3 solid/min; RPM = disk or drum revolutions per minute; N = number of knives in the disk or drum; L = chip length (cm); A = bolt cross-sectional area (cm^2).

In field operations, the actual output is also affected by other characteristics of the machinery, properties of the raw material, work organization, and operator skill. The nominal outputs provided by the machine manufacturers are seldom achieved in the field because interruptions are created by waiting for the off- and on-road vehicles, inadequacy of motor power, idling due to feeding problems, movement of the chipper within the work site, maintenance, and machine breakdowns.

In the northern European landing-comminution systems, about 60% of the work-site time of a chipper operator is spent on chipping. The comminution may be hampered by various *disturbances*, the most common of which are caused by the following factors:

1. Rapid wear and tear of knives due to stones, sand, and metal bodies means not only direct knife costs but also loss of work time due to reduced output and changing of knives.
2. Clogging of the chipper discharge spout occurs particularly when very small, unlimbed hardwoods, branchy softwood tops, slash or oversized trees are comminuted. Thawing snow and blunt knives contribute further to the clogging of the discharge spout.
3. Poor bearing capacity of the ground may hamper the movements of comminuting machines and transport vehicles in the landing area. Terrain chipping in particular requires firm ground.
4. In the northern European system, where the material is decked beforehand on the landing, snow in the piles may make it difficult to grapple the trees in winter. Snow may accumulate under conveyors, choking the in-feed device, especially during a thaw. Freezing of interlacing branches to one another and to the ground may slow the feeding. Snow and freezing problems may be lessened by making large and high landing piles, using protective coverings, and increasing the efficiency of the crane.

5.2.1 Portable Chippers

Because portable chipping units must be readily movable from site to site, the chipper device proper is generally lighter than in stationary machines working permanently at the same location. A lighter chipper means less flywheel action. Consequently, the revolution speed slows down and builds up more rapidly than in stationary chippers. Regardless of the lighter mass, a portable chipper must be reliable and of rugged design, and the maintenance under field conditions must be easy. The power requirement per produced unit should be reasonable, yet the production capacity must be high enough to keep the cost of comminution low. The geometry and other properties of the chips must meet the requirements of the end use under consideration.

A major problem associated with biomass chipping in the woods is feeding. Branches, whether loose in slash or still attached to whole trees or tree sections, cannot be aligned properly and are consequently forced to the chipper knives in nonuniform, random directions, resulting in uneven particle lengths. Another problem is inorganic contaminants, which increase knife costs, reduce productivity, raise energy consumption, and cause interruptions in the work.

More than 80% of all chips used in the world pulp and paper industry are produced with *disk chippers* developed at the end of the 19th century (Fig. 5.6). Because of their relative simplicity, ruggedness, reasonable cost, and ease of maintenance, disk chippers prevail among both stationary and mobile comminution equipment. Forest biomass is forced endwise into a disk chipper through a supporting input spout. Sharp, straight knives, bolted radially on a 400–1000 rpm revolving heavy disk, comminute the raw material progressively into particles of the desired, nominal length. The process is controlled by adjusting the knife height or the anvil, the number of knives, the revolving speed of the disk, and the in-feed speed. The cut of the knives is in a mode intermediate between 90–0 and

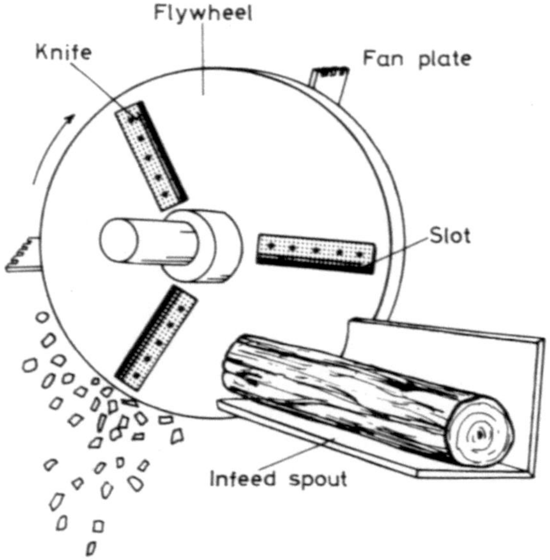

Fig. 5.6. The principle of a disk chipper

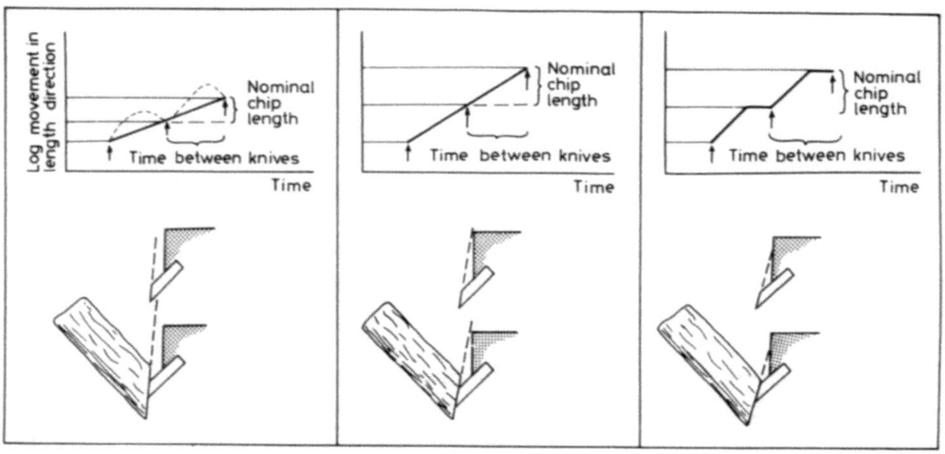

Fig. 5.7. The effect of pull-in angle on the self-feeding of a disk chipper. *Left:* Too small an angle resulting in insufficient feeding. The *dotted curve* represents occasionally observed bouncing. *Center:* Ideal pull-in angle resulting in ideal feeding. *Right:* Normal pull-in angle resulting in accelerations, retardations, and stepwise feeding (redrawn from Hartler 1986)

90–90 (Fig. 5.5). When the knives cut through a log, they exert a pulling force which moves the log against the chipper disk. Proper and continuous suction requires sharp knives. The force that pulls the log through the chipper is exerted by the front surface of the knife. Counteracting forces are friction, mainly from the spout, and reaction from the back of the knife or from the wear plates of the disk. Three cases are considered in Fig. 5.7 (Hartler 1986):

1. In the first case the pull-in or clearance angle is not large enough, leading to short movement and consequently to short chips. A high reaction force from the back surface of the knife has a compressive action on the wood, resulting in damage even in the "undamaged half" of the chip. The reaction force can occasionally lead to bouncing of the log.
2. The second situation represents an ideal case when the movement of the log between two adjacent knives is just sufficient to bring the log up to the disk in the correct position for the next cut. The movement is continuous, as shown in the upper part of Fig. 5.7. The pull-in angle is somewhat larger, and the compressive action from the back surface of the knife is so small that the surface region of the "undamaged half" normally remains undamaged.
3. In the third case the pull-in angle is larger and the inward movement more rapid. The log hits the wear plate of the disk, where it is retarded until the next cut. The compressive action from the back of the knife on the wood is negligible when the log is pressed against the disk. Instead, a reaction force from the disk will act on part of the newly cut surface of the log. The movement will be discontinuous, as shown in the upper part of Fig. 5.7. This case represents a common arrangement in a disk chipper.

The following factors determine the cutting behavior of a disk chipper (Fig. 5.8):

1. The knives progress in the wood along a line of cut that is inclined 1–7° to the plane of the rotating disk. This pull-in angle (α) is determined by the relative velocity of the knife and the rate of feed.
2. The wedge angle of the knife (β), usually in the range 30–37°, affects the particle formation and energy consumption.

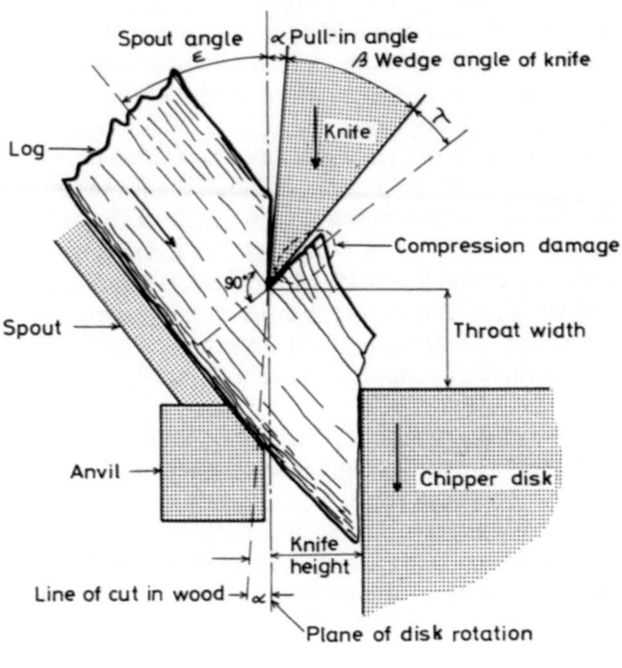

Fig. 5.8. Basic configuration of disk chipping and definition of chipping angles (redrawn from McLauchlan and Lapointe 1979)

3. The feed direction of the raw material is inclined to the disk plane by the spout angle (ε), usually 30–40°, which may be attained by attaching the spout at a horizontal angle only, or in combination with a vertical angle. If the knife height perpendicular to the disk face is h, the chip length is equal to h/sin ε.
4. The angle (λ) between the inner or cleaving face of the knife and the face perpendicular to the grain determines whether the chip formation will be mainly by cleavage or by shearing.

Slots below the knives in the disk permit the particles to pass through the disk. The throat width of the slot should be appreciably greater than the particle length to allow unobstructed flow of the chips. After the chips have passed the disk through the slots, the disk with its knives acts like a fan rotor and blows the chips through an out-feed span. Short plates mounted on the outer edge of the disk increase the blowing action. The blowing power of a portable chipper affects the solid content and bulk density of the chips in a vehicle. Alternatively, the chips are discharged through the bottom of the chamber to a conveyor belt.

Productivity in comminution of forest residue is often limited by the feeding of the raw material through the spout rather than by the reduction action itself. When the raw material consists of very branchy whole trees, unlimbed tops, or slash, the relatively narrow in-feed opening of the spout may seriously slow the operation of a disk chipper. Drum chippers with a larger spout permit smoother in-feeding of slash.

In a drum chipper the knives are on the outer surface of a rotating, horizontal cylinder. Two types of *cylindrical drum chippers* are distinguished on the basis of the in-feeding principle. In side-feed drum chippers, short wood bolts of even length are fed from above in the manner generally used in pulp-grinding machines, so that the knives cut the wood in the grain direction. The method allows thickness control of the particles. In end-feed drum chippers, the raw material is supported by an anvil and forced head-on to the rotor, which produces a cut perpendicular to the grain. Only the latter principle is applicable for comminution of random-length biomass with portable equipment.

In a cylindrical drum chipper the knives are positioned on the surface of the rotor parallel to its axis either in straight lines or in a spiral around the drum (Fig. 5.9). The chip pockets below the knives, corresponding to the length of the chip, drop the reduced wood to a bottom conveyor or to a blowing fan for discharge. Another alternative is to let the chips exit into the inside of the drum through slots in front of the knives. To control the oversized splinters, a drum chipper may be equipped with a built-in screen through which the reduced material passes before leaving the chipper. The size of the screen openings is usually about double the desired chip length.

A *V-drum chipper* consists of two truncated cones facing one another at the narrow ends at an angle of about 90° (Fig. 5.10). The knives are arranged either radially or spirally around the cones. The trees are fed horizontally or by gravity head-on into the V-shaped space between the cones. The reduced chips pass through slots in the conical drums falling onto a conveyor, or they are blown through a spout. A V-drum chipper combines the characteristics of cylindrical drum chippers and disk chippers.

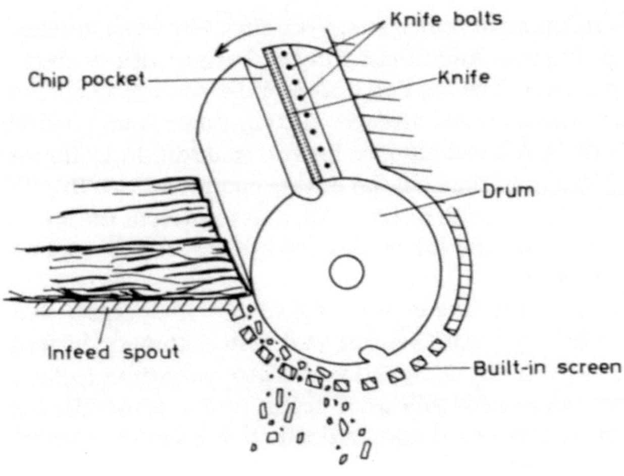

Fig. 5.9. The principle of a cylindrical end-feed drum chipper

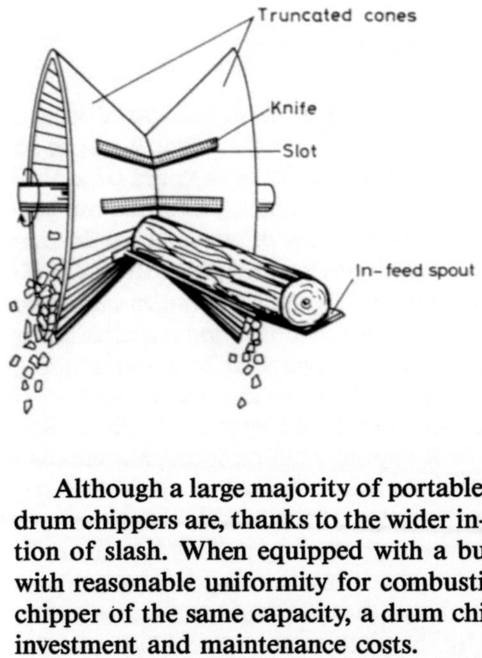

Fig. 5.10. The principle of a V-drum chipper

Although a large majority of portable whole-tree chippers are of the disk type, drum chippers are, thanks to the wider in-feed opening, better suited to comminution of slash. When equipped with a built-in screen, they reduce slash to chips with reasonable uniformity for combustion use. However, compared with a disk chipper of the same capacity, a drum chipper tends to be heavier and has higher investment and maintenance costs.

In a laboratory-scale experiment a cylindrical drum chipper produced considerably less oversized particles from crown material of *Picea glauca* than did a disk chipper. On the other hand, the drum chipper produced substantially larger quantities of essentially pure needles. While the under 16-mm hole-screen fractions from the drum chipper contained 70% of the needles, the same fraction from the disk chipper contained only 37% of the needles. It was concluded that drum chipping facilitates separation of needles for foliage utilization (Young and Hatton 1976).

In conventional disk and drum chippers, the base of the in-feed spout acts as an anvil. The cutting knife pulls a tree with each passage one chip length through the in-feed spout, making the machine in principle *self-feeding*. However, uninterrupted trouble-free operation requires additional *forced feeding* with a chain, belt, or scraper conveyor and rollers. The feeding system must be able to convey the biomass steadily and, to maintain the desired cutting angle and particle length, hold it firmly against the anvil and chipper knives. In addition to forced feeding, rollers prevent repositioning of the material during comminution. To produce good quality chips from the end pieces of the trees, and to prevent the spout from plugging, the dead spot between the forced feeding system and the cutting knives should be as short as possible.

Overfeeding is avoided by adjusting the in-feed rate to 10–15% below that actually required by the nominal chip-length setting, and with automatic in-feed capacity control. The chippers are usually equipped with a heavy flywheel to store energy. If the revolution speed, as a consequence of excess power demand, falls below a given limit, the feeding is interrupted automatically to prevent the chipper and the out-feed spout from plugging.

5.2.2 Portable Chunkers

Chunkwood production and utilization are emerging technologies introducing a new alternative to conventional chipping (Arola 1983a, 1984, 1985; Arola et al. 1982, 1983, 1985). The chunkwood concept seems to offer advantages over ordinary chips if forest biomass is used for the production of energy or structural particle boards. *Chunkwood* is defined as short, thick pieces of wood where the majority of particles have a relatively uniform length of 50–250 mm in the grain direction and a variable cross-sectional area, ranging from about finger size up to entire disks equal in diameter to the material reduced. Chunkwood particles occur only when there is a controlled relation between the material feed and severing of wood by sharp knives or other well-defined cutting tools (Arola 1984).

A self-feeding *spiral-head chunker* was first developed by the U. S. Forest Service in the mid-1970s (Erickson 1976c). In the spiral-head or conescrew chunker the cutter blade is attached to a horizontal shaft in a helical or spiral fashion, somewhat resembling a screw (Fig. 5.11). The edge of the helix is sharpened. As the shaft rotates, the spiraling blade slices through the log, pulls it into the unit, and reduces the wood into predetermined lengths. The cutter depth increases progressively from zero to the maximum diameter of the cut (Arola and Sturos 1982). The length of the chunks is determined by the pitch of the spiral blade and the number of threads. To reduce friction, the pitch should be continuously but only slightly increasing (Erickson 1976c).

A commercial spiral-head chunker was developed more recently in Finland. The first, light model is mounted on the three-point hitch mechanism of a farm tractor. The larger models are constructed on a one-axle chassis, trailer, or truck. Six different interchangeable conescrew heads are available, each producing particles of its own nominal size. In a large model, a conescrew head, weighing up to 300 kg, is fastened to a 1500-kg flywheel with bolts. Also attached to the shaft of the helix are two paddles that throw and blow the reduced biomass through a discharge spout. With a steep pitch and three parallel threads, a conescrew head

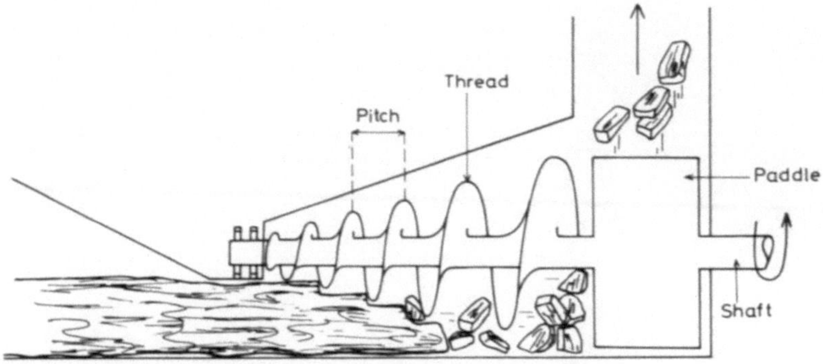

Fig. 5.11. The principle of a spiral-head or conescrew chunker

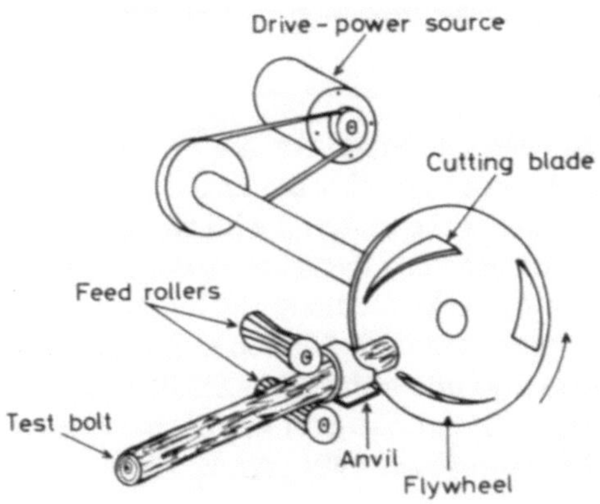

Fig. 5.12. The principle of an involuted disk chunker (redrawn from Arola et al. 1982)

can produce conventional chips, while a head with a slower pitch and only one or two threads reduces the biomass to chunks (Fig. 5.4). In fact, the unit is thus a chipper-chunker.

An *involuted disk chunker*, also developed by the U. S. Forest Service, is another alternative for producing chunkwood (Arola et al. 1983). In the first experimental laboratory machines, two or three cutting blades, bent to a 30-cm radius, were mounted at right angles to the face of a rotating flywheel disk with the trailing end spiraling inward, i.e., involuted. The blades are tapered so that the leading cutting edge projects 13 mm from the disk face and the trailing edge 127 mm, which is equivalent to the maximum depth of the cut (Fig. 5.12). The bite of the knife thus increases from the leading edge to the trailing edge. Since the workpiece does not self-feed, a pair of powered rollers are positioned immediately ahead of the anvil to force the stem into the chunker parallel to the face of the flywheel. The feed speed of the rollers and the rotation speed of the flywheel are interconnected. A larger prototype machine, capable of chunking 30-cm-thick stems, is built around a mobile whole-tree chipper, in which the conventional disk is replaced by an involuted disk-cutter assembly (Mattson et al. 1984).

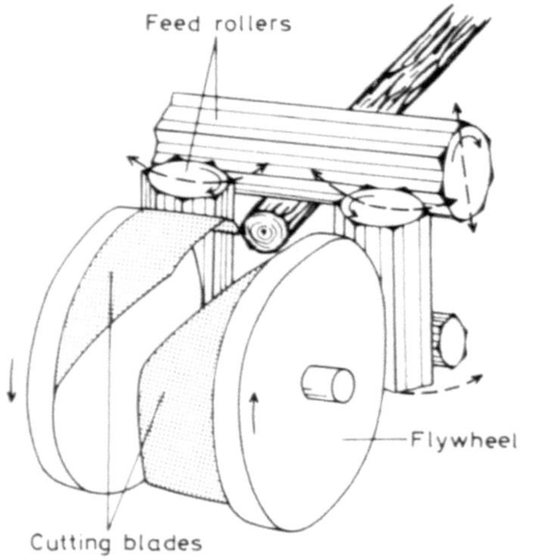

Fig. 5.13. The principle of a double involuted disk chunker (redrawn from Arola 1985)

The involuted disk chunker concept has been developed further by the Swedish University of Agricultural Sciences in cooperation with the U. S. Forest Service. A new prototype double-disk chunker reduces the material with two counter-rotating involuted disks, cutting the stem simultaneously from both sides in a scissor-type action (Fig. 5.13). The counter-rotating disks facilitate the blades acting as anvils for each other, thus removing the problems with friction and manufacture of the complex anvil. Two sets of blades also allow the diameter capacity of the chunker to be increased without increasing the cantilevered length that the individual blades protrude from the disk (Mattson et al. 1984; Arola 1985).

5.2.3 Portable Hogs

Because sand, stones, and metal objects wear and break sharp knives, resulting in excessive downtime, contaminated biomass cannot be comminuted with chippers or chunkers. A feasible solution is offered by low-maintenance hogs which reduce the biomass in a random manner with blunt tools. The resulting pieces have a rough surface, compression damage, an irregular shape, and a wide size distribution. Although the problems of heterogeneous size can be overcome with built-in screens, the product is suitable mainly as fuel for large-scale combustion plants.

Portable hogs are heavy-duty machines with operation limited to landings and terminals. They are mounted on a forwarder, truck, or towable trailer. As in the heavy mobile chippers, the main components are generally a reduction device, in-feed and out-feed systems, operator cabin, and a crane.

The reduction is performed with a single drum or two counterrotating drums around a horizontal shaft at the bottom of an open chamber. In shredder-type hogs the rotor has fixed, strong teeth, while swing-hammer or hammermill hogs beat the biomass with freely-swinging tools hanging in several rows on the rotor

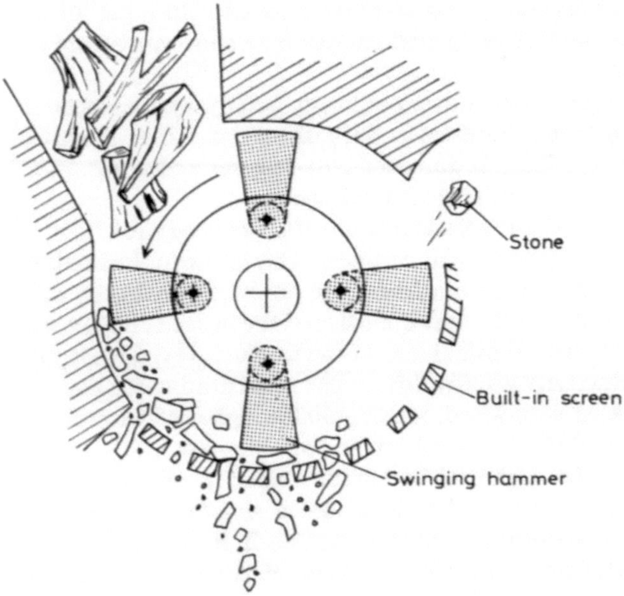

Fig. 5.14. The principle of a hammermill hog

(Fig. 5.14). The incoming material is battered by the teeth or hammers in the chamber. The reduced material is forced, usually through grid bars or a screen, onto a conveyor below the hog bottom, thus reducing the oversized pieces in conjunction with the discharge. The number and shape of the crushing tools, their distance from the anvil or in-built screen, rotor speed, and the screen insert determine the final particle size.

In a *punch-and-die hog* the fixed teeth, located usually in spiral rows, shred the biomass while rotating through a notched anvil. In a *knife hog* the teeth cut the material against a primary anvil. In both machines, the material is further reduced between the teeth and a discharge screen plate. These principles can also be applied to reducing bark residue. In a *hammermill hog* the centrifugal force resulting from the movement of a horizontal rotor keeps the freely-swinging hammers extended. The hammers reduce the biomass by impact and pivot back when encountering a hard foreign object.

Due to their working principles and heavy-duty design, hogs are more resistant to wear and tear and require less maintenance than chippers. Hammermills with freely-hanging impacting tools, and shredders with slowly counterrotating drums are generally the most rugged types of biomass comminutors. They can be used to reduce slash or stump and root mass with a high degree of contaminants. Punch-and-die hogs are somewhat more sensitive to abnormal contamination of biomass. Knife hogs, relying on more or less sharp knives and reducing biomass against a fixed anvil, are more affected by contaminants (Pottie and Guimier 1985).

The biomass is dropped to the hog chamber from above. The feeding takes place by gravity and the rotor itself, or it may be assisted by chain or scraper conveyors. The chamber is wide enough for short slash or presplit stump and root mass. For longer material, a heavy-duty hog can be equipped with a guillotine-type shear which reduces the biomass to desired lengths on a conveyor before in-feed to the crushing chamber.

The hogs are usually protected against overloading caused by large objects. In some machines the rotors can reduce biomass in both directions. When the rotor senses a jam via a preset pressure limit, it automatically reverses and keeps rotating in that direction until another jam and reversal occur.

It is not possible to calculate the theoretical *production capacity* for hogs in the same simple manner as in Section 5.2.1 for chippers. This is due to the random way in which the biomass is fed into the machine. The production capacity of a particular hog is generally defined on the specific basis of the in-feed characteristics of the feedstock and the desired nominal size of the product. If the feedstock is high in wood content, productivity usually drops. The rated production capacities of manufacturers, compared with wood, are 50–300% higher when processing bark (Pottie and Guimier 1985). It is difficult to feed biomass into a strong hog at a rate equivalent to the technical reduction capacity of the device. For example, when the output of a heavy truck-mounted shredder was 9.6 dry tonnes per effective hour for 2-year-old seasoned spruce slash, it was only 5.7 dry tonnes for more elastic, newlyfelled spruce slash (Alexandersson and Johnsson 1985).

A totally different comminution principle is applied in a Finnish 25-t forwarder-mounted stump mass processor prototype designed for combined reduction and decontamination of presplit stump and root mass at the upper landing. The stump and root pieces are sheared randomly in a 2400-mm-diameter, vertical processing cylinder which has a 60-rpm rotating bottom and a fixed vertical shear blade 200 mm from the inside wall of the cylinder. The raw material is dropped into the cylinder from above by crane. The centrifugal force generated by the rotating bottom flings the pieces against the blade which shears them. The reduced pieces are removed from the cylinder through an opening in the wall behind the shear blade.

5.3 Energy and Power Requirement in Comminution

The energy consumption in every forest operation can be divided into five levels: physical, machine, cycle, operational, and system (Gasslander et al. 1979; Sundberg 1981). Figure 5.15 gives an example of the fuel requirement for the harvest and transport of conventional timber in a mechanized and motor-manual system in Sweden. For the purposes of biomass comminution, the five levels of energy consumption can be defined as follows:

1. The *physical level* is a theoretical measure of the minimum energy requirement to overcome the physical resistance of wood when reducing it to the desired sizes. It cannot be measured in field tests.
2. The *machine level* consists of the fuel consumed by a machine to break apart wood material. It does not include energy for driving auxiliary equipment of the comminuting unit.

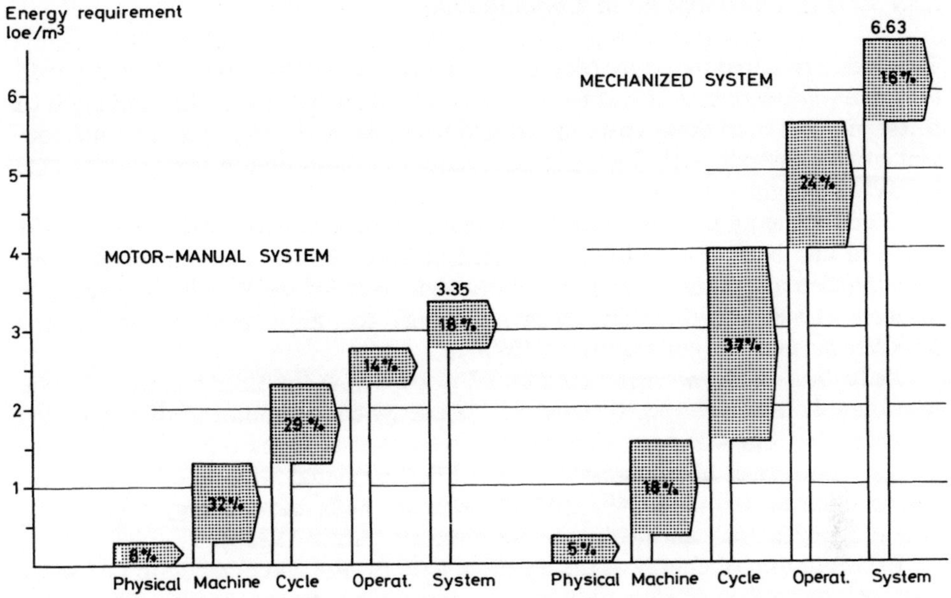

Fig. 5.15. Energy requirement for harvesting and transport of conventional timber (dbh 20 cm, off-road haulage 400 m, on-road haulage 50 km) from stump to mill in liter oil equivalents (*loe*) per m^3 solid under bark (redrawn from Sundberg 1981)

3. The *cycle level* also recognizes the role of material handling in energy consumption while comminuting biomass, including the energy lost to drive pumps, in-feed device, fans, conveyor belts, etc.
4. The *operational level* indicates the total energy used during a normal operational period, including the energy lost during waiting and idling time and for movement within and between work sites.
5. The *system level* also includes the indirect energy used for manufacturing the machine and spare parts, maintenance, and lubricants.

Comminution, in fact, accounts for only a minor part of the total energy consumption in the harvesting of residual biomass. Moreover, when portable comminution machines are used, approximately half of the energy losses occur between the machine and cycle levels (Jones 1981). The actual reduction of biomass is thus not decisive for the total energy consumption of a logging schedule. The differences in energy consumption between various comminution machines may become negligible in relation to the total fuel cost of biomass harvesting.

To allow flexible mobility, portable chippers should not be too heavy. Consequently, it is important to limit the power requirement through proper construction, maintenance, and operation pattern, even though energy consumption per se may not be a critical factor. Because the power requirement is a product of the specific comminution energy in kWh/m^3 and the rate of comminution in m^3/h, it is essential to keep the specific energy requirement as low as possible.

5.3.1 Energy Consumption in Comminution

The energy required to reduce biomass to pieces of the desired size is called *specific comminution energy.* It is directly related to the cleavage and shear strength of forest biomass, and these again are related to the basic density and moisture content of the raw material. The specific energy of comminution is expressed here in kWh/m^3 solid.

Specific energy of comminution increases almost linearly with *basic density* of wood and bark (Fig. 5.16). Hence, the machine power requirement and fuel consumption are usually higher for hardwoods than for softwoods. According to Rodgers (1948), the specific chipping energy for hardwoods is on average 2.6 kWh/m^3 and for softwoods 1.8 kWh/m^3.

Reduction of the *moisture content* of wood below the fiber saturation point considerably increases specific energy. The energy consumption of disk chipping for absolutely dry wood is twice that for green wood. If the wood is frozen, the energy requirement also increases (Fig. 5.17). Compared with wood not frozen, *Quercus falcata* required 24% and *Populus* sp. 64% more energy when comminuted frozen with an involuted disk chunker (Arola 1983a).

Consumption of energy at the machine level for the actual reduction of wood is fairly independent of *stem diameter.* An exception is forced feeding. The energy consumption per unit mass produced of a hydraulic in-feed device decreases with increasing stem diameter at the cycle level (Heikka and Piirainen 1981).

Specific energy is critically dependent on the *side-feed angle* between the cutting knife and the grain direction of the wood. Since the energy requirement is

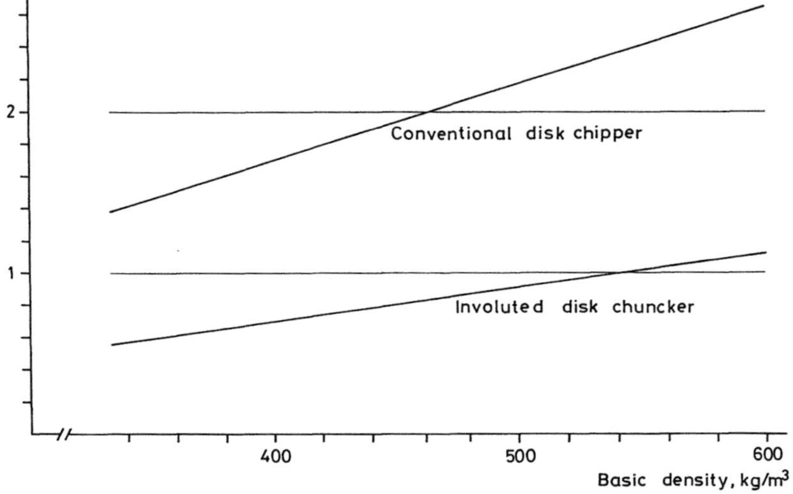

Fig. 5.16. The effect of the basic density of wood on the energy consumption of a conventional disk chipper (Papworth and Erickson 1966) and an involuted disk chunker (Arola 1983a)

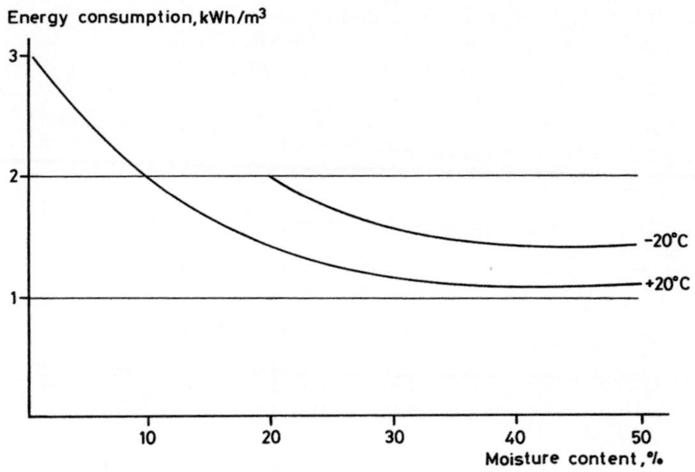

Fig. 5.17. The effect of the moisture content of unfrozen and frozen *Picea abies* wood on the energy consumption of a disk chipper (redrawn from Murto and Kivimaa 1951)

higher for cutting across than parallel to the grain, the 90–0 or 0–90 cutting mode reduces the energy consumption compared with the 90–90 mode (Fig. 5.5). However, the power requirement increases with the surface of the cut.

The compressive stresses in chipping must overcome either the cleavage strength or the higher shear strength of the wood. For small 20–30° *knife wedge or sharpness angles* and large 26–36° *λ-angles* (Fig. 5.8), the energy consumption is low. On the other hand, for a large 50° knife wedge angle and a small 6° λ-angle a reduction definitely occurs by shearing, resulting in excessive compression damage and high energy consumption (McLauchlan and Lapointe 1979). The following table by Papworth and Erickson (1966) illustrates the effect of side-feed angle and knife sharpness angle on the energy consumption of a disk chipper when comminuting maple wood:

Knife sharpness angle	Energy consumption (kWh/m^3) Side-feed angle	
	0°	30°
30°	2.5	2.2
40°	3.4	3.2
50°	5.2	4.2

The energy consumption of comminution is also affected by the *knife or cutting speed,* which is dependent on the revolving speed and diameter of the chipper disk or drum. Figure 5.18 shows the effect of cutting speed on the energy consumption and cutting force of a drum chipper, indicating increasing energy consumption with increasing cutting speed (Pahlitzsch and Sommer 1965). When the revolving speed of a light conescrew chunker was increased from 630 to 1000 rpm, energy

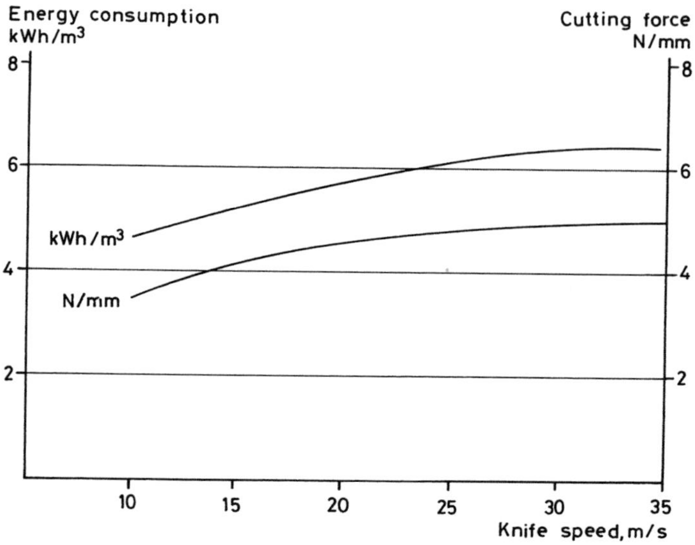

Fig. 5.18. The effect of knife speed on the cutting force per unit length of knife (N/mm) and energy consumption of a drum chipper producing 0.5-mm-thick chips (redrawn from Pahlitzsch and Sommer 1965)

consumption rose by 7–13% (Heikka and Piirainen 1981). In contrast, the energy consumption of an involuted disk chunker dropped from 1.8 to 0.6 kWh/m^3 when the disk speed increased from 92–152 to 339 rpm (Arola et al. 1982).

One of the most important factors affecting the energy consumption of comminution is the *particle size* of the reduced wood. The smaller the particles produced, the higher is the consumption of energy. Hence, the production of 90-mm particles with an involuted disk chunker consumed for aspen 0.6 kWh/m^3 and for *Acer saccharum* 1.0 kWh/m^3, which was only 37 and 33% of the energy used for the production of conventional pulp-sized chips with a disk chipper (Arola et al. 1983). The effect of particle length on the energy consumption of a disk chipper is illustrated in Fig. 5.19. When wood is reduced with a conescrew, the length of the particles is controlled with the pitch angle and the number of threads of the screw. Knives with a slow pitch consume less energy (Fig. 5.20).

The fuel consumption during comminution is considerably below 1% of the total energy content of the wood. Table 5.1, where the chippers produced conventional pulp-sized particles and the conescrew chunker 80-mm chunks, illustrates the fuel consumption in the comminution of 5–10-cm-thick *Alnus incana* stems with light farm tractor-driven equipment. Table 5.1 also shows the respective energy consumption measured at the power take-off shaft of the tractor, excluding the energy used for the crane, idling, etc.

Most of the factors determining chipper energy consumption will also affect hog energy consumption. Knife hogs which use basically similar cutting principles consume approximately the same amount of energy as chippers. Punch-and-die hogs and hammermill hogs have much higher energy requirements due to the less

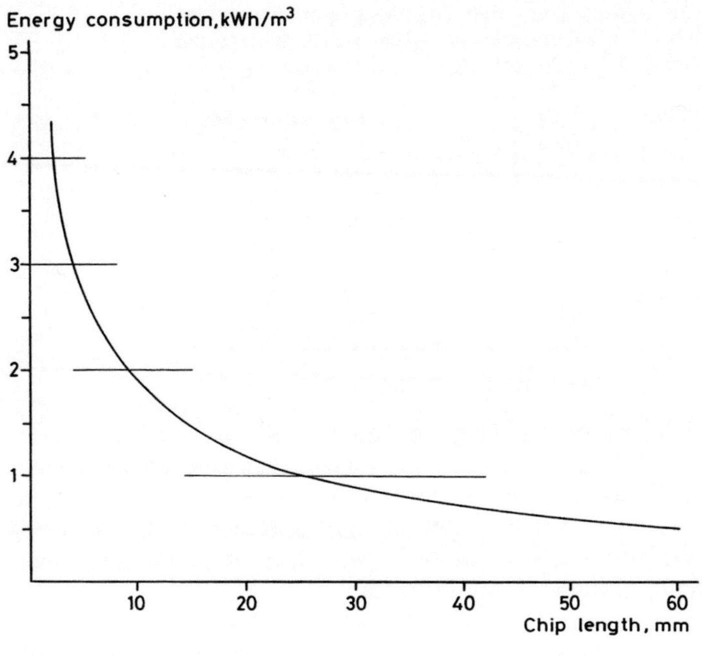

Fig. 5.19. The effect of particle length on the energy consumption of a disk chipper (redrawn from Murto and Kivimaa 1951)

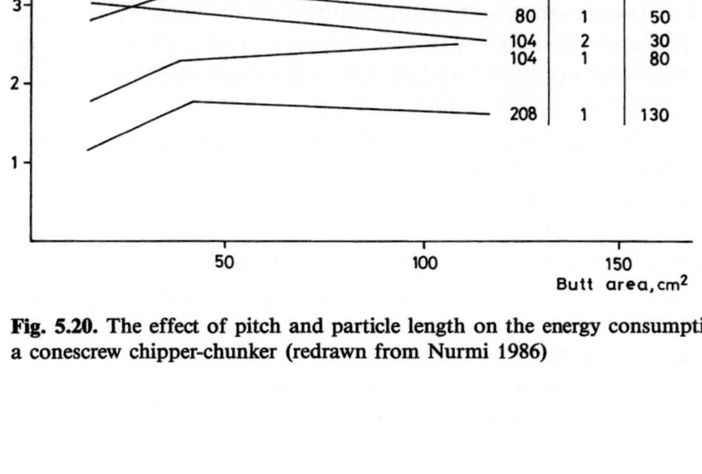

Fig. 5.20. The effect of pitch and particle length on the energy consumption of comminution with a conescrew chipper-chunker (redrawn from Nurmi 1986)

Table 5.1. Consumption of energy (kWh/m^3 solid), and fuel consumption relative to the energy content of the wood, in comminution of 5–10-cm-thick *Alnus incana* stems with light farm tractor-driven equipment (Heikka and Piirainen 1981)

Equipment	Feeding	Energy consumption at PTO (kWh/m^3)	Fuel consumption	
			l/m^3	Wood energy (%)
Conescrew chunker	Manual	1.1	0.4	0.2
Disk chipper	Manual	1.1	0.4	0.2
Disk chipper	Hydraulic	1.9	0.6	0.4
Drum chipper	Hydraulic	3.0	1.0	0.6

efficient technique of comminution. They produce pieces with rougher surfaces and a larger proportion of fines. The greater the surface area exposed, the greater is the energy consumed for comminution (Pottie and Guimier 1985). Contrary to chippers, the energy consumption of a hammermill decreases with decreasing moisture content of the biomass, since as an impacting comminutor a hammermill benefits from the brittleness of dry biomass.

5.3.2 Power Requirement of Comminution

Farm tractor-driven comminution machines tend to place a heavy strain on the drive shaft and power transmission of the prime mover, which can be protected against overloading by a slip-clutch arrangement. The torque developed by comminution is uneven. There are high temporary peaks, depending on the machine type and revolving mass. The strain can be relieved with heavy, large-diameter flywheels and a fast revolving speed, which increase the moment of inertia and dampen the peak torque caused by sudden loads.

Figure 5.21 illustrates the fluctuation of torque on a drive shaft rotating at 630 rpm. The leveling effect of the hydraulic feeding on the fluctuation is distinct. The high peak torque of the conescrew chunker prototype is a result of the light 70-kg mass and small 420-mm diameter of the revolving cone of the test machine. When the conescrew was later equipped with a heavy flywheel and the rotation speed raised, the sharp fluctuation was eliminated (Fig. 5.22).

Figure 5.23 shows the average torque for two diameter classes of six tree species in Michigan as a function of the revolving speed of an involuted disk chunker. Torque increases with the basic wood density and tree diameter, but it can be reduced by using higher revolving speeds. Lower torque results in lower motor power requirement. A heavy, large-diameter disk with a small number of knives revolves at a uniform rate and permits the use of a smaller motor, especially if the trees and bolts to be reduced are not too long. On the other hand, an increasing output per hour usually also means an increasing power requirement.

Power requirement indicates the minimum power of the motor or at the power take-off shaft of a tractor needed for proper functioning of a comminution ma-

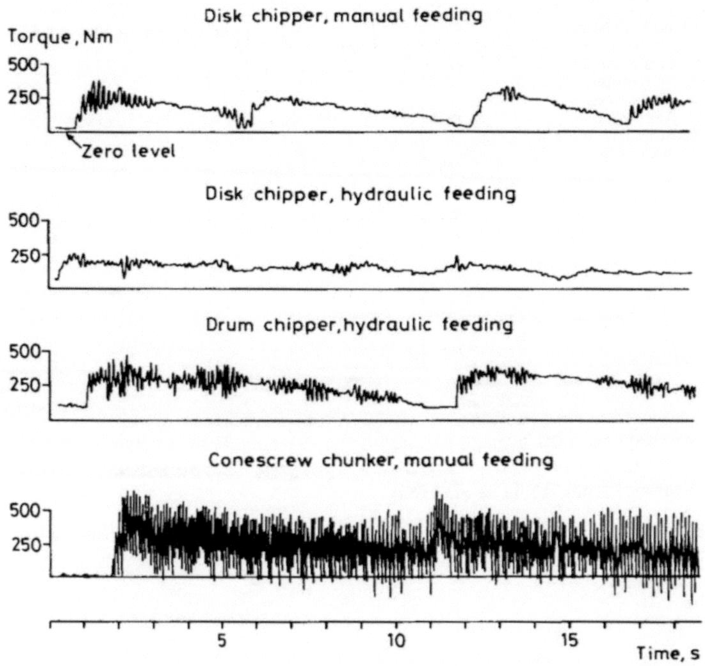

Fig. 5.21. Fluctuation of torque directed by light comminution machines on the drive shaft of a farm tractor. Butt diameter of trees is 5–10 cm (redrawn from Heikka and Piirainen 1981)

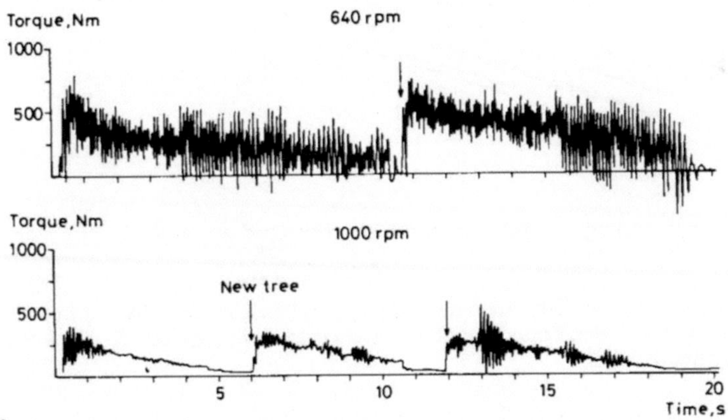

Fig. 5.22. Fluctuation of torque directed by a light conescrew chunker on the drive shaft of a farm tractor when reducing 10 cm (butt diameter) frozen *Alnus incana* stems. Screw speed is 640 or 1000 rpm (redrawn from Hakkila and Kalaja 1981)

chine. Because comminution of residual forest biomass is uneven work, the power requirement is difficult to determine. Butt ends of trees cause power peaks which are much higher than the mean power. A large individual stem requires more power than a bunch consisting of several small trees with the same total cross-sectional area.

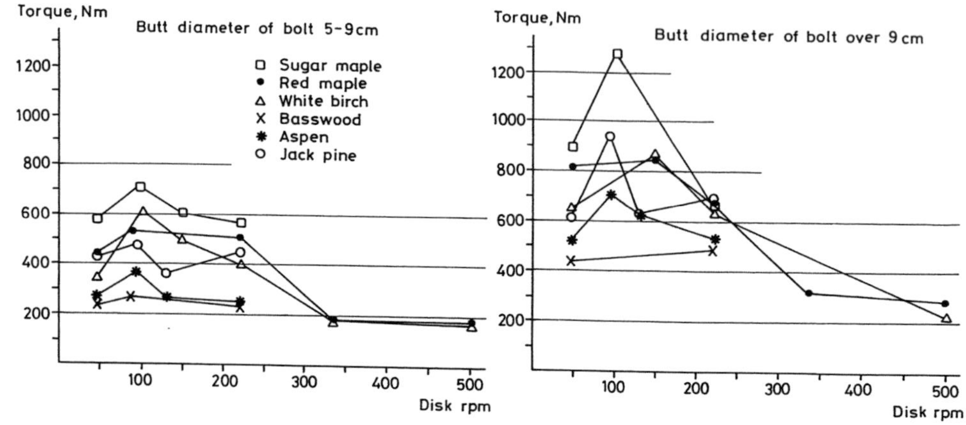

Fig. 5.23. The effect of revolving speed on average torque on the drive shaft of an involuted disk chunker in comminution of *Acer saccharum, A. rubrum, Betula papyrifera, Tilia americana, Populus* sp., and *Pinus banksiana* (redrawn from Arola et al. 1982)

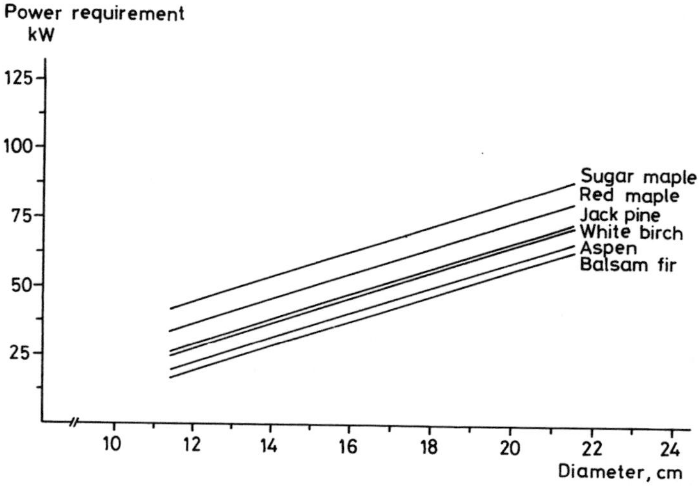

Fig. 5.24. The effect of bolt diameter on the power requirement of an involuted disk chunker in comminution of *Acer saccharum, A. rubrum, Pinus banksiana, Betula papyrifera, Populus* sp., and *Abies balsamea* (redrawn from Arola 1983a)

Figure 5.24 is an example of the effect of bolt diameter on the power requirement when the stem is reduced to about 80-mm pieces with an involuted disk chunker. Depending on the tree species, i.e., on the basic wood density, chunking of an 11-cm-thick bolt with the device under consideration requires in practice a power of 15–35 kW, while a 22-cm bolt requires 65–95 kW.

When wood is reduced to pulp-sized chips and the performance of the machine is kept at the same level, the power requirement is essentially higher, especially for high-density species such as hickories and oaks. Figure 5.25 illustrates the net power demand of a two-knife, 122-cm-diameter drum chipper as a func-

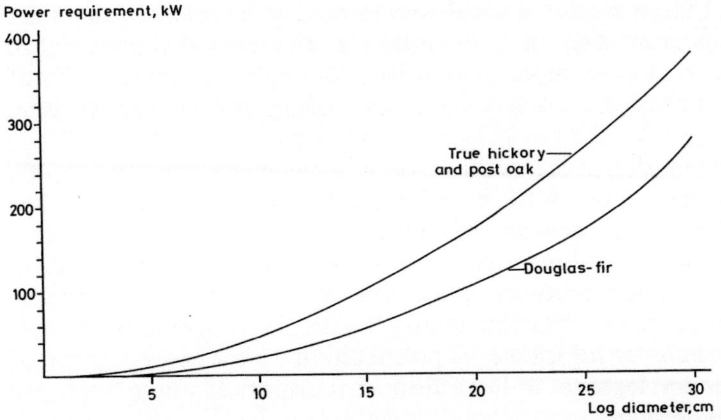

Fig. 5.25. Net power requirement (idling excluded) of a two-knife, 122-cm-diameter, 550-rpm drum chipper fed at 30 m/min with single fresh logs of *Carya* spp., *Quercus stellata*, and *Pseudotsuga menziesii* as a function of log diameter inside bark (redrawn from Koch 1985b)

tion of bolt diameter. For large 30-cm-diameter hickory logs the power demand is 400 kW. Because the two-knife drum design under consideration did not have adequate self-feeding characteristics, the drum was later equipped with three knives. Although the addition of a third knife did not alter the specific comminution energy, it resulted in an increase in the power demand by about 50% (Koch 1985b). If a portable chipper is to be used for size reduction of large trees, the machine must be equipped with a strong power source.

5.4 Position of Comminution in the Harvesting Schedule

Comminution has a central role in a biomass harvesting schedule. The organization and execution of the other phases of biomass delivery are determined partly by the siting, mobility, capacity, and technical characteristics of the machines used for comminution. The equipment is either *portable* or *stationary* at a central processing yard. Most of the portable units are limited to working at an upper landing or on a forest road, but some of them are *mobile*, i.e., self-propelled and able to move independently in the forest.

Depending on the place between stump and mill at which the machine is positioned, three alternatives arise. Harvesting schedules based on comminution with mobile equipment *on the terrain* are used on a small scale in northern and central Europe and also in South America. In clear-cutting areas and in systematic corridor thinnings a terrain chipper may be carried to the stump, while in selective thinnings the comminution has to be accomplished on the strip roads. In addition to comminution and forwarding, a mobile chipper may have other functions such as felling. Since this leads to single-tree handling and slows the productivity of an expensive unit, combining felling and chipping is seldom cost-effective. A specific alternative is a continuously moving swath harvester, which combines felling and comminution (Sect. 4.3).

Comminution of forest residue is usually performed with portable equipment *at the primary or upper landing*, or at the roadside. The method is prevalent in North America and in most European countries. Consequently, the majority of portable chippers are constructed solely for use at landing sites and cannot move to strip roads.

Centralized comminution at the terminal, i.e., *at a secondary or lower landing*, is common in the Soviet Union and some other eastern European countries. The issue is mainly the chipping of stem and branch residue that accumulates in a central processing yard when traditional timber assortments are being prepared from tree lengths or whole trees. Centralized processing has also gained some importance in the Federal Republic of Germany and Sweden. Comminution can also be centralized *at the point of utilization*. A precondition for the generalization of the system is further development of long-distance transport of whole trees, tree sections, and logging residue.

5.4.1 Comminution in Terrain

The mobile comminution machines operating in terrain usually also transport the chips to the landing and are thus actually *chipper-forwarders*. Hauling takes a considerable proportion of the work site time, lowering the productivity of chipping. The comminution productivity of a mobile terrain chipper can be improved by unloading the reduced biomass at the site into a *shuttle forwarder* which hauls the chips to the roadside. Long, off-road haulage distances favor systems in which shuttle forwarders are used. However, since there is likely to be a demand in the foreseeable future for only a small proportion of the residual biomass available for chipping, priority will no doubt be given to sites on which extraction distances are short (Alexandersson 1984).

Compared with landing chippers, a terrain chipper must be more versatile but still lighter in mass. This limits the maximum diameter of the tree and restricts the output. However, the system has many advantages:

1. No supplementary extraction machines are required for off-road haulage;
2. The biomass is kept essentially free of dirt, thus contributing to cleaner chips;
3. Pulling the trees from the side of the chipper with a crane straight into the in-feed device is at least as fast an operation as turning whole-tree bunches parallel to the strip road and loading them onto the forwarder;
4. If chipping is done at the landing, trees must be skidded whole, causing excessive damage in thinning stands, or they must be bucked for forwarder haulage. Neither whole-tree skidding nor bucking is needed in terrain chipping;
5. Branches and other residue break off the trees during chipping. In landing chipping they accumulate in inconvenient piles, but in terrain chipping they are left with their nutrients more or less scattered;
6. Where logging sites are small and the equipment must be relocated frequently, the ability of mobile terrain chippers to move independently from site to site and their low set-up time offer organizational advantages;
7. To reduce the nutrient flow from the forest soil, a terrain chipper can be equipped with a flail delimber or other separator to remove the foliage be-

fore chipping (Sect. 8.6.1). However, the present foliage separators are not satisfactory in continuous operations.

Because of its heavy mass and relatively high center of gravity, a terrain chipper is only suitable for moderately easy terrain. Even with a forwarder as a prime mover, the following restrictions have to be placed on the logging site:

1. The ground must be reasonably level and firm;
2. To limit the time used for the forwarding function, the haulage distance should not exceed 300 m;
3. The bunches gathered alongside the strip roads may be covered by snow in the winter. The snow depth must not exceed 30–40 cm.

If the stand does not meet these requirements, the comminution has to be deferred to the upper or lower landing. A terrain chipper can, of course, be used also at a landing, but compared with the more effective landing chippers its operation costs tend to be higher. However, under the following circumstances the use of a mobile terrain chipper may be feasible also at the upper landing:

1. The landing site is so cramped that a truck, trailer, or chip pallet cannot be placed alongside or behind the chipper.
2. The biomass is scattered in fairly small formations at the landing, which forces the chipper to move repeatedly from pile to pile.
3. The landing site is so soft that trucks cannot drive up to the chipper.

Fig. 5.26. A light farm tractor-driven terrain chipper mounted on the three-point hitch of the prime mover; manual feeding

Light, mobile terrain chippers are used mainly in northern and central Europe for the fuel procurement of chip-fired farms. The chipper is mounted on the three-point hitch of a farm tractor and run by the tractor motor through its power take-off. The chips are blown into a trailer or chip bin behind the tractor (Fig. 5.26). Unloading takes place by dumping into a pallet or directly into the chip storage of the farm. Methods by which the chips are blown into $1-2\ m^3$ sacks have not proved technically satisfactory as yet.

Light terrain chippers are primarily fed manually from the side, which limits the maximum tree size to $20-30\ dm^3$. For taller trees and larger chip quantities the tractor should be equipped with a grapple loader. The in-feeding may be based on the self-feeding effect of the chipper knives, but for a trouble-free flow of raw material many of the light chippers are equipped with one or more pulling rollers. The mass of light, farm tractor-mounted chippers, excluding the tractor and trailer, is 200–600 kg.

Beside chipper characteristics, the output of a light, manually-fed terrain chipper depends on the tractor power, trailer volume, transport distance, tree characteristics and size, and the number of men working in the team. The following average hourly outputs include chipping on the strip road, haulage of the chips, and unloading at storage. The highest hourly outputs foresee a two-man team (Hakkila et al. 1979):

Chipping site	Haulage distance (m)	Output of chipping-forwarding (m^3 solid/operating hour)
Landing	0	3.0–5.0
Strip road	100	2.5–4.0
Strip road	300	2.0–3.5
Strip road	500	1.5–3.0

Heavy, mobile terrain chippers are hauled by a four-wheel-drive farm tractor or mounted on a forwarder (Fig. 5.27). Feeding takes place from the side of the machine by grapple crane. In thinning stands the trees must be bunched perpendicularly to the strip road so that they can be pulled by tractor crane straight into the in-feed device. Since the trees do not need to be turned, the feeding process is simple and fast. The machine can be fed from one side only and, therefore, it has to move in two directions on the strip road. However, with a feller-buncher all trees can be collected on the same side of the strip road.

In a typical forwarder-mounted unit the chipper has its own 100–150 kW engine and a feeding device consisting of a short chain conveyor and one or several powered rollers. The possibility of turning the feeding-chipping device in the horizontal plane speeds up the feeding in thinning stands. In some experimental prototype machines the in-feed device is equipped with a flail delimber for foliage removal.

Heavy 15–25 t terrain chippers usually have a $15-20\ m^3$ dumping bin which is unloaded either backward or laterally onto an exchangeable truck pallet on the ground at the landing. The dumping height is seldom sufficient to permit discharging directly onto an open-top truck. High-tilting bins, which are first lifted vertically and then dumped laterally, are an exception.

Fig. 5.27. A heavy forwarder-mounted terrain chipper dumping the load onto a container

The output of a heavy terrain chipper depends primarily on tree characteristics and size, bunch size, location of the bunches relative to the strip road, biomass yield per hectare, terrain, and haulage distance. The following table is an example of the chipping-forwarding output per effective hour of an 18-t, forwarder-mounted terrain chipper in a young pine stand in Finland when the biomass yield is 30 m^3/ha (Kalaja 1978):

Method of bunching	Output of chipper-forwarder (m^3 solid/effective hour) Haulage distance	
	100	500
Manual bunching	8.8	6.8
Strip road-operating boom-type buncher	10.1	7.5
Stand-operating feller-buncher	10.8	7.6

The larger the bunches are, the better is the output in chipping-forwarding. Consequently, the highest productivity is achieved when the trees are collected on the side of the strip road by a stand-operating feller-buncher. Although trees are carried to within reach of the chipper crane also in manual methods, compared to mechanized methods the piles are smaller and they are left slightly farther from the strip road, which lowers the chipper performance. As a rule of thumb, the hourly output of a heavy terrain chipper in chipping-forwarding is roughly the same as that of an ordinary forwarder in the haulage of small whole trees or tree sections.

Canadian studies of forwarder-mounted terrain chippers show approximately the same production level. The output of a 19-t terrain chipper was 7.0 t fresh whole-tree chips per working hour in a winter experiment, when the off-road haulage distance was 190 m. For fresh slash the output was lower, 5.2 t/working hour, although the study was carried out in summer and the haulage distance was only 120 m (Richardson 1986). On a mixed pine and hardwood site in Georgia, a prototype terrain chipper averaged 5.3 dry t/h when the hauling distance was only 92 m. The output was 4.5 dry t at a 275-m forwarding distance (Stokes and Sirois 1986).

Difficulties resulting from adverse terrain, the need for a separate motor for the chipper, and the high price are some of the obstacles preventing the broader use of forwarder-mounted terrain chippers. In northern countries terrain chippers have to operate part of the time at landings, since the demand for fuel chips is largely concentrated to the winter months during the snow season. Their unit costs are then less competitive.

A new terrain chipping solution under development in Sweden is intended to produce clean industrial raw material from marginal timber from early thinnings. This *pulp chip method*, based on the use of a tractor-mounted delimbing-debarking-chipping device, presents a totally different line of thought in thinning techniques for small-tree stands. The processing of ready-to-use pulp chips is done on the strip road and all other biomass components except the stem wood are left unutilized. The tractor driver controls the processing from the ground, feeding the machine manually or with a light crane. The trees are delimbed, debarked, and chipped in one continuous operation. The chips are collected in a bin at the rear of the tractor, transported to a landing, and dumped into a truck bin (Fröding 1987). The method means abandonment of the principle of whole-tree utilization.

5.4.2 Comminution at an Upper Landing

Comminution at an upper landing presupposes haulage of the material to the chipper instead of moving a mobile chipper to the material. The system may be regarded as the basic organization for biomass harvesting. In addition to portable chippers for whole trees and slash, various kinds of portable chunkers and hogs are available for comminution of biomass at the landing.

Portable comminution machines operating at an upper landing are generally of simpler design than mobile terrain chippers as they do not have to move in forest terrain carrying a dumping chip bin. Consequently, landing chippers are of heavier construction and often equipped with a more effective motor, which provides scope for higher reliability and performance. The maximum permissible tree diameter is also greater for landing chippers.

In the *work organization* widely adopted in North America since the beginning of the 1970s, conventional grapple skidders are used to haul whole trees concurrently with the comminution process from the logging site to within reach of the chipper crane. A continuous supply of wood is essential for full utilization of the capacity of the chipper. The chips are blown directly into a covered semitrailer through an opening in its rear panel. There is no need to move the chipper as work progresses.

When clear-cutting old hardwood stands in the north-central states of the United States, a minimum of 23 t fresh saw- and veneer timber must be available to justify separation of the logs before whole-tree chipping (U.S. Department of Agriculture 1978). Separation may take place prior to off-road haulage, or may be carried out at the landing prior to comminution of low-quality biomass.

The North American landing-chipping system creates a hot logging schedule, requiring careful timing and balancing of the various operations, as skidding, chipping, and chip haulage are carried out simultaneously and are linked tightly together. If chipping is interrupted, the off- and on-road haulage must also be stopped, and vice versa. In a system evaluation in the United States, the cost differences among chipping operations attributable to work organization were primarily a result of inventory delays at the chippers. No inventory delays were recorded and no excess waiting cost attributed to the chipper where comminution was done at landings with available biomass piles. In contrast, on hot logging schedules based on a continuous flow of biomass to the chipper more than half of the total cost of chipping was incurred while waiting for delivery of raw material by the skidders (Kluender et al. 1983).

Although the system as such is prone to disturbances, it is an economical and feasible solution in North America due to reliable heavy-duty machinery, favor-

Fig. 5.28. A light farm tractor-driven landing chunker mounted on the three-point hitch of the prime mover; manual feeding

able technical conditions for logging, and the availability of adequate landing space. In northern Europe, however, forestry conditions are somewhat different. Small-sized trees and logging residue are mainly recovered from small private forest lots with limited landing areas. Due to the low biomass yield per work site, equipment must be moved frequently from landing to landing. Susceptibility to interruptions and downtime must be avoided by all means, and machine operations should not be linked closely together. The interdependence of off-road haulage and comminution is reduced by transporting the raw material to the landing in forwarders, and by buffer-storing it in 3- to 4-m-high inventory piles before starting comminution. Since this means that raw material is not brought within reach of the chipper crane, the equipment must be capable of moving flexibly at the landing as work progresses. This requires good mobility of the equipment and must be taken into consideration when planning the landing arrangements.

Farm tractor-driven landing chippers cover a broad range in capacity and price. The lightest, manually-fed chippers are mounted on the standard three-point hitch of the tractor (Fig. 5.28). Several thousands of these light machines are in use in northern Europe. If the chipper can be placed between the tractor and trailer, the unit may be used as a terrain chipper as well.

More effective farm tractor-driven landing chippers are equipped with a hydraulic feeding device and usually with two wheels. The tractor functions as both the prime mover and the power source of the chipper and grapple loader (Fig. 5.29). The mass of these chippers, excluding tractor and loader, is usually 1 – 4 t.

Fig. 5.29. A light farm tractor-driven landing chipper; feeding by tractor crane

Fig. 5.30. A heavy landing chipper mounted on a towable chassis

Output per operating hour is $5-18\,m^3$ solid. The heaviest machines weigh 8 t and have an output of over $20\,m^3/h$.

Forwarder-mounted comminution machines are designed primarily for terrain chipping. However, due to their heavy mass, some units reducing slash and stump and root mass by crushing operate only at landings. An extra motor is mounted onto the load space of the forwarder to run the comminution device.

In North America, heavy landing chippers are mounted *on a towable chassis* (Fig. 5.30). They are usually equipped with a self-contained knuckle or slide boom crane and an operator's cabin, but some machines require a separate crane for feeding. The trees are loaded onto the feeding table from the rear. The thickest branches of large hardwood trees must sometimes be notched by a chain saw before the feed rolls are able to compress them.

The heaviest of these portable chippers weigh 30 t, have a power source of up to 400 to 600 kW, are able to reduce stems of up to 70 cm in diameter, and have a stated theoretical capacity of over 100 t/effective hour. However, under actual field conditions the output is radically lowered by interruptions in the flow of raw material to the chipper and the downtime caused by waiting for the chip trucks. In the United States, heavy chippers of this type have an average annual output of 29000 green tonnes in in-woods operations and 38000 green tonnes when operated at central yards (Altman 1980).

In Europe, heavy landing chippers are more commonly mounted *on used bogie trucks* for flexible moving within and between landings (Fig. 5.31). Because the feeding table is high above ground, handling of tall trees becomes difficult. However, the material to be comminuted consists mainly of small trees, tree sections, and slash.

The truck serves as a chassis and supplies power to the crane. The source of power for comminution is either the own motor of the truck or a separate 200–300 kW motor. The operator's cabin may be raised for better visibility during comminution, and lowered to attain a low transport profile. Feeding usually takes place from the rear of the machine, sometimes from the side. The reduced product is blown directly onto a truck, trailer, or exchangeable pallet. Heavy, truck-mounted hogs may be used to comminute dirty logging slash at landings.

Truck-mounted comminution machines have a high operational availability. They move rapidly from site to site, and the equipment is quickly assembled. However, a firm, bearing landing area is required. In an extensive system evaluation in Finland, truck-mounted chippers were operational during 60% of the total work-site time. Organizational interruptions accounted for 32% of the time and machine interruptions for 8% (Kuitto 1986):

	Time distribution of truck-mounted chippers (%)
Main time of chipping	58.6
By-time of chipping	1.6
Effective time	60.2
Service	2.9
Knife change	3.3
Breakdown	1.6
Machine interruptions	7.8
Waiting time for trucks	26.7
Personal interruptions	2.5
Other interruptions	2.8
Organizational interruptions	32.0
Total work site time	100.0

The output of heavy truck-mounted chippers is 15–30 m^3 solid per operating hour, and the annual capacity of the most effective units may amount to 40000–50000 m^3 solid. Because it is not profitable to move a heavy machine to a small work site, the minimum chip yield requirement per work site is the full load of a truck and trailer, i.e., 80–130 m^3 loose.

5.4.3 Centralized Processing of Residual Wood

Centralized processing of forest biomass at one place, located as close as possible to the disposal points of the main products, is an alternative for in-woods comminution in the immediate vicinity of the logging site. Within a forest district, raw

Fig. 5.31. A heavy landing chipper mounted on a bogie truck

material is harvested at various logging sites, accumulated at terminals for conversion, and then distributed to adjacent plants or marketed to customers. About 90% of the 250 terminals processing trees or stems to conventional timber in the Federal Republic of Germany operate in conjunction with the wood industry, 5% have been established by forest owners, and another 5% by timber dealers (Wippermann 1985).

Rather than moving a portable processing machine to the woods, the biomass is thus hauled in raw form to heavy-duty *stationary or semistationary machines* for conversion at a terminal. The amount of work at the felling site is minimized by shifting tasks from the woods to a centralized installation, thereby taking advantage of mass processing of the raw material under controlled conditions. Several technical terms are used for the centralized site: *lower landing, secondary landing, timber terminal, centralized processing yard, central conversion yard*, etc. In the United States, the process of dismembering a tree or stem at a stationary or semistationary conversion yard to obtain a maximum product value is termed merchandizing, and the system of machines used for this purpose is called a tree or stem *merchandizer*. In a linear merchandizer the stems move linearly, i.e., longitudinally, past one or more cutoff saws. The system allows great flexibility in selecting product length, but the output is limited to 1.5–2 stems/min and therefore is low with small stems. In a transverse merchandizer the stems move transversely past a bank of saws spaced to cut them into selected lengths, and production rates of up to ten stems with a total linear input of 80–120 m/min are possible (Koch 1985b).

Centralized processing is an integrated operational system. Depending on the raw material and the assortments produced, the tasks consist of measuring,

Fig. 5.32. A centralized processing yard for residual forest biomass in Lycksele, Sweden (courtesy of Jonas Palm)

delimbing, debarking, bucking, comminution, sorting, distribution to adjacent plants, or marketing over longer distances. Much of the work done at the terminal consists of moving the raw material from one suboperation to another.

The existing centralized processing sites in the USSR, Europe, and North America are designed mainly for handling tree-length material, but some are able to process whole trees or various kinds of residue. Centralized processing of wood is a standard solution in the USSR and China. According to Abol (1984), in the USSR as much as 90% of all wood is accumulated as tree lengths and 6% as whole trees in processing yards. Centralized processing is also used to a minor extent in North America, central Europe, and Sweden. Figure 5.32 shows an example of a centralized processing yard for various kinds of forest biomass residues in Sweden.

Some of the key economic forces in support of centralized processing systems are the reduction in harvesting and handling costs, economy of scale, improved value recovery, and complete-tree utilization (Christensen 1986). Compared with mobile processing, the following *benefits* are evident:

1. The dependence of wood processing on heterogeneous terrain, stand conditions, and climate is lessened, and the need for forest machines and workers to move to the trees is reduced. Work safety and ergonomic conditions are improved, productivity is increased, and energy is saved. Centralized processing allows electronic stem scanning combined with computerized decision support systems and the use of heavy-duty, stationary machines, which often have a simpler structure than the mobile ones.
2. The operational and technical reliability and availability of equipment is increased. According to a whole-tree chipper survey in the United States, the machines working at terminals produced annually 29% more chips than the machines in forest locations (Altman 1980).
3. The efficiency of biomass utilization is increased with respect to both quantity and quality. Scaling and bucking are carried out under favorable conditions

by skilled workers, resulting in improved accuracy of cuts, less volume loss, and higher grades of lumber. All conversion residue can be recovered and homogenized as industrial raw material or fuel.

4. Sorting, marketing, and distribution of the widened range of primary products and by-products are improved, and management of activities is made more effective.

The serious *drawbacks* to the centralized processing of forest biomass are connected with long-distance haulage of the raw material to the terminal. The low bulk density of trees and unprocessed residue increases the cost of transport. Traffic regulations and safety measures restrict the haulage of long stems and branchy trees in conventional timber rigs on public highways. Transport of whole trees requires special equipment, and in some cases the investment costs of the processing yards may rise prohibitively high. The system presupposes a concentrated source and steady supply of wood. Moreover, whereas centralized processing lightens the physical strain of forest work, working at a terminal may constitute a higher mental burden.

Centralized processing facilities are often large, stationary systems. Due to the permanence of the design and the large capital investment, they represent long-term commitments. In time they may prove to be rather inflexible in adapting to major changes in supply and market conditions. It may also be difficult to upgrade the system with new technology. Furthermore, an inherent operational weakness of large, integrated terminals is the series nature of the process flow; if one piece of equipment breaks down, it may lead to the entire system falling idle (Christensen 1986).

Wood processing terminals may be equipped to handle several kinds of raw materials or may be restricted to process only a special assortment of biomass. To maximize the utilization of the volume and value of biomass, primarily the following forms of unprocessed material are brought to the terminals (Grammel 1978, 1985):

1. Delimbed, unsorted *tree lengths*. The use of stem wood is optimized according to demand. This is by far the most common type of raw material presently processed at centralized terminals.
2. *Tree sections* which are cut to convenient lengths for transport. The entire above-ground biomass is utilized, the main product usually being raw material for pulp industries. Comminuted fuel is a by-product.
3. *Whole trees*. The entire above-ground biomass is utilized, and the use of stem wood is optimized. However, for ecological and transport-technical reasons, the top of the tree may be left on the site.
4. *Complete trees*, including the stump-root system. All biomass is utilized, and the use of stem wood is optimized. Due to many unsolved problems caused by transport, handling, and soil contaminants, no complete-tree terminals have yet been built or tested.
5. *Logging residue* left behind in the harvesting of merchantable timber, and unmerchantable, small-sized trees from precommercial, silvicultural treatments. The biomass is processed and sorted for industrial raw material or fuel.

The operations of a centralized processing yard consist of the following four principle functions (Grammel 1978):

1. A *collection function*. The forest biomass is concentrated in the processing yards in various forms.
2. A *processing function*. The biomass is processed by delimbing, debarking, topping, comminuting, etc.
3. A *grading and sorting function*. Procedures such as scaling, grading, optimizing the sectioning, screening, etc. are often performed with the help of a computer-aided process-control system.
4. A *marketing and distribution function* related to the main and the secondary products.

Both multistem, high-production units, and single-stem, high-value product units are available. *Market-oriented terminals* are aimed primarily at high value yield of large trees. *Production-oriented terminals* are aimed at cost-efficient utilization of small trees. The third group is terminals for *special tasks*, such as processing stump and root mass or logging slash (Larsson 1986). With all these raw materials, the feasibility of centralized processing is more or less dependent on markets for biomass fuel. The current low application of terminals is, in fact, largely due to the lack of markets for biomass fuels (Bjerkelund 1985).

Terminals based on stationary equipment are usually aimed at deliveries to large consumers. Consequently, they are adapted primarily to the structure of large-scale forestry. On the other hand, terminals with lighter mobile equipment are adapted to the structure of small-scale private forestry. Both types of terminals are *built from components* of which a large share has originally been developed and manufactured for timber handling at ordinary mill yards. Many of these components need to be modified to some extent to function well for a different type of raw material, such as whole trees and logging residue. Another group of components, such as delimbing devices, is actually designed for logging. A third group of components, such as equipment for decontamination of stump and root mass or compaction of bulky biomass assortments, is developed especially for the production of forest energy (Lönner 1985). The interdependence of the operations of various components in the processing line can be reduced by siting buffer storage before and after the most critical function (Hallonborg 1982b).

5.4.3.1 Terminals for Large Trees

Raw material is transported as delimbed tree lengths or whole trees to a large-tree terminal. The goal is to minimize harvesting and processing costs by transferring work from the forest to the terminal and simultaneously maximizing the product value. The economy of the system is largely based on the additional revenues from improved bucking and sorting. Centralized processing makes it possible to adapt the bucking pattern to varying market demands and to increase the number of assortments produced. The terminals are able to process dimension assortments to meet urgent special orders and to deliver quickly desired dimensions to customers. The branches and bark are comminuted for fuel (Kwasnitscka 1978).

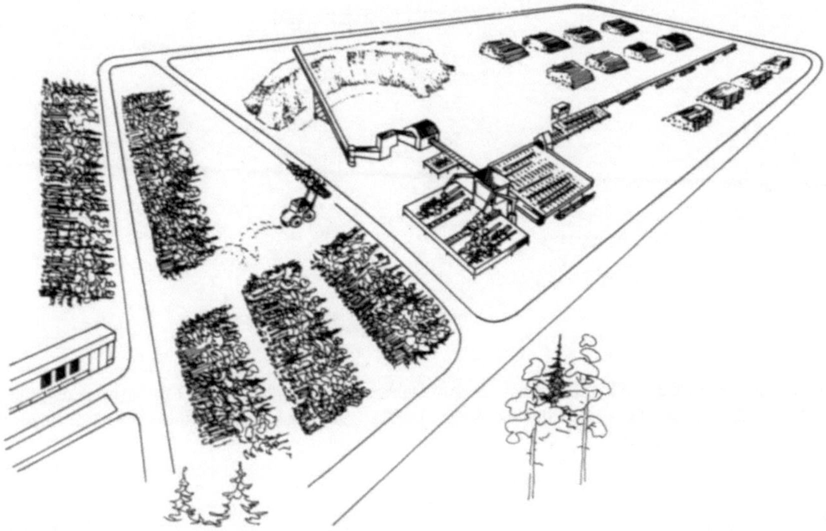

Fig. 5.33. A layout of a terminal in conjunction with a saw mill for processing large whole trees and, as a by-product, small-sized fuelwood trees. The annual capacity is 200000 m^3 of biomass during two shifts, manning six men and a foreman per shift. Investment level is 30 million Swedish crowns (courtesy of Göran Lönner 1985)

For the raw material collection function, a large-tree terminal may have equipment such as stationary and mobile cranes, loaders, and conveyors. The processing function is taken care of by delimbing and debarking machines, chippers, etc. The sorting and grading function is carried out by electrical measuring installations, cross-cutting saws, sorting boxes, different types of directional guiding gears for the specific assortments, and highly developed computer systems with a great variety of tasks. In fact, electronics will be the dominating technical device of the terminals in the future (Grammel 1985).

Figure 5.33 is an example of the layout of a large-tree terminal located at a sawmill. The whole trees are loaded onto the large table and fed into a double line of single-tree delimbers. The delimbed stems are either fed directly to the bucking plant or buffer-stored at the other end of the table. The bucking pattern is optimized by a computer-controlled trimmer. Sorting is also automatic. Top and crown material is fed to the fuel line for chipping and screening. The fuel line has a separate table for in-feeding small-sized trees from precommercial thinnings.

5.4.3.2 Terminals for Small Trees

Most small-tree processing terminals receive the raw material as tree sections. The goal is to minimize the harvesting costs of small-sized pulpwood trees and to utilize the additional biomass from crowns and tops as fuel. The revenues from the fuel are an essential factor in the economy of the terminal, and thus a good market for biomass fuels is a vital precondition of the system.

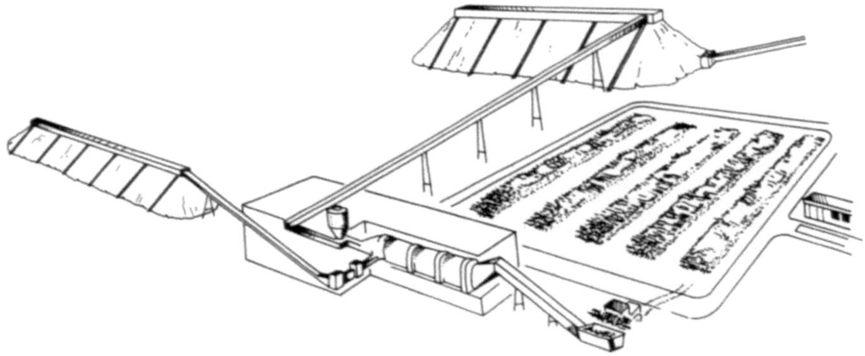

Fig. 5.34. A layout of a tree-section terminal in conjunction with a pulp mill. The annual capacity in 500000 m^3 of biomass in three continuous shifts. Investment level is 50 million Swedish crowns (courtesy of Göran Lönner 1985)

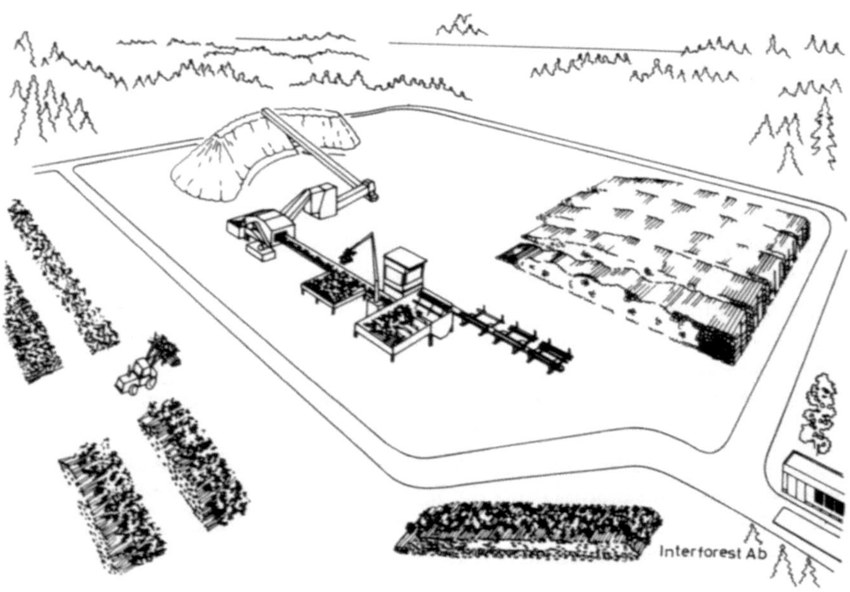

Fig. 5.35. A layout of a separate tree-section terminal for processing unlimbed pulpwood and fuelwood. The annual capacity is 120000 m^3 of biomass in two shifts, manning four men per shift plus a foreman on the daytime shift. Investment level is 16 million Swedish crowns (courtesy of Göran Lönner 1985)

In Sweden, where tree-section terminals were originally developed at the end of the 1970s, many central processing yards are located in immediate conjunction with a pulpmill to guarantee a continuous demand for residual wood and bark for the production of energy. The mass handling technology of tree sections is usually based on drum processing, which can be applied to either delimbing alone or to combined delimbing and debarking (Sect. 8.6.2). Some pulpmills integrate the delimbing of tree sections with the debarking of conventional pulpwood in slightly modified drums, while other mills have special tree-section delimbing facilities based on stationary drums, mobile batch delimbers, or other bunch delimbing principles (T. Jonsson 1985).

In view of the growing need to intensify thinnings of young softwood plantations, tree-section terminals located at a pulpmill are expected to spread in Sweden and Finland. Figure 5.34 represents a case in which a completely new wood room is built for processing tree sections in connection with a pulpmill. The key element of the system is a two-section drum with a conic connecting part. Delimbing is carried out in the first part and debarking in the second one. The clean pulpwood is chipped and stored. The crown, top, and bark residues are mixed, comminuted, and stored in a fuel pile.

Figure 5.35 illustrates a separate tree-section terminal designed to receive random-length pulpwood in tree-section form and unlimbed, small-sized fuelwood from precommercial thinnings. Tree sections are processed in a batch delimber and the logs are then stored. Branches, tops, and undersized trees are chipped, screened, and transported by a conveyor to a pile.

Under good market conditions a small-tree terminal may be aimed primarily at maximizing the product value. The layout is then basically similar to a large-tree terminal and may in central Europe have as many as 100 assorting boxes for various types and grades of products. A terminal may, in fact, be designed for maximizing the value of all size classes of trees, in which case it has a separate line for large and small trees (Dietz 1984).

5.4.3.3 Terminals for Residual Biomass

Terminals designed specifically to process residual biomass are usually located separately or in conjunction with a power plant, without a close industrial connection. Most of them produce comminuted fuels from low-quality raw materials, such as undersized whole trees, logging slash, or presplit stump and root mass. An unbalanced demand of fuel over the year may cause organizational and employment problems. The wide seasonal variation is thus a serious economic burden. When conditions during the warm season are the most advantageous for biomass fuel processing, the demand is low.

The equipment consists usually of loaders, conveyors, chippers or crushers, screens or other sorting devices, and storage facilities for unprepared biomass and comminuted fuel. The comminution machines may be stationary or may move within the terminal from pile to pile or even between several small terminals. Transferring the comminution of residual biomass to central processing yards has important advantages over processing in the terrain or at the upper landing:

1. The close interdependence between the comminution and trucking operations, which is the critical link in in-woods processing, can be eliminated. The reliability and operational availability of the machines increases and the hidden costs due to waiting time decrease. Comminution can be carried out with simple, reliable, energy-efficient machines.

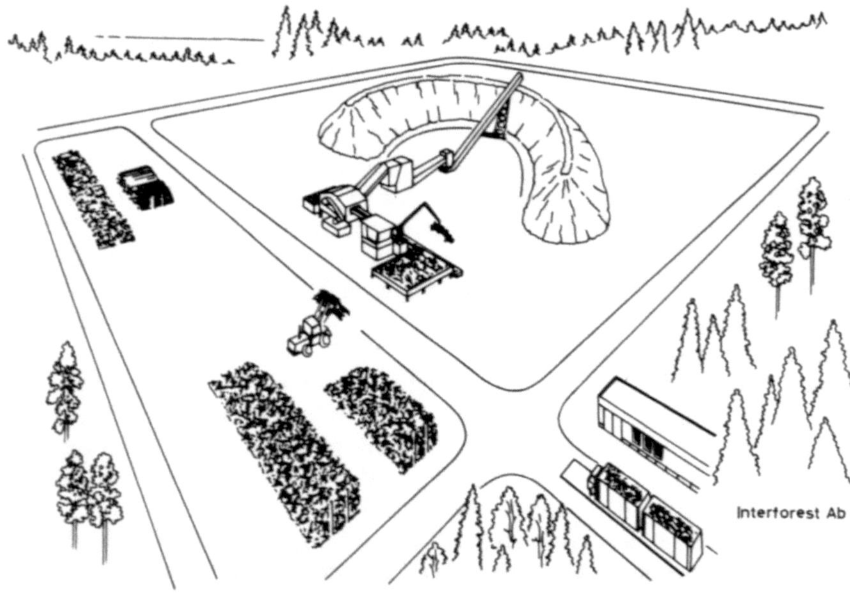

Fig. 5.36. A layout of a separate residue terminal for producing screened, undried fuel chips. The annual capacity is 55 000 m^3 of biomass in two shifts, manning three men per shift plus a foreman during daytime. Investment level is 10 million Swedish crowns (courtesy of Göran Lönner 1985)

2. Chip quality control and upgrading are improved. The comminuted raw material can be divided into industrial and fuel fractions as the prevailing market situation requires. Integrated utilization of forest residue can be improved.
3. Uncomminuted residual biomass can be buffer-stored. As the chip storage time is shortened, problems of dry matter and quality loss are eased. However, if the raw material is unlimbed softwood, harmful insects may propagate in the whole-tree piles during the spring season.

A layout of a typical residue processing terminal is illustrated in Fig. 5.36. The terminal produces screened, undried fuel chips from various kinds of above-ground residues and low-quality hardwoods. A large in-feeding table and a crane determine productivity.

In spite of the potential advantages, centralized comminution of residual forest biomass is only conducted on a restricted scale because of the difficulties of transporting uncomminuted residue. Among the existing installations one of the most interesting is the unique Mackmyra terminal in Sweden for processing stump and root mass, mainly for sulfate pulp mills (Sect. 8.6.4).

Chapter 6

Trucking Residual Forest Biomass

A considerable part of the total price of any assortment of timber is the hauling cost from a landing to the mill. The lower the stumpage price is, the greater is the relative importance of the long-distance transport as a cost factor. Hence, its role is especially important in low-value commodities from residual biomass.

The effectiveness of distribution affects the extent of the economic catchment area around a mill or combustion plant. Both direct costs and indirect cost factors such as reliability under all weather conditions have to be considered when selecting a long-distance transport system. Although moving timber in any form will always have a predominantly economic motive, the method chosen is also greatly influenced by political, environmental, psychological, and physiological factors (Silversides 1981).

Long-distance transport is not a fully independent phase of a delivery system for residual biomass. It is closely tied to logging operations, landing facilities, mill yard handling, and subsequent utilization. The flow of conventional timber and residues from a logging site to a landing affects secondary transport and vice versa. Each stage of the harvesting and transport schedule must be planned to facilitate the execution of the others.

Long-distance transport of residual forest biomass can be done by truck, rail, barge, or pipeline. However, as long as the quantities to be transported are modest and activity is confined to a radius of less than 150 km of the site of utilization, truck transport is the most economical and flexible alternative. Rail, water transport, and pipeline are rarely used alternatives, and trucking is in all events the first stage in any long-distance transport of biomass residue.

Because of its irregular size, form, and composition, residual biomass is a low-density and high-volume product. Contaminants may increase the mass of stump and root material by several tens of percent. These drawbacks make special demands on the equipment and organization of on-road transport. Large load volumes are a prerequisite for effective operation. This is achieved by equipping the truck with a trailer. When using a *semitrailer*, the front end of the load is supported by the truck tractor, while the own axles of a *full trailer* support the whole load.

When considering the choice of vehicle for hauling residual biomass, two different development lines may be discerned. One is represented by truck rigs with an *exchangeable, special superstructure*, and the other by vehicles with a *permanent superstructure*, possibly equipped with a compaction device. Because the seasonal variation in truck transport of both conventional timber and residual biomass is more or less the same, it is difficult to increase the annual usage of equipment through an integrated vehicle technique. On the other hand, one-sided specialization involves economic risks. Flexibility in case of changes in the hauling needs both within and outside the forest sector is, therefore, the strongest argument for exchangeable truck superstructures (Alexandersson et al. 1984).

Achievement of minimum trucking cost per tonne of biomass presupposes that the vehicle can be loaded to its maximum allowable gross mass before the maximum load volume is received. However, as a result of the low bulk density of various assortments made of whole trees, tree sections, and logging residue, the size of a truck load is usually limited by volume rather than mass capacity. There are three alternatives to *improving the capacity utilization* of a truck:

1. Producing *assortments with higher bulk density* through comminution of slash, presplitting stump-root systems, reducing the length distribution of tree sections, or careful stacking of load.
2. Using *large-volume rigs* the maximum allowable load mass may be achieved at least when hauling green biomass of tree species and assortments with a high basic density of wood and bark. In Sweden, the largest truck-trailer combinations may transport a load volume of 145 m^3 on public roads.
3. Improving bulk density by *compaction*, i.e., pneumatic loading of chips using powerful fans, vibration of chips, compression of tree sections with special equipment, etc.

The main objective of designing special rigs is to improve the economics by reducing the tare mass/load ratio and maximizing load capacity. The feasibility of each alternative is, on the one hand, closely linked to transport conditions, such as the quantities of various assortments to be carried annually, distances, quality of roads, and highway regulations specific to each country and, on the other hand, to the harvesting system as a whole. Trucking cannot therefore be unequivocally assessed as a separate operation.

Apart from the efficiency of the mass capacity utilization, the economics of trucking is affected by terminal work when loading and unloading, travel speed, work organization, effective annual working time, and the skill and motivation of the driver. An ideal carrier for a particular case is the result of many factors.

6.1 Transport-Technical Characteristics of Residual Biomass

The trucking cost of residual forest biomass is definitely dependent on the *bulk density* of the assortment under consideration, i.e., its mass per unit of stacked or bulked volume in kg/m^3 loose. Bulk density is a result of moisture content, basic density of the biomass, presence of contaminants, and *solid content* or the solid volume factor of the timber, i.e., the ratio of the solid volume to the stacked volume. For example, if the basic density is 500 kg/m^3, the moisture content on the fresh mass basis is 40%, and solid content of the load is 0.35, then

$$\text{bulk density of the load} = 0.35 \times 500 \times \frac{100}{100-40} = 292\ \text{kg/m}^3 \ .$$

A low solid content results in low bulk density and underutilization of the mass capacity of a truck. When small-sized trees are stacked unlimbed, the *solid content of a pile* may decrease to half of that of delimbed stems, thus increasing the demand for storage area. The following table shows the solid content of five small-tree assortments made of 6–8 cm dbh Scots pine and birch trees, depending on differences in bucking, delimbing, and comminution. The differences between the tree species are mainly due to the crookedness of birch (Kanninen et al. 1979):

Assortment	Solid content of a pile of small timber	
	Pinus sylvestris	*Betula* spp.
Delimbed, bucked to 2-m lengths	0.60	0.54
Delimbed, bucked to 4-m lengths	0.50	0.44
Delimbed, bucked to 3–5-m lengths	0.48	0.35
Whole tree chips before transport	0.41	0.39
Whole trees	0.26	

The solid content of a pile of small whole trees depends on tree variables such as branchiness, crookedness, diameter, length, and length distribution. It is also affected by stacking method, pile size, and weather factors such as frost, snow, and ice.

The smaller the trees are, the lower is the solid content. Larger and heavier trees cause more branch breakage, thus compacting the pile. An increase in pile height has the same effect. Consequently, whole trees and tree sections from early thinnings have a lower solid content and bulk density than those harvested from later thinnings. In both cases the bulk density is too low to allow full utilization of the mass capacity of modern timber trucks. Capacity utilization is even less efficient when hauling logging slash. An essential increase in the solid content may be gained through comminution. The *solid content of comminuted biomass* is dependent on many raw material and equipment variables, the most important of which are listed below.

Particle shape affects the solid content of chips mainly through the *diagonal-to-thickness ratio*. In fact, the inter-chipper differences in the solid content of sawmill chips are partly a result of variations in the diagonal-to-thickness ratio. The higher it is, the lower is the solid content. When this ratio in laboratory chips of Norway spruce increased from 6 to 13, the solid content decreased by one-fourth (Edberg et al. 1973).

Particle size distribution has a definite effect on solid content. With limitations, the mixing of over- or undersized chips with screened accepts increases the solid content. The increase resulting from the mixing of abnormal size fractions with normal accepts is explained by the tendency of heterogeneous particles to fill the spaces between chips during loading and transport. Similarly, whole-tree chips, when compared with uniform industrial pulpwood chips, have a higher solid content due to their more heterogeneous size distribution. As seen in Fig. 6.1, a drastic increase in the dry mass of a load may be obtained through a sawdust mixture of 25–50% in screened pulpwood chips (Edberg et al. 1973).

Tree species appears to affect solid content. Whole-tree chips made from brittle, low-density species contain more fine fractions, which increases the solid content. For the same reason, settling during hauling seems to be greater for low-den-

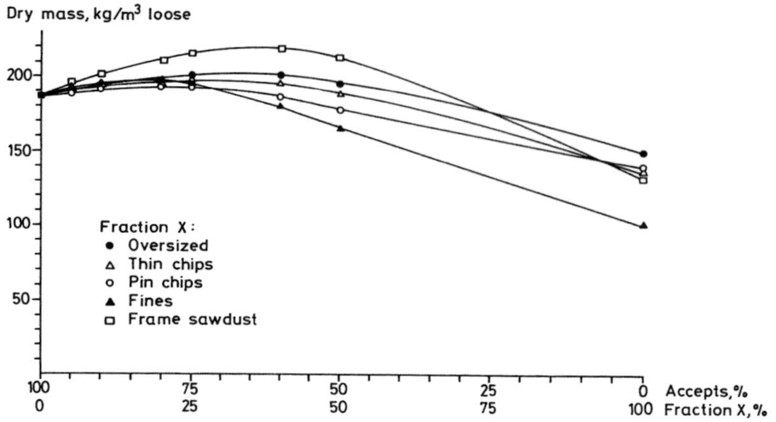

Fig. 6.1. The dry mass content in laboratory chips of *Picea abies* when mixing different screened fractions into the accepts (Edberg et al. 1973)

sity species such as gray alder, as shown by the following results by Kanninen et al. (1979):

Tree species	Basic density (kg/m^3)	Solid content of chips	
		On departure	On arrival
Alnus incana	370	0.457	0.484
Pinus sylvestris	385	0.403	0.421
Betula spp.	470	0.377	0.389

Lowering the *moisture content* of the material prior to comminution results in a slight increase in the solid content of chips. The reason is again the particle size distribution, since seasoned raw material produces more fine fractions when comminuted. Fresh twigs may pass through a chipper unfractioned, causing bridging and large empty spaces in the load. In addition, the increase is partly explained by less friction between chips made of dry biomass.

The *composition of the biomass* has a definite influence on the solid content of the chips, which is positively correlated with the solid content of the raw material before comminution. Branches, especially fresh ones with foliage, tend to produce excessive amounts of oversized particles and pliable twigs, while bark breaks easily into finer particles. When slash is comminuted separately from the stem, and branches cannot be aligned, the reduced mass has a low solid content. For example, when the solid content of birch whole-tree chips was 0.430, the corresponding figure for chips made of green softwood slash was only 0.357, although both were made with the same portable drum chipper (Uusvaara and Verkasalo 1987).

In cold climates the *season* affects the solid content in that frozen biomass, because of its brittleness, produces a higher percentage of fines. The effect is reduced by slower settling of frozen particles during transport. The average solid

Fig. 6.2. Farm tractor-driven TT 1000 TU landing chipper loading a tractor trailer by blowing from above (courtesy of Hannu Kalaja)

content of Scots pine and birch whole-tree chip loads at the power plant was 0.447 in the summer and 0.473 in the winter, as demonstrated by an extensive Finnish study (Uusvaara and Verkasalo 1987).

The *loading method* significantly influences the solid content. Chips blown through the discharge spout directly into a vehicle are denser than chips falling freely from a conveyor, tractor bin, loader, or silo (Figs. 6.2 and 6.3). In the following example, the same forwarder-mounted disk chipper worked first in terrain dumping the chips from its bin to an exchangeable truck container, and then at the landing blowing the chips directly onto a truck (Uusvaara and Verkasalo 1987).

Method of loading	Solid content of whole-tree chips at plant	
	Pinus sylvestris	*Betula* spp.
Dumping from tractor bin	0.464	0.415
Blowing from chipper	0.490	0.477

There are also differences between chippers with respect to the *blowing power.* When a transportable heavy drum chipper was equipped with an extra strong fan, the solid content of whole-tree chips in a truck load was 0.483 even before transport, while the corresponding figure as an average of three blow-loading conventional disk chippers was 0.411 (Kanninen et al. 1979).

Fig. 6.3. Bruks chip shuttle forwarder loading an exchangeable container by side-dumping (courtesy of Bruks Mekaniska AB)

Settling increases the solid content of the load during transport. At first, settling is rapid but slows down significantly at a distance of 20 km (Harris and Phillips 1989). Therefore, the settling of forest chips, although their hauling distances are often short, is almost equal to that of sawmill chips which are generally transported over longer distances. Furthermore, settling of forest chips compared with sawmill chips is accelerated by a higher content of fines, and by the fact that the first kilometers are often over rough roads. The initial solid content, length of haul, evenness of the road, possible freezing of chips, and the type of equipment all contribute to changes of solid content during transport. Settling is also more effective in a trailer than in a truck, as shown by the following results by Uusvaara and Verkasalo (1987):

Raw material of chips	Solid content of chips at plant	
	In truck	In trailer
Delimbed birch stems	0.427	0.450
Whole birch trees	0.451	0.469

From the standpoint of transport efficiency, the solid content before hauling is more important than that after hauling. On the other hand, because the quantity

Table 6.1. Average bulk density of whole-tree chips from three tree species before and after truck haulage for a distance of 50 km in Finland (Hakkila 1984)

Tree species	Basic density (kg/m^3)	Solid content	Moisture content (%)		Bulk density (kg/m^3)	
			Newly felled	Seasoned	Newly felled	Seasoned
Before truck haulage:						
Pinus sylvestris	385	0.41	53	40	336	263
Betula spp.	470	0.39	45	35	333	282
Alnus incana	370	0.44	50	38	326	263
After truck haulage:						
Pinus sylvestris	385	0.44	53	40	360	282
Betula spp.	470	0.42	45	35	359	304
Alnus incana	370	0.47	50	38	348	280

of chips is often determined in terms of loose volume at the plant, the solid content after hauling determines the value of a unit volume of chips.

Table 6.1 is an example of the average bulk density of whole-tree chips made of small-sized trees in Finland. Although the basic density is significantly higher in birch, the differences between the species are evened out by the reverse effects of moisture content and solid content. The bulk density of whole-tree chips made of newly felled trees varies between 320–340 kg/m^3 when settling during truck hauling is excluded. If the trees are well seasoned, the bulk density is 260–285 kg/m^3 or 15–20% less.

The load-space requirement of a vehicle hauling residual forest biomass in one form or another is determined by its mass capacity and the bulk density of the assortment. Figure 6.4 can be used to calculate the load volume necessary for full utilization of the mass capacity of a rig, if the basic density, moisture content, and solid content of the biomass to be hauled are known. Whether these load-space requirements can be met legally, depends on country-specific traffic regulations.

6.2 Improvement of Bulk Density by Compaction

The low bulk density of residual forest biomass results in inefficient trucking. This drawback may be partly compensated by using large-volume rigs or compaction of raw material. The larger the load space, the lower is the bulk-density requirement. If the capacity of a rig is 30 t, for instance, the bulk-density requirement is 353 kg/m^3 for an 85-m^3 rig but only 240 kg/m^3 for a large 125-m^3 rig.

The bulk density of green chips is usually well over 300 kg/m^3 and as such sufficient for large-volume equipment. However, if the moisture content is low and the load space small, the load-carrying capacity cannot be fully utilized. A simple way to compact the load then is to *use a strong fan* in the chipper (Sect. 6.1).

Another possibility is *vibration of the chip load* in connection with truck and railway transport. The compaction effect improves with increasing frequency and

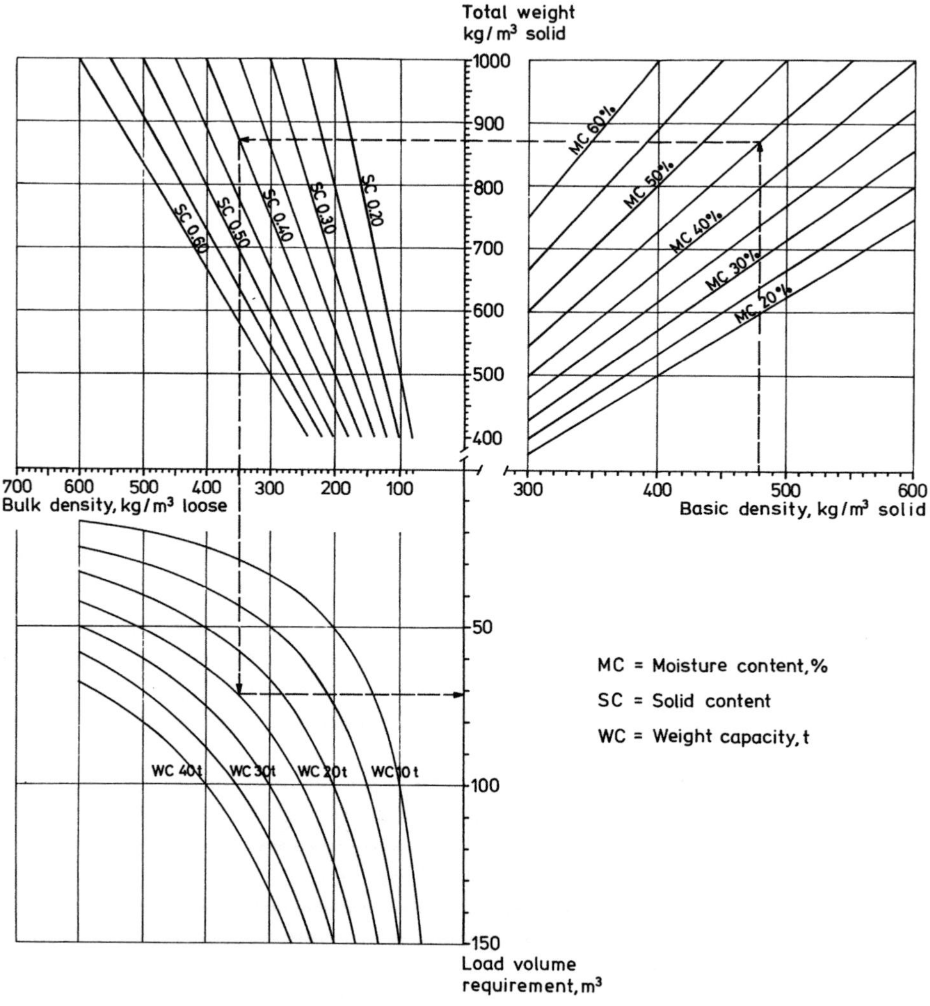

Fig. 6.4. Load-volume requirement for full utilization of the mass capacity of a rig, as a function of the basic density, moisture content, and solid content of biomass

amplitude of vibration. Compaction takes place mainly during the first 30 s, and after 2–4 min the process is fully completed. Compaction is most effective near the source of vibration, as the amplitude and compaction effects decrease with increasing distance. The variation in the moisture content of the chips does not seem to affect the compaction result. In an experiment in the Soviet Union, the maximum increase in solid content was 39–48% when the thickness of the chip layer was only 70 cm, the vibration frequency 1500 Hz, and the amplitude 4 mm. Under practical conditions, a 25-Hz vibration frequency and a 4-mm amplitude produced a 20–25% increase in the solid content of railway cars (Mihailov 1964).

Compressing chips can have little effect, since after compression the chips tend to rebound to their normal size. However, if the chips are dried to the fiber

Table 6.2. The effect of blowing, vibration, and compaction on the bulk density of whole-tree chips under laboratory conditions (Vidrine and Woodson 1982)

Method of compaction	Bulk density (kg/m^3)	
	Liquidambar styraciflua (m.c. 49%)	*Quercus* sp. and *Carya* sp. (m.c. 37%)
Blowing (43 m/s)	474	493
Vibration	419	419
Compaction at pressures of:		
0 kPa	366	373
86 kPa	434	443
172 kPa	461	464
345 kPa	493	491
517 kPa	493	499
690 kPa	493	499

saturation point, compression is more permanent. Volume reductions of up to 33% have been obtained (Silversides 1981).

Vidrine and Woodson (1982) studied the effect of various kinds of treatment on the bulk density of fresh whole-tree chips from low-density sweetgum and high-density oak and hickory in 2.4-m-high, 20–25-cm-diameter plastic pipes under laboratory conditions. Ordinary blown-into-van conditions of in-woods chipping were simulated by using a discharge air velocity of 43 m/s. Other treatments involved pipes filled by dropping from above and then compacted by vibration or compression with a hydraulic piston. Pneumatic loading resulted in as high a bulk density as compaction under 345 kPa pressure. Additional pressure raised the bulk density very little (Table 6.2). Neither vibration nor compression are used in practice in conjunction with truck transport of chips.

Compared with comminuted biomass, the need for load compaction is much more obvious for small-sized whole trees, tree sections, and slash. The following work organizations may be applied:

1. Compaction before loading, i.e., baling;
2. Compaction in conjunction with loading with a grapple crane;
3. Compaction after loading with truck-mounted or separate special equipment.

Baling involves the compaction of feedstock in a compression chamber and produces regularly shaped handling units which are significantly denser than the original feedstock, whereas bundling is simply the grouping together of small pieces of biomass into a larger unit. Bundling does not necessarily include compaction and the resulting product can be of irregular dimensions (Pottié and Guimier 1985).

To *reduce volume by baling*, residual biomass may be compacted and bound to form a cubic, cylindrical, or log-formed bale with uniform dimensions.

Although no commercial forest biomass balers are available, laboratory studies and tests with conventional farm equipment and prototype machines have proven the potential of the concept. The method offers the following operational advantages, which naturally have to balance the costs of compaction:

1. Bulk density may be increased to more than twofold, which would allow full use of the mass capacity in most truck rigs.
2. Conventional bunk-type roundwood rigs would require only minor modifications and little additional fixed mass for transport of residual biomass in the form of bales.
3. Loading bales could easily be handled with ordinary timber cranes, and cross loading from a truck to a railway car would be simple.
4. Bales could be left scattered on the logging site or stored in the landing area or at the terminal without major problems of biomass loss or contamination. This would eliminate the need for uninterrupted movement of biomass from woods to plant, thus preventing a hot logging schedule.
5. Bales could be fed into a chipper continuously without major in-feed problems, which would improve the productivity of the machine.
6. Baling could obviously be carried out with simple, easily maintained, low-investment machines.
7. A baler is a simple, straightforward machine requiring moderate power, containing few moving parts, and promising great durability. However, development of an efficient in-feed system, automatic tying, and suitable banding material is necessary (Stuart et al. 1981).

Danielsson et al. (1977) were able to compress aligned biomass material of Scots pine and Norway spruce to about 35% of its initial volume, using pressure of 0.7 MPa. The pressure requirement for bulk densities of 400 kg/m^3 was almost 0.6 MPa for dry slash but significantly less for fresh (Fig. 6.5). After the pressure had been released, the material partially sprang back, and the fresh biomass more so than the dry. This, however, took place only after the pressure had been reduced to about 0.1 MPa. String, which is adequate to withstand this pressure, seemed to be sufficient for forest biomass bales.

A prototype forest biomass baler consisting of a compaction plate and a shear blade combined in one unit has been constructed at the Virginia Polytechnic Institute and State University in the United States. The 1-t portable machine was equipped with a 74-kW engine, driving a 190 l/min hydraulic pump operating at 13.8 MPa pressure. The baler was able to compact and shear pine, maple, and oak branches and tops into 1.1-m-long, 0.76-m^3 rectangular cubes with a bulk density ranging from 600 to 980 kg/m^3, depending on the moisture content of the biomass. Energy consumption of the baler was 0.7–3.0% of the energy content of the biomass (Walbridge and Stuart 1981).

The strength and dimensional requirements placed on the bales are to a great extent dependent on the transport system and less on subsequent treatment at the destination. If the biomass is carefully built up for compaction, bales become stronger. A technique aimed at producing continuous ropelike bales from short-rotation willow biomass has been studied in Sweden. One possibility is to arrange the butt ends of pliable young willow trees helically around the bale outside,

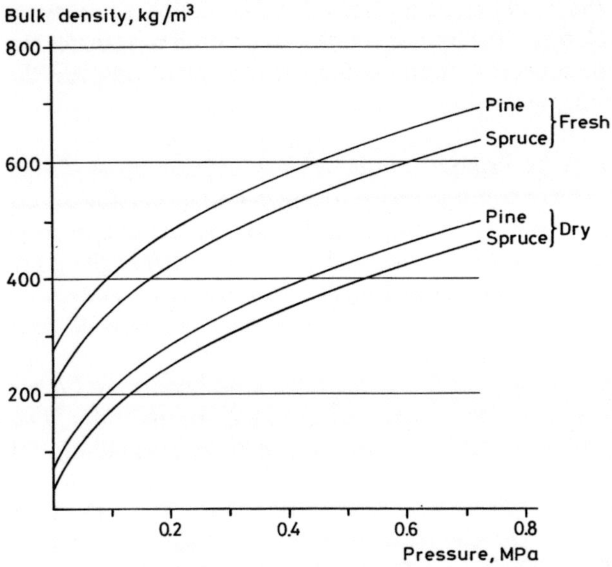

Fig. 6.5. The bulk density of fresh and dry crown mass from *Pinus sylvestris* and *Picea abies* as a function of compression pressure (redrawn from Danielsson et al. 1977)

which would make the bales resistant to bending. Another alternative is to place the butt ends in the middle of the bale, thereby creating more air circulation for better storage properties. An advantage of the latter method is that the short pieces that are lost when bucking the continuous bales to desired lengths consist of thin tree parts only (Säll 1981).

Baling would be more feasible in conjunction with other operations of the harvesting schedule, rather than as a separate operation. For this reason the process should be an integrated part of the continuous flow of material, which in turn will affect the layout of a commercial baling operation (Danielsson et al. 1977). Baling offers the potential of meshing well with small- and medium-sized conventional independent contractor operations in the United States, allowing the contractor the opportunity to recover logging residue which may find a market as an industrial fuel, fiber, or even as fuel for farm and home heating (Stuart et al. 1981).

Walbridge and Stuart (1981) proposed the following alternative organizational applications for a slash baler in the United States:

1. The baler is mounted on the tail section of a forwarder. The machine would move in the stand, feeding slash with its crane into the baler, and leaving the bales scattered over the site for later pickup by a conventional forwarder.
2. The baler is mounted on an old truck or trailer to be used at a landing in conjunction with grid or gate delimbing by skidders. The slash concentrating in the deck area in unsightly, loose piles could easily be recovered using the crane of the baler unit.
3. The baler is incorporated into a loader base at a landing where whole trees are processed with a logging machine. In most two- or three-skidder opera-

tions, only about half of the loader capacity is needed for feeding a processor. Consequently, during slack times the loader could also be used to transfer the branches and tops from the processor to the in-feed of the baler, compress the bales, and stack them onto piles.

A slight compaction effect may be brought about *with a grapple crane during loading.* Certain types of cranes are equipped with lifting cylinders capable of exerting heavy compressive force, thereby causing compaction. Some compaction may be produced by a crane by compressing the load with a log held in the grapple. The effect is only moderate but may be sufficient when the load consists of fairly large tree sections with a high moisture content (Carlsson and Rådström 1984).

Bulk density may also be increased with a *detachable crane-controlled truck compactor.* A Swedish prototype device consists of a simple beam frame, two pulleys, two tie-down cables, and two 19-MPa hydraulic cylinders extending 19 t of power to each cable. After loading the truck, the device is connected to the hydraulic system of the crane and placed across the truck load. The cables are fastened manually to the rig frame and tightened with the hydraulic cylinders for compaction. Finally, the security chains are installed and permanently secured over the load space. The compactor is removed from above the load after releasing the cables. In a field experiment, the bulk density of fresh tree sections increased from the original 250 to 294 kg/m^3 or 18%. The time consumed was about 4 min per compaction cycle. The operation was economically feasible (Larsson and Nordén 1982).

Most truck compactors are permanently mounted on the vehicle. In a Danish compaction rig prototype the load space is enlarged for the duration of loading *by tilting the bunk posts outward.* Once the load is complete, the bunk posts are raised to the original vertical position by a steel wire tightened by a hydraulic motor. The upper ends of the jointed posts are hinged 90° inward by the wire, thus compressing the load from above as well as from the sides. Compared to the legal volume, load space may thereby be enlarged for loading to 2.1-fold. With a power of 4.5 t at the end of each of the eight stakes or 0.27 MPa on both sides, the bulk density of small-sized whole trees could be increased from 165 to 258 kg/m^3 or 56% (Møller 1975). However, climbing onto a load of unlimbed timber to secure the wires was an unattractive and dangerous operation.

The Swedish Exte system utilizes *cross beams* for compaction. The equipment consists of four bunks, each with a pair of telescopic stakes. A hydraulic rotator with a beam is mounted onto the top of each stake on one side of the rig. After the load has reached the conventional height, the stakes are lengthened by 135 cm for continued loading. To compact the overload to legal height, the rotators turn the cross beams, made of strong leaf springs, over to the opposite side. Hydraulically maneuverable grapples then lock the beams onto a clamp at the top of the corresponding stake. The hydraulic cylinders in the stakes pull the beams down and compress the biomass. Depending on the timber assortment, the operation may be repeated two or three times. In Sweden, the 48-t compression power of the device is sufficient for full utilization of the mass capacity of the tree-section trucks, even for hauling seasoned timber with a low 30% moisture content.

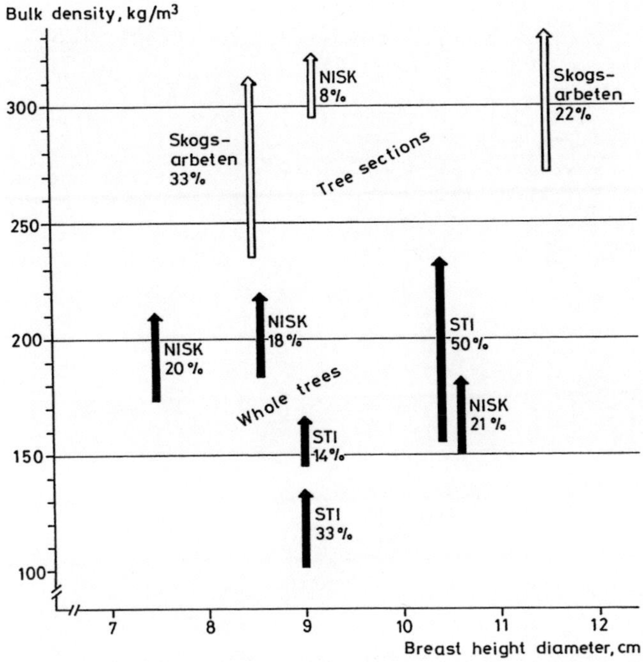

Fig. 6.6. The bulk density of small-sized whole trees (*black arrows*) and tree sections (*white arrows*) in a truck load before and after compaction in various experiments of Skogsarbeten in Sweden, NISK in Norway, and STI in Denmark. Relative increase is indicated (redrawn from Hansen 1975)

The compaction operation raised the time consumption in loading from 1.1 min/t to 2.0 min/t (Kyttälä 1981). The Finnish RKP load-compaction system is based on the same principle.

The Swedish Hydrovåg system uses *telescopic stakes equipped with flaps* for compaction. The flaps automatically turn out to a vertical position when the stakes are stretched to full length. In a field experiment, the solid content of tree sections could be raised by an average of 81%, while the time consumption in loading increased from 3 to 4 min/t (Carlsson and Larsson 1981).

Although load compaction seems to offer good potential for improvement of efficiency in trucking of residual biomass (Fig. 6.6), the preconditions for full capacity utilization must be created when preparing the timber at the site. For example, the length distribution of tree sections has a definite effect on the end result. On the other hand, it is important that the driver has good visibility over the rig when loading it. Since large-volume rigs are cheaper to operate than trucks with compaction equipment, compaction will be feasible only if these measures are insufficient.

6.3 Trucking Unlimbed Timber

The load volume of a modern timber truck with trailer is quite sufficient for hauling conventional pulpwood or sawlogs. The platform need not be extended to

Fig. 6.7. A self-loading truck-trailer unit for on-road transport of tree sections, equipped with RKP compaction device (courtesy of Hannu Kalaja)

allowable maximum dimensions, and the trucks can still take full legal loads. Large-volume rigs guarantee full capacity utilization when hauling forest chips.

Since equipment for hauling of sawmill chips offers a ready and proven alternative for transport of forest chips, development of rigs for other assortments of residual biomass was long neglected. However, systems analyses show that in-woods processing is not the most advantageous solution in all cases of biomass recovery, and that centralized processing may offer an attractive alternative if only the technical-economic problems of on-road transport of unlimbed timber can be solved.

Trucking whole trees and tree sections is still in the early stages of its development. In North America, for example, transport of whole trees is limited to only a few operations. However, where trees are of small size and uniform in length, and have small crowns and relatively high wood density, delivery costs of whole trees over short distances are competitive with those for other more conventional timber forms (Bjerkelund 1985).

During the first half of the 1980s, equipment and methods were developed especially in northern and central Europe; in the Federal Republic of Germany for transport of whole trees and in Sweden and Finland for transport of tree sections (Fig. 6.7). In the Soviet Union, and to a smaller extent in the United States and Canada, whole-tree trucks to central terminals are sometimes run outside the public road network. The restrictions on mass, length, width, and height of load are then naturally less severe.

Branches and tops that protrude from the load not only contribute to low bulk density, but result in other problems as well. If hauling is carried out with conven-

Fig. 6.8. Self-loading ONL-12.2 semitrailer modified for trucking whole trees in the Soviet Union (courtesy of SILAVA)

tional timber rigs, the parts of trees jutting out from the sides or back must be removed with a power saw. This is dangerous and time-consuming work. According to an early study by Hansen (1975), cleaning a load of whole trees from protruding branches by power saw, and the landing area from subsequent slash by hand, required 80–90 min per rig, thus increasing considerably the time consumed at the landing. In another study, Meng (1976) found an increase of 10–20% in the time consumption for loading large whole trees versus tree lengths, and an additional 10 minutes' work with a power saw was required to trim the load.

In order to avoid the costly and hazardous cleaning phase, the rigs must be *equipped with sidewalls.* To keep service mass as low as possible, light constructions of steel wire net, aluminum, and plywood are preferred. The sidewalls must be further strengthened to withstand pressure if load compaction is used. The walls in the North European tree-section trucks extend the entire length of the load space, so that part of them must be opened for unloading equipment at the plant. In Germany, the Soviet Union, and North America, normally only the region of the crowns, i.e., the rear of the semitrailer of a whole-tree truck, is covered by protective sidewalls (Fig. 6.8).

In North America, timber cranes are commonly equipped with a *combination of grapple and heel* to handle logs or tree lengths by the butt. If these cranes are used for whole trees, the mass of the tops and crowns tends to cause trees to spread out at the top during loading, complicating the task of placing the trees on the rig. The trees tend to cross when stacked, which slows down the operation, lowers the solid content of the load, and causes unloading problems. In a Canadian experiment, for example, the duration of a loading cycle was 31 s for a bunch of tree lengths, but 39 s or 26% longer for a bunch of whole trees (Routhier 1982). *A timber crane with a rotating grapple*, grasping trees near their center of gravity, is well suited for handling whole trees. Less effort is then required for lifting, trees caught under adjacent ones can be freed and repositioned more easily, and the risk of bending the trees to the point of breaking diminishes.

A separate crane or wheeled loader which can be used to serve several vehicles is feasible for large work sites. The system allows the use of heavier and more powerful loading equipment, which makes it possible to compact the trees during loading. On small work sites, self-loading trucks offer operational advantages and reduce the size requirement of a landing area. A self-contained crane does not increase the service mass of the vehicle if it can be detached and left at the landing.

Large whole trees are hauled in semitrailers. They may be extendable, articulated, and carried by the truck chassis when no load is being hauled. Butt ends are usually toward the front, but some North American rigs, used primarily on private roads, are loaded *in crisscross fashion* with about half the butts to the front and half to the rear. If the direction of butt and top ends is alternated bunch by bunch, load size increases further, but unloading becomes difficult. A short log is used sometimes as a crosspiece to facilitate unloading.

Lavoie (1980) studied truck transport of pulpwood-sized spruce and fir as delimbed tree lengths and whole trees in Canada. The number of whole trees per load was up to 15% less than that of tree lengths. The quantity of merchantable timber in a load was reduced, but the loss was compensated by the additional biomass from crown material. Loading whole trees required 11 min more per semitrailer truck than loading tree lengths. The difference was 5 min in unloading.

A truck equipped with a full trailer is the most common solution for hauling tree sections. In northern Europe, where the method is increasingly used, one pile of 5-m sections is loaded onto the truck and two piles onto the trailer. Compaction of tree-section loads is feasible over longer transport distances (Sect. 6.2).

Hauling unlimbed timber still presents certain technical and, especially, economic problems, but the efficiency of transport may be increased with careful planning and work organization. Fairly simple modification of conventional roundwood rigs and use of compaction can improve profitability and safety to a considerable degree. The technique may be expected to develop gradually to the point of becoming as reliable as the hauling of conventional roundwood or chips.

6.4 Trucking Unprocessed Logging Residue

Due to its very low bulk density, logging residue may be hauled unprocessed only over relatively short distances. Over longer distances trucking is done with special large-volume rigs, which must be fitted with a deck and sidewalls.

In trucking, as in off-road forwarding, the loading of *crown and top mass* may be rationalized by modifying the conventional timber grapple (Sect. 4.2.2). Compression of the load with the grapple improves bulk density slightly. According to a Swedish study, it was possible to increase the bulk density of a truck load of fresh crown and top mass in this way from 180 to 200 kg/m^3 or by about 10%

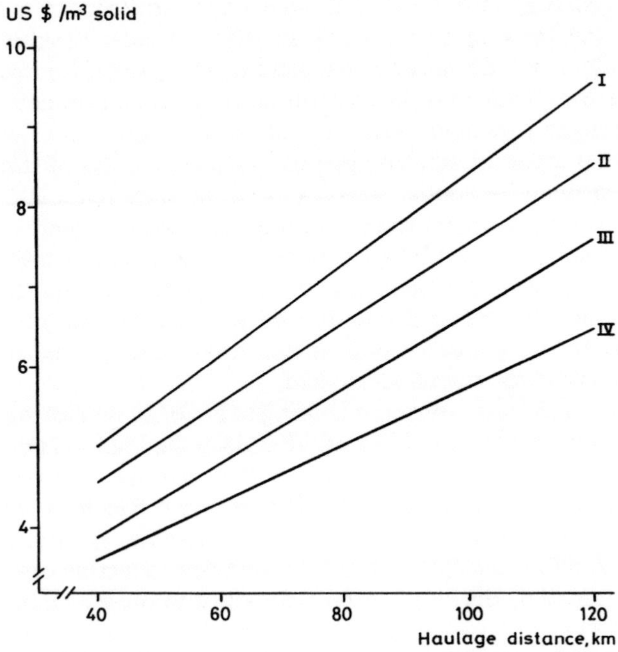

Fig. 6.9. Trucking cost of logging residue as a function of transport distance for four special vehicles in Sweden. Moisture content 40%, bulk density of uncompacted biomass 150 kg/m^3. For explanation of the four alternatives, see the text (redrawn from Alexandersson et al. 1984)

(Larsson 1981). Using a grapple saw for prefractioning crown mass in conjunction with loading does not significantly improve bulk density.

Trucking crown and top mass over relatively short distances does not necessarily require large-volume rigs. However, the longer the distance, the more important is the load size. Figure 6.9 shows the trucking costs for Scots pine and Norway spruce slash in Sweden for four large-volume rigs of various sizes. The uppermost curve represents a self-loading 135-m^3 rig with a net load of 20.3 t. Curve II represents the same vehicle equipped with a compaction device which increases the net load to 25.2 t. Alternative III is a 148-m^3 rig without compaction facilities, but loading is accomplished with a separate crane, compacting the load to the same 25.2-t net mass. Theoretically the most economical solution is represented by curve IV, where the same vehicle is equipped with its own crane and compaction device. While the first three alternatives are designed for a maximum gross vehicle mass of 51.4 t, the fourth one requires an allowable gross mass of 60.0 t. The differences in the economy are significant, the greatest cost benefit being achieved with the largest self-loading rig equipped with a compaction device. Under prevailing Swedish conditions, a switch from option I to option IV increased the maximum economical operation distance by about 65 km. Enlargening the catchment area almost trebled the accessible amount of biomass residue.

In the unmanaged forests of British Columbia logging slash in cutover areas is characterized by the abundance of *stem mass from large overmature trees.* In the case of cable logging systems, most low-grade stem mass remaining after prime logging is found in the landing or within 30 m of the roads. Residue discarded in the landing results from the yarding crew sending in logs that do not meet the utilization specifications, from logs that are too short to load on the conventional trucks, from portions of logs bucked in the landing, and from breakage during yarding. Sinclair (1984) tested integrated systems based on the use of the excess capacity of the conventional log loader to collect residual stem mass into various recovery units. The truck container system showed the greatest potential. The main benefits of the system were the ability to service several landings with one truck, the quick pick-up and drop-off time, the ability to load the container for a longer period of time, and the reduced interdependence of active logging operations. The existing system using a cable skidder and a self-loading timber truck was not competitive against the container system.

Stump and root mass is trucked from the upper landing to a central processing terminal or mill in large-volume, reinforced chip rigs. To reduce the load-volume requirement, stump-root systems must be fragmented before hauling. Because reduction of impure raw material to small chiplike particles is very difficult in the woods, stump-root systems are rendered more easily transportable by rough presplitting on site or at the landing. Presplitting notwithstanding, trucking efficiency is still seriously constrained by the very low bulk density and the high content of contaminants.

Bulk density of stump-root systems is increased by more than 60% through fragmentation into two or three pieces, which in turn may reduce the cost of trucking by 25–40%, depending on the transport distance. Since reduction of stump-root systems improves bulk density and lowers the proportion of contaminants, the degree of prefragmentation is of great significance to transport efficiency. However, gains from the hauling of finer stump pieces must be weighed against the increase in the cost of presplitting, fiber losses caused by in-woods fragmentation, and possible changes in the handling costs at the mill.

An essential prerequisite for flexible stowing and compaction of stump and root fragments during loading is that the load space of the rig is fully visible to the crane operator. This may be achieved by placing the operator high enough to be able to look down on the load, or by constructing the sidewalls of fine mesh, enabling the operator to see through them. The load space of the rig must be strong enough to withstand the rough handling of stowing and compaction with grapple. Furthermore, if the load is to be emptied by side-dumping, heavy demands are also placed on the stability of the rig and the design of the load space. The sides must have no raised edges which might catch the compacted stump wood and hinder its movement (Hansen 1977).

With respect to the crane, the number of joints between the outer boom and the grapple should be kept at a minimum in order to increase grapple stability and thereby compacting effect and productivity. The possibility of locking existing joints would presumably further increase loading productivity and the degree of compaction. A double-acting crane with both lifting and pressing functions obviously improves the chances of achieving a higher degree of compaction. The design of the grapple must be a compromise between the maximum lifting volume, on the one hand, and ease of operation during compaction and stowing, on the other (Hansen 1977).

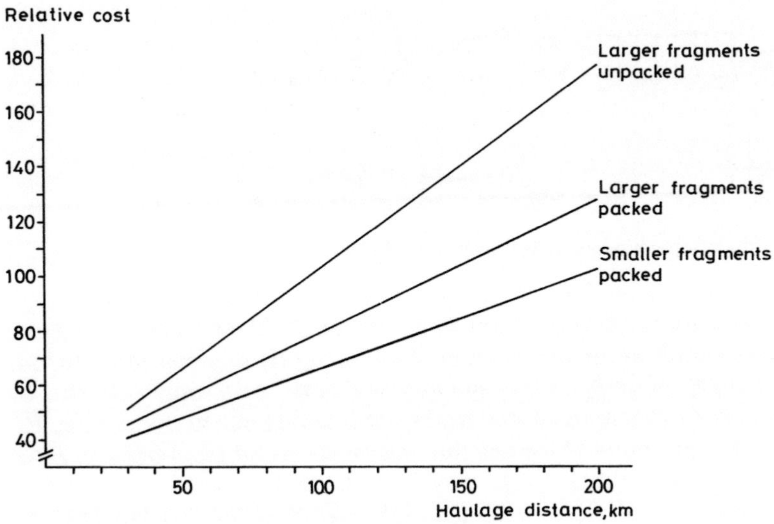

Fig. 6.10. Relative trucking cost of stump and root mass as a function of haulage distance, depending on the degree of fragmentation and packing (Hansen 1977)

In Swedish experiments, the bulk density of stump material of a low fragmentation degree was raised from 160 to 255 kg/m^3 or by 60% by stowing and compaction with a crane. This resulted in about a 30% decrease in the cost of trucking. When the material was reduced into smaller fragments, i.e., six to ten pieces per stump-root system, bulk density was raised to 350 kg/m^3. This lowered the trucking cost by an additional 15% (Fig. 6.10).

6.5 Trucking Comminuted Biomass

Trucking of comminuted forest biomass does not present major technical problems. Rigs originally developed for on-road haulage of sawmill chips are effective and reliable for the transport of forest chips as well. The mass capacity of a rig may be fully utilized, unless the raw material is well seasoned. The direct transport cost of residual biomass is obviously lowest if the material is fractioned in the woods.

In North America, chip transport is done with semitrailers with a load volume of up to 100 m^3 or more. The most feasible solution in northern Europe, taking into account the road conditions, is a combination of truck and full trailer. The largest rigs in Sweden are able to take a chip load of 145 m^3 loose.

Particularly in conjunction with terrain chipping, a full-trailer truck may be equipped with two or three *exchangeable chip containers*. The system requires pallets which can be lowered onto the ground as the upper edge of a rig on wheels is too high for dumping directly from a terrain chipper bin. The pallet system results in a somewhat smaller load size compared with stationary truck rigs. On the other hand, the pallets may be filled and left at the landing for an indefinite period of time, to be picked up when convenient (Fig. 6.11).

Fig. 6.11. The Multilift chip container system (courtesy of Multilift AB)

Although there are no major unsolved technical problems, the hot logging schedules linking comminution and on-road hauling tightly together result in hidden costs due to inherent delays. Consequently, real costs are strongly affected by correct dimensioning of equipment and timing of interdependent operations. Effective haulage of comminuted biomass thus requires careful *planning and organization*, so that the inevitable delays may be reduced to a minimum through smooth synchronization of the operations. Job organization and landing-area layout are of utmost importance for raising productivity and lowering costs of hauling and comminution as well.

Comminuted biomass is a mass article, the loading and unloading of which may easily be mechanized and automated. *Loading a chip truck* takes usually place straight from the chipper onto a truck, trailer, or exchangeable container (Figs. 6.2 and 6.3). During chipping at a landing this is usually done pneumatically by blowing. If the discharge spout is above the load, it is possible to add a chip crown onto the rig, and the truck does not have to be moved every time the chipper changes position as the work progresses. When the chips are fed from above, particles bounce off onto the ground as the load nears completion. Furthermore, to prevent chips from blowing out of the rig during hauling, the load should be covered with a flip top, net, or canvas tarpaulin. If blowing is done from the rear, on the other hand, the load cover need not be removed during loading, and no chips are lost.

Sometimes a whole-tree logging operation produces both softwood and hardwood raw material simultaneously, with the former suitable for pulping, while the latter is used for fuel. If the landing area is large enough for loading two rigs simultaneously, softwood and hardwood chips may be kept apart with an air-powered or hydraulically operated flipper at the end of the discharge spout, allowing the operator to direct the chips into either one of the rigs according to need. Another possibility is a Y-shaped twin discharge spout, which allows the operator to change chip direction by means of a built-in gate inside the chute.

Moving the reduced wood from chipper to rig is technically simple, but the close interlinkage of truck and chipper makes the system operationally "hot" and vulnerable. If comminution is completely separated from trucking, the chips must be stored at the landing for at least a short time and transferred from the ground onto the truck using a wheeled loader or a crane equipped with a chip scoop. This results in additional expenditure, contamination, and biomass loss. To keep the quantitative and qualitative losses in check, the chips must be piled onto a protective ground sheet. Plastic sheets are not durable, and bits of plastic that become

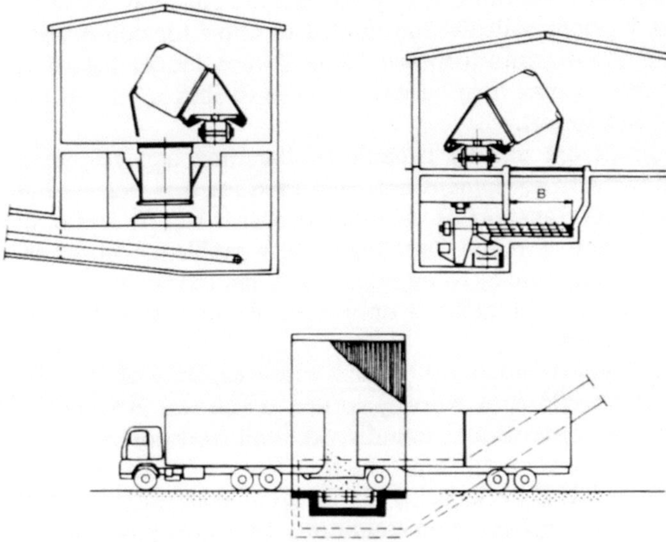

Fig. 6.12. Examples of self-unloading chip rigs. *Top left:* Side-dumping and belt conveyor. *Top right:* Side-dumping and screw conveyor. *Bottom:* Truck with a moving floor and a scraper conveyor (courtesy of Kone Engineering)

mixed with the chips are detrimental to the pulping process. Experiences with buffer storing chips at the landing site have been predominantly negative.

Unloading a chip truck is usually done by dumping from the rear or side with a truck-mounted hydraulic device. Another method of fast self-unloading is a moving rig floor, equipped with a belt or scraper conveyor powered by the hydraulic pump of the truck (Fig. 6.12). The self-unloading feature increases hauling versatility but results in a reduced payload. In North America, many mills receiving large volumes of chips by truck are equipped with a stationary unloading platform, on which a truck-semitrailer unit is dumped from the rear. Chip rigs with hinged sidewalls may be unloaded by pushing with a tractor-mounted blade. Mechanical chip diggers combined with a mechanical or pneumatic conveyor, and vaccuum to suck up the chips are other, albeit infrequently applied alternatives.

In cold climates, frozen green chips may cause unloading problems. Freezing is sometimes prevented with salt or side boards heated with exhaust gas from the truck. A rig may also be equipped with hoses for steam-thawing chips at the destination.

6.6 Alternatives to Trucking

The reserve of unmerchantable forest biomass in all major forest industry countries is many times that of expected short-term demand. This makes it possible to concentrate recovery to sites with the most advantageous logging conditions and placement. With only few exceptions, the catchment area around potential users of residual biomass is geographically too limited to allow for competitive rail transport. However, if the demand increases in the future, longer transport distances must be acceptable. It may then become feasible to shift a part of the biomass transport from road to rail.

Timber *railroad* terminals are usually capable of handling and reloading whole trees and tree sections with equipment normally used for tree lengths and log lengths. The problem in rail transport of unlimbed timber is insufficient bulk density of the load, rather than terminal handling. Safety problems caused by protruding branches may be overcome by using side boards, but the center section of the car side should be left open to facilitate unloading. Another possibility is the use of tree-section containers.

In a Swedish experiment, conventional timber cars with a capacity of 30 t carried a 20.5-t net load of tree sections of Norway spruce, which was 20% below the average load of 26.5 t of conventional roundwood. Rail transport was not competitive for distances shorter than 150 km. The total costs of truck and rail alternatives thereafter run neck and neck, being about equal at distances of 150 to 225 km from terminal to mill, depending on conditions. In reality, however, the longest economical transport distance for tree sections was no more than 100 km, thus barring the possibility of rail transport (Alexandersson 1985).

Rail transport is basically very suitable for moving large quantities of homogeneous, comminuted biomass over long distances, where a direct rail connection exists between a terminal and the receiver. Proven technology is available and commonly used for transport of sawmill chips to pulp mills.

Chip cars are either open or covered. Roofless cars are easier to load and unload, but severe winter conditions of snow, ice buildup, and freezing may cause problems. The choice of loading system depends mainly on the quantity to be transported, storage facilities at the terminal, and the desired rotation speed of the cars between terminal and mill. The following methods are being used for loading saw-mill chips, singly or in combination: pneumatically from a pile or blowing directly from a chipper, dumping from a truck, by gravity from a silo, by belt or scraper conveyor, by crane-mounted scoop loader, or by wheeled loader.

A tractor with a front-end pushing blade is a simple unloading solution requiring a modest capital investment for cars with sidewalls that can be swung open from below. A small tractor with scoop loader or bottom-dumping bucket is another alternative at a relatively low capital investment. Self-propelled digger units combined with mechanical or pneumatic conveyors may also be used, particularly when handling frozen chips or chips from covered cars. Overhead vaccuum unloaders, roll-over rail car dumpers, crane-mounted scoops, and tilting platforms are useful for handling large volumes of chips.

The standard loading and unloading equipment described above is used for rail transport of high-quality industrial chips. Chips made of whole trees, logging residue, or stump and root mass are of lower value and cannot economically be hauled over long distances. Consequently, at present cost levels the catchment area is not large enough to allow for the use of railroads to transport whole-tree chips or other similar products.

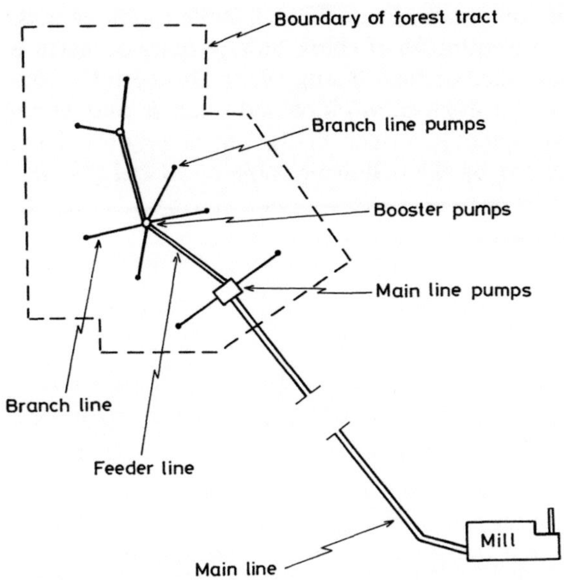

Fig. 6.13. A layout of a chip pipeline (Haug 1972)

A hydraulic *pipeline*, through which solid particles are floated in water or other liquids, is a potential alternative to trucking of comminuted material. The method is currently applied to slurry transport of coal, ore, and phosphate, ground into fine particles, usually of 1.8 – 2.5 mm diameters. However, no proven technology for practical scale long-distance pipeline transport of chips is yet available. Larger particle sizes, wide size distributions, irregular shapes, and the varying densities and moisture content of wood and bark make the operating of pumps, valves and other flow-controlling devices more difficult in chip transport compared with ground coal.

Figure 6.13 illustrates the design of a pipeline system for use under North American conditions. To be competitive, the high-capacity system would require large, contiguous areas with abundant supplies of low-quality wood and a poorly developed infrastructure for timber transport by conventional means. For example, a 20-cm-diameter pipeline with a flow speed of 1 m/s and a chip concentration of 20% could carry 500 t fresh chips in 24 h. Under northern European conditions it would probably be more feasible to adopt a system whereby a pipeline connects two or more large terminals (Haug 1972).

Concentration is defined as the percentage by volume of solid matter in the mixture of chips and water. The optimum concentration is no more than 30% and diminishes with a decrease in the diameter of the pipe. The power requirement for the pumps increases steeply when the concentration exceeds 35% (Elliott 1960).

The critical flow speed is defined as the speed at which the loss of energy due to friction is the least. It increases proportionally with the square root of the pipe diameter. If the speed falls under the critical limit, sedimentation may cause blocking. An economically feasible speed lies between 1 – 3 m/s (Haug 1972).

The loss of energy due to friction varies with different factors such as mass density of wood when saturated, concentration of chips, homogeneity of mixture, diameter and material of pipe, and speed of flow (Haug 1972). However, the consumption of energy for transport of a mixture of chips and water is only about 5% higher than transport of clear water (Gardner 1977). The energy efficiency of pipeline transport is demonstrated by the following table by Mittal (1979):

Mode of transport	Energy intensity (MJ/t/km)
Air cargo	18.00
Highway truck, semitrailer	1.80
Rail transport, freight	0.54
Barging, deep waterway	0.36
Pipeline, crude oil	0.11

A pipeline may be laid out at sharp angles and straight up and down. The line length and length profile determine the effect and location of the pumps. If input and output are at the same level and the ground is even, then only the friction has to be taken into consideration. In uneven terrain, heavier pumps are necessary for carrying the mixture across peaks. Siting of pipes on the ground, underground, or in water, is determined by the weather conditions, and whether operations will be carried out year-round or limited to the summer season. Capital costs will increase considerably if the line must be protected against frost (Haug 1972).

A pipeline permits a high degree of automation, and it eliminates heavy transport from the highways. In warm climates where water is available the operation is not influenced by weather. On the other hand, the system requires high capital investments which are committed for a relatively long period. The requirement of a large and constant throughput, and inflexibility concerning the assortments of timber are serious drawbacks.

Approximately 60–70% of the total costs of pipeline transport consists of capital expenditures. If these costs are compared with truck and rail rates, which do not include the construction of highways or railroads, a pipeline can hardly be competitive. However, in remote and inaccessible areas where the construction of any part of the infrastructure has to be included, pipeline transport may become a feasible alternative.

Chapter 7

Examples of Biomass Harvesting Systems

Succeeding phases of biomass preparation and hauling from stump to the point of utilization are linked together to form a harvesting system. A common way of classifying harvesting systems is based on the form in which the timber is moved from the stump area to the woods landing or roadside. In *shortwood or log-length systems* the tree is delimbed and bucked into timber assortments and hauled typically by load-carrying tractors to the roadside as log lengths. In *tree-length systems* the tree is delimbed and topped at the stump and skidded as tree length to the woods landing. The tree lengths are bucked to conventional roundwood assortments at the woods landing, central terminal, or point of use. In *whole-tree systems* the entire above-ground portion of a tree is skidded to the landing. Intensified recovery of residual biomass, integrated in or separate from the harvesting of conventional timber, introduces totally new or partly modified *biomass harvesting systems.* Pottié and Guimier (1986) reviewed 29 systems for harvesting forest biomass.

A harvesting system may consist of both independent and interlinked operations, which are discussed in more detail in Chapters 4, 5, and 6. The functioning of an entire system is determined in the first place by its weakest link. While it is important not to raise the capital investment by overscaling the capacity of single machines, it is even more essential to avoid fatal flaws in the system. Compatibility of succeeding operations is therefore of great significance.

The first step is typically either felling and bunching of whole trees, collection of slash after conventional stem-only logging or, in the case of below-ground residue, uprooting stump-root systems. Essential phases of the procurement are, naturally, off-road hauling of biomass in one form or another to the landing site and on-road hauling to the point of utilization. A crucial operation specific to most biomass harvesting systems is size reduction of raw material, which may be performed on terrain, at the roadside or upper landing, at a centralized terminal, or at a mill yard (Sect. 5.4).

A large selection of harvesting systems is available for recovering residual forest biomass of different types. The applicability and competitiveness of the various alternatives are determined mainly by the following factors:

1. Properties of the biomass to be harvested;
2. Scale of operation;
3. Logging conditions such as climate, stand environment, terrain, sensitivity of forest soil, etc.;
4. Distances in off-road and on-road hauling;
5. General stage of development and logging traditions;
6. General infrastructure supporting the operations; road network, service facilities for machines, existing logging organizations, etc.;

7. Availability and direct and indirect costs of skilled and unskilled labor;
8. Availability and cost of energy and capital;
9. Structure, location, and energy demand of the forest industries;
10. End use and quality requirements set for the products;
11. Silvicultural and environmental requirements;
12. Role and integration of residue recovery in relation to harvesting conventional timber.

Because crown mass as a rule is only a minor part of the total biomass removed from a certain forest area, at least in larger operations residue recovery must be integrated more or less closely into conventional logging. If the scale of residue recovery is not adequate to guarantee full employment for effective special machines, it is usually more economic to rely mainly on the regular harvesting equipment locally available. In a *one-pass system* all material from a site is harvested simultaneously. In a *two-pass system* the unmerchantable biomass is usually recovered after removal of conventional timber. However, sometimes a feasible order of operations in a two-pass system may be just the opposite. For example, in certain clear-cutting operations the best result may be achieved by harvesting the smallest trees with branches intact in the first pass, in order to improve the productivity of heavy-duty logging machines in preparing conventional timber from large trees and to avoid breakage and soiling of small trees.

Each phase of a harvesting system, either motor-manual or mechanized, is affected by both the preceding and the following operations. The capacity of the entire system is often determined by the least productive phase, including the downtime and idle time elements, which always reduce the long-term performance. If the operations are tightly interlinked, not even the capacity of the least productive operation can be fully achieved. But if the interconnection of the subsequent operations is loose, the system can be balanced by adjusting the shift length separately for each operational function. Since production time is also lost in moving equipment between work sites, the annual capacity of a harvesting system is affected considerably by the size of the tracts. The more effective the machine, the larger is the proportion of time used to move it from site to site.

The profitability of a harvesting system is not determined exclusively by the expected productivity and cost per hour. The *level of annual production* is an important factor in the system selection. Highly mechanized, capital-intensive harvesting systems cannot bear extended periods of downtime and remain economically viable, and therefore they necessitate year-round operation. Since the economy of efficient harvesting systems is adversely affected by the need to move operations frequently, their use creates pressure to provide large harvesting areas. Heavy logging machines and haul trucks also require large landing areas to work efficiently (Mattson 1985).

Eleven proven biomass harvesting systems will be introduced in the present chapter as examples of emerging technology. The equipment referred to is commercially available. Even if the amount of biomass recovered annually with these systems is still small compared with the existing reserves of unmerchantable forest biomass, all the alternatives described are in use somewhere in northern Europe or North America. In practice a procurement organization may employ several harvesting systems concurrently, but here the examples are simplified.

A comparison of costs between harvesting systems applied for different biomass sources in different countries easily results in misleading conclusions. For example, the general infrastructure, relative costs of labor and capital, fluctuating exchange rates, logging conditions, and the scale of operation affect the appropriateness of various alternatives. Hence, the efficiency of simplified harvesting systems is studied here in the light of *labor productivity*, expressed as man-day requirement per m^3 solid procured from stump to mill.

The capital investment for a system, including trucking to the mill, is expressed in US dollars per m^3 solid of annual capacity. The investment need is based on working principally in one shift or 2000 h/a and on the value of the dollar in April 1987. The level of capital investment is often reduced in practice by purchasing second-hand machines and by increasing the annual work time beyond 2000 h/a. Paying off of the investment naturally depends on machine lifetime, which for forest machines is usually 4–6 years, depending on conditions.

Although the feasibility of biomass recovery is greatly affected by a complex of socioeconomic, silvicultural, and environmental factors, the values for labor productivity and capitalization give at least some indication of the applicability and need for modification of a system for recovering a certain type of biomass in a given country. Mutual efficiency comparison between systems in the light of the following sections is not of particular value, because tree size, biomass composition, and technical logging conditions may be totally different. Some systems are designed for whole trees, while others are intended for logging slash or below-ground biomass. Some are designed for small-scale operations, and some for large industrial enterprises.

7.1 Whole-Tree Chipping Systems

Whole-tree chipping systems are aimed at a simultaneous increase of productivity, reduction of logging costs, and improved recovery of forest biomass. A need for the removal of small-sized and low-quality trees may arise as a result of the demand for additional raw material but also for silvicultural reasons. Typical goals are thinning young plantations at the proper time or regeneration of low-value hardwood stands for preferred species.

In northern Europe whole-tree chipping is applied mainly to small-sized trees, generally 5–12 cm in diameter at breast height. Trees thicker than 20 cm dbh are exceptions. Northern European conditions characteristically are small private ownership, close integration of agriculture and forestry, and active participation of farm forest owners in timber harvesting through delivery sales.

In Finland and Sweden thousands of farm owners produce fuel chips from unmerchantable, small-sized trees for their own consumption as a substitute for fuel oil as a source of heat energy. Because the annual need of a farm for fuel chips is too small to allow investment in expensive special equipment, harvesting is based on the use of a farm tractor (Sect. 7.1.1).

Commercial whole-tree chipping operations in northern Europe also have to deal with small trees. Because the annual chip consumption varies from plant to plant within a wide range, a large variety of equipment and harvesting systems is in use. The trees to be harvested are usually bought standing, but a common alternative is a delivery sale in which the forest owner is responsible for cutting and off-road hauling (Sects. 7.1.2–7.1.4).

In the United States and Canada technical logging conditions are more favorable for effective whole-tree chipping systems. Trees are larger, biomass yield per hectare and work site is higher, and the material comes mainly from clear-cuttings. Furthermore, due to larger production series and less strict ergonomic standards, the price level of forest machines is lower. All these factors result in the use of heavy-duty machines and higher productivity.

North American whole-tree chipping systems are rather similar. They are commonly based on transportable chassis-mounted chippers, some of which are able to reduce 40–60-cm-diameter trees of relatively high basic density. An example of these effective systems is given in Section 7.1.5. According to a whole-tree chipper survey of the American Pulpwood Association in September 1979, as many as 732 transportable, heavy-duty chippers were working in the United States for the forest industries, making fuel chips for utilities, institutions, brick kilns, etc., or in operations such as land clearing and right-of-way maintenance. Assuming the same production per chipper as was measured in an earlier survey in 1974, the annual production was estimated as follows (Altman 1980):

	Number of machines	Production (million green tons/a)
Forest industry operations:		
Chippers in woods	488	14.2
Chippers at terminals	115	4.3
Other operations	129	Not known

7.1.1 Chipping Small Trees for Farm Heating

Case. A farm has 80 ha forest land. In addition to conventional timber, these forests produce 0.8 m^3 solid/ha unmerchantable, small-diameter stems and tops annually. To improve the management of young plantations and to reduce heating costs, the farm substitutes chips for 9 t fuel oil annually in space and water heating and grain drying. The annual chip requirement is 50 m^3 solid or 120 m^3 loose.

Chips are burned in a 40-kW boiler equipped with automatic stoker feeding. Uninterrupted operation and efficient combustion require uniform particle size and less than 35% moisture in the chips.

Harvesting Solution (Fig. 7.1). Chips are made of stems and tops less than 10 cm in diameter, because larger material can be sold for pulpwood. Harvesting is timed to fit in with other operations of the farm. Chain-saw felling is started early in the spring and brought to conclusion by the end of the summer. Trees are left

Fig. 7.1. A harvesting system to produce fuel chips from small delimbed stems for domestic farm use

on the ground, branches intact, for transpiration drying for a period of at least 4–5 weeks, depending on the tree species and weather conditions. For better control of particle size distribution and automatic feeding of chips, the trees are delimbed after transpiration drying and collected manually into bunches. The stems are hauled in a tractor trailer to the farm yard and piled manually for subsequent comminution.

Chips are produced by a light chipper mounted on the three-point hitch of a farm tractor. Stems are fed singly by hand. The chips are blown directly into the storage barn and further dried to 30% moisture content by cold air circulation. Due to restricted chip storage capacity and to avoid health problems caused by microfungi, half of the piled stems are covered with a tarpaulin and left for later reduction in the winter.

Productivity depends on conditions and equipment. The manpower requirement is high, especially for the felling and delimbing phase. This is a result of the small tree size, delimbing of trees, and use of nonprofessional labor:

	Manpower requirement (man-day/m^3 solid)
Felling and delimbing	0.285
Hauling to farm yard	0.060
Chipping	0.080
Total	0.425

To produce 50 m^3 solid of chips, the farmer has to work a total of 21 days. In addition to the fuel, such an operation results in silvicultural tending of a 2-ha forest area. Because the tractor, trailer, and chain saw are part of the standard equipment of the farm, the only additional capital investment is US $ 2000–4500 for the chipper. To reduce the capital costs, the chipper can be purchased jointly with neighboring farms. Cooperation is feasible, because the chipper is used for only a few days each year.

7.1.2 Chipping Small Trees for a 1-MW Heating Plant

Case. A rural hospital in northern Europe has a 1-MW heating plant consuming 1500 m^3 solid or 3500 m^3 loose of fuel chips annually. Automatic stoker feeding requires homogeneous chips, free of twigs and other oversized particles. The moisture content should not be higher than 40%.

The peak of heat production and chip delivery to the plant tends to fall in midwinter. Daily chip consumption increases up to 15 m^3 solid during the coldest season. All fuelwood may be purchased from farm forests within a radius of 15 km.

Harvesting Solution (Fig. 7.2). Because few farm owners are fully employed during the year, many of them are willing to sell undersized fuelwood to the heating plant in road side piles. Consequently, the chip procurement system is based on delivery sales from self-employed farmers.

Fig. 7.2. A harvesting system to produce fuel chips from small delimbed stems for a local heating plant

Felling, rough delimbing, and hauling are performed by the farmers in the way described in Section 7.1.1, using the transpiration drying method. To control the moisture content and to speed up the subsequent chipping in the winter, the road-side piles are made as high as possible and covered with a tarpaulin.

Chipping and chip transport to the plant are achieved by a contractor. Both work phases are managed with the same prime mover, a 50–80-kW farm tractor. The chipper, attached to the three-point hitch of the tractor, may be easily dismounted and replaced for on-road hauling by a tipping trailer with a 30–35 m^3 load space. The tractor is equipped with a hydraulic grapple crane to mechanize feeding to the chipper.

Because the farmers perform cutting and off-road hauling with existing chain saws and farm tractors, the only extra capital-investment requirement for the system is the contractor's machinery. The chip operations can employ this equipment only for 90 days annually and that is not enough to pay off the contractor's tractor. Hence, the job is better suited to a tractor owner who has excess machine capacity, particularly in the winter season. The total additional capital investment for a new chipper, trailer, and tractor-mounted grapple crane is US $ 27000–40000.

The total labor requirement of the above simplified harvesting system for 4–12 cm dbh trees is, under average conditions, 0.395 man-day per solid m^3 of chips. In practice, efficiency varies within a wide range, depending on tree size and other technical logging factors:

	Manpower requirement (man-day/m^3 solid)
Felling and delimbing	0.285
Haulage to landing	0.050
Chipping and haulage of chips	0.060
Total	0.395

A prerequisite for major improvement in productivity would be the abandonment of delimbing. This would require modification of chip-handling equipment at the hypothetical plant under consideration.

7.1.3 Chipping Small Trees for a 3-MW Heating Plant

Case. A 3-MW municipal district heating plant consumes 6000 m^3 solid or 14000 m^3 loose of fuel chips annually. The chip handling equipment includes a disk screen and has been designed to permit a moderate amount of oversized particles in the chips. For efficient combustion, the moisture content of the chips should stay below 42%. During the peak season in midwinter the daily consumption of chips increases to 50 m^3 solid.

Unmerchantable, small-sized wood is abundantly available from young stands within a radius of 25 km, but only a few self-employed forest owners are willing to sell timber delivered to the roadside. Consequently, fuelwood has to be bought standing.

Harvesting Solution (Fig. 7.3). To reduce the overhead costs and to secure the availability of forest labor, the procurement of fuel chips is delegated to a local timber company. The organization makes it possible to integrate the fuel chip operations into harvesting of conventional timber. The system is thus based on the use of professional forest workers.

The trees are cut motor-manually in May–August using the felling-piling technique (Sect. 4.1.2.1). The procurement organization supplies the workers with felling frames as auxiliary equipment for the chain saw. Bucking is restricted to trees taller than 8 m, which are cut into two sections. Because some oversized particles are allowed in the chips, delimbing is not necessary. After 4–5 weeks of transpiration drying, the trees are hauled to the landing site and placed into 3-m-high piles with a forwarder equipped with a long-reach crane. A portion of the piles is covered for midwinter operations.

Chipping whole trees is accomplished by a contractor with a 70–90-kW farm tractor-driven landing chipper. Feeding is carried out with a tractor-mounted grapple crane. Reduced wood is blown directly into a tractor trailer with a load capacity of 40 m^3 loose. If other chipping opportunities are not available, the same tractor is used for both chipping and chip transport to the plant. If the con-

Fig. 7.3. A harvesting system to produce fuel chips from small whole trees from thinnings for a local heating plant

tractor serves several customers simultaneously, larger quantities of chips will justify the use of a 40-m^3 truck and exchangeable chip bins for on-road hauling.

When a single tractor is employed for chipping and on-road hauling, the capacity of the farm tractor for combined chipping and transport is 25 m^3 solid per shift. During the peak consumption period, the tractor works in two shifts, or an outside tractor with trailer is hired to help with chip hauling.

For small 4–12 cm trees the average labor requirement of the harvesting system is 0.166 man-day per solid m^3 of chips. Although the trees are not delimbed, felling and piling nevertheless account for well over half of the manpower consumed. The logging organization employs the forwarder contractor during the entire year, but not more than half of the time is used for off-road hauling of fuelwood. Hence, only 50% of the forwarder investment is taken into consideration in the following table. For chipping and chip-hauling equipment investment, the fuel chip harvesting system is naturally charged in its entirety.

	Manpower requirement (man-day/m^3 solid)	Investment (US \$/m^3 solid of annual capacity)
Felling and bunching	0.102	0.3
Forwarding	0.022	14.1
Chipping and haulage of chips	0.042	20.4
Total	0.166	34.8

The hypothetical 3-MW heating plant of the example provides in the forest and transport phase direct work annually for 996 days, corresponding to about 4.5 man-years. As a by-product of the fuel-chip harvesting, silvicultural tending is carried out annually in an area of about 150 ha.

7.1.4 Chipping Small Trees for Pulping

Case. A northern European pulp and paper mill annually consumes 100000 m^3 solid or 230000 m^3 loose of whole-tree chips from small-sized trees. The chips are partly upgraded for additional raw material of sulfate pulp and are partly used for energy production. Moisture content and presence of oversized particles are not of critical importance, but contamination of raw material must be avoided. The procurement of chips is performed by the forestry department of the company in conjunction with conventional pulpwood logging, based primarily on motor-manual methods and use of forwarders. The required amount of small-sized trees is purchased standing mainly from young thinning stands within a radius of 100 km around the mill.

Harvesting Solution (Fig. 7.4). Cutting and forwarding are accomplished by the same methods as in Section 7.1.3, except that transpiration drying is not needed and the felling period is longer. To allow efficient comminution and to reduce

Fig. 7.4. A harvesting system to produce chips from small whole trees from thinnings for large-scale use of a pulp mill

snow problems in midwinter, landing piles of trees are made up to 3.5 – 4.0 m. Eight medium-sized forwarders are required for the entire operation.

Comminution is performed with three truck-mounted chippers, which move flexibly from one work site to another. Chips are hauled by truck-trailer units with a load space of 100 – 120 m^3. Two chip rigs are necessary for each of the three chippers. During peak consumption, comminution is occasionally performed in two or even three shifts. The total investment for the equipment is US $ 43 per m^3 solid of the annual capacity.

This system represents one of the most effective chip harvesting solutions for unmerchantable, small-sized trees under European conditions, but there is a close, "hot" linkage in the schedule between chipping and trucking. The manpower requirement is 0.154 man-day per m^3 solid of chips delivered to the mill. Due to their flexible movement, truck-mounted chippers may accept rather small inventory piles. The minimum size of a work site is a truck load or about 100 m^3 loose of chips.

	Manpower requirement (man-day/m^3 solid)	Investment (US $/$m^3$ solid of annual capacity)
Felling and bunching	0.102	0.3
Forwarding	0.022	14.1
Chipping	0.010	10.0
Trucking	0.020	18.8
Total	0.154	43.2

Chip procurement of the mill provides annually direct work for 15400 days, corresponding to 70 man-years. The silvicultural activity annually covers 2000–2500 ha of otherwise noncommercial, small-tree operations, thus greatly stimulating the silvicultural management of young forests and improving the future availability of high-quality timber.

7.1.5 Chipping Low-Quality Hardwoods for Pulping

Case. Large areas of poorly productive hardwood stands of noncommercial species of inferior quality are not utilized in North America. Yet they could yield significant amounts of biomass per unit area, and for silvicultural reasons they should be clear-cut and regenerated for preferred species.

A hypothetical kraft pulp mill is ready to use 80000 m^3 solid or 185000 m^3 loose of hardwood whole-tree chips annually, partly for pulping and partly as fuel. The material is available within a radius of 100 km.

The diameter range of the trees is wide, from 10–40 cm at breast height. Trees are often crooked and branchy, but technical logging conditions are otherwise favorable. A part of the large butt logs may be used as sawtimber.

Harvesting Solution (Fig. 7.5). Harvesting is based on a fully mechanized whole-tree system employing two independent teams. Each team consists of one or two large feller-bunchers, two grapple skidders, one large whole-tree chipper, three

Fig. 7.5. A harvesting system to produce chips from whole hardwood trees of variable size from clear-cutting areas for large-scale use of a pulp mill

trucks, and nine semitrailer vans. Excluding the supporting machinery, the total capital investment is US $ 2.2–2.6 million, equaling $ 28 per each m^3 solid of annual capacity. The investment cost is lower than for the mechanized systems in Europe, reflecting the larger tree sizes and lower machine prices in North America.

The feller-bunchers sever the trees and pile them directionally into bunches. Bunch size is linked to the capacity of the grapple skidders which haul the trees to the chipper. Skidding is done simultaneously with chipping, which tends to create operational delay time in the system.

The chipper is fed by its own crane. The chips are blown directly into one of the vans. In large landing areas two vans can be loaded simultaneously by changing the blowing direction of the spout according to the utilization properties of the trees. High-quality sawtimber is salvaged by bucking butt logs motor-manually with a chain saw at the landing and stacking them by the heel boom loader of the chipper for further transport with conventional timber trucks.

Compared with the other cases in Section 7.1, tree size is considerably larger and productivity higher. Excluding the role of sawlogs, the labor requirement of chip harvesting is 0.047 man-days/m^3 solid.

	Manpower requirement (man-day/m^3 solid)	Investment (US $/$m^3$ solid of annual capacity)
Felling and bunching	0.010	9.3
Skidding	0.012	4.4
Chipping	0.005	4.2
Trucking	0.020	10.1
Total	0.047	28.0

7.2 Systems for Chipping Slash

Compared with standing, tree-formed raw material, the labor requirement is usually smaller for harvesting on-ground slash, because neither motor-manual nor mechanical cutting is required. The supervision input is also small as purchasing and planning take place in conjunction with the harvesting of conventional timber.

Where slash from unmerchantable stem material is most abundant, transport distances are often very long. This is the case in many remote and mountainous areas in North America and in the northern parts of the Soviet Union. The economic conditions for slash utilization are often more realistic in Europe, although there the on-ground residue is primarily composed of crown mass and its amount per hectare is more modest. This is due to shorter distances and the tendency of the forest industries and municipalities to increase their energy self-sufficiency by using local fuel.

In most cases logging slash is utilized as a substitute for oil. However, chips made from slash may have the disadvantage of heterogeneous composition, uneven particle size, and high moisture content. Moreover, excessive amounts of sand and stones may occur in bare-ground operations. All these factors tend to cause problems when chips are burned in a small heating plant. Consequently, chips of this type are used mainly by large power plants of forest industry companies, because they are less sensitive to the variation in fuel quality.

Fuel chips made from slash from clear-cut areas are significantly cheaper than chips produced from small-sized trees from early thinnings. On the other hand, recovery of on-ground residue does not produce such important silvicultural advantages as selective removal of small-sized trees often does.

Two existing northern European slash recovery systems are introduced below. One represents chipping in terrain (Sect. 7.2.1) and the other at the landing (Sect. 7.2.2). Both are based typically on the use of heavy-duty equipment and full mechanization.

7.2.1 Chipping Slash in Terrain for a Large Power Plant

Case. A large pulp and paper mill burns all of its bark residues, but the capacity of the bark boiler is not fully utilized. To reduce further its dependence on fossil fuels, the mill power plant is ready to purchase chips produced in the woods. Due to the age structure of the forests, thinning of young plantations is not a problem in the procurement area, and the supply of unmerchantable, small-sized trees is limited. Instead, 60000 m^3 solid or 145000 m^3 loose of slash chips is estimated to be readily available annually from company-owned cutover areas within a radius of 100 km around the mill. Chips made of logging slash may thus be substituted for 10000 t fuel oil. When mixed with wet bark, the chips will improve the combustion properties of the main fuel. The operation will also have a silvicultural function, because removal of on-ground residue facilitates preparation of the soil and thus lowers regeneration costs.

Conventional timber is harvested from clear-cut areas with logging machines leaving residue in small piles. Because the minimum diameter of pulpwood is 6 cm, the majority of the slash material is composed of crown mass. The average off-road hauling distance is 250 m.

Harvesting Solution (Fig. 7.6). The harvesting system is based on comminution in the terrain with a forwarder-mounted mobile chipper, capable of working autonomously. No additional machines are needed for the collection of slash or for hauling chips to the landing. To ensure uninterrupted functioning, the unit is equipped with a drum-type chipper with a wide in-feed opening. The volume of the dump bin of the chipper is 17 m^3.

The terrain chipper picks up crown mass and tops with its grapple crane from piles left by a harvester, placing it in the in-feed trough. When the bin is full of chips, the machine carries the load to a landing, lifts and tilts the bin to a height of 4 m, and dumps the load to a 45-m^3 truck container.

Harvesting from stump to truck container is thus basically a one-machine and one-man operation, having an average hourly production rate of 7 m^3 solid of fuel chips. However, to reduce snow problems it may be advisable to haul slash

Fig. 7.6. A harvesting system to produce fuel chips from slash from clear-cutting areas with a mobile terrain chipper for a large power plant

for midwinter operations to landing sites with a conventional or modified forwarder.

Despite the small size and heterogeneous composition of biomass from crowns and tops, the productivity of combined chipping and chip hauling is rather high. This is entirely due to the fact that the material left by a logging machine is ready for comminution as such, and no felling and bunching are required. From an operational point of view the harvesting system is simple. However, despite the possibility of using extra truck containers as buffer storage at the landing site, the close linkage between the operations of the chipper and the chip truck presupposes careful synchronization.

Procurement of 60000 m^3 solid of chips from crown mass and tops with the equipment under consideration requires six forwarder-mounted terrain chippers, four truck-trailer units and 24 exchangeable chip containers. The total capital investment is US $ 2.9 million or $ 49 per each m^3 solid of the annual capacity. The harvesting system is capital-intensive, while the labor requirement is only 0.042 man-day/m^3 solid.

	Manpower requirement (man-day/m^3 solid)	Investment (US $/$m^3$ solid of annual capacity)
Chipping and chip hauling	0.022	31.1
Trucking	0.020	17.8
Total	0.042	48.9

7.2.2 Chipping Slash at Landing for a Large Power Plant

Case. This hypothetical case is fundamentally similar to that in Section 7.2.1, a pulp and paper mill annually burning 100000 m^3 solid or 240000 m^3 loose of chips from logging slash in a large bark-fired boiler. However, the technical logging conditions are somewhat different. Soft sites and uneven ground make the trafficability more difficult compared with the previous example. Moreover, the off-road hauling distances are somewhat longer and the snow cover in winter is thick.

Harvesting Solution (Fig. 7.7). Since the conditions do not favor terrain chipping, comminution is carried out at the landing. Slash is transported to the landing by the same forwarders that first haul the conventional timber from the logging area. A multispiked grapple is installed temporarily on the crane for slash handling. The average output of forwarding is 7 m^3 solid/h.

Comminution is done with a heavy-duty, truck-mounted drum chipper, equipped with an adjustable built-in screen for homogeneous particle size. The output is 20 m^3 solid/h. On-road hauling is carried out with 100-m^3 truck-trailer units. Each unit includes an extra interchangeable trailer to facilitate the linkage between chipping and trucking.

Procurement of 100000 m^3 solid of chips annually requires eight forwarders, three truck-mounted chippers, six truck-trailer units, and six extra trailers. The in-

Fig. 7.7. A harvesting system to produce fuel chips from slash from clear-cutting areas with a portable landing chipper for a large power plant

vestment for the entire equipment is approximately US $ 41 for each m^3 solid of the annual capacity. The manpower requirement is 0.052 man-day/m^3 solid.

	Manpower requirement (man-day/m^3 solid)	Investment (US $/$m^3$ solid of annual capacity)
Forwarding	0.022	13.3
Chipping	0.010	11.7
Trucking	0.020	15.6
Total	0.052	40.6

7.3 Systems for Integrated Recovery of Crown Mass and Conventional Timber

When raw material is procured for the mill in the form of whole-tree chips, segregation of the energy fraction from potential pulpwood raw material is a problem. Upgrading of barky chips has not yet been solved fully satisfactorily (Sect. 8.6). The difficulties in sorting and debarking chips are largely avoided by using integrated, one-pass harvesting systems to deliver the raw material to a processing terminal or mill in the form of whole trees or tree sections. These alternatives make it possible to combine high productivity of forest work and value maximizing of all parts of the forest biomass.

A major drawback is the high cost of long-distance transport due to the low-bulk density of forest material with branches. Integrated harvesting of conventional timber and crown mass thus loses in competitiveness, when the hauling distance increases. It also requires investments in terminals or mill yards, which must be equipped to handle raw material with branches.

Two examples of integrated harvesting are presented below. The tree-section system is gaining in importance in Sweden and Finland as an economical solution for thinning young softwood plantations (Sect. 7.3.1). About 2 million m^3 solid of biomass is expected to be harvested annually with the emerging tree-section system in northern Europe by the end of the 1980s. The other example of an integrated system is the centralized processing of large whole trees (Sect. 7.3.2), actually a modification of the centralized tree-length system. The main goal is the maximum value of the sawtimber, while the recovery of crown material for energy is usually of secondary importance. The system presently has the widest application in the Soviet Union. It is also used in central Europe and North America.

7.3.1 Biomass Recovery Based on the Tree-Section Method

Case. Thinnings are a vital component of the northern European forest management system aimed at the production of high-quality softwood timber. The breast height diameter of the trees removed in the first commercial thinnings is only 6–15 cm, resulting in high costs and, consequently, alarmingly reducing thinning activity. From the forest industry view point it is necessary to stimulate early thinning cuttings by lowering the cost of logging.

Fig. 7.8. A harvesting system for integrated recovery of conventional pulpwood and crown mass from thinnings in the form of tree sections

In order to improve the productivity of timber harvesting in young thinning stands, a sulfate pulp mill modifies its debarking drum to make it suitable to handle unlimbed material. It can accept annually 150000 m^3 solid of 5-m tree sections for simultaneous processing with conventional pulpwood. The tree sections will yield 105000 m^3 bark-free pulpwood and 45000 m^3 mixed fuel from stem bark, crown mass, and broken tops. The average trucking distance is 60 km.

Harvesting Solution (Fig. 7.8). Harvesting is based on partial mechanization. Trees are felled with a chain saw directed in such a way that a forwarder moving on a strip road can grasp either the top or butt of a tree with a long-reach crane. In addition to felling, the cutter tops the crown below 4 cm to improve the bulk density and to reduce nutrient loss from forest soil, and does some light bunching by moving small trees near larger ones so as to improve the productivity of the forwarder. The output of the cutter is 30–90 trees or 3–5 m^3/h, depending on tree size and other logging conditions.

Hauling is performed with a forwarder specially equipped with a slightly enlarged and modified load space, a long-reach crane, and a grapple saw. The forwarder bucks the trees to about 5-m sections in conjunction with loading. The tree sections are hauled to a landing site and placed on 3–4-m-high piles. The forwarder has an average output of 6 m^3 solid/h and an annual capacity of 12000–18000 m^3 solid.

Long-distance transport of tree sections takes place with self-loading truck-trailer units equipped with sidewalls. The load volume of a unit is 110 m^3. To improve the bulk density of the tree sections, the trucks and trailers are equipped with telescopic compaction stakes. The annual capacity of a tree-section truck is 12000–16000 m^3 solid in one shift.

Compared with conventional pulpwood harvesting systems in early thinnings, the productivity of a partially mechanized tree-section system is higher. The manpower requirement is approximately 0.088 man-day/m^3 solid. The procurement of 150000 m^3 tree sections requires about 30 chain-saw workers, 10 forwarders, and 10 truck-trailer units. The capital investment for all the harvesting equipment is US $ 5.0 million corresponding to $ 33/m^3 solid of the annual capacity.

	Manpower requirement (man-day/m^3 solid)	Investment (US $/$m^3$ solid of annual capacity)
Motor-manual felling	0.042	0.3
Forwarding and bucking	0.024	12.2
Trucking	0.022	20.8
Total	0.088	33.3

7.3.2 Biomass Recovery Based on Centralized Processing of Large Trees

Case. A North American integrated forest industry enterprise builds a centralized whole-tree terminal to reduce the costs of harvesting, increase the recovery of biomass, and improve the value recovery of sawtimber trees. The terminal consists of two lines: a main line for sawtimber trees and a smaller side line for pulpwood trees. Residue is comminuted and used as fuel.

The annual capacity of the terminal is 250000 m^3 solid of whole trees, at least 75% of which consists of large trees. All material is harvested from clear-cut areas within a 75-km radius around the terminal.

Harvesting Solution (Fig. 7.9). Fully mechanized harvesting is based on a single machine, a large feller-skidder, which is actually a clam-bunk skidder equipped with a crane-mounted felling head. The machine severs the trees, lifts their butt ends onto the clam bunk, and skids a full load to the landing. It is sometimes possible to do simple sorting according to species or tree size during the felling and unloading phases. Load size and productivity are sharply affected by tree size. In clear-cutting sawtimber trees, the output is 10–16 m^3 solid/h. Under such conditions the annual capacity of a feller-skidder is 25000 m^3 solid in one shift.

Whole trees are trucked to the terminal in semitrailer rigs equipped with sidewalls. Trees are lifted onto the rigs with a separate, mobile crane loader, placing butt ends in front. Trees exceeding the maximum allowable length are topped by a chain saw. Unloading at the terminal takes place with stationary equipment. A whole-tree truck has an average load size of 25 m^3 solid and an annual capacity of 12000–15000 m^3 solid in one shift.

The procurement of 250000 m^3 whole trees requires 10 feller-skidders, 8 truck loaders, and 15 whole-tree trucks. The total capital investment is US $ 5.1 million or $ 20/m^3 solid of the annual capacity.

Fig. 7.9. A harvesting system for integrated recovery of conventional sawtimber and crown mass from clear-cutting areas in the form of whole trees

The total labor requirement of the system is 0.040 man-day/m^3 solid. The high efficiency is partly due to large tree size, since 85% of the total raw material recovery at the terminal meets the dimensions of conventional sawlogs or pulpwood, and only 15% consists of unmerchantable stem portions and crown mass.

	Manpower requirement (man-day/m^3 solid)	Investment (US \$/m^3 solid of annual capacity)
Felling-skidding	0.010	10.4
Trucking	0.030	10.0
Total	0.040	20.4

7.4 Systems for Harvesting Stump and Root Mass

Recovery of stump and root mass for industrial purposes requires heavy-duty machines and full mechanization. Despite high capital investments, the work productivity tends to remain low and the cost of extraction is high compared with stem harvesting.

Stumps and roots of most conifers are excellent raw material for sulfate pulping, but their utilization can be profitable only infrequently. Prerequisites of such conditions are a high stumpage price of conventional pulpwood and a general shortage of fiber raw material, so that marginal raw material from stumps and roots may offer an opportunity for full utilization of the mill capacity. When these conditions were prevailing in northern Europe in the 1970s, research and development of stump and root mass utilization was temporarily very intensive.

Theoretically, an attractive way to reduce the cost of stump extraction is complete-tree logging, i.e., integrated one-pass harvesting of the above- and below-ground portions of a tree. However, application of the principle creates serious technical and organizational problems and presupposes redesigning of the entire harvesting system.

Two examples of proven stump mass harvesting schedules are described below. One represents salvaging of stump-root systems from stony moraine clear-cut sites after stem-only logging in Sweden (Sect. 7.4.1). Another refers to complete-tree harvesting of southern pine from stone-free, sandy sites in the United States (Sect. 7.4.2).

7.4.1 Relogging Stump and Root Mass for Pulping

Case. Due to the shortage of softwood raw material, a sulfate pulp mill in northern Europe cannot operate at full production capacity. No additional conventional pulpwood or saw mill chips are readily available.

The forestry department of the company clear-cuts 4000 ha/a mature coniferous forests within a radius of 120 km around the mill. The yield of conventional roundwood is 200 m^3 solid/ha, leaving behind about 40–45 m^3 solid/ha stump-root systems. The soils are mainly stony moraine. When poor and sensitive soils, steep slopes, and logging sites smaller than 4 ha are excluded, the available area for stump harvesting is reduced to 3000 ha/a. Since only stumps larger than 20 cm in diameter are considered recoverable, and since part of the material from larger stump-root systems is lost in the harvesting operation, the actual yield is 30 m^3/ha. Hence, 90000 m^3 solid can be recovered annually and used as a source of marginal fiber.

Harvesting Solution (Fig. 7.10). A tracked excavator is converted to a stump-harvester by replacing the bucket with an uprooting-splitting head. The stump-harvester is used for combined uprooting, decontamination, presplitting, and collection of the reduced pieces in small piles. Depending on its size and form, a stump-root system is split into two to six parts. The output of a stumpharvester is 3 m^3 solid/h.

Snow and soil freezing prevent extraction of stump and root mass during the winter. A stump-harvester, operating only during six summer months, has an annual capacity of 6000 m^3 solid when working in two shifts.

The split pieces of stump and root mass are carried to the landing site by modified forwarders. The load space is converted to dumping, enlarged to 30 m^3, and equipped with a solid bottom and sides. The grapple is widened and reshaped. The average forwarding output is 7 m^3 solid/h. Stump and root mass is stored in piles for several months to allow weathering and drying to decontaminate it further. Reduction in the soil content lowers the cost of trucking and

Fig. 7.10. A harvesting system for separate recovery of stump and root mass from clear-cutting areas after removal of conventional timber

raises the capacity of the stumpwood processing plant at the mill yard (Sect. 8.6.4). The material is hauled to the mill in self-loading, dumping, 130-m^3 truck-trailer rigs. The grapple of the truck crane is reshaped for stump pieces. The load size is 30 m^3 solid.

Procurement of 90000 m^3/a requires 15 stumpharvesters, 7 stump forwarders, and 7 truck-trailer rigs. The capital investment for the equipment is US $ 6.4 million in total or $ 71/m^3 solid of the annual capacity. In the light of the high capital investment and full mechanization, productivity remains rather low due to the small volume and difficult technical logging properties of the stump-root systems. The manpower requirement is 0.091 man-day/m^3 solid.

	Manpower requirement (man-day/m^3 solid)	Investment (US $/$m^3$ solid of annual capacity)
Uprooting and splitting	0.048	31.4
Forwarding	0.020	15.6
Trucking	0.023	24.2
Total	0.091	71.2

7.4.2 Harvesting Stump and Root Mass with Conventional Timber

Case. A sulfate pulp mill in the southern United States obtains its raw material mainly by clear-cutting mature pine stands. The mill needs 50000 m^3 solid of additional pine raw material without enlarging the traditional procurement area. A potential reserve of fiber is formed by the below-ground portions of the pine trees in the clear-cut areas.

The stump-root system of southern pine is characterized by a deep, large taproot and rather small laterals. The total amount of stump and root mass is 50 m^3 solid/ha, most of it in the central stump-root systems. The soil is sandy or loamy with very few stones.

Harvesting Solution (Fig. 7.11). Recovery of below-ground biomass is integrated with stem harvesting. The felling head of a wheeled, loader-mounted feller-buncher is replaced by a puller-buncher head. Its tubular shear is forced into the ground to sever the lateral roots around the taproot. Removal of laterals makes it possible to lift trees up from the ground and bunch them for hauling. For pines with a breast height diameter of 15–30 cm, the output is 10–16 m^3 solid/h.

Hauling to the landing is carried out by conventional grapple skidders, which also perform rough delimbing with a gate delimber at the landing. The central stump-root system increases the mass per tree and, consequently, the load size in skidding. The output is 9 m^3 solid/h.

The stems are transported to the mill with taproots attached in conventional tree-length trailers. At the mill yard the trucks pass through a washing station, where the driver decontaminates the butt ends of the stems with a fire hose. The stems are then bucked, drum debarked, and chipped for pulping.

In addition to the conventional roundwood, the system produces 35 m^3 solid/ha or 15% below-ground raw material. Approximately 1500 ha of mature

Fig. 7.11. A harvesting system for integrated recovery of the stem and central stump-root system from clear-cutting areas

pine forests, yielding a total of 330000 m^3 stem material, must be harvested to gain 50000 m^3 solid of raw material from stump-root systems. The planned operation requires 15 puller-bunchers, 22 grapple skidders, 28 timber trucks, and 12 loaders for the trucks. The total capital investment for the logging system is US \$ 7.0 million or \$ 17/m^3 solid of the annual capacity working in one shift. The manpower requirement is 0.041 man-day/m^3 solid.

	Manpower requirement (man-day/m^3 solid)	Investment (US \$/m^3 solid of annual capacity)
Pulling-bunching	0.009	4.7
Skidding	0.012	4.5
Trucking	0.020	8.7
Total	0.041	17.9

Chapter 8

Utilization of Residual Forest Biomass

Growth in income and population results in an increasing demand for forest products. The FAO estimates the annual growth in consumption as 1.2% for sawn wood, 2.7% for wood-based panels, and 3.7% for paper and paperboard. The gross wood-in-the-rough required to meet the growth projections of the forest industries is expected to rise to 2388 million m^3 in 2000, without eliminating double counting of industrial residuals used in further processing. The share of mill residues is presently 11.4% of the total consumption of industrial raw material and expected to peak around 1990 at about 12.7%. By that time the industrialized regions should have fully tapped their economically accessible mill residues. The net demand on the forest created by forest-based industries is projected to grow from 1456 million m^3 in 1984 to 2085 million m^3 in 2000. The expected growth is thus 43% (FAO 1982, 1986a).

This consumption for processing purposes by the world forest industries is actually less than one-half of the total world consumption of wood, as in the developing world fuelwood constitutes the bulk of wood consumption. Unfortunately, the statistical basis for fuelwood is weak, particularly for those countries where its use is most significant. The projected increase in world wood consumption in the period 1984–2000 is separated into industrial wood and fuelwood below (FAO 1982, 1986a):

	Consumption (million m^3/a)		
	Fuelwood	Industrial wood	Total roundwood
1984	1595	1456	3051
2000	1820	2085	3910

From the beginning of the present century up to World War II there were few significant changes in the use pattern of forest products. More fuelwood was consumed than any other product, whereas paper, paperboard, and wood-based panels accounted for rather small volumes. During the postwar economic expansion in developed countries, consumption of all the products of industrial wood grew steadily and strongly while that of fuelwood declined even more rapidly. Fuelwood was widely supplanted by more convenient and cheaper fuels, and in developed countries gradually became a marginal energy source. To a certain extent the rise in the consumption of pulpwood was made possible by the decline in fuelwood consumption, as this liberated supplies of small-sized wood for other purposes. However, the decline in fuelwood consumption slowed down after the first energy crisis in 1973–1974 and was reversed in 1978 (ECE/FAO 1986). In the developing world, fuel for cooking continues to be the main use of wood.

Table 8.1. The annual rate of establishment of tropical forest plantations in relation to deforestation of closed forests (UNIDO 1983)

Region	1000 ha/a				
	Forest plantations			Deforestation of closed forests	Replacement ratio
	Industrial	Fuelwood	Total		
Tropical America	283	252	535	4400	1:8
Tropical Africa	64	62	126	1300	1:10
Tropical Asia	233	205	438	1800	1:4
Total	580	519	1099	7500	1:7

On a world basis, wood supplies are judged to be sufficient to meet the prospective demand for industrial raw material in the year 2000, but regionally there are, and will continue to be, differences between production and consumption. Europe, Japan, the Middle East, and North Africa are regions consuming more wood products than they produce. Given certain intra- and interregional transfers, global wood resources may at present look sufficient from the standpoint of the forest industry.

However, when the increasing demand for fuelwood in developing countries is taken into consideration, significant imbalances appear. In some areas, a critical stage has already been reached as regards fuelwood. Deforestation and degradation of tropical forests continue, with effects which are often irreversible since the forest areas remaining are subject to yet more pressure from the population (Marchand 1983). The tropical forest is expected to be reduced by some 12% in the period 1975–2000, largely due to a spontaneously shifting culture and uncontrolled conversion of forest land to extensive grazing. Replacement of deforested tropical areas by artificial reforestation covers only one-seventh of the area simultaneously lost from forest production (Table 8.1).

To secure even the present level of fuelwood consumption in the developing world, 0.5–2 m^3 per capita per annum depending on the circumstances, 50 million ha fuelwood plantations will be required by the year 2000 to keep up with population growth. This foresees an immediate fivefold increase in all forest tree planting programs. Since the work involved and the funds required are huge, research and development in more thorough recovery of forest residue and in more effective combustion and carbonization technology are imperative as well.

In some developing countries increased demands for timber could be met by making greater use of less accepted tree species and tree components and by opening up areas as yet unutilized. In many developed countries inventory data show great potential for expanding the raw material base by using a larger part of the forest biomass. Local shortages, silvicultural goals, rationalization of forest work, and efforts toward increased energy self-sufficiency may result in the generalization of whole-tree logging systems.

Despite increasing use of wood for industrial purposes, the greatest potential use of residual forest biomass throughout the world is expected to be *energy*, primarily through direct combustion (Sect. 8.1) but also through carbonization,

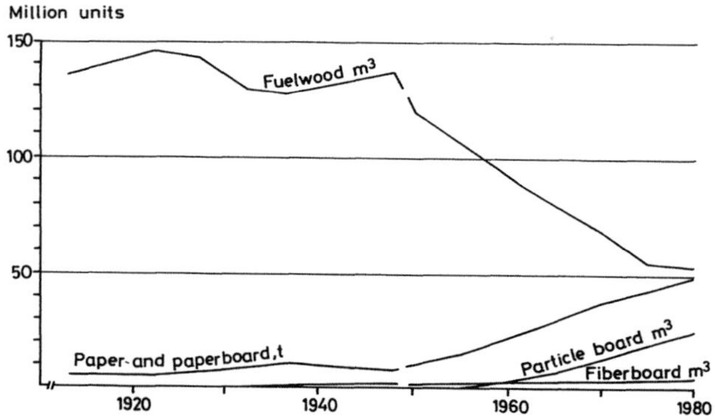

Fig. 8.1. Development of the consumption of fuelwood, paper and paperboard, particleboard, and fiberboard in Europe since World War I (redrawn from ECE/FAO 1986)

gasification, or liquefaction (Sect. 8.4). In the developing world, poor people seldom have access to fuels other than wood or charcoal, which are frequently collected and prepared by the users themselves and burned mainly in simple, inefficient stoves for cooking food. Population growth leads to an increasing demand for fuelwood, but the increase in consumption is constrained by the shortage of the resource itself.

In the developed world, increasing use of renewable biomass for energy could reduce dependence on finite reserves of oil and other fossil fuels. However, the possibilities of forest energy are relatively limited in developed countries. In many densely populated industrialized countries with small forest reserves, only 1–2% of the total consumption of primary energy could be realized through residue from forests and forest industries. On the other hand, there are also countries in Europe and North America in which over 10% of total energy could be produced by forest biomass-derived fuels. An extreme example is Finland where in the mid-1980s about 15% of all primary energy was produced by wood-derived fuels.

Consumption of *paper and paperboard* expanded strongly in the postwar period (Fig. 8.1), and this development is expected to continue. According to the FAO (1982), world consumption of paper will grow from 170 million t in 1979 to 357–411 t in the year 2000.

Paper products may be classified according to end use into three categories: graphic papers for printing and writing, packaging papers, and household and sanitary papers. Each end-use category contains a large number of paper grades. Most of them use a smaller or larger percentage of kraft pulp, which can be produced from softwood and hardwood chips containing residual tree parts. Among the various branches of conventional forest industries, pulp, paper, and paperboard mills are potentially the most important users of forest residue (Sect. 8.2).

Particleboard (Sect. 8.3) is the newest of the mechanically processed wood products, having appeared on the world scene since World War II. The 1970s particularly showed a high rate of expansion of the industry with the Federal Republic of Germany, USA, and the Soviet Union dominating production. Particleboard is likely to be in greater demand as a plywood substitute as veneer logs become scarce. It is also expected that it will find increasing use as core material for veneer-faced panels. The latest development is structural flakeboard, a competitor of structural softwood plywood. Presently, the main source of raw material for particleboard is residue from the mechanical forest industries, but the use of forest residue is also technically possible.

Fiberboard (Sect 8.3) is made of fibers of wood or other lignocellulosic materials with the primary bond derived from the felting of fibers and their inherent adhesive properties. The fiberboard industry is characterized by its ability to use low-quality raw materials with a rather high bark content. The primary source of raw material is residue from sawmills and veneer mills, but chips from forest residue can also be used. However, since the final products are rather cheap, the high cost of harvesting makes the use of forest residue less competitive.

Leading fiberboard producers are the United States and the Soviet Union. The world production of fiberboard grew only 16% during the 1970s, while the production of particleboard more than doubled (UNIDO 1983):

	Consumption (million m^3)	
	Particleboard	Fiberboard
1970	19.2	14.2
1980	40.3	16.2

In the long term, forest biomass has a great potential as a feedstock for *chemical processing* but, the chemical pulp industry excepted, this potential has been realized in only a few isolated instances (Sect. 8.4). Other feedstocks, notably oil and natural gas, have proven more abundant, inexpensive, and easier to handle and process. Oil and gas are extracted from concentrated sources and can be transported relatively cheaply, enabling the construction of very large factories. Units of a similar size for processing residual forest biomass would require enormous raw material catchment areas, entailing high transport costs.

Not even the steep rise in the prices of oil and natural gas in the 1970s brought about a shift to forest biomass or any other new feedstock. One probable reason is the immense investment already made in the petrochemical industry, not only in plants, but also in research and development, and a severe worldwide overcapacity for many bulk chemicals. Under these circumstances, any broad shift toward residual forest biomass as a bulk chemical feedstock is unlikely in the foreseeable future. However, forest biomass has the unique advantage of renewability and, for many countries, it is the only available indigenous raw material for large-scale production of liquid fuels or other chemicals. Furthermore, a few chemical products such as furfural, whose molecular structure makes them easier to manufacture from wood than from oil, or biochemicals extracted from tree foliage, are already being produced from forest biomass. Where sufficient raw material is available, at a suitable price, many more of these chemical products could be manufactured from forest biomass, especially if raw material procurement and processing could be integrated into a larger operation (ECE/FAO 1986).

Forest residue is also a potential source of *fodder* (Sect. 8.5), although under present conditions it has only marginal local importance as animal food. A major change is unlikely as long as there are large agricultural surpluses in many parts of the developed world. However, in the Soviet Union particularly, massive research is being carried out in order to utilize conifer needles as a fodder supplement and a source of vitamins for cattle, chickens, and other domestic animals. The goal of this research is whole-tree harvesting, using other biomass components for conventional forest products and technical foliage for chemical extraction or production of vitamin-rich fodder.

8.1 Use of Chips as a Solid Fuel

The economic viability of various biomass resources is heavily dependent on fossil fuel prices and availability. Before the world-wide energy crisis, wood received very little attention as a source of energy material. The use of cheap fossil fuels and nuclear power was expected to increase almost without limitation, while wood burning was considered to be a squandering of national resources. It is true that the use of forest industry residues to produce energy was expanded and developed, but the techniques of harvesting and processing unmerchantable forest biomass received little consideration in research and development programs.

The steep increase in the cost of oil and the uncertainty of its availability since 1973 suddenly increased the importance of renewable energy and created a worldwide interest in wood as an energy source. A distinctive demonstration of the change in attitudes was the UN conference on New and Renewable Sources of Energy in Nairobi in 1981.

Predictions of oil prices reaching US $ 35–50 per barrel in the 1980s stimulated large research programs during the last half of the 1970s. However, technical advances have not always come up to expectation and, far from rising as expected, oil prices fell radically from over US $ 30 per barrel at the end of 1985 to below $ 10 in mid-1986 and still remained at the low level of $ 16–18 in early 1988. This drop caused some users to cease purchasing forest biomass for energy purposes and resulted in less government concern for research on renewable energy. Nevertheless, it is this very uncertainty over oil prices which should be a spur to the preparation of some oil and coal substitution. A combination of raw material availability, energy demand, and political considerations matched against international pressures and constraints eventually determines the overall national policy toward biomass energy with economics and strategic issues as driving forces (Hall and Coombs 1987).

From an energy point of view, differences between countries are enormous regarding the supply of indigenous energy, population density, standard of living, structure of industry, traffic conditions, climate, sustained yield of forests, and intensity of forestry and wood utilization. *In the developing world*, fuelwood is the only energy source for a large and growing number of poor people. In 1980 about 2000 million people depended on fuelwood or charcoal for their daily energy requirements. These people were faced with an ever-growing shortage and were obliged to overexploit existing resources. Fuelwood shortage affected 1150 million of them, corresponding to an annual deficit of around 400 million m^3 mainly in Africa and Asia. Some 100 million people were unable to satisfy even basic energy needs. If current trends continue, by the year 2000 a total of about 3000 million people will live in deficit areas, unable to satisfy their fuelwood requirements. Yet, because of limited resources, it is estimated that total consumption of fuelwood in the developing world in the year 2000 will not be more than 1700 million m^3 (de Mountelembert and Clément 1983).

In developing countries, urban populations with medium and high incomes depend mostly on commercial energy supplies, especially in the form of oil and electricity and, to a varying extent, on commercialized forms of traditional fuels such as charcoal. The same is true for industrial activities, except for small-scale industries, where large amounts of traditional fuels are used. People belonging to lower income groups and rural residents, in general, cannot afford using commercial fuels. They use a mixture of traditional fuels, such as wood, charcoal, agricultural and animal wastes, and some kerosene (Renewable ... 1985).

Although fuelwood and charcoal in developing countries are commonly burned at a very low efficiency in open fireplaces or simple stoves, the problems are primarily socioeconomic and ecological rather than technical. In the deficit areas, all available woody material around population centers is collected for fuel, and practically no logging residue is left on site. Since this book is technically oriented and meant to review the short- and long-term utilization potential of unmerchantable forest biomass, the fuelwood problems of the developing world are omitted.

The situation is very different *in the developed world*, where forest residue is only one among many available energy alternatives. While fuelwood in developing countries is used mainly for cooking food, in developed countries the use is primarily for production of heat and steam and, to a much lesser extent, electricity. In many developed countries a large proportion of wood-derived fuels is used by the forest industry. Due to the high cost of harvesting and transport of forest biomass and low prices of competing fossil fuels, a significant increase in residue consumption for energy is unlikely in the short term but very possible in the long term.

Although the utilization of wood for fuel is typically a characteristic of developing countries, a surprisingly large proportion of the annual drain from the forests ends up in the production of energy in many developed countries as well, if the secondary use of processing residues from the forest industry is taken into account. Table 8.2 shows an example of the development of the consumption of wood-derived fuels in the United States after the energy crisis in 1973. Total consumption in 1983 corresponded to 155.5 million t of wood equivalent at 15% moisture content. Two-thirds of this amount was burned in solid form and almost one-third in the form of black liquor from the chemical pulp industry. Wood-derived fuels may thus be of great significance even in industrialized countries.

Heat is the lowest common denominator among fuels, but measuring the heat content alone misses the concept of energy quality, which in turn affects energy efficiency. For example, for most purposes petroleum is a more efficient fuel than biomass, as measured by both thermal and economic criteria, and electricity is

Table 8.2. Development of consumption of wood-derived fuels in the United States after the worldwide energy crisis in 1973 in million tonnes of wood equivalent at 15% moisture content (Koning and Skoog 1986)

Consumer group	Million tonnes of wood equivalent per annum			
	1972	1976	1980	1983
Pulp and paper industry				
As hog fuel	2.7	4.8	7.4	13.2
As bark	5.4	5.4	6.2	7.1
As black liquor	42.5	45.4	49.2	49.8
Mechanical forest industry	21.5	24.1	27.1	25.0
Other industry	2.3	2.4	2.8	2.9
Residential	24.2	30.7	52.1	55.9
Others	0.7	0.8	1.6	1.6
All consumers	99.3	113.6	146.4	155.5

an even higher-quality fuel. The higher quality of electricity is particularly useful in manufacturing where its conversion to mechanical work and heat at the point of application can be controlled exactly with respect to time.

In the 1970s a search for cost-effective and efficient systems for the utilization of residual biomass for energy was given a high priority in many countries with a forest industry and an advanced technology based on *chip combustion* was developed. A *whole-tree combustion* concept, headed for generation of electricity in large power plants, burning whole trees or unlimbed tree sections, was introduced in the United States in the mid-1980s (Ostlie 1988 a, b), but under prevailing conditions this new operation model does not seem to offer immediate or economic advantages over existing chip technology.

Chip systems continuously have a heavy emphasis in most national research and development programs as well as in many of the multilateral research projects based on the IEA Forestry Energy Agreement (Gilliusson 1984). Accordingly, the present chapter is concerned only with direct combustion of comminuted biomass. Although traditional fuelwood technology still prevails in small-scale operations, it is not reviewed in this conjunction, because it is closely tied to local conditions and is generally well known and documented.

In older European literature *the unit of energy* is the calory (cal) which is the quantity of heat energy needed to warm 1 g of water 1 °C. In the United States, energy is expressed in British thermal units (Btu). In the SI system the basic unit of energy is the joule (J) and its 1000-fold multiples kilojoule (kJ), megajoule (MJ), gigajoule (GJ), and terajoule (TJ). Another unit frequently used in this book is the kilowatt-hour (kWh) which is equivalent to 3.60 MJ. All these units are interconvertible. Conversion factors are given at the end of the book.

8.1.1. Techniques for Direct Combustion of Chips

In order to release thermal energy, chips are burned in a combustion chamber or furnace which is a part of a boiler system. In single-family houses the chip furnace is fed either manually or by a small-diameter screw conveyor through a fuel gate which usually requires fine chips with even particle size and a moisture content below 30%. In larger installations feeding is handled by means of a screw, piston, scraper, or belt conveyor.

The boiler has two functional parts: the furnace and the convection section. In the furnace, heat is transferred from the flame through radiation. The walls of the furnace generally consist of brickwork and various types of water-cooled steel surfaces or a water jacket. In a furnace which is partly cooled by a water jacket, and where the rest of the containment surfaces consist of brickwork, heat radiation is also obtained from the hot brickwork surfaces. The hot flue gases from the furnace are passed on to the convection section of the boiler, sweeping past steel pipes and steel surfaces on the opposite sides of which water circulates. In this part of the boiler, the heat from the flue gases is transferred to the water by means of convection. The radiation and convection surfaces are collectively referred to as the heating surface of the boiler (Dickens et al. 1983).

Dry chips can be burned directly in the boiler furnace, but moist chips are usually burned in a precombustion furnace. Since the precombustion furnace is not water-cooled, the combustion temperature rises high enough to dry moist chips and gasify them in the presence of limited primary air. The volatile gases are led to the boiler furnace for complete combustion with excess secondary air. The precombustion furnace can be a separate unit or an integral part of the boiler. An old oil-fired boiler can be retrofitted for chip combustion by installing a precombustion furnace in front of the boiler and connecting it to the existing equipment with a gas duct. The system accepts chips with a rather high moisture content.

Basically, there are three categories of furnace design in use in larger installations: grate or fixed-bed burners, fluidized-bed burners, and suspension burners. Most installations use a *grate burner*, which can be fed from the top countercurrent to the airflow, or from the bottom concurrent with the airflow. Combustion may be accomplished in one or successively in two chambers. In Dutch ovens fuel is fed through a chute in the oven roof and forms a conical pile on the grate in the oven or precombustion furnace. In inclined-grate furnaces fuel enters the burner at the top of the grate and is dried as it slides down to the lower horizontal section where combustion occurs. In spreader stokers, fuel is fed either pneumatically or mechanically into an even, thin bed across the grates. When the fuel is added above the grate, smaller particles and volatiles burn in suspension, while the large pieces of fuel fall into the grate and burn in the fuel bed. The grate may move to assist transfer of the bed material from the feed end to the ash removal end (Karchesy and Koch 1979).

In the *fluidized-bed burner*, biomass is combusted with the aid of an inert, granulated, solid material such as sand or limestone. When a stream of gas is blown at a low rate through distributors located at the base of the bed through granulated material, the bed behaves as a porous medium and allows the gas to pass through. With increasing gas flow the particles of sand and biomass start to float and the bed expands from its original static depth. With an adequately high rate of gas flow, the bed behaves as a liquid, boiling violently. The biomass undergoes dehydration, carbonization, gasification, and combustion in the bed, and overfire secondary air is not generally required. The large particle surface area promotes heat transfer between gas and solids. The behavior of the bed depends on the rate of gas flow and, accordingly, three main types of beds are distinguished: bubbling bed, expanded or partial circulating bed, and full circulating bed.

The high thermal inertia of the fluidized bed levels the fluctuations in heat input and permits the use of heterogeneous fuels with varying heating values. Fuel moisture content can be as high as 60–65%. The method can be used efficiently to reduce harmful flue-gas emissions of sulfur dioxide and oxides of nitrogen, the former by means of a suitable absorbing matter, such as limestone in the bed, and the latter by means of the low, 750–950 °C temperature of the bed and furnace. Fluidized-bed burners are suitable for efficient multifuel firing of low-calorific-value solid materials such as chips, bark, peat, pulp sludge, and municipal waste for steam or hot water generation and even for gasification. However, the technique requires fuels of small particle size. Jahkola (1983) presented the following conclusions based on results of research and development activities and commercial applications regarding fluidized-bed burners:

1. Combustion efficiency is high at low air excess;
2. Multifuel firing is possible, either one by one or combined;
3. Turn-down ratio is good;
4. No supporting fuel is required;
5. No predrying is required;
6. The absence of movable parts improves the reliability of the burner;
7. The method is environmentally safe, since it results in low NO_x formation and has the capability for sulfur removal.

In *suspension burners*, biomass of very fine particle size, such as sander dust or sawdust, is burned in suspension, usually in the presence of an auxiliary fuel. Fuel feeding is accomplished pneumatically. Combustion takes place in a very short time and in a compact chamber. Due to the short contact time, the fuel must be rather dry. Since the requirements regarding the particle size and moisture content of the fuel are critical, the method is not suitable for residual forest biomass.

Farms and rural single-family houses are among the most natural users of fuelwood. Traditionally, wood is burned in the form of short, split bolts, but the technology is labor-intensive and very difficult to rationalize. Comminution of forest residue, based on light farm tractor-mounted equipment, offers a feasible solution for effective biomass recovery, transport, and automatic fuel feeding even for small-scale operations.

Use of chips for heating farm houses and single-family homes is most common in northern Europe, but even there it is modest compared to the use of traditional fuelwood. For example, while wood was the main fuel for space heating for 120000 farms in Finland in the early 1980s, the number of chip-fired systems was estimated at 5000. Due to the relatively high cost of automatic equipment and the strict requirements of particle-size distribution in stoker feeding, manual handling of chips and chunks is still common in small-scale systems. Nevertheless, development of handling equipment and improved quality of chips has resulted in increased use of stoker boilers, particularly in Sweden.

To decrease handling and storage problems of comminuted forest biomass, the particle size may be uniformed and bulk density improved by *densifying* the material. Densified products include disk- or log-formed briquettes, generally 50–100 mm in diameter, and smaller 10–25 mm diameter pellets. Both are pressed from fine particles, such as sawdust, shavings, or milled peat, usually without a binder. Use of homogeneous pellets greatly improves the reliability of automatic stoker feeders, particularly in small units of single-family houses. However, since densification requires dried, fine-ground starting materials, production costs of briquettes and pellets from residual forest biomass tend to become prohibitively high for use for normal space-heating purposes.

Owing to technical development, changed price relations, the search for new job opportunities, and public investment subsidies, the number of small- and medium-sized *chip-fired heating plants* increased rapidly in the early 1980s. Since not all planners were sufficiently familiar with the problems inherent in the procurement and utilization of fuel chips, the rapid expansion did not always proceed without difficulties. Although many plants functioned fully satisfactorily, inadequately tested plant designs were also accepted and had to be corrected later. The experience gained during the first half of the 1980s helped to establish better technical and organizational conditions for reliability, economy, and ease of utilization. Development was particularly distinctive in northern Europe. For example, at the end of 1979 Finland had only 14 forest-chip fired 0.5–10.0 MW heating plants, but during the following 3-year period, 1980–1982, as many as 87 new plants of this size class with a total capacity of 180 MW were started up. Moreover, apart from the forest industry power plants, eight plants of over 10 MW with fuel selection which included forest chips were also put into use concurrently.

The category of 0.5–20 MW chip-fired heating plants in various countries primarily serves municipal district heating, garrisons, hospitals, educational institutions, greenhouses, dairies, and small industrial enterprises. The profitability of these plants is often burdened by a low capacity utilization rate. For example, in Finland in 1982 utilization of forest chips, calculated per effective megawatt of the installed plant capacity, was 1310 m^3 solid, corresponding to an annual peak utilization time of only 2850 h. Underutilization was due to temporary overcapacity of new district heating plants, problems customarily inherent in the start-up phase, and especially due to a low heat requirement during the summer season.

In chip-fired plants producing process steam for industrial purposes, dairies for example, the degree of utilization is generally fairly even throughout the year, which is an appreciable advantage to both the plant itself and the chip-procurement organization. In contrast, seasonal variation is very great at municipal district heating plants. Their heat production and chip delivery peaks tend to fall in midwinter. For rational dimensioning of storage facilities, handling equipment, chip boiler, and procurement organization, and for smoothing the work load during the critical season, it is generally beneficial to design the chip boiler for a certain basic load only and meet the temporary peak need by means of an oil-fired, auxiliary stand-by boiler. During the summer period of low heat requirement, the additional oil-fired boiler can be used as the sole producer of heat since a small load in the chip boiler is uneconomical due to poor efficiency and the relatively high demand of labor per unit of energy produced.

The *forest industries* are both a consumer and a producer of energy. Although the forest industries have sometimes expressed concern over the impact of the increasing demand for fuelwood on the price development of industrial wood, this effect can actually be beneficial as it strengthens the long-term availability of high-quality wood through improved forestry practices. Increased use of forest and process residues, together with improvements in the energy efficiency of the operations, may make some forest industry plants self-sufficient in their energy consumption. However, the use of solid biomass residues requires additional labor due to handling problems and wear and tear on the equipment.

Compared to oil- and gas-fired power plants, chip-fired plants must sustain considerably higher capital investments per unit capacity because of more complex materials handling, a larger storage yard, and a more expensive boiler. These high capital costs must be compensated through lower fuel costs. Since the forest industries have the organization, labor, equipment, and expertise needed for fuel-chip harvesting, and since they have ready access to forest residues, they are in a position to procure large amounts of fuel chips using the most economical methods. According to a survey of fossil fuel using industries in the United States, the primary barriers to using wood energy are the conversion costs associated with changing to wood fuel and the use of wood fuel handling system (Ostermeier et al. 1988).

In the forest industries, the heat produced by biomass combustion is used to generate steam. This steam may be used for drying pulp and lumber, to turn turbines that provide mechanical energy, or to turn electricity-producing turbines. *Electricity generation* requires high-pressure, superheated steam. The cost of the boiler increases with the steam-pressure requirement.

Efficiency here is the ratio of the electrical energy produced to the thermal energy of the fuel input. The efficiency for converting residual biomass into electricity is low compared to conversion of biomass into thermal energy, for small plants usually less than 20% and for large power stations 30–35%. However, in remote areas self-sufficiency provided by small local biomass-fired power plants may be of more importance than economic efficiency. At 25% efficiency, approximately 0.7 t of dry biomass is needed to generate 1 MWh of electricity.

If steam is used only for electricity generation, the used steam can be condensed and recycled to the boiler as feed water. In a cogeneration unit, electrical power and a useful form of thermal energy such as steam are produced simultaneously, steam usually being the main product. Superheated, high-pressure steam is produced to run a back-pressure turbine and exhausted steam from the turbine is then used for process applications or heating. The overall thermal efficiency of the system can be increased significantly, e.g., to 75%, but this is done at the expense of the reduced output of electricity. Hence, integration of the cogeneration plant into the utility distribution system offers significant advantages that may not be available to an isolated industrial cogeneration plant.

The term cogeneration is used normally in the context of coincident production of electrical and thermal energy in one facility. This *electrical/thermal cogeneration* utilizes high-pressure steam through a turbine generator to produce electrical power and low-pressure exhaust steam for mill or district-heating uses. However, cogeneration can apply to the coincident production of mechanical and thermal energy as well. The *mechanical/thermal cogeneration* also utilizes high-pressure steam to drive a turbine and to produce low-pressure steam for mill or heating purposes, but instead of using the turbine as a power source for a generator, the turbine drives some mechanical piece of equipment directly.

For mechanical/thermal cogeneration, the turbine output is used directly as input to the driven equipment, whereas for electrical/thermal cogeneration the turbine output is first utilized to drive a generator to produce electrical power, which must be converted back to mechanical power in an electric motor to drive the desired equipment. The losses involved in converting mechanical energy to electrical energy and back again to mechanical energy can average from 5 to 10% or more. In addition to energy efficiency, mechanical/thermal energy also offers cost advantages in capital investments and maintenance. Consequently, this system can be an economically feasible solution for biomass utilization if the application meets certain criteria regarding the power requirement, sufficient demand for turbine exhaust steam, and need to drive the equipment with variable speed. The mechanical/thermal cogeneration concept has received only limited application, mainly in the power-plant portion of pulpmills. Other possible applications in the forest industry could be mechanical turbine drive of fans, refiners, chippers, or hogs (Beebe 1980).

Direct combustion of biomass has *environmental consequences*, which are partly negative and partly positive. Gaseous combustion emissions include particulates and organic compounds which result from incomplete burning, nitrogen oxides, metals, and sulfur dioxide. These emissions contain mutagenic and carcinogenic compounds and are therefore an issue of concern in many countries, particularly in population centers. The impact of the emissions vary from country to country as well as regionally within national borders. For instance, mountainous regions where air stagnation may occur are more susceptible to emissions than flat, open regions. The larger the combustion plant, the better the control of emissions, small residential equipment tending to pollute relatively more than large installations (IEA Forestry ... 1987).

The low sulfur content of wood (Sect. 3.3.3) results in low sulfur emissions from biomass combustion into the atmosphere. Forest biomass produces more ash than gas and oil but less than coal and peat. Disposal of this ash at landfill areas may cause leaching problems but, in fact, ash from forest biomass is a valuable soil-improvement agent and should be recycled in agriculture and forestry (Sect. 9.4.2).

8.1.2 Effect of Chip Properties on Direct Combustion

High plant investment costs limit the use of forest chips as fuel. In Finland, an average of 70% of investment costs is for equipment, 20% for buildings, 3% for planning, and 7% for other items (Toropainen 1982). The greatest part of investment costs relates to the chip handling and storing systems. The more advanced the automation, the higher are the equipment costs. On the other hand, without automation, a larger work force is needed, which is a heavy cost factor, especially for small plants.

The reliability of a heating plant depends essentially on the chip-handling equipment. The operations of the following devices are especially critical:

1. Chip receiving equipment;
2. Chip silos and hoppers;
3. Chip conveyors;
4. Furnace feeders with their automatic control.

The significance of chip properties for operation economy is often ignored in plant design. While one plant may be well fitted for a variety of chip types, the efficiency and uninterrupted operation of another plant may allow no variation in chip properties, so that a wrong or reduced chip quality creates a risk for disturbances in the feed and combustion cycle. Different adjustment of plants to contrasting chip assortments demonstrates the need for accomplishing the fuel specifications before building the plant. At least the following chip characteristics should be determined in the planning phase of the heating plant: nominal size and size distribution of particles, occurrence of oversized particles and fines, moisture content, heating value, and bulk density (Gärdenäs 1986).

First of all, the sensitivity of the handling equipment to disturbances is frequently the result of unnecessarily complex systems or incorrect dimensions. Secondly, disturbances are generally a result of adverse variation of chip properties. As long as the fuel is faultless the equipment functions well, but it is hardly possible to keep chips made from forest residues reliably within a narrow quality range. Consequently, it is imperative in equipment planning to guard against irregularities that do not come up to the assessed requirements.

The aim must naturally be homogeneous and clean chips, but costs rise with demands. For instance, the costs of stem chips are nearly 10% higher than those of whole-tree chips at the heating plant. The following general observations characterize the effect of fuel quality on the operation of 0.5 – 10 MW chip-fired heating plants in Finland (Kirjavainen and Lahdenperä 1983; Hakkila 1984; Kalaja 1984).

Average particle size may be of significance to control the air required for combustion. In most chippers it is possible to regulate the nominal particle size with the position of the knives or anvils, the number of the knives, feeding speed, or screen construction (Sect. 5.2.1). Variation of average size is, however, no major disturbing factor in practice.

Oversized particles and twigs may cause clogging in the gratings of the receiving devices, bridging in the silos, breaks in the unloaders and conveyors, and disturbances of the automatic control devices. Stoker feeders in particular are sensitive to twigs and large particles. The smaller the stoker diameter, the more easily the system becomes clogged. The diameter of the stokers of 0.5 – 10 MW heating plants varies from 10 to 60 cm. Experience of dog or scraper conveyors is, for the most part, positive.

Problems caused by oversized particles and twigs can be reduced by making the conveyors sufficiently wide, avoiding long moving distances, and through correct coating and shaping of the silos. It is possible at larger plants to place a simple disk screen at the head of the handling system.

Fine material, even when abundant, constitutes a minor problem compared with oversized particles. However, it may slide below the grate or cause clogging so that the air required for combustion is distributed unevenly in the chip bed. More unusual disturbances are clogging of the stoker feeder, disturbance of the photocells that function as sensory organs, controlling automatic feeding, and problems caused by particulate emissions.

Foliage contains large amounts of respiring live cells and is rich in nitrogen, thus accelerating the deterioration of the chips. Consequently, together with fines it increases the danger of energy loss, microfungal spores, and spontaneous ignition if chips are stored for longer periods than 2–3 weeks (Sect. 8.1.4.2). When profuse, as in chips produced from fresh slash in the winter, needle clusters may also increase freezing and bridging and may clog conveyors and the grate.

An *excessively high moisture content* lowers the effective heating value of chips and, above all, the efficiency of combustion, thus diminishing the available energy (Sect. 8.1.3). When the moisture content varies irregularly, control of combustion becomes difficult, resulting in further loss of energy. A high moisture content is harmful also to the flow of chips and tends to lead to bridging, especially if moist chips freeze. Freezing of chips may prevent the start up of the automatic silo discharge. Decaying fungi thrive in moist chips, the optimum moisture content being about 50%.

Exceptionally low moisture content may result in too rapid combustion and escape of emission gases which carry heat with them. Very dry chips may catch fire before the boiler. Too fierce combustion may also cause excessive particulate emissions in the environment, thus resulting in pollution problems.

The amount, properties, and behavior of *ash from combustion* depend mainly on furnace controls, the ash separation apparatus, and the composition of chips. High proportions of bark and foliage result in excessive ash formation. Decomposition of organic matter during chip storage increases the relative content of inorganic materials correspondingly. Inorganic contaminants also increase the ash production and reduce the nutrient content and fertilization value of ash. High silicon content, i.e., the presence of sand, decreases the fusion temperature of the ash and may therefore cause slag formation, resulting in handling and recycling problems (Sect. 9.4.2.1).

The noncombustible components of the chips are either retained as bottom ash in the furnace, which may interfere with the combustion process, or are entrained as fly ash with the stack gases, tending to cause erosion of heating surfaces and ducting and emission problems. Special equipment is therefore required for separating solid particles from the gases leaving the furnace.

Chip properties are thus of great importance for continuous and efficient operation of small- and medium-sized heating plants. The smaller the installation, the higher is the risk of disturbances when low-quality chips are used. Large power plants in conjunction with forest industry enterprises are much less sensitive to erratic variation in moisture content, particle size, and other properties of fuel chips.

The average properties of the fuel and the range of variation must be taken into account in the preliminary planning of a chip-fired heating plant. It is necessary to know the types of forest biomass and the methods of fuel preparation to be used for the production of chips. However, frequently insufficient attention has been paid to fuel quality in plant design. This has resulted in faulty chip handling and boiler adjustments, leading to vulnerable and uneconomic operation. Dimensions of system components are, for example, dependent on whether the chips are received unscreened or screened. Under certain conditions simple upgrading of chips by disk screening or other means may be feasible to ensure uninterrupted functioning of the fuel handling chain. While the basis for conveyor dimensions is the maximum feeding capacity required by the boiler, the roughness of this equipment is determined by the possible occurrence of oversized particles, twigs, frozen lumps, stones, etc.

8.1.3 Energy Efficiency of Chip Combustion

Energy is used in economic production in two general steps. First, solar, geologic, and other natural energies produce a wide variety of resources and environmental services that are essential for human existence and economic production. In the second step, human-directed production energy, such as human muscle or fossil fuels, is used to locate, extract, refine, process, assemble, transport, and otherwise upgrade the organizational state of natural resources to useful goods and services. All energy that humans control and direct for their own purposes can be called *economic energy* and can be used to do economic work. The amount of economic work possible depends on both the quantity and *quality of energy* and the efficiency of the process. High-quality fuels exist in more concentrated forms, are more flexible, require less processing, and can be used with high end-use efficiency. For example, 1 kWh of electricity, generated from any source of primary energy, can do more economic work than 1 kWh of biomass, which would be reduced to about one-third when converted to electricity (Hall et al. 1985). Most of the energy consumed in harvesting of forest biomass is provided by diesel fuel and gasoline and is difficult to replace with alternative fuels.

Energy planning sometimes overlooks the fact that energy must be expended to obtain another type of energy. Thus, coal must be mined, processed, stored, and transported, and all these phases of work consume energy in one form or another. On an energy-use basis, forest biomass is not a cost-free product either. Harvesting, preparation, storage, and transport of residual biomass for fuel require energy. In the case of short-rotation energy plantations a significant amount of energy is consumed in the production phase, especially when fertilizers are applied. In addition to the energy needed for operating the machinery, energy used for manufacturing the machinery must also be taken into account in energy-balance calculations. According to Blankenhorn et al. (1978), energy consumption in the production of steel is 79.5 MJ/kg and in the production of tractors 158.9 MJ/kg. Manual labor consumes 1.34 MJ/work hour and truck transport 1.0 MJ/t/km.

Disregarding the solar energy, usually considered as free energy, the input/output ratio in forestry is around 1:20 or inputs are around 5% of the output. The inputs include direct energy as fuel, indirect energy as embodied in the machines, and energy for transport and food for labor. Looking at forestry as an energy system, its efficiency is very high, provided solar energy is free (Sundberg and Silversides 1988).

The energy cost of extracting fuel varies with conditions and evolving technology. Thus, an important aspect of fuel quality is the amount of energy required to locate, extract, and refine the fuel to a socially useful state. This aspect of fuel quality is measured by *energy return on investment*, which is the ratio of the gross amount of fuel extracted in the energy transformation process to the economic energy required to make that fuel available to society. The denominator of the ratio includes both direct fuel use and other less obvious indirect economic energy. Net energy is actually a more relevant measure of energy supplies than gross energy because it is the energy available to produce final goods and services (Hall et al. 1985). As explained in Section 5.3, several levels of energy consumption can be differentiated in biomass harvesting: physical, machine, cycle, operational, and system levels, the latter including the indirect energy used for manufacturing the machines and spare parts.

The energy return on investment for a typical North American whole-tree chipping operation is 32 : 1 (Hall et al. 1985). Because the performance and energy efficiency of machines are reduced with decreasing tree size, the corresponding ratio is lower under European conditions. For example, in Finnish whole-tree chipping operations it is estimated to be about 25 : 1. Nevertheless, not more than 3–5% of the heating value of biomass is theoretically consumed for the system energy of preparation and transport of chips. However, the quality of liquid fuel consumed in the delivery phase is essentially higher than that of the chips.

The net heat recoverable in the combustion process is less than the potential heating value of the biomass. The main cause of heat loss in chip combustion is the vaporization of the water initially present in the fuel or generated when hydrogen and oxygen combine, and by raising the temperature of the steam to that of stack gases which escape from the system. The relative energy output from a boiler, expressed as a percentage of the input, is called *boiler efficiency*. The output usually means the steam generated in the boiler, multiplied by the difference in heat contents of the feed water and steam leaving the boiler. The input equals the dry mass of fuel multiplied by its calorific heating value.

Moisture content of the biomass is the main fuel factor affecting boiler efficiency. First, an increase in the moisture content both reduces the effective heating value of the biomass (Sect. 3.5) and results in increasing heat loss with escaping vapor, which always carries heat. Secondly, a high moisture content may have a detrimental influence on the efficiency of combustion by reducing the flame temperature, which again may result in incomplete combustion and loss of energy in the form of escaping carbon monoxide and hydrocarbons. Since there is more water in biomass fuel than in fossil fuel, and since the moisture content and other important fuel properties of biomass tend to vary erratically, boiler efficiency is generally less in chip combustion.

Small furnaces in which it is difficult to achieve a sufficiently high combustion temperature react to changes in moisture content more sensitively than large ones. Medium-sized heating plants are usually designed so that an optimum result is achieved with 30–35% moisture content. When the moisture content rises to

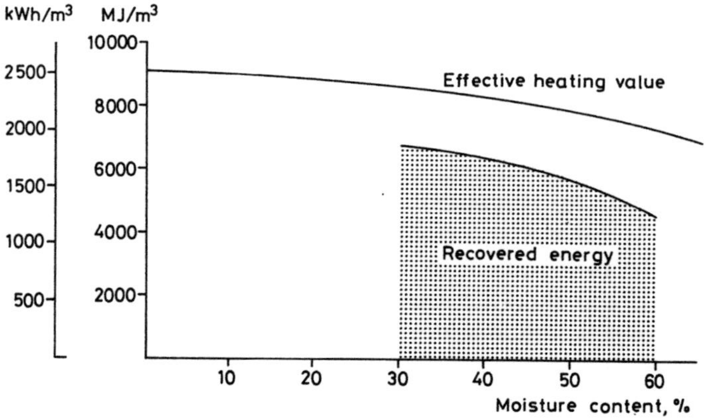

Fig. 8.2. An indicative example of the effect of moisture content on the recovery of thermal energy (MJ/m^3 solid) from forest biomass in a typical 1-MW chip-fired heating plant. Assumed boiler efficiency is 80% for a moisture content of 30%, and 70% for a moisture content of 50%

50–60%, combustion begins to occur with great difficulty and efficiency falls drastically. Chips with such a high moisture content are classified at many small- and medium-sized heating plants as unfit for use. When the moisture content exceeds 60–70% the heating value of the fuel does not suffice to maintain combustion. Excessive moisture content also reduces the overall steam-producing capacity of a heating plant.

On the other hand, very dry chips may burn too explosively. This results in increasing escape of heat and particulate emissions into the atmosphere. However, since the effect of moisture content is plant-specific, no universally applicable correlation can be postulated. Figure 8.2 illustrates the share of the thermal energy of chips recovered at a typical medium-sized heating plant in Finland when the furnace is run at nearly full load.

Removal of water from biomass by means such as transpiration drying of whole trees or forced-air drying of chips is an important measure to improve boiler efficiency (Sects. 8.1.4.1, 8.1.4.2). Heat of stack gases may be utilized for predrying chips to obtain a better overall efficiency of the operation.

Boiler efficiency depends not only on the fuel properties but also on the furnace and its controls. No burning system is capable of recovering the energy potential of a fuel in its entirety. Energy loss in combustion is caused primarily by the following factors:

1. Stack gases carry heat with them;
2. Stack gases escape unburnt;
3. Incompletely burnt carbon is left as charcoal among the ashes;
4. Heat radiation and convection from the furnace to sites where it is not needed.

To ensure complete burning of chips, the combustion chamber is supplied with more air than is theoretically needed for burning. A chip furnace usually requires 25–50% excess air, which increases the amount of stack gases and heat loss. This loss can be reduced by improving the heat exchanger system of the boiler to capture more heat from escaping gases, by using the gases for predrying chips, or by preheating the combustion air and boiler feed water with stack gases.

A commonly used type of combustion equipment in the forest industries is the spreader-stoker boiler. The fuel stream is projected into the furnace by mechanical or pneumatic means and spread uniformly over the grate. A portion of the fuel particles falls to the grate and forms a bed, while another portion is entrained and travels upward through the furnace with combustion gas. The latter portion, called *char carry-over*, results in incomplete combustion and loss in boiler efficiency of 2–10 percentage points, and places a severe strain on the pollution control equipment. This particle entrainment is dependent on the shape and density of the chips, especially on their thickness. Carry-over of char from the furnace is attributable to particles larger than 1 mm, since particles less than 1 mm usually burn completely under normal furnace conditions. The gas velocity increases with the fuel firing rate. This increases the amount of large fuel particles entrained and reduces their residence in the furnace. The particle size may thus affect combustion and boiler efficiency, depending on the furnace type (Spurrell et al. 1987).

Heating plants operate at undercapacity for a considerable part of the year. Efficiency falls when the load diminishes. For this reason, and also because of furnace interruptions, *annual efficiency* is always poorer than with full-load operation using the same fuel. According to Immonen and Seppälä (1984), the annual efficiency of medium-sized chip-fired heating plants in Finland is 73% on an average.

If forest biomass is burned in order to generate electricity, the overall conversion efficiency is considerably less. Based on a boiler efficiency of 69%, a steam-cycle efficiency of 39%, and auxiliary power requirements of 9%, the overall efficiency of electricity generation from wood can be estimated at 25%. Cogeneration of steam for industrial processes or district heating offers both cost and energy efficiency advantages (Love and Overend 1978).

The maximum transport distance from the viewpoint of energy utilization can be determined by balancing harvesting, transport, and processing energy against recoverable energy from forest biomass. Under North American conditions, net energy can be achieved if truck round-trip haul distance is kept within 2900 km for production of heat, within 2500 km for production of steam, and only 173 km for production of electricity (Blankenhorn et al. 1978).

8.1.4 Drying and Storage of Fuel Chips

The moisture content of biomass varies between species, between trees, within a tree, and during the season. In a live tree, the moisture content increases from stem base to stem top and from branch base to branch top, and is generally highest in the foliage (Sects. 2.1.1.2.1, 2.1.1.3.1). This within-tree variation of moisture content appears to be more distinct in conifers. In a single stand, the moisture content of a dominant tree is usually slightly higher than that of a suppressed tree. Under conditions prevailing in the boreal forests, the moisture content of conifers is highest during the dormant season. The moisture content of hardwoods is at its highest in the spring just before the leaves appear, then drops below the annual average after bursting into leaf, and rises again in the autumn to the higher winter level (Hakkila 1962).

Approximately half of the total mass of a newly felled tree consists of water. However, considerable differences occur between species, partly due to differences in the basic wood density. The moisture content decreases with increasing basic density if the amount of water per unit volume of biomass remains constant. For example, Heding (1984) reported the following average moisture contents for Danish tree species: *Picea abies* 51 – 58%, *Betula* spp. 47%, *Fagus sylvatica* 45%, *Fraxinus excelsior* 40%, *Quercus robur* 45%, and *Salix* sp. 50%. In Finland the moisture content range of small-sized trees varies as follows (Hakkila 1984):

Species	Moisture content range of newly felled, small-sized trees (%)
Pinus sylvestris	50 – 60
Picea abies	50 – 60
Alnus incana	45 – 55
Betula spp.	40 – 50

The moisture content of newly-felled trees tends to be too high for efficient combustion and utilization of boiler capacity. Small furnaces, particularly, are sensitive to excessive moisture content of fuel, since this reduces both the effective heating value of the chips and the efficiency of combustion and, consequently, the available heat. Beyond a certain point, which can be called the *critical moisture content*, the energy efficiency of a chip furnace shows a remarkable decline. The temperature at the zone of combustion drops, the furnace begins to soot heavily, and the emission gas darkens. When the moisture content increases further, the combustion will eventually stop. For furnaces designed for fresh chips, the critical moisture content must be as high as possible, preferably around 60%, but this goal is very difficult to achieve. Typically, the critical moisture content is around 50% (Heding 1984).

Because of past experience, increasing attention is being paid to the moisture content of wood as a cost factor in chip-fired heating plants. Figure 8.3 presents the variation of chip moisture content from October 1981 to April 1983 as the mean of a large number of 1 – 10 MW heating plants in Finland. The average moisture content was 40% in the winter and significantly less in the summer. Since operation at maximum boiler capacity is often necessary during the cold season, and since a large majority of the total consumption of chips is concentrated in the winter season, the moisture content of chips in the summer is of less importance.

In addition to the average moisture content, uniformity of moisture is another crucial quality factor, as irregular variation makes combustion control difficult and results in loss of efficiency. Figure 8.4 illustrates monthly load-to-load variation of the moisture content of chips from September 1980 to March 1983 in a medium-sized heating plant in Finland. The standard deviation between the trac-

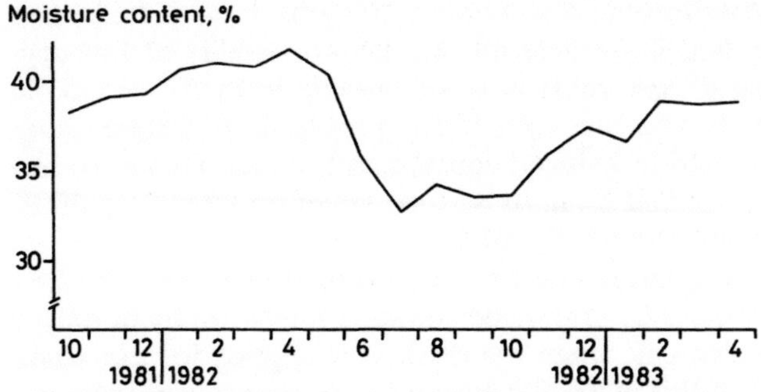

Fig. 8.3. Monthly variation of the average moisture content of forest chips from October 1981 to April 1983 according to measurements of a large number of heating plants in Finland (Hakkila 1984)

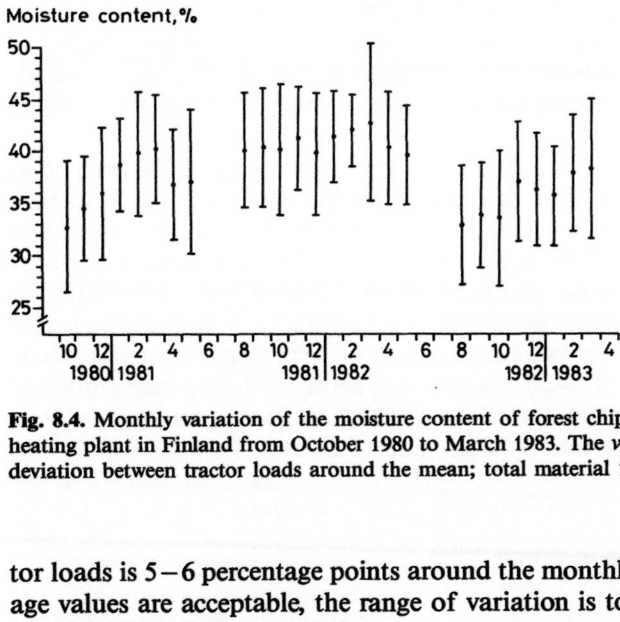

Fig. 8.4. Monthly variation of the moisture content of forest chips received by a municipal district heating plant in Finland from October 1980 to March 1983. The *vertical lines* illustrate the standard deviation between tractor loads around the mean; total material 1766 loads (Hakkila 1984)

tor loads is 5–6 percentage points around the monthly mean. Although the average values are acceptable, the range of variation is too wide from the viewpoint of optimal operation control.

The efficiency of combustion can be improved by reducing and leveling the moisture content of the biomass. Lower moisture content also results in reduced cost of transport and storage, improves the reliability of chip handling systems, and facilitates furnace maintenance. Furthermore, the activity of microorganisms becomes difficult below a certain moisture level. This reduces the dry matter loss caused by wood-destroying fungi, spore concentration, exposure of workers to mold dust, and danger of spontaneous combustion in the chip piles and silos during prolonged storage.

Energy loss due to moisture is compensated in practice by increased fuel consumption. If the biomass is very wet, achievement of sufficient flame temperature may require use of oil as an added fuel. The direct cost caused by the moisture can be evaluated through the additional consumption of chips or oil. Kärkkäinen (1981) used three alternative models for determining the profitability of fuel drying. In all cases, the burning of wet chips was not feasible from the economic point of view. Low-cost drying of biomass was highly profitable at the prevailing price level, especially if it resulted in reduced consumption of oil. Hence, to improve the competitiveness of residual biomass as fuel, attention must be paid to the moisture content in the operational planning.

Drying of conventional fuelwood in covered or uncovered piles is established traditional technology. However, even prolonged storage usually results in only a slight reduction of moisture content, when the method is applied for unbarked stems and long bolts. Although this method continues to be usable in small-scale operations, it is not well suited to large-scale fuelwood procurement using modern logging technology. Consequently, fuelwood drying in piles is not discussed in this book.

Another approach is transpiration drying, which is also based on drying before comminution of fuelwood. This method allows low-cost reduction of moisture content to a level determined by physiological tree factors. Since integration of transpiration drying in whole-tree logging under summer conditions is generally simple, the method is in common use, especially in small-scale operations (Sect. 8.1.4.1).

Drying is much more complicated and costly after the biomass has been comminuted, particularly in large-scale operations. In fact, the goal in chip storage may often be to protect the chips against increased moisture content rather than to reduce the moisture. Nevertheless, a number of artificial drying methods have been developed and are being used by farms, single-family houses, and small- and medium-sized heating plants. These methods are applied not only because they improve the energy efficiency of combustion but also to reduce the activity of microorganisms in chips (Sect. 8.1.4.2). Natural and artificial drying and storage of chips and chunks are reviewed in the proceedings of an IEA Conference on Production, Storage, and Utilization of Wood Fuels (Gislerud 1988).

8.1.4.1 Transpiration Drying of Small Trees

An effective and economical solution for the reduction of the moisture content of fuelwood is summertime *transpiration drying*. In this method the felled tree is left to season on the ground with its crown intact. Drying takes place through transpiration as long as the foliage maintains its activity. This ability of foliage to remain physiologically active as the tree dries is species-specific and determines the lower limit of the moisture content that can be achieved. Transpiration drying is also called biological drying, leaf seasoning, or sour felling.

As other natural wood-seasoning methods, transpiration drying is dependent on weather conditions. Drying is minimal in a humid, rainy climate, and during winter when the trees are inactive. The following climatic variables have an effect on the rate of evaporation from an open water surface and from the surface of leaves and needles as well (Visser 1987):

1. Wind;
2. Air humidity;
3. Air temperature;
4. Temperature of the evaporating surface;
5. Amount of moisture in the surface available for evaporation;
6. Nature of the surface.

Visser (1987) studied transpiration drying of 19-year-old *Pinus radiata* trees in the Republic of South Africa and 35-year-old *Pinus sylvestris* trees in the Federal Republic of Germany. An A-pan, consisting of an open tank filled with water and exposed to the atmosphere in the plantation, was used as *an evaporation meter.* The mean mass loss in 48 radiata pines after 34 days was 21.9% of the original total fresh mass. At the same time, 21.1 cm water evaporated from the A-pan. Thus, for the first month 1% mass loss equaled 1 cm water evaporated from the A-pan. At this stage of the process, the trees lost their ability to transpire, and a breakaway developed between the mass-loss rate and evaporation of water. No significant differences were found in the rate of total mass loss of trees lying in direct sunlight and those lying in the shadow of the plantation. In Germany the mass loss of Scots pine was also closely correlated with evaporation of water from an A-pan. The correlation lasted 19 days after which the breakaway began. At this stage a mass loss of 22.8% existed and 28.4 cm water had evaporated. The author concluded that measurement of the amount of water evaporated from an A-pan is a handy tool to determine the mass loss of pines in transpiration drying, and the method could be used to provide some indication of the time when the effective transpiration period expires within a certain season.

Transpiration drying of small-sized conifers felled in the winter starts in northern Europe as soon as the trees are free of snow, but it is not really significant until May. Conifers require the whole summer season to dry well. Hardwoods felled in the winter begin to season after they have burst into leaf, which occurs only after the leaves appear on growing trees. In the summer, transpiration drying of hardwoods takes place usually within a month, under favorable conditions even in a fortnight. The more rapid the drying, the more uniform is the moisture content both between and within the trees. In all stages of the seasoning process, the moisture content is more uniform in hardwoods than in conifers. The smaller the trees, the better is the result of transpiration drying (Fig. 8.5).

Provided that felling takes place sufficiently early, the moisture content of 5- to 10-cm diameter Scots pine trees may exceptionally fall to 30%, but it usually remains around 35–45%. Norway spruce achieves a somewhat lower moisture content, 30–40%, and only under very favorable weather conditions will the moisture content fall below 30%. Birch dries rapidly to 30–35% moisture content (Fig. 8.6). The moisture content increases by several percentage points if the trees are not removed from the forest before the relative humidity rises with the onset of autumn (Hakkila 1962). In a Danish experiment, 30-day transpiration drying had little effect on the moisture content of Norway spruce, which was assumed to be the result of the high density of the remaining stand (Heding and Jakobsen 1985).

Rogers (1981) studied transpiration drying of three tree species during the cool fall and winter season in eastern Texas (Table 8.3). In *Pinus taeda*, a large propor-

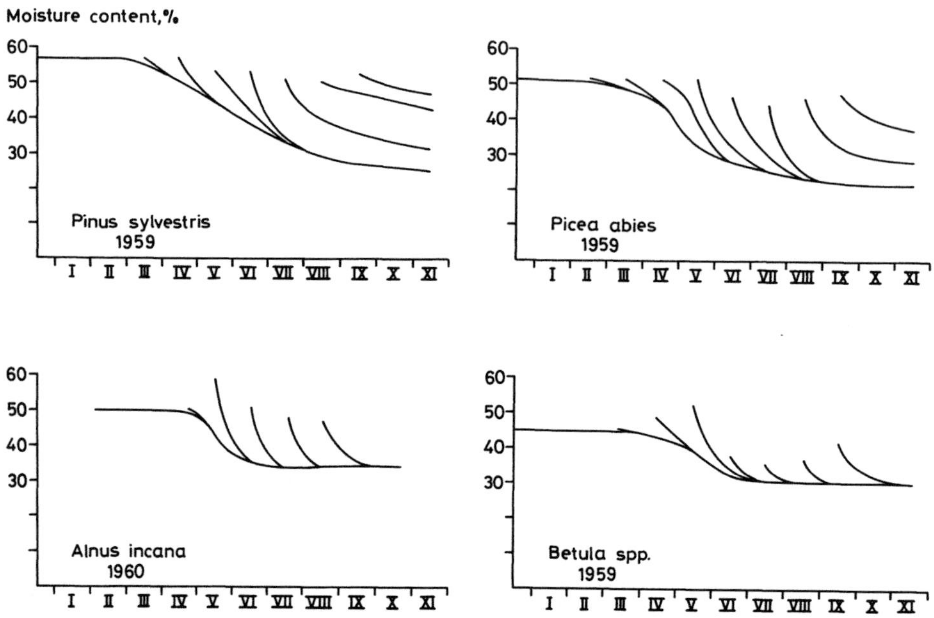

Fig. 8.5. Development of the moisture content of 5–10-cm-diameter *Pinus sylvestris, Picea abies, Alnus incana,* and *Betula* spp. felled during different months and left in the forest for transpiration drying in southern Finland (Hakkila 1962)

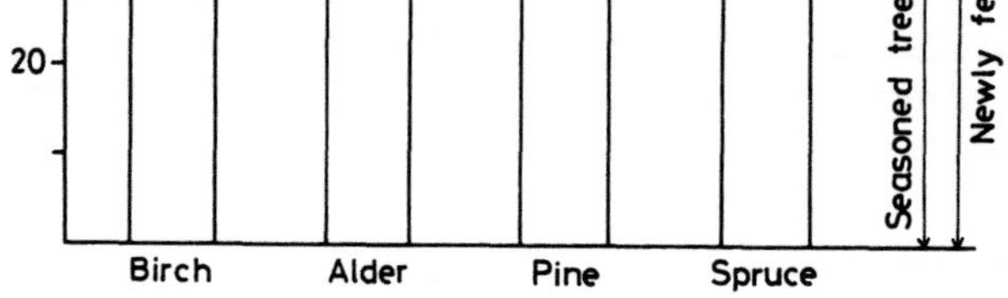

Fig. 8.6. Seasoning profit in transpiration drying of small-sized *Pinus sylvestris, Picea abies, Alnus incana,* and *Betula* spp. (Hakkila 1984)

tion of moisture loss occurred during the first 3 weeks after felling. This indicated that only a short delay in whole-tree harvesting can have a great impact on the fuel potential of biomass. In *Quercus alba,* moisture loss was relatively constant for the duration of the study. This was attributed to the presence of tyloses which slow moisture movement in white oak wood. After 9 weeks of transpiration drying, the moisture content of sapwood was 35% for loblolly pine, 37% for white

Table 8.3. Development of moisture content in the sapwood and heartwood of *Pinus taeda*, *Quercus alba*, and *Liquidambar styraciflua* during transpiration drying in the fall and winter seasons in Texas (Rogers 1981)

Tree species	Stem part	Moisture content (%)					
		Weeks after felling					
		0	1	2	3	6	9
Pinus taeda	Sapwood	56.7	54.0	51.2	48.4	40.5	35.6
	Heartwood	41.1	38.7	36.3	34.1	28.7	25.8
Quercus alba	Sapwood	42.0	41.3	40.6	39.9	38.2	36.9
	Heartwood	41.7	41.6	41.5	41.4	41.0	40.7
Liquidambar styraciflua	Sapwood	53.4	52.0	50.6	49.4	46.4	45.1
	Heartwood	54.3	53.7	53.3	52.8	51.8	51.3

oak, and 45% for *Liquidambar styraciflua.* The corresponding figures for heartwood were 26, 41, and 51%, respectively. The author assumed that drying occurs in Texas at a higher rate in the summer, and pointed out that drying rates vary between geographic regions due to weather conditions. According to Watson et al. (1986b), a 40% moisture content is attained in Alabama by transpiration drying of hardwoods during the months in which foliage is present.

In Finland, transpiration drying is commonly integrated in fuelwood harvesting systems. The method covers approximately 40% of all fuel chip production from small trees. This is due partly to the easy applicability of the method to the felling-piling technique used for small-tree logging (Sect. 4.1.2.1). Transpiration drying, haulage of seasoned trees to the landing site in a sufficiently early phase, high and regular piles, and the avoidance of snow in winter chipping suffice to keep the moisture content of chips safely below 40%.

Piles can be *protected against rain* by use of tarpaulins. In a recent study the moisture content of chips made from whole-tree piles covered in the autumn was 36.1% during the following winter, whereas from corresponding uncovered piles it was 39.1%. Covering did not only slow down the reabsorption of moisture but also facilitated winter chipping. The costs of covering, including materials and work, accounted for 2–4% of the total costs of forest chips (Uusvaara 1984).

Since transpiration drying helps to leave foliage at the site, it also improves the composition of whole-tree chips for most utilization purposes, reduces respiration and microbial activity during chip storage, and reduces the loss of nutrients from the forest soil. However, a portion of the nutrients is probably translocated from foliage into branches and stem before the abscission of leaves. According to Watson et al. (1986c), drier trees require more knife changes in chipping, reducing productive time and driving up the cost per productive hour and the cost per tonne of chips.

8.1.4.2 Storage of Fuel Chips

Uninterrupted operation of combustion systems requires stockpiling of fuel to assure a continuous supply during adverse harvesting conditions. Depending on the chip procurement organization and distribution of the annual consumption throughout the year, fuel chips may be stored for months prior to use. Since wood has a low bulk density and heating value, in large plants chips are typically buffer-stored in exposed outside piles. Only in small consumption points can chips be protected against rain.

Wood, bark, and foliage are susceptible to biological and chemical reactions and decomposition by fungi during storage. Distinct moisture gradients develop in the piles, and the average moisture content of the chips may change. As these changes tend to degrade the fuel by reducing its heating value and may even cause fire hazards and health problems, serious attention must be paid to the duration, technique, and management of storage. The effect of storage on the fuel value must be taken into account in operation planning.

Fuel chips that are not properly dried tend to release heat when stored in a pile or silo. During the first days of storage the temperature increase in chip piles can be attributed to respiration in live cells of the biomass. More long-lasting heat release is caused mainly by respiration of the microorganisms which always colonize any kind of biomass. These reactions typically result in an increased carbon dioxide concentration in whole-tree chip piles. Bergman and Nilsson (1979) measured as much as 26% carbon dioxide in the air of the interior of a whole-tree chip pile about 2 weeks after storage, which was much more than in piles made from debarked chips. In overheated piles extra heat is assumed to be generated by chemical oxidation reactions. The process of deterioration in chip piles proceeds during storage as follows (Assarsson 1969; Bergman 1972):

1. Small amounts of nutriment are constantly consumed through *respiration reactions* in live parenchyma and cambial cells of a tree, releasing carbon dioxide, water, and heat. Owing to the limited diffusion of oxygen into the wood and of carbon dioxide from the wood, respiration proceeds slowly in the living tree. When the tree is reduced to small particles, diffusion increases and respiration reactions accelerate, especially in materials such as whole-tree chips which contain most of the live, above-ground components of the tree. Respiration starts when the temperature exceeds 0 °C, and increases with temperature up to about 40 °C. At somewhat higher temperatures live cells begin to become disorganized and die. Most of them typically cease functioning at about 60 °C.
2. *Microbial reactions* are caused by four basic types of microorganisms. Staining and molding fungi (*Ascomycetes* and *Fungi imperfecti*) are able to cause soft rot in wood, but are not very active decomposers. They are harmful in pulpwood chips because they cause discoloration problems in the production of unbleached sulfite pulp and mechanical pulp and spread spores which can cause allergic reactions. The second category contains wood decay fungi (*Basidiomycetes*) which are mainly responsible for decomposition of chips during storage. They are either white-rot fungi, which attack both carbohy-

drates and lignin, or brown-rot fungi, which leave the main part of lignin alone. The latter group rarely has been isolated from stored chips. The third category, soft-rot fungi (*Ascomycetes* and *Fungi imperfecti*), attacks the surface layer of chips which becomes softened. Although the attack proceeds slowly, the damage can be extensive, especially since they can grow at both low and high moisture levels which may be inhibitive for white-rot and brown-rot fungi. The fourth category of microorganisms contains yeast and bacteria, which appear to be of less importance in chip storage.

3. Most of the decaying fungi have their optimum temperature between 20–30 °C and are inhibited at temperatures above 40 °C. In large chip piles, temperature conditions generally permit decaying and staining fungi to attack the wood only in the outer, colder areas in a compacted pile, whereas the main center part of the pile is usually too hot for these fungi to thrive. As regards moisture conditions, fast growth of decaying fungi occurs at 25–60% moisture content, which is typical for fuel chips. These fungi also need oxygen, which is usually available. Furthermore, they require bound nitrogen for growth, for instance in the form of inorganic salts or protein. In bark-free wood, which contains little nitrogen, this may limit the activity of the fungi. Since foliage has a high concentration of nitrogen and protein, its presence in biomass accelerates fungal activity and deterioration of the chips during storage.
4. In addition to respiration in live cells and microorganisms, various *organic chemical reactions* take place in the chip piles. Live cells die and unsaturated extractives in these cells easily react with atmospheric oxygen in exothermic reactions. These direct oxidation reactions become of some importance at about 40 °C and of primary importance at about 50 °C. When the temperature rises to the point of ignition, which may occur in pockets formed by fines and foliage in whole-tree chip piles, the excessive heat is assumed to be generated by chemical oxidation reactions.

Biomass is hygroscopic, absorbing and losing moisture in response to the environmental conditions. When fuel chips are stored in exposed, outside piles, it is necessary for fuel management planning to include the factors affecting the moisture changes and deterioration of chips during storage.

Incident rainfall tends to increase the moisture content of chips during storage unless the pile is covered. This increase can be affected through the size and shape of the pile. Saucier and Phillips (1985) recorded rapid increases in the moisture content in 3- and 6-m-high piles of hardwood whole-tree chips during a 12-month storage experiment in Georgia. In an uncovered flat pile, the average moisture content increased in 12 months from an initial 44% to a high of 63%. Although substantial surface evaporation of moisture presumably took place, the flat pile shape caused precipitation to perculate and saturate the pile increasingly deeper with time. A covered flat pile, on the other hand, decreased significantly in average moisture content with storage time. In an uncovered conical pile, the moisture content increased considerably but clearly less than in the uncovered flat pile. The cone shape thus definitely provided some shed effect from incident rainfall. A central vent installed in another uncovered conical pile provided increased ventilation and evaporation of interior moisture, but the effect was only slight:

Storage time (months)	Average moisture content (%)		
	Uncovered flat pile	Covered flat pile	Uncovered conical pile
0	43.8	44.4	42.2
3	54.5	41.9	51.9
6	62.0	39.8	54.8
9	63.1	33.8	53.7

Covered storage may be economically attractive to companies that currently store fuel chips in open piles. This could be particularly true if some of the operational benefits of drier fuel, such as the possibility of using the boiler at low fire or at full peak capacity, are important in the operation of the combustion system. Depending on the length of the storage time and other conditions, covered storage may be economical in many cases. A cost-effective storage building in large-scale operations is a wooden, pole-type shed with open sides and a tin roof (Harris 1988).

The microbial reactions start in the interior of the pile. This results in a temperature rise and evaporation of water from the heated particles of the biomass. When the water vapor passes through the surface layer of the pile it is cooled off and precipitated. The water leaving the interior thus concentrates partly in the surface of the pile. However, if the pile is protected against incident rain, its average moisture content decreases slowly, depending on weather conditions and ventilation. The result is an uneven moisture content, and in practically all parts of the pile moisture conditions allow the growth of microorganisms. The slight increase in the effective heating value of chips through drying during storage is often negated by loss of dry matter through the decomposition activity of fungi. Consequently, extended storage of chips should be avoided, and as much as possible of the buffer stores should consist of uncomminuted biomass.

In general, uncovered chip piles can be divided into two regions which exhibit unique moisture characteristics and abrupt *moisture gradients*; the wet surface region and dry inner regions. White et al. (1983a) found the depth of this wet surface layer to be similar for three piles of fresh hardwood whole-tree chips in Virginia, independent of pile height, which varied from 3 to 6 m. After 1-year outside storage the wet layer was confined to the outer 0.9–1.5 m of the pile. The rise in surface moisture was due partly to rainfall and partly to condensation of water vapor from hot, moisture-laden air moving toward the pile surface. The spontaneous heating in the pile interior resulted in temperatures about 20 °C above the ambient temperatures. This moisture gradient gave rise to the movement of warm air from the center region toward the surface of the piles. Within 60–120 days of storage, the saturated outer layer of whole-tree chips had an average moisture content of 60%. The average moisture content under the saturated layer was 29%, instead of the initial 40% at the beginning of the experiment. The weighted average moisture content increased during storage.

Since the depth of the saturated surface layer appears to be independent of pile height, the tendency of exposed chip piles in temperate climates to increase in moisture content can be reduced by shortening storage cycles and building higher piles. However, since piling to greater heights is known to increase the likelihood of spontaneous fires, internal temperatures of large whole-tree chip piles should be closely monitored.

White et al. (1983b) also studied the development of the calorific heating value of whole-tree chips in the storage experiment referred to above. Heating value was not a function of location within the pile; nor was it affected by pile height. The average calorific heating value of chips was relatively unchanged during the first months, although a slight initial increase was observed. After 4–6 months the heating value decreased drastically, and after 1-year storage it was 9% lower than initially. Since the average moisture content of the chips increased, the reduction in the effective heating value was even greater. Variation in the calorific heating value increased with storage time, obviously due to pockets of conspicuous decay which were found randomly throughout the piles. These overall trends in the calorific heating value, i.e., initial increase followed by a decline, were reported to be confirmed in the literature.

Whole-tree chips typically evolve more heat and decompose more rapidly than conventional debarked chips. Loss in dry mass during storage increases in the following order: clean debarked chips, whole-tree chips, bark, foliage. However, the mass loss also depends on the tree species. Aspen, for example, deteriorates very rapidly and oak quite slowly. The species differences seem to be much greater for whole-tree chips than for debarked chips (Springer 1980).

A 6-m-high, uncovered pile of hardwood whole-tree chips constructed during the winter in Virginia was relatively inert until the ambient temperature increased to about 20 °C in the spring. The pile did not begin to increase in temperature until about the 17th week of the study in early April. The degradation of the chips appeared to begin at the same time, and thereafter a uniform dry mass loss of 1.5% per month was measured in biologically active piles. No change was noted in the level of acidity in the water-soluble matter in either the biologically active or inactive portions of the storage cycle of 1 year. The average pH value for the solubles in the bark fraction was 3.8 and that in the wood fraction 4.1. These levels of acidity are corrosive to storage and handling equipment. Further, the leachate from storage piles resulting from incident rain was also considered to be acidic at a level that may be harmful to the environment (White et al. 1986).

Springer et al. (1978) investigated the storage characteristics of whole-tree chips made from 10-cm-diameter *Pinus taeda* trees in simulators. Unscreened whole-tree chips generated significantly more heat than screened whole-tree chips or bark-free chips. The bark and needles in the whole-tree chips were assumed to be responsible for the higher temperatures. Unscreened whole-tree chips thus appeared to present increased likelihood of spontaneous ignition. Chemical treatment with formaldehyde or p-nitrophenol solutions was highly effective in preventing chip heating throughout storage. The cost of treatment chemicals was considered economically acceptable for pulpwood chips but not for fuel chips.

Unscreened whole-tree chips suffered significantly higher mass losses than the screened bark-free chips. Surprisingly, the greatest losses were suffered by the screened whole-tree chips. After 6 months in simulators, the loss of dry mass was 5.4% for unscreened whole-tree chips, 10.3% for screened whole-tree chips, and 4.3% for screened bark-free chips. Chemical treatment reduced the loss to 1.2% for screened whole-tree chips and 0.8% for bark-free chips. According to Springer et al., a likely explanation may be that the higher temperatures experienced by the unscreened whole-tree chips destroyed or inhibited the growth of certain aggressive wood-destroying fungi which continued to thrive in screened whole-tree chips. Large quantities of bark fines in the unscreened whole-tree chips may also have been partially responsible for somewhat inhibiting the attack of the microorganisms, as pine bark is known to contain large quantities of polyphenolic materials. No significant changes in the calorific heating value occurred during storage.

Jirjis and Theander (1986) studied the effect of storage on the chemical composition of fuel chips prepared from coniferous logging residue. In compacted outside piles, the lignin/carbohydrate ratio increased during storage in the foliage and bark, while wood particles changed very little. In uncompacted outside piles, the change in the foliage and bark fractions was similar to that in compacted storage, whereas the wood fraction consistently showed a declining lignin/carbohydrate ratio. Compacted inside piles exhibited only slight changes in the surface layers but a clear increase in the lignin/carbohydrate ratio in the pile interior. When the indoor pile was left uncompacted, the results were generally similar for the foliage and bark, but for wood fractions the lignin/carbohydrate ratio declined. The results reflect differences in the species composition of active fungi, depending on the treatment of the chip piles.

Significant changes occur during storage in the content and composition of extractives in chips. Typically the total resin content decreases rapidly in outside storage of chips, and in the remaining resin components the fats are hydrolyzed and further oxidized. These reactions are the result of respiration activity in live cells. They make the resin more hydrophilic and more easily soluble in chemical pulping. A few weeks of chip storage equals about 1 year of roundwood storage in terms of the decrease in resin content, which is of great advantage in sulfite cooking. For kraft cooking, however, these seasoning reactions are not beneficial since they decrease the yield of by-products (Dillner 1972).

The Department of Forest Products, Swedish University of Agricultural Sciences, has conducted a large number of storage studies on various kinds of raw materials intended primarily for the production of pulp or for fuel. Logging slash from *Pinus sylvestris* and *Picea abies* plays an important role in these studies. Some of the major findings can be summarized as follows (Thörnqvist 1984, 1985b, c, 1986a, b; Table 8.4):

1. Temperature development in logging slash depends for the most part on the moisture content and size of the pile, degree of comminution, composition of the biomass, and ambient temperature. The optimum moisture content for a temperature increase is considered to be 50%. In small piles of slash built by logging machines on site, the pile temperature follows the ambient temperature. In the interior of 6–7-m-high roadside piles of uncomminuted slash, the temperature is about 20 °C above the ambient temperature. For comminuted

Table 8.4. Examples of the moisture content, dry-mass loss, and change in energy content of biomass fuels under various storage arrangements in Sweden (Thörnqvist 1986b)

Type of logging slash	Moisture content (%)	Loss of dry mass (%)	Change in energy content (%)
Winter-felled slash in small piles on site			
Stored till August	25	10	0
Stored till October	45	25	−23
Winter-felled slash in well-ventilated 4-m-high pile			
In pile from May till September	45	1	+4
In pile from May till December	50	2	+3
Fresh biomass in 3-m-long, 0.5-m diameter bundles in 4-m-high pile from May till January			
Slash	45	2	0
Pine from early thinning	55	8	−8
Spruce from early thinning	55	5	−6
Birch from early thinning	40	2	0
Fresh, comminuted slash in 60-m^3 pile			
Covered storage from May till November	20	17	−7
Uncovered storage from May till November	50	20	−18
Uncovered storage from May till January	55	24	−23
Chips from fresh slash in 7-m-high, 5000-m^3 piles from June till January			
Compacted, covered pile	40	10	−6
Compacted, uncovered pile	45	12	−12
Uncompacted, covered pile	25	7	−2
Uncompacted, uncovered pile	30	8	−5
Comminuted slash, dried with cold air fan in June to 15% m.c. and then stored in 60-m^3, covered pile till January	15	2	+12

logging slash, which is piled up in 3-m-high piles at temperatures above the freezing point, the temperature increases during a 2-week period to 60−70 °C. In the pile interior the temperature remains constant at this level for 6−7 months. However, if the comminuted slash is piled up at below-freezing temperatures, several additional weeks are required before this level is achieved. The continued temperature development in the pile is rather independent of the ambient temperature.

2. The moisture content in fresh logging slash lies between 55 and 60%. When the slash from winter logging is allowed to remain on the site in small processor-made piles, the moisture content sinks during the summer to 25−30% only to rise again in the autumn and winter to 45−50%. If the slash is collected during winter and stored in 4−5-m-high piles, the moisture content sinks to 40−45% during summer and autumn. If fresh slash is comminuted and stored in piles, a minor amount of drying occurs initially. After 6−7 months, two well-defined regions have been built up as a result of the heat development in the pile center. The inner region has a moisture content of 20−25%, while the moisture content in the surface part is as much as 65−70%. The average moisture content of the pile lies at the initial level, but if the storage time is extended, the average moisture content of the pile increases.

3. Dry-mass loss is caused by two factors: needles and thin twigs fall off during handling and storage, and microbiological and chemical decomposition occurs. If slash remains on the site in processor-made small piles, the dry mass losses in storage reach about 25% after one vegetative period or 3.8% per month, and 45% after two vegetative periods or 2.2% per month. If slash is stored uncomminuted in 4–5-m-high piles, the dry mass loss is 0.2–0.3% per month. If fresh slash is comminuted and stored in piles for 6–7 months, the loss of dry mass is 2.5–3.0% per month. If the chips are first dried to 25 or 15–20% moisture content, the monthly loss of dry mass is 0.8 and 0.25%, respectively.

Thörnqvist (1987) summarized experiences of 13 *spontaneous fires* in piles of comminuted wood fuels. Most fires originated in the border area between different assortments of fuel or in the border area between compacted and uncompacted fuel. The probable cause of the fires was assumed to lie in the different permeability in various parts of the piles. In two cases the centers of the outbreaks were localized to areas containing foreign metal objects which were assumed to function as a catalyst. According to the author, the following instructions should be followed to avoid self-ignition of large chip piles:

1. Store separate assortments of wood fuel in different piles;
2. Attempt to minimize the dispersal of the moisture content in the pile;
3. Never compact piles made from comminuted logging residue or similar material;
4. Avoid metallic objects in the pile;
5. Store wood fuel in oblong piles with the width of the base equal to twice the height of stack;
6. Avoid pits or humps along the side of the stack;
7. Build the pile along the direction of the prevailing wind;
8. In order to obtain good drying and to minimize energy losses during the storage period, the pile height should not be less than 6–7 m. However, to prevent overheating and spontaneous self-ignition, avoid exceeding the following pile heights:

	Maximum height (m)	
	Uncompacted pile	Compacted pile
Unbarked stem chips	15	12
Hardwood whole-tree chips	12	9
Softwood whole-tree chips	10	7
Comminuted logging slash	7	–
Bark	7	4
Sawdust	6	4

When supplied with a suitable moisture content and temperature, microorganisms multiply rapidly in organic materials. Not only do they decompose and stain the biomass, but they also produce vast amounts of *spores which may cause allergic reactions* in workers exposed to infected materials (Fig. 8.7). In acute form

Fig. 8.7. Microfungi multiply rigorously during storage of whole-tree chips and produce vast amounts of spores which may cause allergic reactions in workers (courtesy of Thomas Nilsson)

this disease is characterized by an acute onset of respiration difficulties, often in combination with a hacking cough, as well as chills, fever, headache, nausea, and pain in joints and muscles. The symptoms often disappear when exposure ceases, but further exposure may give rise to another acute episode with a latency period of a few hours. In more serious cases, a restrictive diminution in lung function which influences gas exchange is present, and lung fibrosis may occur (Belin 1984).

The disease, *allergic alleovitis,* has long been known in the sawmill industry as *wood-mold disease.* It is identical with the "farmers lung" syndrome which is caused by exposure to materials such as moldy hay. Handling of infected fuel chips is known to cause similar problems. Castrén and Lukkala (1986) found fewer symptoms among farmers using softwood chips than among those using hardwood chips for space heating.

During storage of whole trees and slash on the logging area or in piles, the increase in total microfungi is very small. Even if small trees or slash are tied in bundles, little increase occurs in the amount of microfungi (Flinkman and Thörnqvist 1986). If, instead, wet biomass is reduced and stored in a pile or silo in the form of chips, the amount of microfungi increases rapidly.

When chips are fresh and contain foliage and other live tissues, and the size of storage formation is only some tens of cubic meters, the fungi multiply drastically during a storage period of a few months. Up to 100 billion spores in 1 kg of chips is not uncommon. In larger, e.g., 400 m^3 storage formations, the number of germinative fungal spores is usually less, apparently due to the high heat development in combination with high moisture content. However, high concentrations of germinative spores may occur locally in large piles as well. In a large-scale storage experiment in Sweden the amount of germinative spores increased in chips from conifer slash and *Quercus robur* stem mass 3- to 150-fold during a storage period of 8 months, depending on the treatment of the chips. A covered pile of tractor-compacted oak stem chips was a distinct exception, showing 10000-fold increase in the amount of health-hazardous microfungi (Thörnqvist 1985b, c, 1986a).

Storage properties are generally better in chunks than in pulp sized chips. Chunks dry somewhat faster and to a lower moisture content under natural convective air drying due to their bulky character. The form of chunks allows larger air spaces and easier air flow as opposed to the flat character of chips, which form thin, close layers, restrict air movement, and trap moisture (Sturos et al. 1983; Sturos 1984, 1989; Nurmi 1987).

Nilsson (1987a, b) found both the total number of microfungal spores and the number of viable spores after an 8-month storage period to be lower in chunk wood than in chips. The amount of spores was smaller in 60–80-mm chunks than in 120–140-mm chunks. Cold-air drying reduced the development of viable spores during storage.

There are several possible approaches to *inhibiting the growth of fungi* and deterioration of biomass during prolonged storage. The first possibility is storage under anaerobic conditions. Laboratory studies show that no heating and deterioration occur under the 1% oxygen level in chip piles, but large-scale anaerobic storage of chips is impractical. The second alternative is chemical treatment prior to storage, but no cost-effective treatment has been developed for long-time storage of fuel chips. Consequently, the only practical approach for protecting the fuel from bacterial and fungal attack is to dry it below the fiber saturation point of wood, i.e., 20–24% moisture content, and to shelter it from rewetting (Springer 1979).

A prerequisite for fuel chip storage in single-family houses and farm buildings is that the chips are properly dried. Although transpiration drying of fuelwood under favorable climatic conditions is generally sufficient to allow efficient combustion of chips, it is not fully satisfactory from the storage point of view, since there is enough moisture left to allow the microorganisms to multiply in chip piles. Consequently, if artificial drying of chips is not possible, comminution of both fresh and partially dried fuelwood to chips or chunks for small-scale use should be carried out in small quantities according to need, so as to avoid long-time storage.

Gislerud (1983) found the dry matter loss during 3- to 5-month indoor storage of green and semidry fuel chips made from unbarked birch stems to be 1–1.5% per month. After such storage, the chips released a large number of fungal spores. To keep the amount of microfungi as low as possible, artificial drying should begin within a few days of chipping, and drying should be rapid. If molding develops before or during the drying, handling of dried chips will actually be accompanied by a strong exposure to mold dust, since spores escape more easily from dry than from wet material. Use of dry chips may thus result in a high spore concentration of the air even if molding has not advanced far (Gislerud 1982; Pellikka and Kotimaa 1983).

Chips intended for domestic consumption can be dried artificially to about 20% moisture content by blowing cold outside air through the chip silo or pile with the aid of fans, perforated floors, air vents, etc. Turning and mixing the chips during drying, as well as use of heated air in blowing accelerates the drying. Warm air can be obtained from combustion gases, or it can be accumulated by using simple equipment, such as plastic shelters or solar panels. The storage space must be covered, well ventilated, constructed from inorganic materials, and kept as clean as possible. Nevertheless, a breathing mask should be used when handling the chips inside (Alhojärvi 1981; Thörnqvist 1982; Gislerud 1982, 1984; Thörnqvist and Gustafsson 1983).

8.2 Use of Residual Biomass for Pulp and Paper Products

Research in forest residue utilization was long headed primarily toward pulp and paper industries, until the worldwide energy crisis of the late 1970s changed the emphasis to energy. Among the scientists whose studies meritoriously stimulated development in the pulp and paper branch were Crossley (1938, 1952), Keays (1968, 1971a–e, 1973, 1974) and his co-workers (Keays and Hatton 1971a, b, 1975; Hatton and Keays 1971, 1972; Hunt and Hatton 1975), Chase and Young (Young and Chase 1965; Chase and Young 1976a, b; Chase et al. 1971, 1973), Eskilsson and Hartler (Eskilsson 1969, 1972, 1974; Eskilsson and Hartler 1973, 1974; Hartler et al. 1977), and Lönnberg (1975a, b, 1976a–e; Lönnberg et al. 1975). Later, Timell (1986c) thoroughly reviewed the literature of compression wood and softwood branches in pulp and paper manufacture. Simultaneously with laboratory studies, mill-scale production runs in the United States, Canada, Sweden, Finland, and other countries were a necessary contribution to make a breakthrough possible.

The separation of wood into fibers can be accomplished by either mechanical or chemical methods, or their combinations. The choice of the process is determined by the raw material availability and quality requirements of the end product together with economic and environmental conditions and the cost of energy. When the justification for residue utilization is evaluated, two aspects are essential from the pulping point of view. First, will the process accept low-quality raw material with a high bark content and heterogeneous composition. Second, will pulp and paper properties, such as strength and appearance, be acceptable for the end use concerned?

Mechanical pulping of roundwood is conventionally carried out by grinding short bolts under specified pressures against a revolving cylindrical stone. The preferred raw material is fresh spruce wood. Since the method requires aligned, straight, debarked bolts of equal length and at least 7 cm in diameter, it is not suitable for residual forest biomass. The requirements regarding the form and dimensions of raw material are overcome in the disk refining method, which is designed for mechanical grinding of chips. Defibration of chips takes place between two textured disks and may be preceded by various pretreatments. For example, in the production of thermomechanical pulp (TMP) the chips are softened by heating prior to defibration for more complete refining. The resulting refiner pulp is usually superior to conventional groundwood pulp in both tear and bonding but is lower in brightness and opacity.

In the most common mechanical processes, the yield of pulp is 91–98% of the dry mass of wood. Raw material loss is restricted mainly to very short cells and cell sections, and some water-soluble extractives. The chemical composition of mechanical pulp thus differs little from the original composition of the wood. Bleaching of bark remnants is not possible in mechanical pulp. Raw materials with a high content of extractives, lignin, or contaminants such as bark, twigs, foliage, and grit are not suitable for mechanical pulp. Furthermore, short-fibered hardwoods and branches yield excessive amounts of fines in mechanical pulping. Consequently, this method cannot be used for the production of paper-grade pulps from residual biomass. However, disk refining is used when fiberboard is produced from residual biomass (Sect. 8.3).

Semichemical pulping processes consist of a mild chemical treatment and subsequent mechanical refining. The chemical stage can be conducted with almost any known pulping reagent. In the most common process, neutral sulfite semichemical pulping (NSSC), the major chemicals are sodium sulfite and sodium carbonate or bicarbonate. The pulp yield is inherently high, 60–80%. Almost all of the semichemical pulp is used for coarse paper grades. The main product is unbleached corrugating medium used as fluting material for combination with paperboard to make corrugated boxes.

The requirements set for the raw material regarding cleanliness and debarking are rather light. Raw materials with low content of lignin and extractives give the highest yields. Low-quality temperate-zone hardwood is the main source of raw material, low-density hardwoods generally producing better pulps than dense hardwoods. Sawdust from conifers is also used as a raw material for semichemical pulp. The strength and stiffness of the end product are dependent on the tree species rather than on the tree component. Therefore, where proper hardwood species are available, whole-tree chipping systems appear to have a high potential in the semichemical pulp industry. Several NSSC mills, especially in the United States, use whole-tree chips regularly.

In the *acid sulfite process*, chips are digested in a solution of sulfurous acid. The cooking solution contains a calcium, magnesium, sodium, or other bisulfite and a greater or lesser proportion of free sulfur dioxide. The yield of pulp is 45–55% of dry mass of wood. This process produces pulp for high-grade, bleached printing papers. Due to an unfavorable energy balance, waste-water problems, and strict requirements for raw-material quality, many sulfite mills, especially in North America and northern Europe, have been converted into kraft mills. Consequently, world production of sulfite pulp is decreasing. On the other hand, tightening requirements of air protection may improve the competitiveness of the sulfite process in the future, particularly because of its lower need of chlorine as a bleaching medium.

Spruce wood is the preferred raw material, whereas many hardwoods, pine heartwood, and resinous pine wood are unsuitable. Very little bark can be allowed in the chips, since bark tannin components are condensed in the sulfite process with the wood and bark lignins and thus decrease delignification. Excessive amounts of bark and other contaminants, presence of harmful extractives, undesirable species composition, and other adverse quality factors prevent the use of residual forest biomass for sulfite pulping. *Dissolving pulp processes* for the production of special pulps with a high content of pure α-cellulose also exclude the use of rough forest residues.

The *sulfate or kraft process* uses sodium hydroxide or sodium sulfide as the main chemical of the cooking liquor. Important characteristics of the process are pulp with a high strength and an effective system of chemical recovery. Spent or black liquor is concentrated to about 65% solids and then fired in the chemical recovery furnace for recycling the cooking chemicals. Bark and fuel chips from forest residues together with the black liquor may make an efficient kraft mill self-sufficient in energy. The yield of pulp, 40–55% depending on the cooking process, tree species, and end product, is somewhat lower than in sulfite pulping. In addition, the process can recover valuable by-products from the tall oil and is suitable for almost all common softwoods and temperate hardwoods.

Wood residues from debarked logs from sawmill and veneer plants are a common raw material of kraft mills. The use of whole-tree chips or residual wood from stumps and roots is also technically possible. For example, softwood and hardwood whole-tree chips are accepted by many North American kraft mills as partial raw material of unbleached and bleached linerboard used as facing material in corrugated or solid fiber boxes. The kraft process is of special interest from the residue-utilization point of view for the following reasons:

1. It is the most common pulping method;
2. It can produce pulp with high strength properties from different softwood and hardwood species;
3. It may accept larger concentrations of bark and resin than other common pulping processes without being seriously affected by pitch problems;
4. The resins and fats in pine wood can be recovered as by-products;
5. The cost of increased chemical consumption is greatly reduced by efficient chemical recovery;
6. The heat energy from nonpulpable components of the biomass can be recovered efficiently from the chip-screening rejects or black liquor.

Within the pulp industries, the kraft and semichemical mills are presently the potentially most important users of residual forest biomass. Substantial quantities of additional marginal raw material could be obtained even if the acceptable admixture of whole-tree chips were kept relatively low, e.g., 10–15%. Although raw material shortage alone only seldom motivates the recovery of additional biomass components for pulping, increased productivity of forest work and reduced cost of raw-material procurement together with silvicultural advantages argue in favor of the whole-tree logging principle.

However, many barriers to increased consumption of whole-tree mass for pulp do exist. The APA/TAPPI Whole Tree Chip Utilization Committee listed various kinds of general, energy, and fiber barriers to the increased use of whole-tree chips in the pulp and paper industries. Some of the major fiber barriers which must be offset by the benefits, or overcome through technical development, are as follows (Altman 1982):

1. Decreased production capacity of the mill;
2. Increased cost of chemicals due to high bark content;
3. Increased cost of pulp-mill maintenance;
4. Decreased pulp quality;
5. Varying uniformity in pulp blend caused partly by species swing;
6. Inability of many installations to screen and store whole-tree chips separately and to meter at an acceptable level;
7. Need for a process to improve chip quality;
8. Market surplus of roundwood and bark-free chips;
9. Lack of concrete information regarding amount of grit already in conventional raw materials;
10. Built-in reluctance by pulp mill management to permit the use of any fiber form which may add to the uncertainty of the pulping process.

These barriers are basically either qualitative or operational. Sections 8.2.1 and 8.2.2 deal with some quality problems of pulp and paper products in conjunction with whole-tree chip pulping. Sections 8.2.3 – 8.2.6 are concerned with operational problems, including capacity reduction of the mill. Both barriers are closely connected with economic aspects.

8.2.1 Strength Properties of Pulp and Paper

The desired strength properties of pulp are dependent on the paper or paperboard product to be made from it. For example, tear strength is especially important in newsprint and stiffness in corrugating medium, whereas both tensile, burst, and tear strengths must be moderate to high in fine printing papers. Pulp strength is primarily a result of raw material properties such as fiber morphology and chemical composition and, on the other hand, of the method of fiber separation and subsequent beating of the pulp. Among the important morphological properties are cell length, ratio of cell length to diameter, cell-wall thickness, and occurrence of reaction wood. The tendency of thin-walled, large-diameter cells to collapse and become ribbonlike, versus the tendency of thick-walled tracheids with a narrow lumen to retain the rounded cross-section form on sheet formation, significantly affects the strength of paper (Sect. 3.2).

Gleaton and Saydah (1956) made kraft pulp from debarked wood components of a 46-year-old *Pinus palustris* tree. The strongest pulp was produced by stem wood, but pulps from top, stump, and root wood were only slightly weaker. Branch-wood pulp was inferior to all other pulps in burst and tear factors. There was a distinct correlation between the length/width ratio of the fibers and overall physical strength of paper produced from various components of the tree.

Kurrle (1963) and Young and Chase (1965) pulped unbarked chips from unmerchantable tops, branches, and roots of Norway spruce, using nitric acid and ammonium bisulfite processes. The strength characteristics of top and root pulps compared favorably with those of stem pulp. In contrast, branch pulps were inferior in tensile strength, burst, tear, and folding endurance.

In a study by Alestalo and Hentola (1966), kraft pulp from unbarked tops of Scots pine and Norway spruce was comparable with good stem-wood pulp. Stump and branch pulps were on a level with low-quality stem-wood pulp. A special characteristic of branch pulp was its remarkably high stretch.

Matjusjkina et al. (1966) made kraft pulps from fresh stump, root, and stem woods of Scots pine. Wood from the central stump required a longer cooking time for proper delignification, and a longer time for beating. Stump and root wood produced pulp with 20–30% lower strength. However, due to *synergistic effects*, a mixture containing 20% stump- and root-wood pulp and 80% stem-wood pulp, both beaten separately, improved tear strength and folding endurance, slightly lowered tensile strength, and gave equal burst strength compared to normal stem-wood pulps. A mixture containing 50% stump- and root-wood pulp and 50% stem-wood pulp produced handsheets with all strength properties 3–5% below those of stem-wood pulps.

Dyer (1967) found kraft pulp from branches of *Thuja occidentalis* inferior to stem pulp in burst and tensile strength, a little higher in tear strength, and superior to all other pulps in stretch properties. Root pulp was also characterized by inferior burst and tensile strength (Fig. 8.8).

Hosia et al. (1971) examined kraft pulping of Scots pine and Norway spruce crown material, inclusive foliage. The tensile strength, tear, and burst were inferior to those of stem-wood pulp, but the stretch was 4–7% above that of the stem-wood pulp level. The presence of needles significantly affected the yield and chemical consumption but only slightly the papermaking properties of the pulp. This was due to the almost total dissolution of the needles during pulping.

Hakkila and Palenius (1971) examined kraft pulping of whole-tree chips from softwood thinnings. The presence of stem bark, branches, and foliage reduced the pulp strength slightly. The loss in overall strength was 15% for Scots pine and 5% for Norway spruce. The proportion of branch wood was not large enough to increase the stretch.

Keays and co-workers examined kraft pulping of the stem, top, branches, stump, roots, whole tree, and complete tree of *Tsuga heterophylla* (Keays and Hatton 1971 a), *Picea glauca* (Keays and Hatton 1971 b), *Pseudotsuga menziesii* (Hatton and Keays 1972), and *Pinus contorta* (Johnstone and Keays 1973). Unbleached pulps from unmerchantable stem tops were comparable with stem pulps in burst and tensile strength but lower in tear, probably due to the presence of juvenile wood. Pulps from stumps and roots were lower in burst and tensile strength but comparable in tear to stem pulps. Branch pulps were inferior to all other pulps in all strength properties. On an average, the burst, tensile strength and tear of branch pulps from the four species were only 61, 65, and 79% of the corresponding values of stem pulps. On the other hand, branch pulps typically showed high stretch values. Beating times to a given freeness for branch pulps were short, but the increase in burst factor and tensile strength was slow.

Keays (1973) concluded that the specific characteristics of branch-wood pulps were mainly a result of the high proportion of compression wood tracheids. Pulps from softwood branches were comparable in strength to normal hardwood pulps

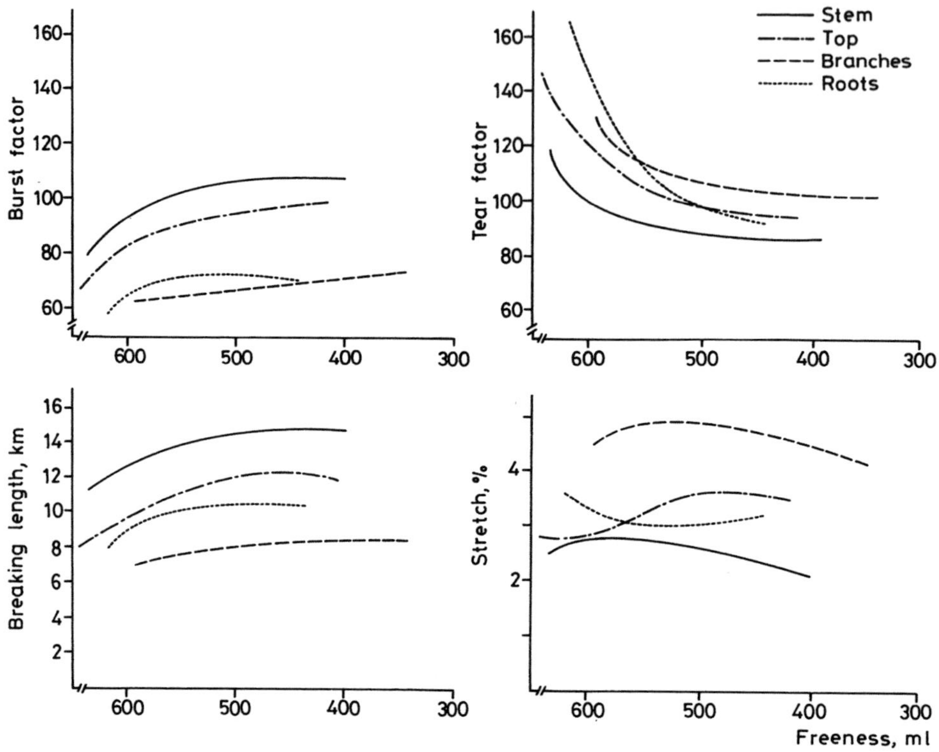

Fig. 8.8. Strength characteristics of unbleached kraft pulps from unbarked tree components of a *Thuja occidentalis* tree with a dbh of 21 cm, as a function of Canadian standard freeness of pulp (redrawn from Dyer 1967)

and might find specialty uses. It was pointed out that in whole-tree pulping of softwoods either branch wood is undercooked or stem wood overcooked. The same problem but reversed is associated with beating.

Chase et al. (1971) evaluated the potential of separate and mixed components of unbarked, small-sized puckerbrush trees in Maine as a raw material for kraft pulp. The tree species studied were *Acer rubrum, Alnus rugosa, Betula populifolia, Populus tremuloides, Prunus pensylvanica*, and *Salix gracilis*. The maximum strength values of kraft pulps were attained generally after 30 min of beating, except for willow pulps which usually started to deteriorate early in the beating cycle. Stem wood produced the strongest pulps with the best yield. Branch pulps were superior to stump and root pulps in burst and tensile strength but inferior in tear. Whole-tree pulps, containing all above-ground wood and bark components of a tree, compared in strength characteristics favorably with standard commercial hardwood pulps made from debarked stem wood. Bleached pulps were generally stronger than unbleached pulps, most significantly in tear and burst strength (Fig. 8.9). The species with the longest fibers, i.e., birch and aspen, produced the strongest pulps. The study established that at least four of the six puckerbrush species can be pulped by the kraft process with the bark intact to

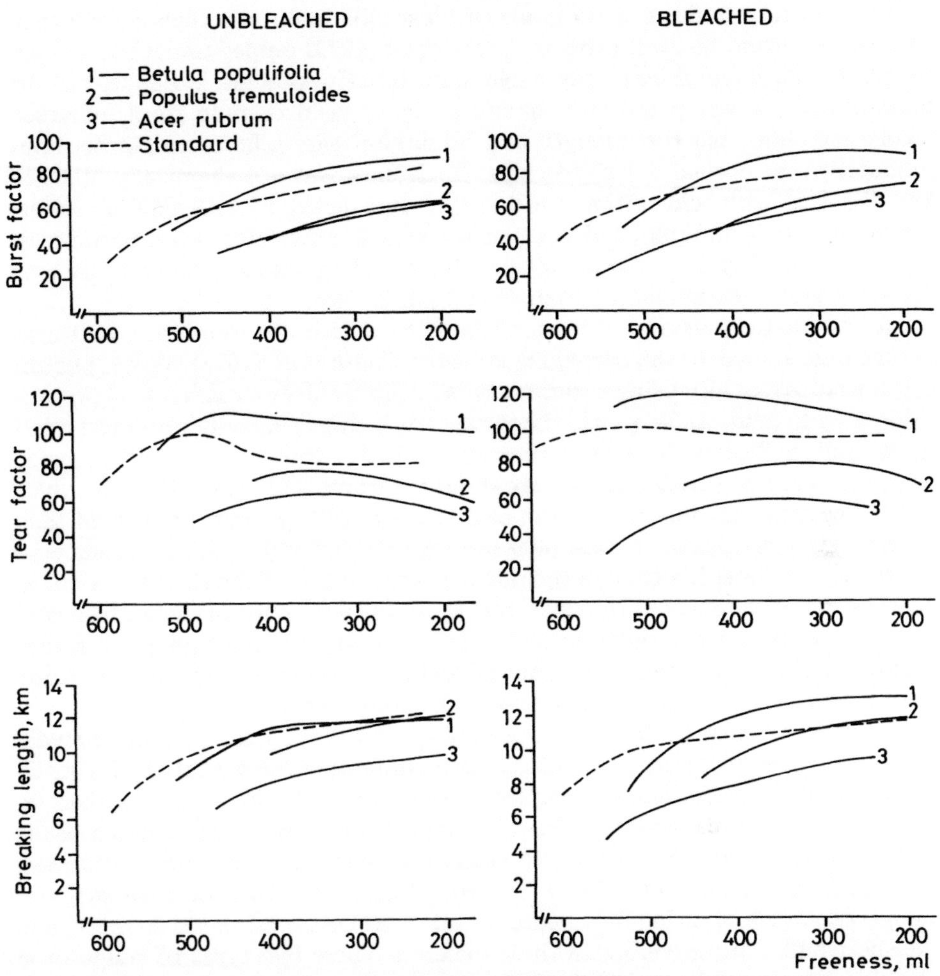

Fig. 8.9. Strength characteristics of unbleached and bleached kraft pulps from whole-tree chips of *Betula populifolia, Populus tremuloides*, and *Acer rubrum*, and those of standard hardwood pulps from debarked stem wood, as a function of Canadian standard freeness of pulp (redrawn from Chase et al. 1971)

produce a pulp that is equal to commercial hardwood pulps in strength characteristics.

As it would be almost impossible to segregate small-sized hardwoods according to species, Chase et al. (1973) continued the whole-tree pulping studies with species mixtures from the same six puckerbrush species. Kraft pulps obtained from the mixtures were slightly lower in physical strength than would be normally expected of pulps similarly produced from standard hardwood chips. A series of cooks of various combinations of mixed whole-tree chips and standard hardwood chips resulted in essentially the same burst, tensile, and tear strength for the mixtures and pulps made exclusively from normal, debarked chips.

In order to produce a good grade of bleachable pulp at a higher yield than was realized from the kraft process, Chase et al. (1973) pulped unbarked and debarked *Betula populifolia* by the magnesium bisulfite process. Compared to the kraft process, it was possible to obtain a higher yield of pulp from unbarked whole-tree chips, but the strength and brightness were inferior to sulfite pulps produced from normal debarked chips. Birch bark was not chemically attacked and dissolved with acid sulfite liquor to the same degree as with alkaline sulfate liquor, and the acid liquor did not penetrate the bark-covered wood particles as rapidly and uniformly as did the alkaline liquor. The pulp had potential use only for some grades of unbleached paper and paperboard.

To explore the potential of unmerchantable puckerbrush stands as a corrugating-medium source for the packaging industry, Chase et al. (1973) cooked puckerbrush whole-tree chips using a sodium-based neutral sulfite semichemical process. Compared to commercial pulps, the puckerbrush chips generally produced a less dense, bulkier sheet with superior strength properties. Burst was 30–50% greater, tensile strength 20–30% greater, tear about the same, and crush strength, which is a critical property for corrugating medium, 10–25% greater. Because of their superior strength characteristics, puckerbrush pulps proved an attractive material for corrugated board, although the physical appearance of the sheets was not as good as that of commercial samples due to undissolved bark particles. In a continued study, the neutral sulfite semichemical process produced from puckerbrush whole-tree chips pulps that compared favorably with commercial material in important strength characteristics (Chase and Young 1976b).

Chase and Young (1976b) examined the suitability for kraft pulping of small-sized trees of seven conifer species from early thinnings. The several species generally attained their maximum strength at different beating times because their responses to the beating action differed. The table below summarizes average results of unbleached pulps from unbarked stems and whole trees of *Abies balsamea, Larix laricina, Picea abies, Pinus resinosa, P. strobus, Thuja occidentalis*, and *Tsuga canadensis*, showing the maximum values of burst and tensile strength, and the sheet bulk corresponding to these maxima. These two types of pulps from thinning raw material were inferior to standard softwood pulp in burst and tensile strength, but the differences were moderate and attributable to the presence of bark, branches, and needles. Bulk, being the reciprocal of sheet density, was well correlated with sheet strength properties. Low bulk indicates a compact, well-bonded sheet, usually meaning good burst and tensile strengths.

	Raw material of pulp		
	Standard debarked stem wood	Unbarked stems from thinnings	Whole trees from thinnings
Bulk (cm^3/g)	1.45	1.60	1.61
Breaking length (km)	14.7	13.4	12.9
Burst factor	99.0	96.3	91.7

Fellegi et al. (1974) made laboratory and mill-scale studies of the suitability of softwood and hardwood branches for raw material of pulp. Both kraft, bisulfite, and thermomechanical pulps from unbarked and debarked Norway spruce branches showed inferior strength properties and could be used only for low-grade wrapping paper or roofing felt. Inferior pulp quality was attributed to the anatomy and chemical composition of branch wood. In *Fagus sylvatica* and *Carpinus betulus* the differences between the morphological properties of branch and stem woods were minor and, consequently, it was not surprising that unbarked hardwood branches produced a good-quality NSSC pulp. Although chemicals consumption was higher in the cooking of branches owing to their large bark content, hardwood branches were graded as suitable raw material for corrugating medium. Tensile and burst strengths and folding endurance were either on the level of values for stem wood pulps or higher.

Eskilsson (1974) characterized unbleached and bleached kraft pulps made from different components of Scots pine and Norway spruce. The very high content of compression wood in softwood branches led to inferior pulps with low strength. Debarked branch wood more than 5 mm in diameter produced pulps with 70% of the strength of stem-wood pulps. The strength of bark pulp was 50% and that of pulps made from needles and under 5 mm diameter twigs only 30% of that of normal stem-wood pulps. Branch-wood pulp was considered to be of low economical value for "strong" papers. For bark, twig, and needle pulps, neither the yield nor strength were sufficient to cover the cost of processing. The author therefore recommended that branch components should be at least partially removed from whole-tree material before pulping, especially since they give rise to various operational problems and are thus unsuitable in today's mill practices.

Due to the high proportion of juvenile and compression woods, pulps from unmerchantable tops possessed the same characteristics as branch-wood pulps, albeit to a significantly lesser degree. These pulps were probably acceptable for most purposes. Pulping of wood from stump and over 25-mm roots gave rise to no special problems. Fibers from stump and root woods were smaller and slightly weaker than those from stem wood, resulting in a somewhat lower sheet strength. However, the difference was assumed to be insignificant in a mixture of pulps. The fibers in under 25-mm roots had quite another character and produced inferior pulps, but since a majority of small roots is left behind in stump extraction, this component may be disregarded.

Table 8.5 summarizes the results of Eskilsson's study on strength characteristics of kraft pulps made from various tree components. The last column shows the product of relative cooking yield and relative strength, illustrating the relative strength obtainable from a unit mass of raw material. In addition to the component pulps, Table 8.5 also presents relative strength data for two pulp mixtures. The first mixture represents the natural composition of complete-tree mass with small roots excluded, and the second partially cleaned complete-tree mass with only half of the original bark, needle, and twig mass left.

While pulps from unmerchantable tree components were inferior to stem-wood pulps in the most important strength properties, they were characterized by higher stretch and, possibly partly related to this, a comparatively low bending stiffness. In branch-wood pulps the stretch was 6–8% before beating and as much as 8–10% after beating, while the corresponding values for stem-wood pulp were only 3.5 and 5.5%. Eskilsson assumed that the general trends found in

Table 8.5. Relative strength obtainable from a unit mass of raw material in kraft pulping biomass components of *Pinus sylvestris* and *Picea abies* (Eskilsson 1974)

Biomass component	Relative cooking yield	Relative pulp strength	Strength obtainable from unit mass of raw material
Stem wood	100	100	100
Top wood	93	80	74
Stump wood	96	90	86
Root wood	95	90	86
Branch wood	76	70	53
Unbarked twigs	38	30	11
Bark	40	50	20
Needles	49	30	15
Complete tree	89	93	83
Complete tree, partially cleaned	93	97	90

the study are probably valid also for conifers other than Scots pine and Norway spruce as variation in the morphology and chemical composition of wood among different tree components follows more or less the same pattern in most softwood species. The results were not considered applicable to hardwoods which differ in several respects from softwoods.

Lönnberg (1976b) illustrated the effect of bark upon the strength of kraft pulps from debarked, 3- to 5-year-old *Alnus incana* and *Betula pubescens* stems by adding a 10–20% charge of bark pulp from the same trees. The bark pulp fines raised the initial drainage resistance of the pulp, resulting in a significantly shorter beating time for the attainment of a given degree of freeness. As a consequence, the fibers were insufficiently fibrillated, and the strength obtained was inferior. Moreover, it was evident that the bark fines hindered smooth fiber-to-fiber contact and, therefore, fiber bonding. The author concluded that, from the standpoint of pulp strength, 25% marginal raw material from small-sized hardwoods containing significant quantities of bark might be substituted for conventional birch chips.

Blasinska (1977) made kraft pulp from unbarked and debarked chips from trimmed Scots pine branches and chips from crown mass, containing all of the branch wood, branch bark, twig, and needle components in their natural ratio. When used alone, trimmed branch-wood chips gave inferior pulps. However, when they were blended with standard pine chips the result was satisfactory. With a mixture of 30% branch wood and 70% stem wood, the pulps obtained were acceptable. When the proportion of branch wood was increased to 50%, the quality was lowered considerably. Chips made from crown mass produced unacceptable pulps. The laboratory results were confirmed by a commercial trial.

Bublitz (1976) made laboratory kraft pulps from a mixture of *Pseudotsuga menziesii* and *Tsuga heterophylla* whole-tree chips, from a mixture of *Acer* spp. and *Alnus rubra* whole-tree chips, and from standard *Pseudotsuga menziesii* chips from sawmill residues. Whole-tree chips came mainly from young thinnings, whereas the control sawmill chips originated primarily from the outer growth rings of large stems. The shorter, thin-walled fibers in softwood whole-tree pulps improved internal bonding, and denser paper resulted from the ease with which they collapsed. Therefore, whole-tree pulps were stronger than or equivalent to control pulps with regard to physical properties dependent on good fiber bonding, such as burst and tensile strength, folding endurance, and stretch. On the other hand, control pulps exhibited higher tear strength. The small amount of bark in softwood whole-tree chips, only 5–7%, did not seriously degrade strength properties. Hardwood whole-tree pulps were weaker than the softwood control pulp in the early refining stages at bulk values of 2.00–1.75 cm^3/g, but gained strength quickly during refining. With the exception of tear strength, hardwood whole-tree pulps were equal to or stronger than control pulps at bulk values of 1.50–1.25 cm^3/g. However, in the hardwood whole-tree chips used in the study, the proportion of bark was unusually low, only 3%.

Murakami et al. (1979) made kraft pulp from mature and juvenile stem woods, over 2-cm debarked branch wood, and under 2-cm unbarked branches of a *Pinus densiflora* tree. Unbeaten branch pulp produced dense handsheets with a comparable tear factor and tensile strength and folding endurance superior to that of mature and juvenile stem wood. However, beating affected the strength more favorably in stem-wood pulps.

Heimola et al. (1980) made bleached kraft pulp from unbarked and debarked branches of Scots pine. The greatest difference between branch-wood and stem-wood pulps was found in the shrinkage. Potential shrinkage of unbeaten branch-wood pulps was 11–12%, whereas that of the stem-wood pulps was about 5%. The difference remained after beating. The high potential shrinkage of the branch-wood pulps was probably related to the high proportion of compression wood fibers and their unusually great fibrillar angle. Due to these shrinkage properties and poor dimensional stability, the authors did not consider pine branch-wood pulps as a very suitable raw material for wood-free printing papers. Instead, they were considered as possibly suitable for grades such as sack paper which require good stretch properties. However, relatively low strength level and poor porosity will probably be limiting factors.

Law and Koran (1982) cooked kraft pulps by the same schedule from debarked stem wood, debarked branch wood, and unbarked branch mass of *Picea glauca*. Handsheets made of unbeaten bark-free branch pulp showed higher bulk and stretch, comparable burst and tensile strengths, and considerably lower tear strength than sheets made from stem-wood pulp. In the case of beaten pulps, branch pulp gave comparable tensile strength and stretch and lower tear strength. The presence of bark cells in branch pulp reduced the sheet strength and caused a dirt problem. Pulping mixtures containing up to 20% branch wood reduced the tear, burst, and tensile strengths of stem-wood pulp by about 10%. It was concluded that the inclusion of branch wood in the furnish within the natural amount limit would be possible, but branch bark should be allowed only in chips used for paper products where cleanliness is not a critical factor.

The studies referred to above show quite unanimously that pulps produced from branch wood, both unbarked and debarked, are inferior in their strength to all other pulps made from various wood components of a tree. The low physical strength of branch pulps has been shown in several other studies as well (Hägglund 1925; Sproull et al. 1957; Kubínek and Šedivý 1959; Ivanova 1960; Yonezawa et al. 1962; Worster and Vinje 1968; Starck and Palenius 1970; Ageeva et al. 1978). Nevertheless, when branch wood is present as a mixture in whole-tree chips in the natural ratio, it does not seriously affect the strength characteristics of the pulp.

Stump and root woods may also be blended in normal softwood chips in significant proportions without any major effect on the strength properties of kraft pulps. A mill-scale trial in Sweden with a chip mixture containing 80% normal stem wood and 20% stump and root wood of Scots pine and Norway spruce did not result in reduction of pulp strength except in tear index, which was reduced by 6% (Projekt ... 1977 a). A linerboard mill of Interstate Paper Corp. in Georgia was reported to obtain 30% of its raw material from used wood of mature pine stumps after extraction of chemicals by another plant. The reclaimed stump fiber had no adverse effect in the properties of linerboard (Davis 1971).

8.2.2 Appearance and Optical Properties of Pulp and Paper

In addition to strength, appearance and optical properties are of primary significance in many grades of paper, particularly in fine papers made of bleached pulps. Among the appearance factors, brightness, dirt specks, and opacity of the paper are especially important. Contaminants such as bark, twigs, foliage, and grit increase the content of poorly cooked shives and dirt in the pulp, adversely affecting the brightness and cleanliness of paper.

The kraft and NSSC processes are tolerant regarding the composition variation and purity of the raw material, but excessive amounts of contaminants are of serious concern for them as well. According to Hartler et al. (1977), in Scots pine and Norway spruce chips the layer of inner stem bark is especially hard to digest. Bark limits for chips are determined by both operational disturbances and the appearance of the paper. Measures such as chip upgrading, prolonged cooking to a lower Kappa number, effective screening and centricleaning of pulp, and increased chemical charge in bleaching are applied to improve the appearance of the end product in conjunction with the use of rough chips.

Horn and Auchter (1972) made kraft pulp from 12 softwood species from the western region of the United States in order to establish the consequences of a high content of bark in the raw material. A comparison was made between clean, debarked and rough, unbarked chips from normal stem wood. The average content of bark for the 12 species was 10.7% on a dry mass basis. By efficient centricleaning and bleaching, the amount of *dirt specks* could be lowered to an acceptable level, comparable to bark-free pulps. However, chips with a very high bark content resulted in nonacceptable dirt levels in the pulp, requiring more efficient centricleaner systems. When black spots with an area of 0.04 mm^2 or more were counted by Tappi Standard Method T-213 ts-65, the average results were as follows, expressed in terms of parts per million of surface area (see Scheme p. 397).

Wawer (1975) found the amount of bark specks in bleached hardwood kraft pulps extremely large during high bark periods in chip furnish in conjunction with

	Dirt level of pulp (ppm)	
	Debarked wood	Unbarked wood
After screening	64	461
After centricleaning	20	45
After bleaching	6	9

mill-scale pulping of aspen whole-tree chips despite extensive screening and centricleaning. This pulp could not be accepted for the manufacture of fine papers. Microscope examination showed minute shives and pieces of inner bark in the pulp. Powell et al. (1975), when using hardwood whole-tree chips in mill-scale operation, found it necessary to lower slightly the target Kappa number of unbleached kraft pulp to keep the dirt level under control. The target number was 12.0 for standard chips, 10.9 for screened whole-tree chips, and 10.7 for unscreened whole-tree chips.

In addition to the amount of dirt specks, *brightness* and its stability is another important appearance factor affected by chip contaminants. Alestalo and Hentola (1966) found the brightness in kraft pulps from unbarked tops and branches from Scots pine and Norway spruce to be considerably lower than in pulps from debarked stem wood. Inclusion of needles in the chips further reduced the brightness. Hosia et al. (1971) found the brightness difference in corresponding pulps from the same species much smaller. The reason was assumed to be a higher chemical charge.

Young and Chase (1965) compared the optical properties of pulps made from unbarked stem, top, root, and branch components of *Picea glauca* by two different acid processes. Among pulps produced by the nitric acid process root and top pulps showed the highest brightness. Branch pulps were less transparent than the other pulps, i.e., they had the highest opacity. The ammonium sulfite process gave the same results.

Eskilsson (1974) made kraft pulp from separate components and natural component mixtures of Scots pine and Norway spruce. Of the unbleached pulps, the branch-wood pulps had a slightly lower brightness than other wood pulps, but this could be explained by a higher Kappa number. Pulps from twigs and pine needles had a rather low, and those from spruce needles an extremely low brightness. *The purity of color* was approximately the same for most pulps. The bark pulps were an exception, showing a lower spectral purity so that the color appeared more gray. The low brightness of the pulps made from the nonwoody components adversely affected the brightness of a component mixture, even when these components constituted only a moderate percentage of the raw material.

In unbeaten bleached pulps from bark-free top, branch, stump, and root woods, the brightness was at least as good as that of stem-wood pulps. Although the fiber material from branches was slightly more colored, the higher light-absorption coefficient was compensated for by a higher scattering coefficient due to the smaller fibers. When bleached pulps were beaten to a given sheet structure, pulps from branch and top woods had a significantly higher brightness than the pulps from stem wood. This was fully explained by the very high scattering coefficient, caused by the small and narrow fibers. Pulps from twigs, needles, and bark had a much lower brightness than those from wood raw materials. This was mainly due to light absorption in the fiber material itself and only to a lesser extent to the sheet structure. In a pulp mixture, however, the effect of those colored components was hardly noticeable.

The *brightness stability* was roughly equal for all pulps except for those from needles. The color of the needle pulps was very sensitive to heat. An accelerated aging test caused these pulps to turn almost brown. Instability to heat treatment was also quite noticeable in the pulps prepared from mixtures of the tree components. According to Eskilsson, the behavior of needle pulps was explained by the high content of extractives in foliage and their ability to survive both pulping and bleaching.

Starck and Palenius (1970) studied the effect of stem-bark and crown components on the suitability of chips from small-sized Scots pine and Norway spruce trees from early thinnings for the production of kraft pulps (Table 8.6). The brightness of unbleached pulps was affected more readily by bark than by crown mass. In pine pulps, an increase in the bark and crown mass proportion in the chips resulted in an increase in the amount of dirt specks in the pulp. This tendency did not appear in spruce pulps. Dissolving of needles seemed to be complete for both species.

Lönnberg (1976c) studied the suitability of bleached kraft pulp made from hardwood whole-tree chips for wood-free printing papers. The whole-tree pulp studied was a blend of equal amounts of *Alnus incana, Betula pubescens, Populus tremula*, and *Salix caprea* pulps, all prepared from leafless, 10- to

Table 8.6. The effect of bark and crown mass on the brightness of unbleached kraft pulps (pH 4.5) made from *Pinus sylvestris* and *Picea abies* thinnings; Kappa number 32 (Starck and Palenius 1970)

Tree species	Relative composition of chips			Brightness of pulp SCAN (%)
	Stem wood	Stem bark	Crown mass	
Pine	100	–	–	30.8
	100	4	–	28.3
	100	8	–	26.8
	100	–	8	30.0
	100	–	16	29.4
	100	8	16	26.5
Spruce	100	–	–	31.2
	100	3	–	29.9
	100	6	–	28.9
	100	–	32.5	28.1
	100	–	65	26.4
	100	6	65	25.6
Pine	Whole tree			25.4
Spruce	Whole tree			23.5

15-year-old whole trees. This pulp was mixed in various proportions with a commercial, bleached pine kraft pulp. For comparison, similar procedures were achieved with mixtures of commercial, bleached birch and pine kraft pulps. Whole-tree pulp provided a denser sheet and a better opacity than normal birch pulp. At a low, 10%, proportion of hardwood pulp among pine pulp, the effects on opacity and sheet-surface roughness were not significant. However, a blend with 25% hardwood pulp clearly indicated a lower degree of roughness for the fine paper made from whole-tree pulp as one component. The effect of whole-tree pulp upon opacity was also favorable. The wire side of the "whole-tree paper" showed distinctly longer oil-absorption time than normal fine paper, but the top side of the two paper qualities was equivalent.

Since the strength properties of branch-wood pulps from softwoods are comparable to those of stem-wood pulps from hardwoods, and since they generally have good optical properties, they are of interest as a raw material for fine papers. Heimola et al. (1980) studied the suitability of bleached kraft pulps from unbarked and debarked branches of Scots pine for wood-free offset printing papers normally containing 20–60% birch pulp and 40–80% pine pulp. The light-scattering coefficients of the branch pulps were significantly higher than those of the standard pine pulp. In the case of bark-free branch wood, they were also higher than those of standard birch pulp. Bark had a definite impairing effect on the light-scattering ability, and after the initial stage of beating, the branch pulp with bark reached light-scattering coefficient values below that of standard birch pulp.

When the proportion of bark in *Populus tremuloides* chips exceeded 3%, the effectiveness of the screening and cleaning system of a Canadian bleached kraft pulp mill to remove bark shives and specks diminished greatly (Wawer 1975). Hunt and Hatton (1975) found an increase of three percentage points in screen rejects of *Acer saccharum, Betula alleghaniensis*, and *Fagus grandifolia* pulps, when the bark content of chips was 10%. Similarly, a bark content of 20% gave an increase of six percentage points in screen rejects. For *Populus tremuloides* the corresponding figures were even higher. An effective pulp cleaning system is thus of great importance in conjunction with whole-tree pulping. Nevertheless, the authors concluded that most modern kraft mills have suitable chip-screening and centrifugal cleaning equipment to process satisfactorily unbleached pulps from chip furnishes containing up to of 20% hardwood or softwood whole-tree chips, or bleached pulps from furnishes containing up to of 10% of such chips. They emphasized that whole-tree chips should be metered into the digester or chip-screen feed at a uniform rate in order to guarantee optimum results.

8.2.3 Pulp Yield

Pulp yield shows the proportion of the original mass of chips recovered, on the dry mass basis, in the pulping process. The pulp referred to may be unscreened or screened and also unbleached or bleached. In chemical pulping, the yield is determined by raw material factors, such as the cleanliness and chemical composition of the wood, and numerous variables of the pulping process. The major technical factors affecting the unbleached kraft pulp yield are the alkali charge and the Kappa or permanganate number of the pulp. The *Kappa number* is a measure of the lignin content of the pulp, estimated from the consumption of potassium permanganate by the pulp. According to Tasman and Berzins (1957) and Kleppe (1970), the percent of residual lignin in pulp is equal to the Kappa number multiplied by 0.15.

Effective alkali varies from about 15 to 23% of the dry mass of the chips charged. According to Legg and Hart (1960) and Aurell and Hartler (1965), an increase in the effective alkali charge of 1% NaOH on wood will reduce pulp yield by 0.10–1.15 percentage points. An increase in sulfur in the cooking liquor increases the yield slightly.

When pulping whole-tree chips, it is very difficult to ensure that all biomass components cook to the same degree of delignification. Pulps from unbarked chips tend to have higher Kappa numbers than pulps from bark-free wood. However, to make pulp fibers acceptable, cooking must continue to a given Kappa number prior to bleaching. Since branch wood contains more lignin than stem wood, its delignification also requires a longer time. Therefore, some components of whole-tree chips will overcook and some will undercook. If the cooking is scheduled to minimize the percentage of poorly cooked screen rejects, overcooking of part of the chips will cause yield losses.

Pulp yields are low for any bark, but considerable differences occur between tree species. In a study by Horn and Auchter (1972) of 12 softwood species from the west coast of the United States, the yield from stem bark was about 20%. When stem chips were pulped unbarked, the total yield was reduced from 46.1 to 43.6%, and the screened yield from 45.4 to 42.9% due to an average bark content of 10.7%.

Hartler et al. (1977) made kraft pulp from bark of different tree components of Scots pine, Norway spruce, and birch. For all species, the yield from stem bark was highest and that from root bark lowest. However, the crusty stem bark was more difficult to pulp than other bark types, as indicated by more specks in the pulp. Stump bark and branch bark were intermediate in yield:

Raw material	Yield of kraft pulp (%)		
	Pinus sylvestris	*Picea abies*	*Betula* sp.
Stem bark	26.1	17.4	27.1
Root bark	10.0	14.5	17.5

According to Hunt and Hatton (1975), the screened kraft pulp yield from the stem mass of *Betula alleghaniensis, Fagus grandifolia*, and *Populus tremuloides* was four percentage points lower for every 10% bark in the raw material. For *Acer saccharum* the reduction was three percentage points. Hardwood barks were considered less detrimental to pulp yield and cooking conditions than the barks from softwoods (Table 8.7).

Alestalo and Hentola (1966) studied the suitability of unbarked stump and root mass of Scots pine and Norway spruce for the production of unbleached kraft pulp. In comparison with pulps made from debarked stem wood of pine,

Table 8.7. Effect of wood/bark mixture on screened yield and alkali consumption in kraft pulping of stem mass of *Betula alleghaniensis* (Hunt and Hatton 1975)

Raw material composition		Percent of dry mass of chips			Kappa number
Wood (%)	Bark (%)	Screened yield	Screen rejects	Alkali consumed	
100	0	51.6	0.8	11.1	12.3
80	20	44.0	6.6	11.1	14.5
60	40	35.3	12.1	10.9	16.2
40	60	26.5	21.0	10.7	20.2
20	80	19.0	26.8	10.3	21.4
0	100	15.7	26.8	10.9	17.3

the yields were lower. In the case of stumps, the difference could be attributed to larger amounts of bark and extractives. In the root mass of spruce these variables did not explain the yield difference:

Raw material	Dichloromethane extractives in wood (%)		Total yield of kraft pulp (%)	
	Pinus sylvestris	*Picea abies*	*Pinus sylvestris*	*Picea abies*
Standard chips			46.8	
Unbarked stumps	5.8	1.1	44.0	45.0
Unbarked roots	3.0	1.0	44.4	41.0

In pine wood the lower yield of kraft pulp from stumps and roots is well compensated by a higher yield of by-products. For example, the average yield of resin products in Poland is 48 kg/m^3 for fresh stumps and 121 kg/m^3 for mature stumps of Scots pine. In Sweden, a 20% admixture of rough-screened chips from pine stumps doubled the recovery of tall oil soap simultaneously with a slight decrease of the resin content in the pulp (Projekt ... 1977a).

Keays and Hatton (1971a, b) prepared unbleached kraft pulps from bark-free biomass components of *Picea glauca* and *Tsuga heterophylla*. The possible synergistic or antagonistic interaction between tree components was studied in detail by use of a microdigester assembly. The effect of any one wood component on the pulping characteristics of any other wood component were, if existing, quite small. At Kappa number 20, all components, except branch wood, gave pulp yields of 94% or higher of that of stem wood. The yield from branch wood was considerably lower (see Scheme p. 402).

Hosia et al. (1971) found the yield of unbleached kraft pulp from the crown mass of Scots pine and Norway spruce to be 31.2 and 26.7%, respectively. When all needles and bark were removed from the chips, the yields from bark-free branch wood were 39.5 and 36.9%, respectively. The Kappa number varied from 39 to 49. Fellegi et al. (1974) examined the possibility of producing packaging ma-

Raw material	Unscreened yield of kraft pulp (%)	
	Picea glauca	*Tsuga heterophylla*
Stem wood	47.7	45.4
Top wood	44.8	43.3
Branch wood	36.8	33.7
Stump wood	45.4	43.4
Root wood	46.5	44.7

terials, such as wrapping paper, linerboard, and corrugating medium from unbarked branches. Laboratory studies gave a 35.0% kraft pulp yield from unbarked branches of Norway spruce, a result confirmed by a mill-scale experiment. Unbarked branches of *Carpinus betulus* and *Fagus sylvatica* gave kraft pulp yields of 40.9 and 40.7% and NSSC pulp yields of 65.8 and 71.3%, respectively.

Eskilsson and Hartler (1973) made kraft pulp from various biomass components of Scots pine and Norway spruce trees 20–30 cm in diameter. The recovery of cellulose was 86–97% from the wood components, about 60% from needles, and even less from bark. Accordingly, the total yield of pulp was very low from needles and bark, 22–25 and 19–20%, respectively (Table 8.8).

Table 8.8. Recovery of cellulose and total yield of pulp from kraft pulping of biomass components of *Pinus sylvestris* and *Picea abies* (Eskilsson and Hartler 1973)

Biomass component	Total pulp yield (%)		Recovery of cellulose (%)	
	Scots pine	Norway spruce	Scots pine	Norway spruce
Stem wood	47.2	49.7	93	96
Stump wood	46.8	44.0	86	94
Coarse root wood	46.2	44.0	87	97
Slender root wood	37.9	35.6	93	97
Needles	22.3	25.2	61	58
Bark	19.3	19.8	58	42

Chase et al. (1973) made unbleached kraft pulps from unbarked stem and branch mass of small-sized *Acer rubrum, Alnus rugosa, Betula populifolia, Populus tremuloides, Prunus pensylvanica*, and *Salix gracilis*. Compared to pulps from debarked commercial hardwood chips, the average yield from these unbarked puckerbrush stems was only slightly lower. On the other hand, unbarked branches gave a considerably lower yield despite higher Kappa numbers. All puckerbrush pulps exceeded commercial pulp in alkali solubility, indicating a higher content of hemicellulosic material. This was assumed to account for the fact that some of the puckerbrush pulps were stronger than the commercial pulps. The average yields for the six puckerbrush species were as follows:

Raw material	Total pulp yield (%)	Kappa number	Alkali solubility (%)
Commercial chips	50.5	30.1	5.4
Unbarked puckerbrush stems	49.4	25.0	8.9
Unbarked puckerbrush branches	43.0	35.8	8.2

Chase and Young (1976b) found that the screened yield of kraft pulps from seven conifer species from thinned stands in Maine was strongly affected by the presence of bark and needles. The average yield from unbarked stem wood was 87% and that from whole-tree chips only 78% of the yield obtained from a unit mass of bark-free standard chips, although the degree of delignification remained higher:

Raw material	Kappa number of pulp	Yield of pulp (%)
Bark-free standard chips	23.5	46.1
Unbarked stems from thinnings	29.6	40.3
Whole trees from thinnings	31.3	36.0

Lönnberg (1976a, b) made kraft pulp from whole-tree material from *Alnus incana, Betula pubescens, Populus tremula,* and *Salix caprea* trees 3–5, 10–15, and 20–25 years of age. The yield from leaves ranged from 13 to 19%, depending upon the tree species. The presence of leaves tended to increase the Kappa number and the amount of screening rejects. The yield from bark was generally 25–30%, but only 15% for alder. The author concluded that 4% bark in the chips is obviously acceptable from the standpoints of fiberizing and bleaching. For aspen and

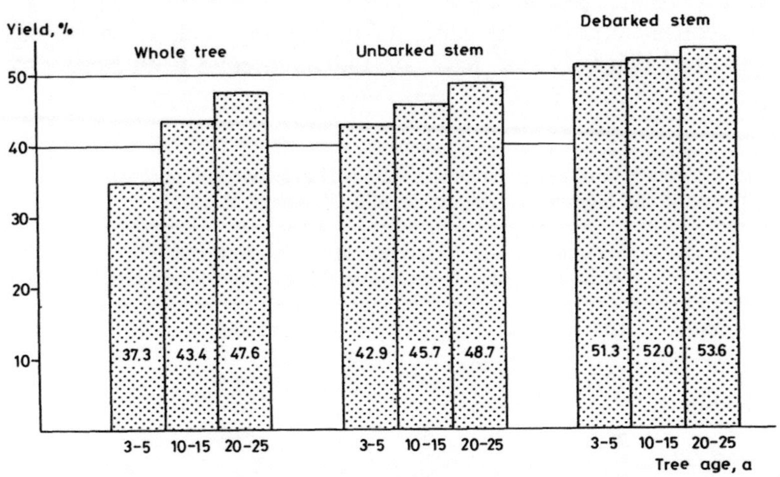

Fig. 8.10. Total kraft pulp yield from debarked stems, unbarked stems, and whole trees of 3–5, 10–15, and 20–25 years of age. Average values for *Alnus incana, Betula pubescens, Populus tremula,* and *Salix caprea* (Lönnberg 1976a)

willow, a bark content as high as 8% in the chips did not seem unrealistic for kraft pulping. Inclusion of both foliage and bark reduced the average yield from whole-tree chips significantly, especially in biomass from young trees with a high content of nonwoody components (Fig. 8.10).

Heimola et al. (1980) prepared bleached kraft pulps from unbarked and debarked 1- to 8-cm-thick branches of 60- to 70-year-old Scots pine trees. Yields from bark-free branch wood were about ten percentage points lower than those from stem wood because of differences in chemical composition. The yield from branch bark was only 20% and, consequently, the pulp produced from unbarked branches contained about 5% bark fibers. The total yield at Kappa number 33 was as follows:

Raw material	Total yield (%)	
	Unbleached pulp	Bleached pulp
Unbarked branches	37.0	34.2
Debarked branches	38.5	35.8

It is quite clear that fiber yield, when expressed in percent of raw material cooked, is reduced in conjunction with whole-tree pulping. The reduction in yield is primarily a result of the high lignin content in softwood branches, high contents of bark and other contaminants, and uneven cooking. On the other hand, if the final yield is calculated as a percentage of the unbarked raw material arriving at the mill, yield differences between conventional bolts and whole-tree chips are rather small (Table 8.9).

Pulp yield together with the basic density of the raw material determines *the specific raw material consumption* in pulping, generally expressed in m^3 solid per tonne of pulp at 10% moisture content. Large differences occur between biomass components. For example, the production of a tonne of bleached kraft pulp at 10% moisture content required 5.1 m^3 stem wood, 4.6 m^3 stump wood, 7.4 m^3 branch wood, 9.1 m^3 unbarked twigs, 17.2 m^3 bark, or 14.0 m^3 needles from Scots pine

Table 8.9. Final yield of bleached kraft pulp from oak-dominated hardwood raw material, as calculated from the unbarked biomass arriving at the mill (Powell et al. 1975)

Yield factor	Standard hardwood chips	Unscreened whole-tree chips	Screened whole-tree chips
Unbleached Kappa number	12.0	10.7	10.9
Material to digester, percent of material to mill	90.0	100.0	94.4
Screened yield, percent of material to digester	46.8	43.7	44.7
Bleached yield, percent of material to digester	44.5	40.8	42.6
Final yield on material to mill	40.1	40.8	40.2

Table 8.10. Relative basic density, yield, and raw material consumption on a volume basis in kraft pulping of biomass components of *Pinus sylvestris* and *Picea abies*; Kappa number 35 (Projekt ... 1977a)

Biomass component	Scots pine			Norway spruce		
	Basic density	Yield	Raw material consumption	Basic density	Yield	Raw material consumption
Stem wood	100	100	100	100	100	100
Top wood	89	93	121	100	93	108
Stump wood	116	96	86	100	96	104
Over 5 cm root wood	115	95	92	121	95	87
Branch wood	112	76	117	159	76	83
Twigs	106	38	248	148	38	178
Bark	64	40	390	85	40	294
Needles	85	49	240	112	49	182

(Stade 1976). Table 8.10 shows the relative raw material consumption on the volume basis when pulp is made from various biomass components of Scots pine and Norway spruce. The consumption is least for stump and root wood of pine and root and branch wood of spruce due to their high basic density, and most for needles due to their low basic density and low yield (Projekt ... 1977a). The capacity of a digester is determined by the raw material consumption in solid m^3/t, the solid content of chips in the digester, and the time required for cooking. All these variables may also affect the mill capacity, depending on the dimensions of succeeding operations in the process.

8.2.4 Beatability and Drainage of Pulp and Paper

In a sheet of paper the fibers are kept together because wet hemicellulose and cellulose surfaces become attached to one another by hydrogen bonds between hydroxyl groups on drying. The strength of these bonds may be enhanced by mechanical treatments which alter fiber features such as cross-sectional shape, void volume, and length. In this process, referred to as *beating*, fibers are crushed, cut, and macerated by refining them in the presence of water. The original tubular cross-sectional form disappears when the fibers collapse, thus improving fiber flexibility. The fibrils of the cell wall are partly loosened or fibrillated which increases the fiber-to-fiber contact area. The fibrillation particularly and, to some extent, the flexibility are further improved by removal of the constraining primary cell-wall layer. All these effects of beating enhance interfiber bonding and produce a denser, stronger, and smoother paper.

The density, fold endurance, burst, and tensile strength of paper typically increase with beating. On the other hand, tear strength decreases simultaneously. If beating is carried too far the fibers will suffer from excessive physical damage, and strength will start decreasing after reaching a maximum. As beating proceeds, the *freeness* of the pulp decreases, i.e., *drainage* of water from the pulp becomes slower. The degree of freeness thus indicates indirectly the development of pulp strength resulting from the beating.

Eskilsson (1974) examined the impact of beating on kraft pulps from various tree components of Scots pine and Norway spruce. Stump- and root-wood pulps deviated only slightly from the stem-wood pulps with respect to beating and drainage, and these deviations were completely explained by the differences in the cross-sectional dimensions of the fibers. Branch wood, especially from pine, and top wood gave pulps that were easy to beat to a dense sheet structure. In a pulp mixture where the beating energy was adjusted to the main components, the pine branch-wood pulp tended to be overbeaten and was seen as a potential source of drainage problems. When the beating was adjusted to give a predetermined sheet density, the dynamic drainage rate was only slightly lower for branch-wood pulps than for stem-wood pulps.

Pulps from the nonwoody parts of the tree gave paper sheets with an unusually high density and with accompanying difficulties in drainage of the pulp slurry. The most extreme pulp was the bleached pine twig pulp, which in the unbeaten state gave a high sheet density of around 1 g/cm^3. It was obvious that even small amounts of twigs, bark, and, to some extent, needles in the raw material affect the sheet structure and drainage properties considerably. Accordingly, a complete-tree pulp, prepared from a mixture of all tree components in their natural proportions, required only half as much energy when beaten to a given sheet structure and had a 70% longer drainage time than the stem-wood pulp. According to Eskilsson, the drainage behavior of the pulp was partly a result of the flexibility of the fibers and partly caused by the excessive amount of fines in the pulp slurry.

Heimola et al. (1980) found the properties of branch-wood kraft pulp from Scots pine to be satisfactory for use as partial raw material of fine paper with the exception of dimensional stability. Simultaneously beaten furnishes containing both branch-wood pulp and stem-wood pulp also did not fulfill the requirement of stability. Dimensional stability could be improved by beating the two pulp components separately and treating the branch-wood pulp very gently. In this way it was possible to meet the requirements with mixtures containing 30% branch-wood pulp. However, it was questionable whether gently beaten or unbeaten branch-wood pulp could be used without impairing the surface strength of the paper.

Hardwood pulps are generally beaten faster and require less energy than softwood pulps. Lönnberg (1976b) found the drainage significantly affected by the bark fines in whole-tree pulps from small-sized hardwoods. Bark fines actually gave a higher drainage resistance to whole-tree pulps than was provided by fibrillation on beating, thus resulting in reduced pulp strength. Pulping chips from unbarked stems or whole trees of 20- to 25-year-old hardwoods required at least 20% more washing capacity than did pulping the same material after debarking. The drainage values achieved with 10- to 15-year-old whole trees were definitely too low.

High drainage resistance results in dewatering problems and, consequently, reduced speed on the drying machine in papermaking. Mill-scale runs show that difficulties are not significant when the share of screened whole-tree chips in the raw material is not more than 15%. However, the effect of crown material is more severe. In a Swedish trial, a 10–20% mix of rough-screened chips from softwood logging residues from clear-cut areas resulted in drainage problems on the drying machine. The wet end did not dewater the pulp sufficiently and operational problems arose in the press section. This necessitated a reduction in the machine speed (Projekt ... 1977a).

8.2.5 Consumption and Recovery of Chemicals

The consumption of cooking chemicals is governed largely by the percentage of raw material which is dissolved in pulping. Bark and foliage require more alkali than debarked wood per unit mass of chips, and especially so per unit mass of pulp produced. However, as pointed out by Chase et al. (1973), the term *chemical consumption* is something of a misnomer since most of the active chemicals in the cooking liquor in the kraft process are recovered from the spent or black liquors and reused in what is essentially a cyclic closed system. On the other hand, the term does give an indication of the amount of chemicals required for cooking various types of chips to a given Kappa number. Deficiency in liquor charge results in coarse pulp and, consequently, in an excessive amount of screening rejects.

Horn and Auchter (1972) studied the effect of stem bark in kraft pulping of 12 west-coast softwood species in the United States with an average bark content of 10.7%. A slight increase from 11.0 to 11.2% in the consumption of active cooking chemicals such as caustic soda was required to reach a bleachable pulp grade at 30–34 Kappa number, when stem chips were cooked unbarked under laboratory conditions. The increase was attributed to the bark constituents. Relating these data to the reduced pulp yield, the total amount of active chemicals consumed for a tonne of screened unbleached pulp increased by 6.9%. In bleaching, the use of chlorine was increased from 9.6 to 9.8%. Reduction in the dirt level was given as the reason for the need of increased chlorine.

Hosia et al. (1971) found the chemicals consumption of debarked branch wood from Scots pine and Norway spruce to be considerably higher than that of stem wood. For unsorted crown mass containing a large percentage of foliage, the consumption of chemicals per tonne of pulp was twice as high compared to stemwood pulping. Worster et al. (1976) made kraft pulp from southern pine for linerboard. The requirement of active alkali to reach a given Kappa number was 13.8% for regular chips, 14.6% for whole-tree chips, and 17.0% for chips prepared from crown mass with foliage. Bublitz (1976) found the consumption of alkali at 50% kraft pulp yield 2.5–3 percentage points lower for whole-tree chips from *Pseudotsuga menziesii* and *Tsuga heterophylla* thinnings than for sawmill chips from Douglas-fir. Young and Chase (1976b) made kraft pulp from small-sized softwoods of seven species from early thinnings in Maine, using an alkali-to-wood ratio of 0.20. Chips from unbarked stems consumed 76% of the alkali charge when cooked to Kappa number 30. Whole-tree chips from the same trees consumed 79% of the alkali at Kappa number 31. Hakkila and Palenius (1971) kraft pulped Scots pine and Norway spruce chips from early thinnings. The active alkali requirement for Kappa number 33 increased with the proportion of nonwoody components in the chips. The consumption was highest for whole-tree chips of spruce due to the large proportion of crown material:

Raw material	Requirement of active alkali (%)	
	Pine	Spruce
Debarked stem	17.4	17.1
Unbarked stem	17.9	18.1
Whole tree	18.5	20.3

A 4–6% mix of whole-tree chips from southern pine in the total furnish of a linerboard mill did not cause any chemical problems. However, as higher rates of whole-tree chips were fed to the digester, problems appeared. To reduce harmful effects, and to improve pulping control, 0.5 percentage points alkali were added arbitrarily (Conner 1978). According to Temler (1980), whole-tree chips require a slightly higher alkali charge than regular chips in order to keep pulp rejects at an acceptable level and to maintain sufficient residual alkali in the spent liquor to prevent lignin precipitation. Hardwood whole-tree chips containing 7% bark and 2.5% twigs required about one percentage point higher alkali charge but gave three percentage points lower yield than standard hardwood chips with about 2% bark.

In a study by Powell et al. (1975) of kraft pulping of hardwood whole-tree chips, the Kappa-value mill target was obtained with the normal 16% alkali charge. This was surprising because pure bark required much more alkali to reach a given Kappa number. Presumably the additional active alkali needed to pulp the bark in the whole-tree chips was compensated by the smaller alkali requirement in pulping the woody portion of the whole-tree chips. It was assumed that the lignin in the wood of small stems, tops, and branches in whole-tree chips was more easily removed.

Chase et al. (1973) found that whole-tree chips from small-sized puckerbrush trees consumed more chemicals than standard debarked hardwood chips. For each puckerbrush species, the branch mass showed a higher chemical consumption than the stem mass. The figures below are average results for six puckerbrush species in kraft cooks with an alkali-to-wood ratio of 0.20. The last column shows the organic-to-inorganic ratio in the black liquor:

Raw material	Total pulp yield (%)	Alkali consumed (%)	Ratio of organic to inorganic in black liquor
Standard hardwood chips	50.5	66	1.66
Unbarked stems of puckerbrush	49.4	75	1.81
Unbarked branches of puckerbrush	43.0	81	2.01

The black liquor from pulping unbarked chips thus has a higher ratio of organic to inorganic materials than that from bark-free standard chips, if the same alkali-to-wood ratio is used in the cooks. However, if a higher alkali-to-wood ratio is used for unbarked chips, the black liquor may have the same organic-to-inorganic ratio. In the kraft pulping process, energy is recovered from the organic portion of the liquor by burning the solids in a recovery furnace and producing steam. The inorganic solids pass through the recovery process and are reused in the cooking liquor. The higher the organic-to-inorganic ratio, the greater is the heating value of the liquor. The black-liquor solids are an important source of energy for a kraft mill.

Depending on the dimensions of the pulping equipment, an increase in the black liquor may cause overloading of evaporation and recovery boiler capacity. For example, use of hardwood whole-tree chips containing 7% bark and 2.5% twigs produced 20% more solids in black liquor than the standard hardwood chips with 2% bark (Temler 1980). If the full capacity of the recovery furnace of a mill is needed when standard chips are pulped, use of whole-tree chips will ultimately cause a reduction in pulp production because of insufficient evaporation and recovery-boiler capacity, and liquor loss may occur from boil-over. Thus, under certain conditions, use of whole-tree chips may require modifications in the chemical recovery system of a kraft pulp mill.

Bark and foliage contain much greater amounts of cations such as calcium, potassium, and magnesium than bark-free stem wood (Sect 3.3.3). Excessive *concentrations of cations* may accumulate in liquid circulation, evaporation, and chemical recovery systems. Consequently, raw material from residual tree parts may cause problems in a closed waste water circulation system (Virkola 1976). In a mill-scale experiment in Sweden, a 10–20% mixture of rough-screened whole-tree chips from softwood thinnings did not cause any chemical problems. However, it was concluded that where problems exist they may be intensified by the use of whole-tree chips. The largest deposit to accumulate was calcium carbonate, but the increase in silica and aluminum was also considered harmful. A 20% mixture of stump and root woods increased the silica content in kraft pulp and black liquor by 40% (Projekt ... 1977a). Conner (1978) reported chemical problems caused by an increase in calcium scale in the digesters, recirculation lines, and heating surfaces in a linerboard mill due to the use of whole-tree chips from thinnings of southern pine. More frequent boil-outs and acid-cleaning, and careful attention to the residual alkali in the waste liquor fed to the evaporators permitted fairly normal operations.

Since bark and foliage lower the brightness and increase the dirt in a pulp, the chemical requirement on bleaching increases correspondingly. A further increase in chlorine consumption may be caused by black liquor carryover resulting from a higher content of solids and drainage resistance of whole-tree pulps. According to Wawer (1975), the average chemical requirement on bleaching in a kraft mill increased by 0.2 percentage points for each percent of bark in aspen-dominated whole-tree chips. Increase in chemical requirement is generally associated with a lower bleaching yield. For example, Powell et al. (1975) found a bleaching yield as high as 95% for pulps from standard hardwood chips and screened whole-tree chips from oak-dominated hardwoods, but only 93% when whole-tree chips were pulped unscreened.

8.2.6 Wear and Tear of Equipment

Biomass always contains inorganic compounds. The concentration of minerals is greatest in bark and foliage (Sect. 3.3.3). Minerals are embedded in cells partly in the form of large crystals, often as silica or calcium oxalate. During harvesting, trucking, storage, and mill-yard handling, additional sand and grit accumulate in the biomass, particularly in the bark.

Conner (1978) developed a classification system of three categories for sand and grit in forest biomass. Surface sand and grit that will settle consists of the heavier, abrasive particles, which can be separated by washing, agitation, or air flotation. Suspended sand and grit contains lighter particles that remain suspended in a fluid system, and can be determined by a solids test. Embedded sand and grit contain particles which cling to freshly cut wood as a result of moisture, sap, and resins. The amount is determined for chip ash after removing the settled and suspended inorganic materials. The total amount of sand and grit is roughly 0.2% in conventional chips but is substantially higher in forest residue, depending to a high degree on the percentage of bark, foliage, and twigs as well as on logging conditions and methods. Shifting of bark fines and dirt during storage and handling causes accumulation in the base of chip piles which shows up when chips are reclaimed from low-pile levels.

Excessive amounts of sand and grit cause rapid wearing of equipment throughout the entire chain, from chipping to papermaking. Typical problems are reduction in chipper knife sharpness and wearing of chip feeders, pipes, conveyors, pumps, digesters, refiners, centricleaners, paper-mill rolls, fourdrinier fabrics,

Fig. 8.11. Raw material remains clean when the off-road transport is performed by a load-carrying tractor; Valmet 832 forwarder (courtesy of Hannu Kalaja)

Fig. 8.12. Bruks system for processing tree sections with transportable equipment at landing site. Bunch-delimbing tree sections for pulp raw material and drum-chipping residue for fuel (courtesy of Bruks Mekaniska AB)

size presses, etc. Sand and grit together with bark and foliage increase the calcium and silica levels in the cooking liquor, causing scale deposition and increased cleaning costs. They also cause more ash and slag formation in burning of bark and chip-screening rejects and reduce the fertilization value of the ash.

To avoid problems of wear and tear, biomass should be kept as clean as possible during harvesting and storage (Sect. 8.6.1). In this respect forwarders offer an important advantage over skidders (Fig. 8.11). In boreal forests, snow cover helps in winter operations. Separation of fuel and fiber components with bunch delimbers prior to comminution improves the cleanliness of the chips used for pulp production (Fig. 8.12). Careful removal of fines by screening chips is of great importance. If the amount of sand and grit in chips remains excessive, chip washing or special equipment such as a sand separator on the digester and a liquor flushing system may be required.

8.2.7 Concluding Remarks

Differing fiber-morphological and chemical properties, compared to conventional pulpwood together with inorganic impurities, prevent the use of residual forest biomass in most pulping processes and places many barriers on its use even in kraft and neutral sulfite semichemical mills. Because chips from residual tree

Table 8.11. Relative yield, chemicals consumption, and strength index of kraft pulps from tree components of *Pinus sylvestris* and *Picea abies* (Projekt ... 1977a)

Biomass component	Relative pulp yield, percent of dry mass	Relative chemicals consumption	Relative strength index
Stem wood	100	100	100
Stump wood	96	100	90
Root wood	95	97	90
Branch wood	76	130	70
Bark	40	130	(50)
Needles	49	130	(30)

parts tend to contain unusually large amounts of bark and other organic or inorganic contaminants, many of the operational and qualitative problems are actually a result of the impurity and heterogeneity of the raw material.

Table 8.11 shows the relative pulp yield, chemicals consumption, and strength index of kraft pulp from biomass components of Scots pine and Norway spruce. A marked difference occurs between stem, stump, and root woods, on the one hand, and branch wood, bark, and needles, on the other. The latter occupy a proportionally larger part of the digester capacity in relation to produced pulp. Against the background data (Table 8.11), the Swedish pulp and paper industries evaluated stump and root mass from clear-cutting areas and debarked stems from precommercial and early commercial thinnings as the most interesting reserves of unmerchantable forest biomass (Projekt ... 1977a).

Of the different biomass components of a conifer, the merchantable stem contains the wood most suited for conversion to pulp and paper, whereas conifer branches constitute a poor raw material, largely because of its high content of compression wood. With only one or two exceptions, all investigators agree that branch wood affords kraft and sulfite pulps in lower yield and with inferior strength properties than does wood from the merchantable stem or top. Therefore, branch wood can only be used when mixed with stem wood (Timell 1986c).

Recovery of residual tree parts in an integrated operation together with the stem, using methods such as whole-tree chipping or tree-section harvesting, offers a rational solution for simultaneous logging mechanization and intensified biomass utilization. However, although whole-tree chipping systems have many advantages from the standpoints of logging and trucking, the use of this raw material for pulp and paper production is not without problems.

Chips made from forest residues and small-sized whole trees do not meet the normal quality requirements of kraft and NSSC mills. Since the mill machinery is designed and scaled for a cleaner and more homogeneous raw material, operational problems occur frequently when these rough chips are pulped. While laboratory experiments make it possible to forecast properties of pulp, paper, and paperboard from whole-tree chips and similar materials with a high accuracy, only continuous, commercial processing can detect the seriousness of expected and unexpected problems. These problems are closely dependent on a large number of variables such as tree species, tree size, methods of harvesting and chip preparation, proportion of residual tree components in the final chip mixture, upgrading and storage of chips, and the end product. Their seriousness also depends on possible bottlenecks of the successive phases in a pulping process and the utilization intensity of the mill capacity. Thus, the findings of one mill are seldom directly applicable to another, although they may use the same chemical process.

Experience has shown that the problems relate primarily to processing rather than to the quality of the end product. The operational difficulties, again, arise largely from the fact that the existing mill machinery was designed for a bark-free raw material. When the chips contain more contaminants than was planned for, bottlenecks arise in the production process, which may reduce the overall capacity of the mill. Consequently, and also for ecological reasons, methods should be developed to leave the twigs and foliage in the forests, or alternatively to segregate these tree components from the main furnish at the mill prior to the digestion. Bunch delimbing (Sect. 5.2.4) and chip upgrading (Sect. 8.6) are examples of methods for rational quality improvement of whole-tree material.

Pulping of unbarked wood is technically difficult and uneconomical in a mill that is designed for debarked wood. The situation is different if a new pulp mill is built. All operations can be specially designed with dimensions that will allow a high content of bark in the chips. Although the presence of bark results in reduced digester yield, the total yield of fiber from a given quantity of timber increases when unbarked wood is pulped. First, no wood is lost in debarking, and secondly, additional fiber is gained from the bark. The table below presents a flowchart for dry bark mass in kraft pulping of unbarked stem mass from west-coast softwoods in the United States (Auchter 1972). Three-quarters of the bark is lost during the cooking operation. This part of the bark goes into the black liquor and further to the recovery boiler, where its energy is recovered in the form of steam:

Flow of bark	Proportion of bark
Loss on chipping	5.4
Loss in cooking	73.9
Loss in pulp screening	0.8
Loss in centrifuging	1.2
Loss in bleaching	2.0
Total loss	83.3
Accepted fiber	16.7
Total bark	100.0

A kraft mill originally designed to use unbarked wood recovers the bark heat energy by burning black liquor solids. Elimination of the drum-debarking plant and bark boiler reduces the costs of investments and labor, but additional capacity and investments are required for chip upgrading, digesting, washing, centricleaning, bleaching, and liquor evaporation and recovery. All these modifications can be taken into consideration when a new mill is built. Existing mills, on the other hand, can generally tolerate whole-tree chips only in rather modest proportions without reducing the mill capacity. The allowable limit is due to whole-tree chip properties, chip upgrading facilities, bark content of the standard chips, dimensions of equipment, and quality requirements of the end product.

According to Powell et al. (1975), shifting from standard debarked hardwood chips to 100% whole-tree chips, while maintaining bleached pulp output, brightness, and dirt level constant, would increase the digester loading 5–7% because of lower yields and the need to produce pulp with a lower Kappa number. Caustic demand and loading of the recovery system would increase substantially. If the mill is presently running at digester or recovery capacity limitations, the production capacity would be reduced without additional equipment investments. On the other hand, if the production is limited by wood supply, the use of whole-tree chips as a substitute for normal debarked chips could increase the total pulp production of the mill.

In a well-designed kraft pulp and paper mill a 15% mixture of whole-tree chips, or chips from stump and root wood, does not cause any major problems. If operational disturbances, such as large variations in Kappa number or excessive impurities, occur in the use of standard chips, these problems must be solved before mixing chips from residual biomass. Whole-tree chips make the pulping process more sensitive to disturbances and reduce the operational scope. Therefore, increased requirements are set for normal raw material and equipment. Possibilities of accepting other low-quality raw materials, such as hardwood chips in a softwood line or recycled fiber in a paper mill, are reduced when the role of whole-tree chips grows. However, upgrading of raw material reduces the problems considerably (Projekt ... 1977a).

Lönnberg (1976b) suggested that a bark content of about 5% in the furnish should be adopted as a criterion for chip mixtures from standard hardwood chips and whole-tree chips from young hardwoods. The most profitable result is attained with high debarking yields, high chip-screening yields, and high pulping yields. The implication is that, to some extent, the low-yielding bark should be removed without wood losses, and that screening should be applied for chip classification rather than for chip removal. According to Lönnberg, a 25% mixture of whole-tree chips from young hardwoods should be acceptable with normal hardwood chips for many kraft mills.

When whole-tree chips are mixed with debarked standard chips, any cooking schedule will be a compromise of the optimum conditions for a given tree species and biomass component. If the proportion of whole-tree chips in the mix varies over wide extremes, some portions may overcook and other may undercook. The situation is even more complex if the species composition of chips varies erratically. Overcooking will result in yield loss and heavy loads on the caustic and recovery processes. Undercooking, on the other hand, will result in a large amount of screen rejects, low brightness, and excessive dirt in the pulp. To avoid variation in the pulp quality, uniform metering of whole-tree chips into the furnish is necessary. Isolating the whole-tree chips and treating them in a separate line would allow the process conditions to be fixed at the optimum balance of economics and final product properties (Powell et al. 1975).

According to Nyholm (1976), an admixture of up to 20% stump and root wood had no detrimental effect on the quality of kraft pulp from Scots pine.

However, taking into account the limited supply of stump raw material from clear-cutting areas, 5–10% was regarded as a more realistic figure.

To summarize, whole-tree chips tend to create process and quality problems in a pulp mill designed to use bark-free wood. These problems will probably cause operation and investment costs that must be weighed against the broader raw material base, additional fuel for production of energy, reduction in harvesting costs, and improved possibilities of carrying out early thinnings in young stands. Efficient upgrading of unbarked chips would solve many of the problems presently constraining whole-tree utilization. In fact, development of chip upgrading is the critical factor that will ultimately determine the feasibility of whole-tree chipping systems in raw material procurement of the pulp and paper industry (Sect. 8.6). Alternatively, the same raw material may be recovered, and the same forest-end advantages achieved, with the tree-section method (Sect. 7.3.1), where the raw material is simultaneously delimbed and debarked prior to chipping at the mill. The technical suitability and economic viability of various alternatives must be holistically evaluated in each specific case separately.

8.3 Use of Residual Biomass for Panel Products

Throughout the world, the fiber- and particleboard industries were developed to utilize residues from mechanical conversion of wood to products. The principal sources of raw material of these forest industry branches are green and dry planer-mill shavings, sawdust, slabs, edgings, trimmings, veneer-log cores, waste veneer, etc. For example, a survey in the United States in 1973 showed that 65% of wood raw material for particleboard consisted of planer shavings. Other sawmill waste accounted for 12%, plywood-mill waste 10%, roundwood 7%, pulp-type chips 5%, and other materials 1% of the total raw material base (Dickerhoof 1976). The proportion of planer shavings is considerably less in Europe.

On a world scale, the fiber- and particleboard industries are not great consumers of wood, and much of the raw material is obtained as mill residues from other units of an integrated enterprise. Therefore, although they are modest in the quality requirements of raw material, their possibilities to absorb forest residues is rather limited. However, they may have considerable local significance as users of unmerchantable forest biomass. By using small-sized hardwoods and softwoods from precommercial and early commercial thinnings these mills could stimulate silvicultural activities within their catchment area and thus improve prerequisites for the production of high-quality timber for other branches of the forest industries.

Fiberboard is composed of defibrated fibers produced by refining or exploding. Fiberboards are classified according to density into insulation board, medium-density fiberboard, and hardboard, which is the prevailing fiberboard product. Except in insulation board, the density is determined by the degree of compaction of the fiber mat during pressing. High compaction results in good fiber-to-fiber contact and, consequently, high strength.

Particleboard is made in a dry process of small wood particles and adhesives which are in either one or three layers. The desired particle form is usually achieved by cutting wood parallel to the grain in the 0–90 mode (Fig. 5.5). Thin particles produced in this way are called flakes. Large, square flakes are called wafers, and *structural particleboard* made from them is called *waferboard.* Strands are flakes whose length is at least three or four times as great as the width. Strands are used for manufacturing *strandboard,* i.e., oriented-strand particleboard. Both waferboard and strandboard are relatively new products specially developed for structural purposes.

It is technically possible to make almost any kind of fiber or particle panel material from any species of wood in almost any form. Some forms or kinds of raw material, however, are little used because they result in a finished product that is undesirable from a merchandizing standpoint, are impractical from a manufacturing standpoint, or are uneconomical from a handling viewpoint. This is particularly true when residues are used (Lewis 1971).

An economical evaluation of properties that can be achieved with a new raw material has to be based on the definition of the specified use of the panel product in question. For example, wall-building panels have four main barrier functions: a mechanical, an optical-aesthetic, a climatic, and an acoustic barrier function. Secondary functions are grouped into that of security and safety, including fire, water, and fungi resistance, and that of stability and durability, including resistance to long load and dimensional stability. There are also separate panel functions which do not directly affect the properties of the wall. One group is assembling and joint formation, where screw holding and internal bonding strength are important properties. Another group is convertability with properties such as the wear of cutters in joinery plants (Back and Lundqvist 1976).

For the mechanical wall-barrier function, especially bending stiffness, bending strength, and impact resistance are important. Internal bonding strength is important mainly with respect to assembling the panels. The optical-aesthetic barrier function requires properties such as surface smoothness, surface regularity, brightness, and moderate paint consumption, to achieve a predetermined surface coverage and gloss. Acoustic barrier qualities depend on the density of the panel. The board properties together with possible process disturbances, raw material consumption, and the cost of delivery determine the suitability of residual forest biomass for the production of various kinds of panel products.

Much of the trouble in a board factory, when using chips from whole trees or logging residue, is connected with the large amount of fine particles. Together with oversized chips the fines may cause plugging and bridging, which tend to reduce production capacity. The fine fractions of chips, containing high percentages of bark and foliage, are also responsible for various kinds of process disturbances, such as slow drainage and increased wastewater load in fiberboard mills or dust in particleboard mills. Fines may also adversely affect the strength and surface appearance of the board. On the other hand, pulverized coniferous foliage may be suitable for use as extender and filler in adhesives in the particleboard and plywood industries (Barton 1981).

The fine fractions of the chips contain excessive amounts of inorganic contaminants, which are very harmful in all panel products. Thus, chip upgrading is of great importance in the production of composite boards. Differences in raw material screening are a common source of apparent contradictions in results of both laboratory trials and mill runs. The use of whole-tree raw material in the panel-product industries was reviewed by Nordlinder and Tufvesson (1975).

8.3.1 Use for Fiberboard

In fiberboard manufacturing, wood is reduced to fibers and fiber bundles by mechanical methods, namely by frictional force or by steam pressure. Fiber separation may be promoted by steaming, but chemicals are not used to dissolve intercellular lignin. Consequently, the yield of fiber is high.

The sheet is formed by depositing the fibers on a fourdrinier-type wire. To produce hardboard or medium-density fiberboard, the sheet is either pressed and dried between platens of a hot press, or first dried and then pressed. Insulation board is produced when the fiber sheet is dried without further hot pressing.

The fiberboard industry is the most modest of all the branches of the forest industry in its raw material requirements but in technical preparedness it ranks first as a user of rough forest chips. The chemical composition of the wood, the morphological properties of the fibers, the presence of bark, and the particle size of the chips are less important than they are in the pulp and paper industries. However, the chips must be free from sand and grit which wear the refiners and other equipment and are particularly harmful in the final product.

Hosia and Kortelainen (1971) made 3-mm hardboard in the laboratory using screened chips from debarked stem wood, screened chips from unbarked branches, and unscreened chips from the crown mass of Scots pine and Norway spruce. Bark and needles reduced the yield, but the presence of needle extractives remarkably improved the water-resistance properties of the hardboard. Unbarked branch chips gave the greatest tensile strength or internal bond. In agreement with earlier observations by Kehr (1976), inclusion of needles reduced the strength of the board made from branches. Needles and fines also decreased the brightness of the board (Table 8.12). A mill-scale trial with unscreened chips from spruce-dominated logging slash confirmed the laboratory results of the technical properties of

Table 8.12. Technical properties of 3-mm hardboard made in the laboratory from stem and crown components of *Pinus sylvestris* and *Picea abies* (Hosia and Kortelainen 1971)

Tree species	Biomass component	Yield (%)	Density (kg/m^3)	Tensile strength (MPa)	Water absorption (%)	Thickness swelling (%)	Brightness (%)
Pine	Debarked stem	86.7	1090	22.0	78.4	57.2	22.9
Pine	Unbarked branches	85.1	1170	26.5	67.9	49.1	13.1
Pine	Crown	80.1	1130	22.0	65.9	42.3	12.8
Spruce	Debarked stem	87.5	1080	26.5	77.5	55.5	24.7
Spruce	Unbarked branches	85.5	1160	33.4	69.1	50.6	11.7
Spruce	Crown	81.6	1150	28.9	60.0	40.5	8.9

the board. With the exception of chip handling and increased wastewater solids, the use of logging slash did not cause major disturbances in the manufacturing process.

Hosia and Kortelainen concluded that softwood branches are much better suited for fiberboard than for pulp and paper production, because in fiberboard manufacturing the chemical and fiber-morphological properties of the biomass are of less importance. No insurmountable technical barriers seemed to exist in the use of branches, but the costs of branch raw material were considered excessive compared to mill residues. Certain differences in technical properties from normal boards might also cause marketing problems.

In the Soviet Union, large amounts of crown material and stem residues accumulate in timber terminals in conjunction with whole-tree and tree-length logging. Since the off-road transport is carried out by ground-skidding, and the distances may be rather long, in summer the slash may contain up to 10% inorganic contaminants. Consequently, the slash must be washed before further treatment. Voronicyn and Pavlov (1968) explored the possibilities of using such residue for the manufacture of 3–4-mm hardboard. Mill-scale trials showed that it is technically possible and economically feasible to use this residue as board raw material. The bending strength of hardboard was improved when processing residue from both hardwoods and softwoods was substituted for conventional stem mass. Board made from birch residue did not meet requirements. A precondition of its acceptance was treatment with steam or chemicals before defibration.

Lajunen (1975) reported a mill-scale experiment manufacturing 3.2-mm hardboard from unscreened, softwood-dominated whole-tree chips. The water absorption, thickness swelling, and internal bonding increased, while the bending strength and surface smoothness remained at a normal level. The only process disturbances were chip bridging and a considerable increase in the solids content of the wastewater from the sheet-forming machine. Whole-tree chips were considered suitable raw material for hardboard, but their use at the mill under consideration was not possible without increasing the wastewater cleaning capacity.

Helge (1976) reported two mill-scale experiments using whole-tree chips as a substitute for unbarked sawmill chips in fiberboard manufacturing. In the first experiment whole-tree chips from thinnings of Scots pine and Norway spruce were substituted for unbarked sawmill chips. The chips were screened before feeding into the defibrator. The experiment covered three different chip mixtures with the proportion of whole-tree chips 33, 50, or 67%. The only operational disturbance occurred in a build up of chip plugs before the screw feeder of the preheater of the defibrator, which reduced the defibrating capacity by 15–30%. Although the chips contained large quantities of bark and needles, there was no increase in the oxygen demand in the wastewater. The influence of whole-tree chips on wastewater pollution naturally depends on the normal raw material of the plant, and since unbarked sawmill chips contain a high proportion of bark, the pollution contribution of whole-tree chips was not significant. The use of whole-tree chips did not adversely affect the density, strength, water absorption, and thickness swelling of the hardboard. A high sand content, however, reduced the workability of the board.

In the other experiment, unscreened whole-tree chips from Norway spruce thinnings were defibrated and refined separately and then mixed with sawmill chips in proportions varying from 0 to 100%, to make medium-density fiberboard. There were no operational disturbances except on the sheet-forming machine, where serious problems arose because of wear and tear on the wet lap knives. When the machine was supplied with pulp from 100% whole-tree chips, the knives had to be sharpened almost constantly. The high sand and grit content was obviously associated with the off-road transport, which was accomplished with skidders (Sellaeg 1974). At a preheating time of 4 min, the fiber yield was 93.5% from debarked slabs, 90% from unbarked slabs, and 91.5% from spruce whole-tree chips. No impact of whole-tree chips on board quality could be noted with regard to traditional criteria. The working quality was, however, seriously reduced because of the high sand content. Presupposing that the sand difficulties can be overcome, Helge (1976) evaluated whole-tree chips as an acceptable raw material in fiberboard production.

If whole-tree chips are used for top coating pulp in fiberboard manufacturing, they tend to give a dark board surface. Furthermore, the needles may cause resin stains and burning marks as a result of hot pressing. However, most mills use a special surface pulp for coating board, thus eliminating the problem (Nordlinger 1976).

Technical foliage of *Picea glauca*, including twigs up to 6 mm in diameter, was processed by Law and Koran (1979) in a disk refiner. The main purpose was to extract protein and chlorophyll-carotene material from the liquid fraction of the product. Once the liquid was removed by centrifugation, the remaining pulp contained 70% of the original dry mass, consisting of 18% extractives, 23% lignin, and 62% holocellulose. This solid fraction had three principal cell constituents: long and rigid bands of epidermis-hypodermis, granular and parenchymatous mesophyll, and rigid strands of fibrous core.

The pulp was used as the raw material for medium-density, 600 kg/m^3 fiberboard with liquid phenol-formaldehyde resin as an adhesive. The strength of the board was considerably below the minimum requirements. Water absorption was particularly high, and the thickness swelling also exceeded the permissible limit. Although the technical properties improved with increasing resin content, foliage pulp was considered unsuitable for the manufacture of medium-density fiberboard.

A thorough evaluation of the suitability of various softwood materials for the production of fiber- and particleboard was carried out by a joint research and development project of the Swedish board industries (Projekt ... 1977b). Bark and foliage characteristically had a low yield. Consequently, chips prepared from small-sized whole trees and logging slash gave a lower yield than chips from debarked stem wood, but the yield was nevertheless almost 90% of the dry mass of chips, and comparable to that from unbarked sawmill chips. A low yield was associated with increased oxygen consumption of wastewater from the process (Fig. 8.13). Foliage, fines, and bark particularly had high drainage resistance.

The raw material source had only a slight impact on board properties. However, whole-tree chips from early thinnings tended to produce better fiberboard than chips from logging slash. Low bending strength was associated with a high

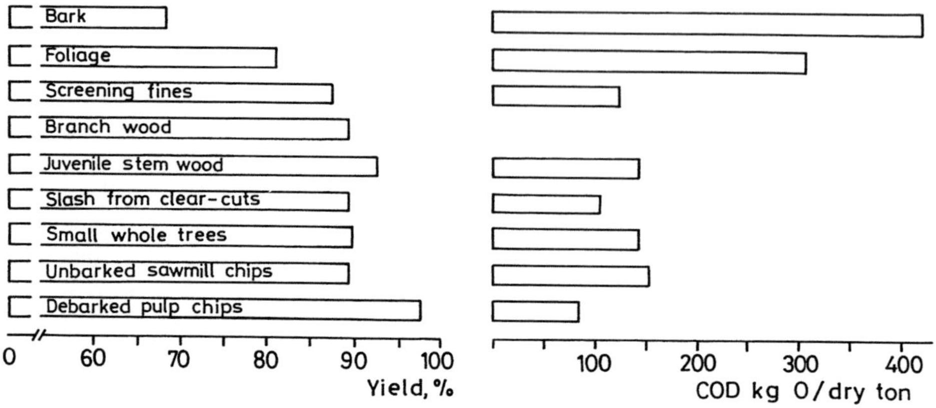

Fig. 8.13. Fiber yield from various softwood materials and the oxygen consumption of wastewater in hardboard manufacturing. Laboratory results from Sweden (redrawn from Projekt ... 1977b)

amount of branch wood in the raw material, whereas the presence of foliage reduced the internal bonding of the board. Excessive amounts of bark, foliage, and fines reduced the impact resistance. Variation in component ratios did not significantly affect water absorption or thickness swelling of the board (Fig. 8.14).

Although all other properties of hardboard in the Swedish experiments were generally fully satisfactory, sand and grit had an adverse effect on workability, which is one of the most critical board properties. The proportion of sand was 0.02–1.2% in whole-tree chips from precommercial thinnings and 0.07–3.0% in chips made from logging slash from clear-cut areas. The highest values were the result of improper treatment such as temporary storage of chips on the ground at a landing site. Despite screening and washing, fiberboard produced from chips with a large amount of inorganic contaminants contained 0.1–0.5% sand, while the allowable amount was only 0.02–0.04%.

The dulling effect of sand on tools depends on the particle size, hardness, and form of the impurities and on the density of the board (Klinga and Back 1964). In the Swedish experiments, about 20–50% of the sand could be segregated with fines from the chips. Screening of fines completely removed sand particles larger than 0.8-mm, 75% of the 0.5–0.8 mm particles, and 35% of 0.25–0.5 mm particles. The drier the chips, the easier was the removal of sand (Projekt ... 1977b).

When residual forest biomass is used for the production of fiberboard, another poor-quality raw material is usually substituted. Nevertheless, the use of forest residues or whole-tree chips in fiberboard manufacturing may be feasible for other reasons. Depending on the mill facilities, the use of these raw materials may be associated with operational disturbances and inferior board quality, but the problems can often be overcome with rather simple measures. The following *conclusions* can be drawn from laboratory trials and mill runs made in different countries.

Depending on the amount of bark and foliage, the fiber yield is smaller than for bark-free wood. A better reference, however, is unbarked sawmill chips to which whole-tree chips generally are equal. If chips are not upgraded, twigs,

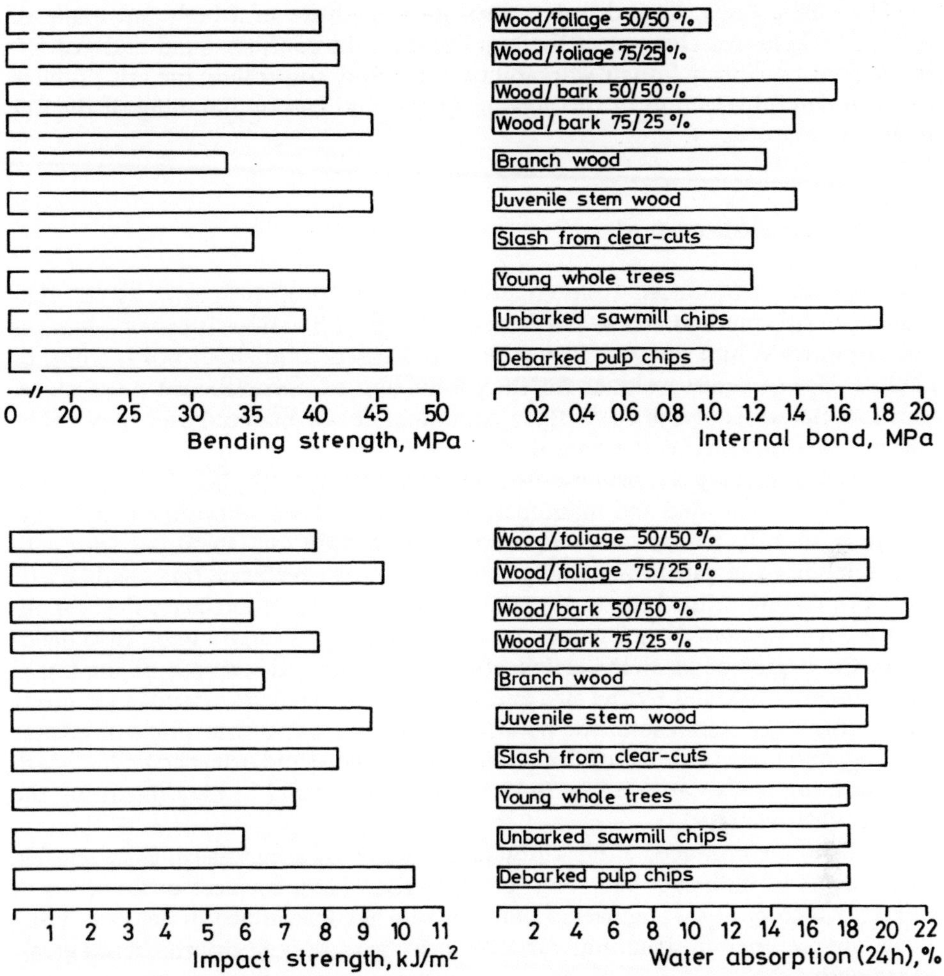

Fig. 8.14. Strength and water-absorption properties of hardboard made from various softwood materials. Average results from Swedish laboratory and mill-scale experiments (redrawn from Projekt ... 1977b)

needle clusters, and fines impede chip handling on screw and belt conveyors and in silos. Feeding disturbances may lower the output of the defibrator, which interferes with formation of the fiber web. These difficulties disappear when fines and oversized fractions are segregated from the chips before the process begins.

Once the chips are screened, web formation, dewatering, pressing and hardening of the sheet cease to be problems. If whole-tree chips are substituted for cordwood, the biological oxygen demand of the effluent increases. This adverse feature gains importance from the fact that the effluent cleaning systems of fiberboard mills are often primitive. However, there is no essential change compared with unbarked sawmill chips.

The color, strength, and water absorption qualities of fiberboard made of screened whole-tree chips meet the requirements. In contrast, sand and grit left in the board seriously impair workability. Therefore, controlling the grit problem calls for careful selection of the logging system and careful handling of the raw material.

8.3.2 *Use for Particleboard*

Particleboard is produced from small particles of wood primarily by the flat-platen pressing method. These particles are mixed with various types of adhesives and compressed into boards. They may be processed from either bolts, chips, or chunks. Use of forest residues for the production of particleboard presupposes that the factory is equipped to flake comminuted raw material and, preferably, can use this furnish for the core portion of three-layer board.

Particle geometry is a prime consideration affecting both the important properties of the board and the manufacturing process. Face smoothness and edge tightness are affected by particle geometry. Since thin and small particles with their pliability and gap-filling ability generate gap-free surfaccs, fine dustlike particles or pressure-refined fibers are generally used for board surfaces. Particle pliability not only directly influences appearance but, to a large extent, also determines the degree of particle-particle contact. Mechanical strength of the board is thus greatly affected by the size and shape of the particles. Particle geometry also controls the response of the board to water, i.e., water absorption, thickness swelling, and linear stability of the board. Furthermore, the behavior of the board to machining is also affected by the type of particle used in manufacturing the product (Moslemi 1974).

Particles are produced either through the action of a hammermill, generating small pieces of indistinct geometry, or more commonly by a set of knives as in flakers, which produce engineered particles of a predetermined dimension. Long flakes are preferred in structural particleboards because they give the board greater strength.

Flakes can be produced from both roundwood and comminuted materials. When forest residue is used for particleboard production, it is customarily delivered to the mill in the form of chips, which are then reprocessed into flakes. The geometry of the resulting flakes depends on the form and size distribution of chips produced from the residual biomass. Hence, great attention should be paid to the cleanliness of the raw material in the harvesting phase and the maintenance of the comminution equipment.

Use of forest residue for particleboard is, to some extent, constrained by the presence of bark, since large amounts of bark have a negative impact on board properties. Over 10% proportions of bark produce significant adverse effects on the strength and dimensional properties of the board. At present, the particleboard industry occasionally uses bark primarily in the core layer and even then only in quantities normally occurring in roundwood. The following disadvantages are associated with large amounts of bark in the raw material (Moslemi 1974):

1. Bark reduces the board strength. Increasing board density or core resin content can, in part, offset the drop in the internal bond, but such measures may be uneconomical.
2. Thickness swelling and linear expansion of the board increase.
3. Surface smoothness of the board decreases.
4. Bark handling during processing can become a problem.
5. Bark creates maintenance problems for wood-reduction equipment and other cutting facilities due to the existence of dirt.
6. The maintenance of a relatively constant furnish pH becomes difficult.

Bark thus adversely affects most particleboard properties. Lehmann and Geimer (1974) observed a 20–30% reduction in bending strength and stiffness, internal bond, and linear expansion when 25% *Pseudotsuga menziesii* bark was included in the conventional softwood raw material of structural particleboard. Wisherd and Wilson (1979) reported a general decrease in the strength of particleboard at a linear rate as the substitution level of bark from *Pseudotsuga menziesii, Pinus ponderosa*, or *Alnus rubra* increased. On the average, a 20% bark substitution reduced the bending strength 24%, internal bonding 21%, and the modulus of elasticity 17%. Thickness swelling and linear expansion were not significantly affected. The authors concluded that a high-quality, underlayer-grade, medium-density particleboard can be fabricated in the laboratory by substituting as much as 20% Douglas-fir or ponderosa pine bark, or 10% red alder bark, for wood furnisch. Muszynski and McNatt (1984) studied 19-mm particleboard made of different admixtures of Scots pine and Norway spruce bark. An increase in bark content resulted in an increase in water absorption, hygroscopicity, and thickness swelling. The strength and the resistance to nail, screw, and dowel withdrawal decreased simultaneously. Nevertheless, the study indicated that homogeneous particleboards with a spruce-bark content up to 30% can be used, for example, in the furniture industry. The best strength properties were obtained with a bark particle size of 3–8 mm.

Carre (1980) considered whole-tree chips, from the technological point of view, a mixture of particles from stem wood, branches, and bark, with board properties related to component ratios and particle geometry. Bending strength, internal bond, and shear strength were all adversely affected by the presence of bark. Small branches were only slightly better raw material than bark. The reason was assumed to be the lower compression resistance of bark and small branches and, consequently, lower pressure during the formation of the glue joint. On the other hand, boards made from bark or small branches showed considerably less thickness swelling during water immersion, and less thickness swelling and linear expansion under exposure to humid air. This behavior of the board was also explained by greater compressibility of particles of bark and small branches, resulting in less internal strain development during pressing and weaker reaction to water. Carre concluded that all treatments by which the fine fraction can be reduced increase the technological value of chips, making the whole-tree chips more like standard stem chips.

Statkov et al. (1969, 1970, 1976) carried out mill-scale experiments manufacturing particleboard from unbarked branches of *Fagus sylvatica*. One-layer board was made exclusively of branch mass, whereas in three-layer board only the middle layer was composed of branch mass. The strength and screw-holding properties complied with official standards. In one-layer board, the strength increased with branch diameter due to the decreasing percentage of bark, but no impact of branch size was observed on the three-layer board. The authors concluded that unbarked beech branches larger than 2-cm diameter are well suited for the production of particleboard. However, in one-layer board unbarked branches can be used only if discoloration is allowed. Branches from *Abies alba* and *Picea abies* with over 2-cm diameter were acceptable raw material as well.

Liiri et al. (1972) studied the suitability of crown and whole-tree chips of Scots pine, Norway spruce, and birch as raw material of the core of three-layer particleboard. Particle preparation with a ring-type flaker displayed no special difficulties as far as resin, needles, and bark were concerned. As the bending strength is determined mainly by the faces of the board, it was hardly affected by the presence of these components. Only when the middle layer was composed entirely of crown mass, was a slight decrease in the bending strength noted. When particleboard was manufactured to a density of not less than 700 kg/m^3, all tested raw materials except crown mass from spruce could be used as core material up to 100% without seriously reducing the internal bond, a critical property of particleboard. Not more than 25% of spruce crown mass could be mixed in the core furnish, mainly because as much as 40% of it was composed of needles (Fig. 8.15). A large proportion of needles reduced the adhesiveness of the particles and, therefore, the internal bond. Generally, the more wood and less needles the residue materials contained, the more suitable they appeared to be for the middle layer of particleboard.

Chen and Paulitsch (1974) analyzed the effect of bark and needles of Scots pine and Norway spruce on the properties of particleboard. The strength decreased and water absorption increased with increased amounts of needles in the raw material. Fatty substances, waxes, and oils in the needles caused difficulties in gluing. However, formaldehyde liberation improved compared with normal particleboard due to the large amount of extractives.

Chen (1979a, b) carried out laboratory-scaled experiments in the use of whole-tree biomass for single-layer particleboard in Taiwan. Board made from whole-tree chips of young *Cunninghamia* sp. reached the standard level. Board made from branches of 7-year-old *Trema orientalis* showed satisfactory properties. The bark and leaves reduced the board strength but improved water absorption and thickness swelling. Single-layer particleboard made from whole-tree chips of the same species met the standard requirements. The use barriers were economical rather than technical.

Kehr and Drechsler (1978) made one- and three-layer particleboard with core particles from crown material of Scots pine. Special attention was paid to the effect of branch diameter and the presence of needles on critical board properties. The bending strength increased slightly and the internal bond significantly with branch diameter, due to the decreasing proportion of nonwoody components (Fig. 8.16). Thickness swelling after 2 h in water was greater for boards made from large branches. Bark decreased but needles increased the swelling. After a 24-h immersion the effect of branch diameter was negligible. The authors concluded

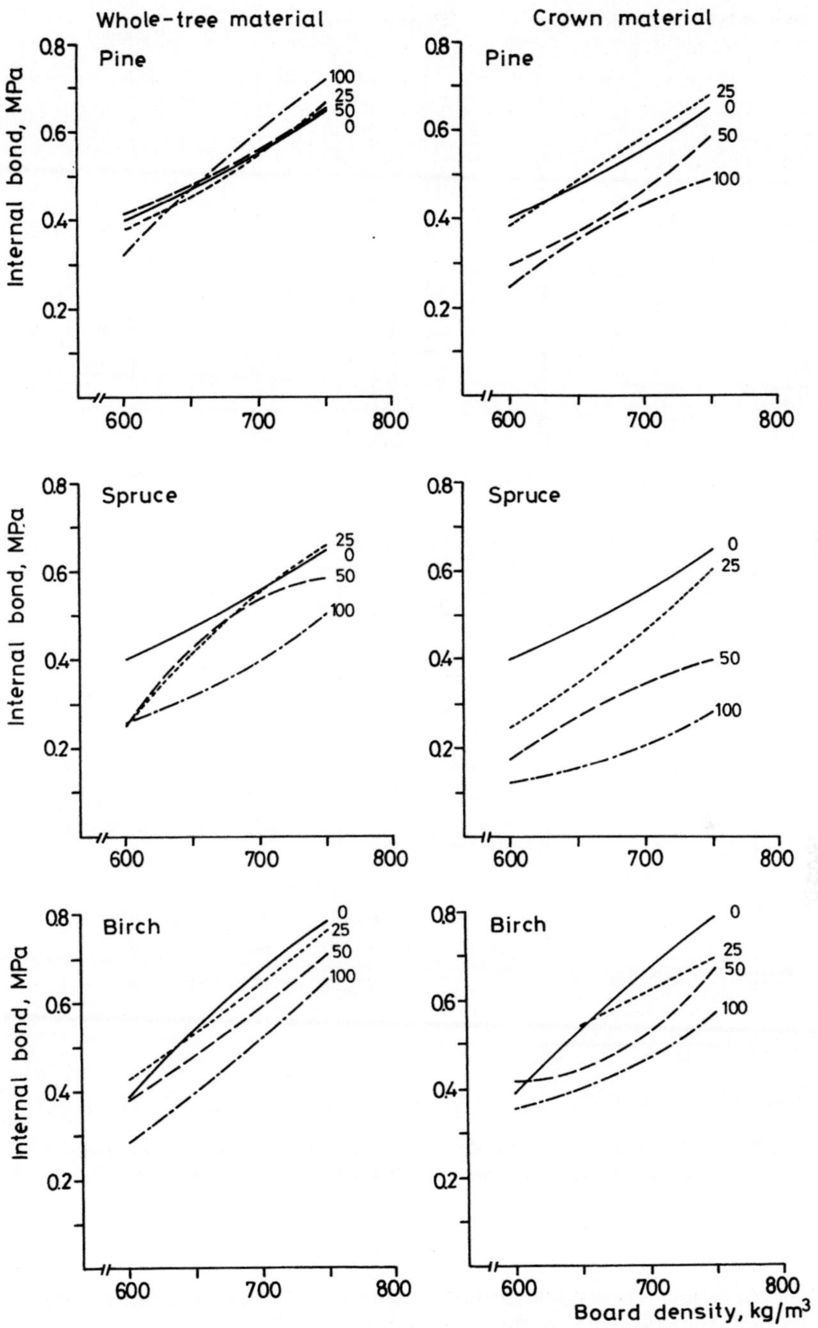

Fig. 8.15. Internal bond of three-layer particleboard from *Pinus sylvestris, Picea abies,* and *Betula* spp. with 0, 25, 50, or 100% residue material in the core, as a function of board density. Primary raw materials were debarked birch, residue material in the core whole-tree chips (pine and spruce), or unbarked stem chips (birch) on the *left* and crown chips (pine and spruce) or leafless branch chips (birch) on the *right* (redrawn from Liiri et al. 1972)

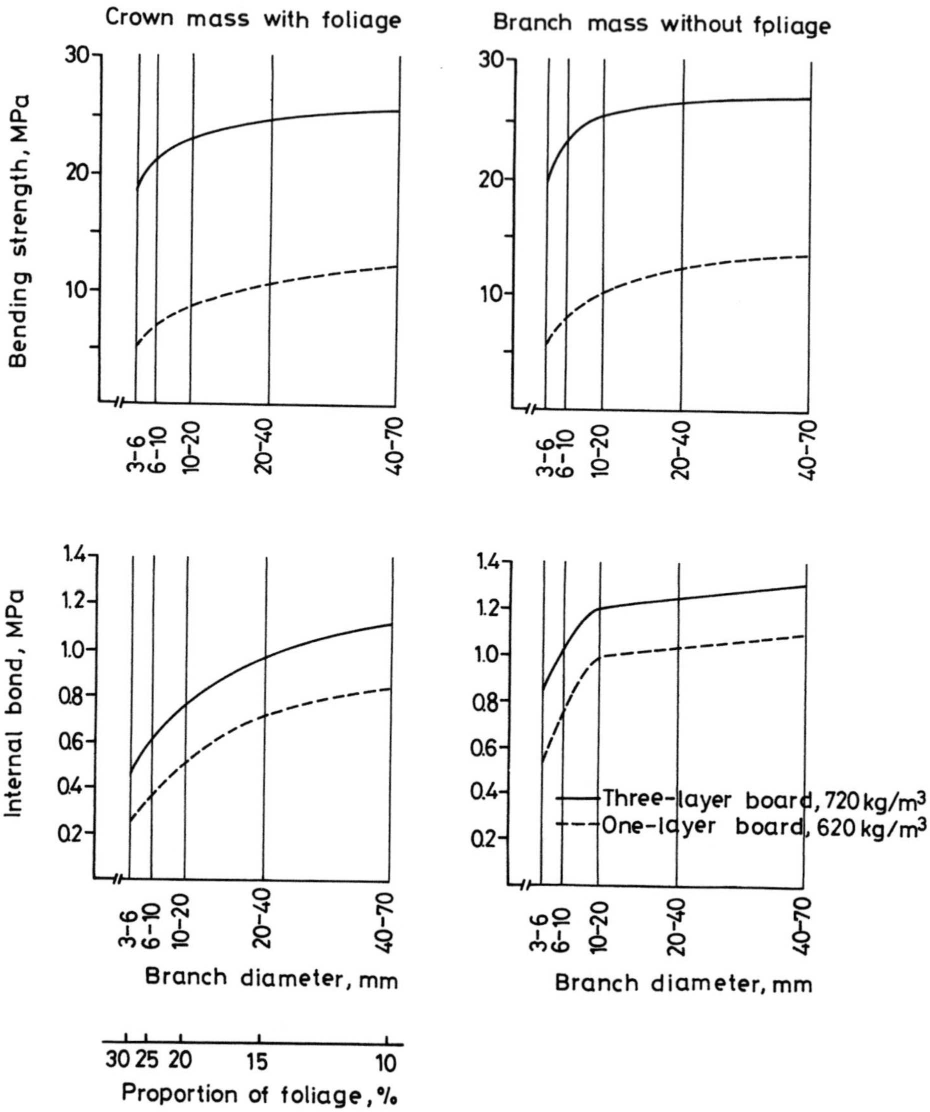

Fig. 8.16. Bending strength and internal bond of one- and three-layer particleboard made from crown and branch material of *Pinus sylvestris*, as a function of branch diameter (redrawn from Kehr and Drechsler 1978)

that some of needles and loose bark should be eliminated from the crown material in order to maintain the high quality of particleboard.

Philippou and Collett (1978) prepared three-layer, medium-density particleboard from crown and top residue of 38–51-cm diameter *Abies concolor* trees. Board quality was affected by the branch diameter and presence of needles, bark, and reaction wood. Board with satisfactory mechanical and physical properties for conventional uses was obtained from branch sections and tops with a diameter greater than 13 mm. Mainly because it included a large proportion of needles, crown material of thinner diameter failed to yield adequate mechanical properties. Mixtures of logging residue and conventional bark-free furnish gave products with properties that met commercial standard requirements for medium-density boards. Better properties were achieved when residue was used in the core and conventional furnish in the outer layers.

Paulitsch (1977) prepared on laboratory scale three-layer particleboard with the faces from conventional softwood material and the core from leafless branch mass of *Fagus sylvatica* and *Quercus robur.* Even when the proportion of fines in the middle layer was as high as 40%, the board met DIN requirements as regards internal bond and thickness swelling. However, some adjustments to binder formulations, hardeners, and buffering additives were needed.

Lehmann et al. (1981) carried out several mill-scale experiments using various mixtures of residual biomass and normal raw material for the production of fiber- and particleboards in East Germany. When the middle layer of the particleboard was composed exclusively of screened whole-tree chips from Scots pine thinnings, the result was adverse. Core material containing 40% unscreened whole-tree chips from understory trees also gave negative results, largely due to problems caused by long particles. However, when these particles were removed and the proportion of whole-tree chips was reduced to 20%, the result was entirely satisfactory. Raw material mixes containing 15–30% unscreened whole-tree chips or crown chips from Scots pine gave positive results in the production of medium-density fiberboard. A positive result was also obtained when the hardboard raw material contained 25–50% unscreened chips from pine crown material.

The joint research and development project of the Swedish board industries found operational disturbances caused by the use of forest residue rather minor. Problems occurred in feeding chips to flaking machines. Flaking knives suffered from sand and grit and, in the case of fresh chips, from resinous materials in foliage and bark. The fines content of the chips was generally large and increased further in flaking, giving rise to dust disturbances. These problems could be reduced by removing fines and oversized fractions by screening.

Particleboard strength generally decreased with increasing content of bark and foliage. Nonwoody components affected the bending strength only slightly when located in the middle layer, but the influence was much greater when they were used in the outer layers of the board. Fines decreased internal bonding. Nonwoody components also increased the thickness swelling of board made from Scots pine, but for Norway spruce the relation was not clear. These adverse effects of nonwoody materials were partly associated with a lower board density. Consequently, they can be reduced by increasing the board density which, however, results in a higher consumption of flakes and adhesives. Particleboard made from forest residue was also characterized by an excessive sand content (Projekt ... 1977b) (see Scheme p. 428).

Inorganic contaminants and high logging costs make the use of underground tree components for board production presently impossible. Nevertheless, it is interesting to observe that clean stump and root wood has been found to be an excellent raw material for the production of particleboard. Biblis and Lee (1978)

	Sand content in chips (%)	Sand content in board (%)
Standard hardwood chips	0.2	0.06
Whole-tree chips	1.3	0.10
Chips from logging slash	3.6	0.19

investigated the possibility of utilizing bark-free stump and root wood from 16-year-old *Pinus taeda* for fabricating mobile-home decking and industrial-grade particleboard. Strength properties and dimensional stability of the board met or exceeded commercial standards. Screw-holding ability and internal bond were considerably above the standards.

It can be *concluded* that chips from whole trees or crowns do not generally cause serious technical problems in the particleboard industry, but the raw material must be as clean as possible. Problems caused by sand and grit are not as serious as they are in the fiberboard industry. Removal of fines and oversized fractions permits smoother processing.

The possibilities of using residual biomass depend, however, on the manufacturing technique. A mill must at least have equipment to transform the chips into flakes. Branch and crown material may be used but in lower proportions than whole-tree chips among conventional raw material. When restricted to the core, bark does not affect the color or smoothness of the face. The share of the middle layer in the total raw material depends on the board thickness but is generally over 50%.

The bending strength of the board depends primarily on the properties of the outer layers. As forest residue is placed in the core it has little effect on bending strength. However, internal bonding also depends on the composition of the middle layer; foliage and fines tend to weaken it. Forest residue may detract from the hygroscopic properties of particleboard, but investigations to date are contradictory.

If the share of fine fractions in the flakes increases, glue consumption increases and the strength of the board suffers. There is a positive causal relationship between the size of the chips and that of the flakes made from them. Consequently, the particle size of chips or chunks should be as large as possible to prevent the flake size from becoming too small.

8.3.3 Use for Structural Flakeboard

Structural flakeboard is an exterior-grade panel for use as construction sheathing in walls, floors, and roofs as a substitute for plywood. In this use it is likely to be subjected to moisture conditions more extreme than those usually found in the interior use of panels. The major concentration of manufacturing capacity of structural flakeboard is around the Great Lakes in North America. The main source of raw material at the present time is good-quality poplar stems.

Structural flakeboard is a panel whose surfaces are composed mainly of 40- to 80-mm-long and 0.5-mm-thick flakes bonded together with a weather-resistant or highly durable resin. It must have three key properties not required in conventional particleboard commonly employed as furniture or floor underlay: durability, dimensional stability, and capability of carrying building loads even under adverse environmental conditions. Aligning wood flakes or strands in one direction produces structural flakeboard of even higher strength and stiffness. Flakes may be produced directly from roundwood or from small chunk particles called fingerlings. Flakes made from roundwood are usually of better quality and contain less fines (Gardner et al. 1978).

Structural flakeboard which is manufactured with all flakes placed randomly is called *waferboard*. In another type of structural flakeboard, *strandboard*, the flakes are aligned in the outer layers parallel to the major panel axis. Core flakes may be either randomly disposed or aligned at right angles to those in the outer layers.

Important properties of structural flakeboard are, to a large extent, determined by flake geometry, flake quality, and flake alignment, together with wood density and resin content. Flake geometry and quality, on the other hand, depend greatly on the size and form of the raw material used for preparing the flakes.

Numerous researchers have shown bending properties of structural flakeboard to increase directly with flake length and inversely with flake thickness. Improved bending performance has been demonstrated for flake lengths up to 75 mm and thicknesses as low as 0.15 mm. Increase in slenderness, i.e., the ratio of flake length to flake thickness, has been shown to affect bending properties favorably up to ratio of 300:1. Below 13 mm, flake width is of minor importance with regard to bending properties. In contrast, the internal bond, i.e., the tensile strength perpendicular to the surface, increases with increasing flake thickness and decreasing flake width and length. Flake geometry also affects durability by its influence on the irreversible thickness swelling which occurs typically after wetting and is attributed to the release of stresses accompanied by some loss of gluing bonds. Decreasing flake thickness improves thickness swelling, springback, and linear expansion properties of the board. The optimum flake thickness for durability and dimensional stability thus becomes a compromise. A practical size range from the standpoint of flake production, handling, and mat formation is 50–75 mm in length, 0.4–1.3 mm in thickness, and 12–25 mm in width (Geimer and Price 1978).

In the mid-1970s the U.S. Forest Service conducted a national research program to develop a structural panel product that could be manufactured from unmerchantable softwood stems in the western parts of the country. The raw material studied consisted mainly of large pieces of broken logs with diameters varying from 12–60 cm. More than 80% of the raw material was sound stem wood, while the rest consisted of bark or wood with decay. The goal was to produce a structural flakeboard that would be fully adequate as an exterior-grade construction material.

Acceptable strength and stiffness were found to be attainable using random flake construction. Aligned face construction produced high strength and stiffness more easily, except that linear instability perpendicular to face-flake alignment was seen as a potential source of problems in service. However, this linear instability can be reduced in face-aligned board by cross-aligning the core flakes.

Random face-flake construction was found adequately stable in both panel directions. Thickness stability remained a problem except at an uneconomically high resin content level. Board properties could be improved by increasing board density, increasing flake length, removing bark and decayed wood fractions, and increasing the amount of face flakes (Ramaker and Lehmann 1976). Behavior at various relative humidities of the panels was comparable with that of commercial and laboratory-made boards previously tested. Fire penetration, fire endurance, flame spread, and smoke yield characteristics were comparable with the fire performance of the commercial flakeboard. The target goals were not met for basic properties such as bending strength and stiffness and retention of internal bond strength after accelerated aging, but were surpassed for other properties such as nail holding, shear, and hardness. Generally, the panel made from large-diameter, Douglas-fir forest residue was acceptable as an exterior sheathing product (McNatt 1975, 1978; Holmes et al. 1979).

Heebink (1974) made structural flakeboard from unmerchantable *Pinus contorta* stems left unutilized in conventional logging operations when the minimum top diameter of timber was 15 cm. The residual stem mass originated partly from live and partly from dead trees. There were no technical difficulties in producing acceptable board from stem sections greater than 75 mm in diameter. All sizes of residue could be used to make three-layer boards with the core made of slivers from chunked residue and the outer layers of flakes produced on a conventional disk-type flaker. It was pointed out that, if necessary, board properties could be improved by an additional 1–2% resin.

Lehmann and Geimer (1974) prepared structural flakeboard from forest residue of *Pseudotsuga menziesii* utilizing various combinations of sound and decayed stem wood from live trees, stem wood from dead trees, bark, and branches. Adequate strength and dimensional stability were maintained as long as extreme amounts of bark or branch wood were not included in the panels. Board quality was significantly lower than that of control panels when more than 12% bark was included, when panels were composed entirely of tops and branches, or when panels contained a combination of residue and were composed of less than 75% sound wood. The most detrimental effects were found in panels composed of small unbarked branch material which, due to the high proportion of reaction wood, resulted in poor dimensional stability.

The authors concluded that with minimal care in selection and combination, Douglas-fir residue from clear-cut areas containing only small amounts of bark, branch wood, or badly decayed wood is an excellent raw material for strong, stable, and durable structural flakeboard.

Calvé et al. (1986) evaluated the possibility of including bark, branches, and unmerchantable stems in the manufacture of structural flakeboard using conventional waferboard technology. Four diameter classes of aspen trees from 2 to 36 cm at stump height were compared. For each class, board was made from debarked stems, unbarked stems, and whole trees. A drum-type laboratory flaker with a single rotating knife was employed for the production of wafers 76 mm long, 0.69 mm thick, and random widths.

The 2-cm-diameter trees were flaked in bundles, with and without branches and bark. Small trees naturally produced narrow wafers. Although narrow wafers will generally require a slight increase in adhesive as they present a larger surface area, they may still produce good-quality structural flakeboard. Particularly in the production of oriented strandboard, long narrow wafers are desirable. The wafers produced from the 20-mm specimens were irregular in shape with gagged ends and a noticeable amount of cross grain, and they contained a large quantity of fines. One problem encountered in processing very small trees was their lack of rigidity. They were very flexible and vibrated under the action of the flaker knives. The bark separated quite easily from the wood tissue during flaking and, consequently, a large part of it could be removed with fines rejects in screening. Calvé et al. made the following observations regarding the properties of structural flakeboard made from various diameter categories of aspen raw material:

1. Board produced from debarked wood showed lower dry bending strength and higher linear expansion, when the flakes were prepared from small stems. Other board properties were not adversely affected when furnish from small stems was employed, except for the 20-mm specimens which clearly produced a lower-quality panel.
2. Adding bark reduced board quality in all stem-diameter classes. However, even the furnish from the 20-mm stems, although it contained 23% bark, still produced a board of acceptable quality.
3. Whole-tree material yielded a board with lower quality than panels produced from debarked and unbarked stems, but the adverse effect of bark and branches was less than expected. Even the board produced from small-diameter whole trees met the standard requirements.
4. When the lower-quality material was concentrated in the board center and the outer layers were composed of high-quality furnish, strength problems associated with the bark and branches were reduced significantly. Three-layer board with a face-core-face ratio of 15:70:15 yielded board with similar bending, modulus of elasticity, and linear-expansion properties as board made from debarked stems. However, the internal bond of the three-layer board was 10–24% lower than in structural flakeboard made entirely from corresponding clear material.

Structural flakeboard relies on engineered, high-quality, bark-free flakes to develop superior strength properties at low resin levels and medium densities. For the most part, the flakes for these products are made from peeled logs or pulpwood bolts. The logs are directly converted with disk-, drum-, or lathe-type flakers. This standard equipment is not suitable for producing flakes from whole trees or forest residue of irregular form. If these raw materials are to be used for structural flakeboard, they must first be reduced to chunks and further to fingerlings which are 5–11 cm long and about 2 cm in cross-section. The fingerlings are processed into flakes in a ring flaker (Haataja et al. 1984).

In this process, the size and geometry of fingerlings are important for several reasons. The length of the fingerling controls the flake length, which should be about 50 mm or more. Since ring flakers rely on centrifugal forces to position the long dimension of the fingerling parallel to the cutting edge of the knife, the length-to-thickness ratio of the fingerling should be about 4:1 to ensure a proper positioning with the knife. Particles that are long enough but too thick in cross-section do not feed well through a ring flaker without plugging the machine. Furthermore, the cross-section of the particle controls flake width. Although structural flakeboard requires an engineered flake with the size restricted to fairly close tolerances, Haataja et al. (1984) anticipated that these flakes could be produced from small trees and residue material using a combination of equipment such as a U.S. Forest Service prototype chunker (Sect. 5.2.2), IWR chunkwood splitter, and standard ring flakers.

8.4 Use of Residual Biomass as a Feedstock for Chemicals

Forest biomass contains three primary constituents: cellulose, hemicelluloses, and lignin. In chemical processing, cellulose and hemicelluloses are the primary raw materials for the manufacture of improved products in solid, liquid, or gaseous form.

It is technically possible to process a great variety of valuable chemicals from forest biomass, but from the standpoint of large-scale production of bulk chemicals, forest biomass suffers from certain disadvantages. It is widely scattered and bulky, which results in a high cost of collection and transport. Since it is heterogeneous in composition, a given chemical is obtained in a rather low yield and usually in a mixture of several reaction products, from which it can be isolated only at considerable cost. Economics are controlled by the cost of raw material, cost of investment and processing, competition from alternative raw materials, and product markets (Klemola 1976).

Technology for making liquid or gaseous fuels from wood has been available for more than a century. However, gasification has been limited to the generation of low-caloric-value producer gas, and liquid-fuel technology has been limited to ethanol. Presently, few wood gasification, hydrolysis, and fermentation plants are in operation, although the use of producer gas is slowly increasing. Characteristically, improved fuels from forest biomass, particularly liquid fuels, have been of major concern only during times of national emergency. After the worldwide energy crisis in the 1970s, when petroleum dwindled and became a major source of balance-of-payments deficits in many countries, alternative fuels derived from renewable biomass feedstocks once again became important. This popularity rises and falls with fluctuations in oil prices and availability on the world energy market. But oil depletion is sure and continuous, and eventually mankind must turn increasingly to substitutes from renewable resources. Thus, technology for replacing at least a portion of petroleum-derived fuels with forest biomass must be developed through research and demonstration (Zerbe 1982).

There is a wide variety of possible conversions for biomass into improved secondary fuels. Many of these conversions can be made at individual, community, or industrial levels. In fact, the making of charcoal, distillation of alcohol, extraction of oils from plants, and many other conversions have been traditional practice in many areas of the world, but typically are restricted to small-scale operations. However, bulk production of fuel-grade alcohols and oils at competitive prices implies activities on an industrial scale. In some cases, for example with the production of ethanol, positive net energy balances and acceptable economics require sophisticated chemistry and systems engineering (Renewable ... 1985).

A great diversity of approaches exists for conversion of biomass into improved fuels or other chemical products. While some technologies are commercially available, some are only in the early stages of research and development. Future utilization of the vast chemical potential of renewable forest biomass reserves will depend on the successful development of merging technologies.

Thermochemical conversion technology employs elevated temperatures to process biomass to chemicals and improved energy forms. Biomass thermal reactions are highly dependent on both temperature and heating rates. Consequently, a wide range of products can be selectively manufactured using appropriate reaction conditions. Since biomass is highly reactive, thermochemical conversion processes are in many cases limited by heat transfer rather than by reaction kinetics. Careful attention must therefore be paid to the engineering aspects of these processes in order to exploit fully this reactivity (IEA Forestry ... 1987). Thermochemical process approaches include gasification (Sect. 8.4.1), pyrolysis (Sect. 8.4.2), and direct catalytic liquefaction (Sect. 8.4.3).

Biological conversion technology uses microbial action to process chemical products from biomass. An example is anaerobic digestion which, with the aid of microorganisms, produces a gas mixture from biomass containing 60–70% methane and 30–40% carbon dioxide. This biogas can be burned to produce heat or, for example, it can power internal combustion engines to produce electricity. Anaerobic digestion is used to process feedstocks such as wet manure and solid municipal waste. Since it is not a suitable method for untreated forest biomass, partly because of the high content of lignin in wood and bark, it is not discussed further in this chapter.

Biochemical conversion technology is a combination of biological and chemical processing steps. Since forest biomass is not readily accessible to enzymatic reactions, biological, chemical, or physical pretreatment is used to disrupt its lignocellulosic structure and allow release of carbohydrates for further conversion with microorganisms. A combination of acid hydrolysis of biomass and fermentation of the hydrolysis product to ethanol for use as motor fuel is an example of an attractive approach for large-scale biochemical conversion of forest biomass (Sect. 8.4.4).

Numerous chemicals can also be produced from forest biomass by *extraction or pressing*. Since extraneous materials rather than primary cell-wall constituents of biomass are recovered, the quantitative potential of extraction is small compared to the bulk products referred to above. However, some of these biochemicals may be very valuable. The products are often specific to genus, species, and tree component, extractives from conifer needles being perhaps of the greatest interest (Sect. 3.3.2.2).

Examples of promising approaches for the production of improved fuels from biomass are listed below. Some agricultural plants are included for comparison. Kelp is characterized by its high content of water. Sugarcane and beet represent plants rich in sugar, whereas corn and potatoes are characteristically rich in starch. Oil palm contains various kinds of oil.

Plant material	Conversion process	Primary product
Tree biomass	Gasification	Synthesis gas, methanol
Tree biomass	Pyrolysis	Charcoal, biocrude oil
Tree biomass	Catalytic liquefaction	Biocrude oil
Tree biomass	Enzymatic liquefaction	Biocrude oil
Tree biomass	Hydrolysis + fermentation	Ethanol
Corn, potatoes	Hydrolysis + fermentation	Ethanol
Sugarcane, beets	Fermentation	Ethanol
Kelp	Anaerobic digestion	Methane
Oil palm	Extraction, pressing	Oils

8.4.1 Gasification of Forest Biomass

The primary method of extracting thermal energy from biomass is direct combustion. To ensure complete burning, combustion is normally accomplished in excess air. If heat is applied to biomass at a high temperature with a deficiency of oxygen, incomplete burning of the volatile components results in the formation of gas. This process is called *gasification*. In addition to thermochemical gasification, conversion of biomass to gaseous products may also be accomplished with biological processes.

Compared to other solid fuels, such as coal, biomass feedstocks have unique properties that offer potential advantages for gasification processes. These advantages include (Schiefelbein 1985):

1. High volatility. Forest biomass possesses 70–90% volatile material versus 30–45% for typical coals. A large fraction of most biomass feedstocks can be devolatilized rapidly at relatively low temperatures.
2. High char reactivity. Biomass chars gasify rapidly in the presence of steam at relatively low temperatures. Consequently, biomass can be converted to gaseous or liquid fuel at faster rates and lower temperatures than most coal.
3. Low ash content. Incombustible residue remains as ash or slag and must be removed to avoid plugging the reactor. Ash also causes particulate emissions. In biomass gasification ash removal systems are simplified and disposal costs reduced.
4. Low sulfur content. Gas clean-up costs are reduced and biomass can be reacted in the presence of catalysts without sulfur poisoning problems. The corrosive effect of sulfuric gases is less serious.

On the other hand, certain properties of biomass feedstocks are less desirable in gasification. High moisture content results in energy loss and may make predrying necessary. Low bulk density causes handling and storage problems and can result in particle entrainment. Economy of scale is difficult to achieve because biomass is widely dispersed. Although large concentrations are not very important for gasifiers themselves, they are of great significance if synthesis gas is to be used as feedstock material for methanol production (White 1983b).

Requirements regarding the cleanliness of the product gas in terms of tar and particulate content depend on the intended use of the gas. The main end uses are as follows (Kohan 1985):

1. *Heat applications.* Low-caloric-value gas is burned in a boiler, primarily to produce heat. Gas cleanliness will generally be determined by local environmental requirements or by the desire to avoid fouling heat-transfer surfaces. Possible utility applications include new or retrofitted power plant boilers, where the gas produced from biomass may provide all or a portion of the fuel.
2. *Power applications.* Low-caloric-value gas is sent to a diesel or internal combustion engine. Cleanliness requirements are more stringent than in heat-only uses. This is due to the fact that tar can foul low-temperature inlet valves and controls on the engines, and because particulates can abrade or corrode internal engine moving parts. During World War II, cars, trucks, tractors, buses, and other vehicles were fueled in many countries with biomass gas produced from on-board gasifiers burning comminuted wood or charcoal. Other applications include fuel-retrofitting existing diesel-engine or combustion-engine generators with biomass-derived gas.
3. *Synthesis applications.* Medium-caloric-value gas is produced from biomass by removing virtually all sulfur, tars, and particulates, and by directing the very clean gas to subsequent catalytic processing steps. End products can be methanol, ammonia, etc.

The gasification process includes several competing chemical reactions. Greatly simplified, carbon reacts with water steam and carbon dioxide, producing carbon monoxide and hydrogen. Involvement of air in gasification results in a high content of nitrogen and, consequently, in a *low-caloric-value producer-gas*. Nitrogen can be eliminated if oxygen and steam are used instead of air and steam for gasification. The product is then a *medium-caloric-value gas*. The exact composition of a gas mixture depends on process variables such as temperature, pressure, heating rate, time, and presence of a catalyst as well as on the moisture content and other properties of the biomass processed. The following table illustrates a typical composition of gas of low and medium-caloric value, water vapor excluded (Benemann 1978):

	Proportion (%)	
	Low caloric-value gas	Medium caloric-value gas
Hydrogen (H_2)	15	30
Carbon monoxide (CO)	20	40
Carbon dioxide (CO_2)	10	20
Hydrocarbons (C_mH_n)	5	10
Nitrogen (N_2)	50	–

At a temperature of 15 °C and a pressure of 101.3 kPa, the effective heating value is 10.2 MJ/m^3 for hydrogen, 12.0 MJ/m^3 for carbon monoxide, and 33.9 MJ/m^3 for methane. For a mixture of low-caloric-value producer gas the average figure is 4–6 MJ/m^3 and for a medium-caloric-value gas usually 10–14 MJ/m^3, both of which are low compared to the heating value of natural gas, about 39 MJ/m^3.

The efficiency of gasification is defined as the ratio of the total energy of gas to the energy fed into the gasifier. Efficiency is reduced by factors such as heat loss from the gasifier, production of steam or moist air for gasification, and electricity consumption during the operation. Total efficiency is generally over 90%. If the gas is to be used immediately after production in a combustion furnace, its heat of formation can be recovered. If, on the contrary, the gas is used as a fuel for internal combustion engines, it must be cleaned, which results in loss of heat.

Filén et al. (1984) processed peat pellets, sod peat, 35-mm chips from unbarked stems, 60-mm chunks from unbarked stems, and chips from logging slash with a variable particle-size distribution in a 1.5-MW updraft gasifier. Gases produced from peat and wood fuels did not differ greatly. Typical contents of combustibles in the dry gas were 19% hydrogen, 24% carbon monoxide, and 2.5% methane. The calorific heating value of the raw gas, steam and tar included, was $3.5-5.0\,MJ/m^3$. The corresponding value of dry gas was $4.5-6.2\,MJ/m^3$, depending primarily on the moisture content of the feedstock, which could exceed 50% without any process disturbances occurring. Total efficiency was 92–97%, the most significant efficiency factor being the gasifier load rather than the moisture content of fuel.

Gasifiers are commonly classified by the method used to transfer fuel, heat, and gas in the reactor. Method selection depends mainly on the fuel, use of the gas, and the required production capacity. A gasifier may use either a fixed-bed, fluidized-bed or *entrained-flow technique*. The last mentioned alternative requires very fine, generally less than 0.1-mm fuel particles and is applied only for gasification of coal.

In a *fluidized-bed gasifier* gas flows upward through a bed of fine particles, such as inert sand or reactive limestone. The gas flow results in turbulent lifting and a bed with an expanded surface area, thus promoting heat transfer or chemical reaction. Fluidized-bed gasifiers are claimed to produce a very low tar content, but high gas velocities tend to carry ash and char out of the reactor, and these products must be separated in cyclones or bag houses. Fluidized-bed gasifiers appear to offer substantial advantages over fixed-bed designs in unit throughput and reactor control, which allows lower capital investment. While fixed-bed gasifiers are most suitable for piece sizes larger than 6-mm, fluidized-bed gasifiers can operate with a range of sizes, but are typically intended for 3–6 mm particles. Particle size and shape can have a marked influence on the performance of gasifiers (White 1983b).

Two alternative *fixed-bed techniques* are in common use: updraft gasifiers and downdraft gasifiers. The oldest and simplest principle is the counterflow or *updraft gasifier*. Comminuted biomass enters the generator at the top, whereas air enters at the base of the system. A series of zones of thermochemical destruction forms within the gasifier while the fuel gradually moves downward against an ascending stream of gases. In the top zone, moisture in the fuel evaporates and the dry solids heat up to a temperature at which devolatilization begins. The fuel descends through the succeeding pyrolysis zone, where the temperature rises gradually to 700 °C. Decomposition and devolatilization continue as the solids descend through the pyrolysis zone. Finally, the solid material enters the gasification zone at the bottom of the generator, where only char and ash remain. Here carbon is consumed in combustion reactions with oxygen at a temperature of about

1000 °C to form carbon oxides, and in the endothermic reaction with steam to form carbon monoxide and hydrogen. A grate at the bottom of the reactor supports the biomass material and allows ash to drop through to the base. Chemicals, tars, and oils are produced in the pyrolysis zone. They tend to condense in the cooler upper region and remain in the outlet gas. Depending on the feedstock and conditions of combustion, the concentration of tar is 40–50 g/m^3 of gas. For this reason, the gas and entrained material are usually burned directly in an adjacent furnace for heat. An updraft gasifier makes if possible to process relatively fresh biomass, but vaporized water is mixed with producer gas, reducing its heating value.

In a coflow or *downdraft gasifier*, air or oxygen is fed into the central part of the generator. The products of combustion and oil vapors from pyrolysis pass through a bed of glowing charcoal. The oil is cracked to a simpler producer gas which exits through the base of the reactor. Tar and oil are reduced to less than 10% of the amount produced in updraft gasifiers, and an almost tar-free gas is obtained. Due to the low content of tar, the gas is suitable for pipe transport and engine fuel. The maximum allowable moisture content of wood is 25–30%. A variation of the downdraft system is the *crossdraft gasifier*, which gives a more controllable performance but is limited in size.

Producer gas generated from solid fuels such as wood, charcoal, coal, peat, or agricultural residues, has been used *to power internal combustion engines* for more than 100 years. Producer gas does not require major modification of existing engines, and can be produced locally with a simple stationary or vehicle-mounted generator. On the other hand, producer gas generates less engine power than petroleum, the fuel is bulky, the gasifiers require careful and continuous maintenance, and they make the vehicle cumbersome. A gas producer requires 2.5–3 kg of dry wood to generate roughly the same amount of energy obtained from 1 liter of gasoline. Although these limitations are substantial, they are not insurmountable. The established technology can be applied in case of fuel emergency, and more efficient and economical designs are being developed. During World War II more than 1 million vehicles around the world operated on producer gas (Producer ... 1983). Modern wood-gasification technology and the economics of its application to internal combustion engines was reviewed by FAO (1986b). Figure 8.17 shows a modern wood gasifier mounted on a farm tractor.

The main components of the system are a generator to make producer gas from solid fuel, a cleaner to filter soot and ash from the hot gas, a cooler to condense tars and other liquid impurities of the gas, and valves to mix the cleaned producer gas with air and meter the mixture into the engine intake manifold. The heart of a gasifier is formed by the generator, i.e., a cylindrical or rectangular metal tank containing space for a 30-min to 2-h fuel supply, a combustion chamber, and an ash pit. The fuel falling into the combustion chamber is kept burning with air drawn through the chamber section. The updraft generators, with air entering the combustion chamber below the fire zone, are best suited for use with tar-free fuels such as charcoal, especially for stationary engines. The downdraft generators, with air entering the combustion chamber above the fire zone, produce less tar and are well suited for vehicle use. The crossdraft generators, with air entering through a nozzle projecting into the side of the combustion chamber, are suitable

Fig. 8.17. Producer gas-operated Valmet 702 farm tractor (courtesy of Vakola)

for motor vehicles using dry, low-tar fuels. In each type of generator, air drawn through the combustion chamber keeps the fuel burning and produces a bed of red-hot charcoal. This bed is sufficiently compact so that the gas streams flowing through contact the glowing carbon surfaces to form producer gas (Producer ... 1983).

Gas generators used for driving various types of engines require comminuted wood. Chunks may give a better result than conventional chips. Fines and overlarge particles should be removed. To prevent plugging and ash problems, the fuel should be devoid of inorganic contaminants and have a rather low bark content and uniform particle size. Since excessive moisture cools the generator and lowers the quality of the gas, efficient operation usually requires air-dry wood with a moisture content less than 25%. Hardwoods are preferred to softwoods due to their lower tar content. Hence, small-sized hardwoods are potentially a suitable raw material for the production of fuel for vehicle gasifiers. However, the requirements for seasoning, preparation, upgrading, and storage of chips and chunks are rather high.

A number of manufacturers are developing gasification systems for heating purposes, but the majority of existing systems are at a research or pilot-plant stage. Few systems can be classified as commercial on the basis of multiple unit sales and documented operating experience. One of the most interesting applications is *retrofitting oil-fired boilers* for gas. There are two options to adjust an oil-fired heating plant to solid fuels. One alternative is to supply the oil boiler with a precombustion chamber and burn the solids as completely as possible before feeding the gases to the boiler for heat exchange. The other alternative employs a separate gasifier, which generates good-quality, low-caloric-value producer gas in the presence of primary air only. The gas, at a temperature of 200–300 °C, is fed directly into the boiler and burned in the presence of secondary air. The gasifier alternative allows efficient combustion with very little excess air and good control of the operation, resulting in improved efficiency and reliability.

Leppälahti et al. (1984) studied an industrial updraft-gasifier application in a 5.2-MW district *heating plant* in Finland. The system was suitable for chips and chunks meeting the following basic requirements: moisture content less than 55%, deformation temperature of ash above 900 °C, and a particle size allowing sufficient permeability of the solids layer to gas. When sod peat with a 40% moisture content was used as fuel, the total efficiency of the plant was 84.9%, taking electricity consumption into account. The gasification heating plant was competitive with direct-combustion plants fired with corresponding solid fuels. Competitiveness with oil required a high annual peak operating time of 3000–5000 h.

A modern kraft pulp mill may produce more wood-derived fuels than needed for its process-steam production. However, certain operations in the process are presently bound to use oil or natural gas as their source of energy.. The most important of these units is usually the lime kiln of the mill. The target of energy self-sufficiency has led to a search for new methods using biomass fuels in *lime-kiln firing*. Among these are mixing of wet biomass fuels with lime mud and pulverized firing of dried, fine-ground biomass in the kiln. However, these alternatives are not energy-efficient and tend to raise the concentration of aluminum, silicon, and other impurities in the lime. This causes plugging and scaling problems at the evaporation plant and other phases of the process, especially if the biomass is rich in bark or foliage.

Most of these drawbacks can be avoided with a new gasification system, which allows the use of coarser particle size, facilitates flame stabilization, and makes it possible to separate the main part of the ash from the gas in the gasifier. The method is based on the circulating fluidized-bed combustion process. To reach sufficiently high flame temperature the biomass must be dried to 20–25% moisture content with flue gases or by firing a side stream of the producer gas at the drying plant. The existing 17–35-MW installations have shown that a biomass gasifier can be successfully integrated in the lime kiln to make a kraft pulp mill independent of the availability of oil (Kiiskilä 1987).

One of the most attractive application options of gasification is the *production of methanol*. Biomass-derived methanol can be used to power automobiles as a neat fuel, blended with gasoline, or converted to gasoline. The large amount of nitrogen in the producer gas in conjunction with air gasification is a disadvantage, since it increases not only the size of the gasification equipment and piping but also the size of the methanol-synthesis compressor, reactor, and associated heat exchangers. Therefore, despite the added cost of an oxygen plant, oxygen combined with steam is preferred to air as the gasifying agent in the production of *synthesis gas*, which is the first stage in the conversion process.

Natural gas, where available, is a preferred starting material for methanol. Much consideration has also been given to gasification of coal for the production of methanol, but wood has some distinct advantages over coal. If feedstock costs are not taken into consideration, for plants of the same capacity, processing costs are cheaper than for coal, since wood is easy to gasify and contains much less sulfur, heavy metals, and ash. Environmental and safety concerns favor renewable biomass feedstocks, but the advantage of coal is economy of scale, i.e., the advantage of being available in high concentrations. Although liquid fuels are produced from coal on a large scale in the Republic of South Africa, wood-to-methanol plants have not been built yet (Zerbe 1985). Examples of other bulk products which can be produced from synthesis gas are ammonia, ethanol, and hydrocarbons.

When the final product of gasification is indirectly methanol, thermal-efficiency data for the gasifier alone are of limited usefulness. Account must be taken of the exit-gas composition, energy input for oxygen production, and ability of downstream operations to utilize any heat recovered from the gasifier. Therefore, more meaningful thermal-efficiency values are associated with the total biomass-to-methanol conversion process. In a comparison of four alternative processes, White (1983b) estimated the efficiency as 40–50%, with the highest estimate for the process that did not require oxygen. Coal-to-methanol efficiency based upon an advanced, undemonstrated gasifier was projected to be near 60%. In another feasibility study of a 1000 t/day wood-to-methanol plant using a Winkler gasifier, the efficiency was 55%, cooling of gas and water being the major sources of energy loss (Tomaselli 1982). It must be emphasized that these estimates are not proven values from industrial operations.

Since the cost of biomass feedstock is a large part of the selling price of methanol, the overall efficiency has a direct impact on the production economics. The potential supply of biomass-based methanol is large and several processes for biomass-to-methanol conversion exist, but such methanol will have to compete in cost with methanol from coal and natural gas. The large capital investments required and the high financial risks involved will act as barriers to implementation of this new technology, even if production feasibility could be demonstrated at a pilot-plant level (White 1983b).

8.4.2 Pyrolysis of Forest Biomass

The physical and chemical decomposition of biomass to combustible gases, liquids, and charcoal by the action of heat in the absence of oxygen is called *pyrolysis*. Gasification (Sect. 8.4.1) differs from pyrolysis in that the deficiency of oxygen is less severe, the reaction temperature is higher, and the only usable product from the process is gas. In older literature, pyrolysis is referred to as destructive distillation of carbonaceous materials.

Liquid and solid pyrolysis products retain the "memory" of the original polymeric constituents of forest biomass. Differences in the chemical structure and proportional distribution of cellulose, hemicelluloses, lignin, and extraneous materials give rise to different products from hardwoods and softwoods. A rough example of an average mass and energy balance of the products from the slow pyrolysis of wood is presented below (Overend 1982). The exact result of the process depends decisively on raw material and operational conditions (see Scheme p. 441).

Several processes of pyrolysis are in use and under development. They differ in the method and rate of heating, and these differences influence the yield of the

Pyrolysis product	Distribution (%)	
	Mass balance	Energy balance
Pyrolysis gas	20.0	5.9
Pyrolysis liquid	23.5	31.4
Charcoal	30.0	52.2
Water	26.5	–
Process energy	–	10.5
Wood	100.0	100.0

three product categories. *Rapid heating* to high pyrolysis temperatures, in the range of seconds, increases the yield of lower molecular-weight liquids and gases, and reduces the yield of charcoal. *Slow heating*, in the range of hours, leads to lower gas yields, higher molecular-weight liquids, and higher charcoal yields (Baker 1983a). In most processes, pyrolysis occurs at 250–650 °C where fairly rapid devolatilization of various kinds of gases and tars takes place, leaving a solid charcoal composed mainly of carbon. In the gasifier environment at a higher temperature and in the presence of some, albeit deficient, oxygen, the pyrolysis gases and liquids such as tar, methanol, ammonia, and acetic acid undergo further reactions known as cracking and reforming. In gasification, these reactions continue until either the equilibrium composition is obtained, or their proportions are stabilized by the gas stream being cooled to low temperatures and thus being quenched (Overend 1982).

The effect of different ways and rates of heating on the yield and product composition have an important role in current research on biomass pyrolysis. In principle, the process is determined by the desired product category as follows:

1. If the primary product target is charcoal, the pyrolysis process is based on a low heating rate and a moderate process temperature.
2. If the primary product target is biocrude oil and subsequent refining into improved fuels, the pyrolysis process is based on a high heating rate and a moderate process temperature.
3. If the primary product target is producer gas for direct combustion or synthesis gas for indirect production of liquid chemicals such as methanol or ammonia, decomposition of forest biomass is carried out by gasification at high temperature in the presence of deficient air or oxygen.

Production of charcoal or *carbonization* has been practiced as long as history has been recorded. Unlike other forms of chemical conversion of forest biomass, carbonization can and still is often being carried out by very primitive methods. A substantial portion of the world's charcoal is produced in developing countries in smouldering piles of earth-covered wood, called earth kilns or pits. Although this method can also be used to produce tar from resinous pine wood, by-products are not generally captured, and it is not possible to restrict combustion to only that fraction of the fuelwood which is needed to generate the heat for carbonization. Due to poor process control, and because a part of the product is left on the ground in the form of fines, the thermal efficiency of carbonization in earth kilns is very low.

Charcoal was produced for industrial purposes hundreds of years ago on such a scale that the activity resulted in forest destruction over large areas. The primary use was in iron manufacture, where about 1 t of charcoal was needed for each ton of pig iron. In the United States, charcoal production continued to increase until 1880 when about 700000 t were manufactured. As larger iron blast furnaces were built, the use of charcoal in the iron industry declined, in part because charcoal lacked the strength adequate to support the overburden in the higher furnaces, and in part because transportation distances and costs of charcoal increased. The iron industry preferred charcoal with a high crushing strength, e.g., that produced from dense hardwoods such as maple, birch, beech, oak, and hickory (Baker 1983b).

While the efficiency of earth kilns tends to remain at about 10% or less, a better process control is achieved by using stationary brick and concrete kilns or portable steel kilns. Most of these kilns represent simple technology, characterized by labor-intensive work methods, batch-operated equipment, loss of liquid and gaseous pyrolysis products, and problems of air pollution. Externally heated, continuous carbonization retorts make it possible to collect the vapors and liquids as by-products from the carbonization, thus greatly improving the efficiency of the operation. By-products include acetic acid, methanol, acetates, soluble tar, creosote oil, pitch, etc., and gases may be burned for heat production. The energy efficiency of modern retorts is about 40%, the yield of by-products included.

The yield and fuel properties of charcoal depend on both feedstock properties and process conditions, particularly the operating temperature. For example, Satonaka (1982) reported the following values for charcoal produced from Japanese woods:

1. Ultimate analysis shows the following rough proportions for the main elements of wood: carbon 50%, oxygen 44%, and hydrogen 6% (Sect. 3.5). When wood is heated to 600 °C, the content of carbon in the solid material increases rapidly to 90%, whereas the content of oxygen decreases to 7% and that of hydrogen to 2.5%. At a temperature of 1100 °C these proportions are 96, 2, and 0.5%, respectively.
2. The average calorific heating value of wood in 12 softwood species was 20.8 MJ/kg and that in 59 hardwoods, 19.8 MJ/kg. When *Quercus mongolica* wood was heated, its calorific heating value per unit mass increased rapidly and attained a maximum of 33.5 MJ/kg at 600 °C. Above this temperature it decreased slowly to 31.9 MJ/kg at 1100 °C. Of the original heating value of the feedstock 48 and 41% was left, respectively.
3. The volume of the solid fuel decreased rapidly to 50% of the original at 400 °C and then slowly to 40% at 600 °C and 30% at 1100 °C.

Since fuelwood transport costs over long distances tend to be too high for poor people in urban areas of developing countries, upgrading of fuelwood to charcoal is often feasible. By carbonization, the mass and volume of the biomass are reduced and the heating value of fuel per unit mass is increased. Thus, charcoal can be transported and handled more economically than an equivalent amount of wood, and, although much of the energy in wood is lost in conversion to charcoal, the loss is largely compensated by the higher efficiency of charcoal stoves compared with wood stoves. Moreover, when burned in simple stoves, charcoal does not produce smoke.

The importance of charcoal as a fuel and, on the other hand, the inefficiency of production in developing countries have been widely recognized in recent years. The problem has been addressed by innumerable projects directed toward raw material supplies, manufacturing, distribution systems, and associated economic considerations. Much work has been done in designing kilns appropriate to the prevailing socioeconomic conditions of the region concerned. This has led to the development of a wide array of inexpensive, efficient designs in a range of sizes, based on local construction materials, technology, and available wood. Conversion efficiencies of 25–35% for high-grade charcoal are not uncommon (Preston 1982).

Although the largest demand for charcoal is for cooking food in developing countries, charcoal is used commonly in barbecues and restaurants in developed countries as well. Charcoal is also used as a heat source for drying tobacco and other agricultural products, melting metals, and for manufacturing lime, cement, metallic silicon, etc. An especially large user of charcoal for refining iron is Brazil, where large areas of eucalypt plantations are managed primarily for the production of charcoal. Among the many industrial uses of charcoal are products such as activated carbon, polishing carbon, and certain types of pencils.

Traditionally, charcoal is made from roundwood. Continuous, multiple-hearth Herreshoff furnaces or rotary kilns can convert reduced materials such as sawdust, shavings, chips, and bark to charcoal of fine particle size at about 25% efficiency. The product can be used in lump form, ground into powder for use as feedstock for activated carbon or fuel, or briquetted. The briquetting process consists of grinding, mixing the powder with a starch solution, pressing, and drying. Nurmi (1986) studied carbonization of screened, over 80-mm chunks from whole birch trees in a simple portable, 4.1-m^3 Mark V metal kiln. Compared to conventional 1-m firewood bolts, chunks required longer carbonization time, produced a slightly lower yield, and had a higher content of carbon. The results of this limited orienting experiment were considered promising for the use of chunkwood in the charcoal industries. The main advantage over conventional feedstock was that the wood preparation and kiln loading could be mechanized, thus reducing the cost of production.

Although charcoal is a very important domestic fuel for cooking in many developing countries, the industrial demand for charcoal is only local and small on a world scale. In the long term, the production of biocrude oil by pyrolysis obviously has a larger potential, since biocrude oil could be refined to liquid fuels and other valuable chemicals.

Until the 1920s during the *wood distillation* era, most industrially important organic chemicals were distilled from wood, primarily from hardwood. Cast iron and steel retorts made it feasible to recover by-products of pyrolysis in the form of crude "pyroligneous acid" and noncondensable gases. The acid from hardwood pyrolysis was refined to make acetate of lime, methanol, and tar. Acetate of lime was used in the manufacture of wood vinegar, acetic acid, many commercial acetates, and other products. Methanol was used widely as a solvent in the production of varnishes, dyes, formaldehyde, perfumery, and explosives, and was mixed with grain alcohol to produce denaturated alcohol, etc. Tar was burned to heat the retort and wood gas to dry and preheat the wood. Softwoods, especially stump and stem wood from pine, were also distilled, but the products were different from those obtained from hardwoods. In addition to charcoal, pine tars, oils, and turpentine were produced, but the softwood distillation industry was never as large or as significant as the hardwood distillation industry (Baker 1983b).

Industrial application of pyrolysis for the *production of liquid fuels* and other chemicals is limited by the large number of products. The traditional methods of wood distillation are no longer energy-efficient and economically competitive, and no proven modern technology is yet available. The methods presently investigated and developed in many countries are based on the conversion of forest biomass through rapid pyrolysis to biocrude oil and, as a by-product, to charcoal.

Pyrolytic oils can be produced by many processes. Although there are several possible mechanisms for producing low-value char and tar, the production of high-value olefins appears to be limited to conditions of rapid heating rates and relatively high temperatures. In recent years, the concept of rapid pyrolysis has emerged as a promising alternative. Using rapid heating rates, yields of gases and liquids as high as 95% can be obtained, including up to about 20% high-value olefinic products such as ethylene, benzene, toluene, and xylene, which are potentially useful as fuels and octane enhancers. Pyrolysis research is presently directed toward capitalizing this potential, paying special attention to the following aspects (Preston 1982):

1. Using extremely rapid pyrolysis to maximize yields of high-value olefins and aromatics. A central problem is obtaining high heat fluxes.
2. Improving the yields of rapid pyrolysis reactions by using reactive atmospheres, which would allow less stringent heating rate requirements, thus making the reactor simpler.
3. Using rapid pyrolysis to produce low-cost fuel oil substitutes. An example is an entrained-flow pyrolysis prototype unit built by Georgia Tech in the United States. The oil products are chemically quite different from petroleum liquids, obviously requiring further upgrading. Oil yields approaching 40% have been obtained, and yields as high as 60% by mass may be possible as operating parameters are optimized.

Despite active research by many organizations throughout the world, large-scale industrial utilization of rapid pyrolysis is not economical at the present state of development and with present oil prices. Another possible approach for the future production of biocrude oil from forest biomass is direct catalytic liquefaction, which involves the heating of biomass slurries under a reducing atmosphere of a mixture of carbon monoxide and hydrogen. This alternative is discussed in Section 8.4.3.

8.4.3 Direct Catalytic Liquefaction of Forest Biomass

Thermochemical liquefaction of forest biomass may be based on either direct or indirect processes. Indirect liquefaction involves the production of synthesis gas through oxygen gasification and conversion of the gas to methanol and, possibly, further to gasoline (Sect. 8.4.1). In direct thermochemical liquefaction, biomass is converted to liquid in a single step.

In a broad sense, *direct liquefaction* is defined as any thermochemical conversion process resulting in liquid products from biomass feedstocks without going through a separate intermediate gas phase. According to this definition, two different direct approaches are available; rapid pyrolysis at a low pressure (Sect. 8.4.2) and catalytic liquefaction at a high pressure. In a narrow sense, direct liquefaction of biomass is commonly defined as a particular technology of reductive, catalytic liquefaction of the feedstock at high pressure to produce biocrude oil. Characteristically, the biomass is mixed either with water or with recycled product oil to form a slurry which is subjected to 20 MPa pressure and 300–400 °C temperature in the presence of a catalyst in a reducing atmosphere (Stevens 1985). The latter definition of direct liquefaction is used in this book.

The feasibility of converting biomass by direct catalytic liquefaction at elevated temperatures and pressures was demonstrated at the U.S. Bureau of Mines' Pittsburgh Energy Research Center (PERC) in the early 1970s. Later, the Lawrence Berkeley Laboratory (LBL) of the University of California introduced a modified concept of the PERC process. Several other processes for direct catalytic liquefaction are being studied as well.

High pressure and moderately high temperature are required in catalytic liquefaction. Catalysts such as alkali metal carbonates, a reducing gas atmosphere of carbon monoxide and hydrogen, and water favor the liquefaction reactions. Innumerable complex reactions take place simultaneously. Cracking, polymerization, charring, dehydration, and volatilization all occur. Gas and solid residues are produced in addition to the crude oillike liquid, *biocrude oil.* By using a gas chromatography mass spectrometer more than 80 compounds have been identified in the oil. The liquefaction mechanisms are not well understood, but oxygen removal by reaction with carbon monoxide to produce carbon dioxide and hydrogenation by hydrogen with cleavage of polymer bonds have been postulated. However, there is some controversy over the role and requirements of carbon monoxide in the reaction system (Technical ... 1980).

The PERC process requires dried, finely-ground feedstock that is made into a slurry with recycled biocrude oil. The LBL process accepts a feedstock in the form of green chips. In the LBL process, chips are transformed into a slurry in a 0.1% dilute sulfuric acid solution and heated at 180 °C temperature for 45 min to prehydrolyze the hemicelluloses and weaken the structure of the wood. Sodium carbonate is added, and the slurry, containing 20–30% organic matter, is homogenized by passing it through a refiner. The slurry is then heated to 330–370 °C at high pressure under a synthesis-gas atmosphere. After a 10–30 min residence time, the effluent is cooled and passed into a pressure let-down vessel. Finally, the oil and aqueous phases are separated. Compared to the PERC process, the principal difference is that an aqueous slurry rather than a recycled oil slurry is used. Since water has a high heat capacity, the slurry must be as concentrated as possible to reduce the energy consumption of heating (Klausmeir 1985).

The catalytic liquefaction processes are not well understood, and the biocrude oil products are not well characterized. Reliable data on correlations of product yields and compositions with feedstock properties are not available. Nevertheless, particularly in direct catalytic liquefaction, the composition of the biocrude oil appears to be rather dependent on the chemical composition of the feedstock and, on the other hand, on conditions in the liquefaction reactor. The lignin component of the biomass produces significant amounts of phenolic and other aromatic materials. Cellulose liquefaction products are derived from D-glucose building blocks and include pyrans, olefins, ketones, and alcohols. Hemicellulosic liquefaction products are derived from five- and six-carbon sugars and include the same cellulose derivatives and furans.

The biocrude oils tend to be inferior to petroleum crude oils. They are corrosive due to acidity, unstable, and foul-smelling. They have a high moisture content and low heating value, and many of them are solids or semisolids at room temperature. Most of these undesirable properties are the result of high levels of oxygenated organic compounds such as phenols and carboxylic acids. Although they are suitable for some uses as produced, it is necessary to refine and upgrade the biocrude oil to increase the product value, for example by separating it into phenols and phenol-free oil (Klausmeir 1985).

Comparative tests of the IEA program for research, development, and demonstration on energy from forestry indicate that product yield and quality vary from process to process. Energy yields of over 70% can be obtained in the form of biocrude oil for both high-pressure direct catalytic liquefaction and low-pressure rapid pyrolysis liquefaction approaches. The rapid pyrolysis process results in products with higher concentrations of oxygen and organic acids and a lower viscosity. The high-pressure products from catalytic liquefaction contain larger quantities of phenolic derivatives. These differences influence the value and end use of the crude oils and will have a significant impact on the upgrading procedure. An economic analysis for hypothetical plants manufacturing 1000 t of biocrude oil per day suggested that high-pressure processes may be more costly. But since the products from lower-cost processes are less stable, highly oxygenated, and probably more difficult to upgrade into gasoline substitutes, production and upgrading must be considered as an integrated concept. At the present state of development, neither liquefaction approach could be considered as necessarily superior (IEA Forestry ... 1987).

Biocrude oils can be upgraded into hydrocarbon fuels. Several conceptual upgrading approaches exist, but research is at a very preliminary stage. Upgrading to hydrocarbon fuels must remove oxygen, reduce molecular weight, and must address cetane/octane considerations. Tests in the IEA project indicate that high-pressure products could be selectively hydrotreated to form gasoline-like hydrocarbons with an energy content of 44 MJ/kg. However, very great uncertainties regarding even basic data, such as mass and energy balance, prevent a detailed economic analysis of the upgrading step (IEA Forestry ... 1987).

Processes and equipment for direct liquefaction of biomass and upgrading of the resulting biocrude oils are not technically ready for industry-scale applications, and the estimated costs of the products are not competitive compared to petroleum products at the 1988 oil-price level. A breakthrough is not probable in the near future. However, intensive research and development work is being conducted in many countries, and the long-term potential of this approach continues to be of great interest.

8.4.4 *Hydrolysis of Forest Biomass*

Approximately 40–50% of the dry cell-wall mass in woody cells is cellulose, 20–35% hemicelluloses, and 20–35% lignin. The proportion of hemicelluloses is largest in hardwoods and that of lignin in softwoods. Of the two main classes of hemicelluloses, galactoglucomannans predominate in softwoods and xylans in hardwoods. All these polysaccharides are composed of large molecules formed by the combination of small repeating structural sugar units (Sect. 3.3).

In *hydrolysis*, cellulose and hemicelluloses are depolymerized to form simple sugars. Cellulose dissolves and is degraded in strong mineral acids by hydrolysis to dextrins and oligosaccharides, and ultimately to glucose. Hemicelluloses dissolve readily in dilute acid to their monomeric components, yielding hexoses (glucose, mannose, galactose) and pentoses (xylose, arabinose) together with a monomethyl glucuronic acid or oligosaccharides containing this acid. Lignin remains insoluble, can be removed from the solvent by filtering, and is used primarily for fuel. Sugar solutions from acid hydrolysis of wood have been used on a commercial scale primarily for the production of ethanol and fodder yeast through *fermentation.*

The yield and composition of the hydrolysis products depend on both the operational conditions and the raw material properties. For example, sugars obtained from softwoods differ distinctly from hardwood sugars. Nonetheless, hydrolysis makes the sugars readily accessible for microorganisms and, consequently, for further biological conversion. Since the enzymatic reactions are highly specific, the subsequent conversion of simple sugars by fermentation generally leads to chemical products of high purity. The primary products can be used for fuel, chemicals, or starting materials for a large variety of other chemical products.

Once monosaccharides are economically obtained in either an acid or enzymatic hydrolysis, the way is open to commercial chemicals. The hydrolysis liquor can be concentrated to molasses for animal feed. Crystalline glucose can be isolated from solution for the food industry, and glucose in solution can be isomerized to food-grade syrup or, quantitatively, into fructose. Hydrogenation yields sorbitol, also used in food products. In other reactions antibiotics and vitamin C can be synthesized. Lactic acid, enzymes, and steroid hormones are examples of valuable glucose fermentation products. Pentoses can be converted into yeast and protein products to be used as animal fodder. Mild acid hydrolysis of pentosan-rich hardwoods releases into the solution substantial amounts of xylose, whose hydrogenation yields xylitol for use as a sweetener, plasticizer, and starting material for various chemicals. Under more severe process conditions, established with steam hydrolysis without any added mineral acid catalyst, pentosans can be converted into furfural, using the organic acids liberated from wood as catalysts (Klemola 1976).

In addition to the examples mentioned above, a large number of other commercial chemicals can be obtained from forest biomass through hydrolysis and fermentation. However, even if no stumpage price is involved, the cost of forest biomass is presently, with few exceptions, too high to permit economical manufacturing. Although technology development will probably improve the competitiveness of many hydrolysis products in the long term, large-scale utilization of residual biomass in the biochemical industries presupposes manufacturing of products in bulk. Throughout the world, therefore, the focus in research and development programs is in *liquid fuels.*

Two biomass-derived alcohol products attract wide interest as future options to petroleum products. One is methanol from synthesis gas through thermochemical gasification of biomass (Sect. 8.4.1), and the other is *ethanol through biochemical conversion* techniques, using hydrolysis and fermentation. Presently, ethanol is used primarily for making alcoholic beverages and as a raw material and solvent in the synthetic chemicals industry, but like methanol, it is also well suited for use as a motor fuel. Ethanol and methanol both have good antiknock properties and therefore enable increased engine efficiency by raising the compression ratio. Problems caused by exhaust-gas emissions are reduced compared to a conventional gasoline engine. However, cold-starting is still a major difficulty, as are water-phase separation, gas-lock formation, and corrosion problems. Both ethanol and methanol can be used as a motor fuel either in pure form or as a 10–20% mixture in gasoline (Riekko and Anttila 1980).

Ethanol can be produced from three different types of biomass constituents, i.e., sugars, starches, or celluloses, in order of increasing difficulty. *Sugar-rich biomass,* such as sugarcane and sugar beet, can be fermented directly into alcohol through the action of enzymes provided by yeast. In the process, ethanol, carbon dioxide, and heat are produced. *Starch-rich biomass,* such as wheat, corn, and potatoes need to be hydrolized before fermentation. Although this is relatively easy, it does represent an additional step, and thus increased complexity and cost. Ethanol technologies based on sugar- and starch-containing biomass are well developed. Their feasibility was established in Germany during World War II, and ethanol is presently produced from sugarcane on a large industrial scale in Brazil.

Recently, researchers have begun looking at the production of ethanol from *cellulose-rich biomass.* While the use of sugarcane is limited by seasonal cropping, forest biomass is available year-round. Since cellulosic biomass has to be hydrolized before fermentation, the process is more difficult. Nevertheless, the technology has developed considerably in recent years, and pilot plants now exist. Advanced high-technology systems may include not only sophisticated processes but also genetically engineered enzymes and feedstocks (Renewable ... 1985).

The world production of ethanol from biomass was 8.4 billion liters in 1981 and was expected to rise to 20 billion liters in 1985. Brazil and the United States accounted for more than 95% of the production. In Brazil, ethanol was produced from sugarcane for use in automobiles as a substitute for imported petroleum products. In 1981, approximately 20% of petroleum consumption was displaced by ethanol blended in gasoline, and the displacement forecast for 1985 was 10.7 billion liters, corresponding to half of the country's gasoline demand. Industrially processed alcohol fuels continue to remain expensive, but the Brazilian program shows the great potential of the concept. Although the feedstock is presently composed of sugarcane, research and development efforts are resulting in new, advanced biomass-to-ethanol conversion technologies which will probably also make processing of residual forest biomass more efficient and economical (Renewable ... 1985). Presently, the following techniques are available (Zerbe 1985):

1. Ethanol and yeast are produced by *fermentation of sugars from the spent liquor* which is obtained as a by-product in sulfite pulping. The liquor is first treated to remove sulfur dioxide and lignosulfonate products prior to fermentation. Approximately 95 liters of 95% ethanol are obtained per ton of sulfite pulp. As the number of sulfite pulp mills has decreased, the importance of ethanol as a by-product of sulfite pulp has decreased simultaneously.
2. The process that has been most successful for the conversion of wood to ethanol is *dilute acid hydrolysis* at a relatively high temperature, followed by subsequent fermentation of the sugars to alcohol. Such plants operated in Germany and Switzerland during World War II, and several plants are still operating in the Soviet Union, producing ethanol, yeast, furfural, and crystalline glucose. Most dilute acid hydrolysis processes are a single-stage type. In two-stage operations, hemicelluloses are hydrolyzed in the first stage, for example by heating with steam, and the more refractory cellulose in the second saccharification stage with acid, which facilitates the recovery of both five- and six-carbon sugars.
3. Another process for conversion of wood to ethanol is *concentrated acid hydrolysis* at low temperature followed by fermentation, but the method is not used in practice. Industrial applications of this process have been very limited.
4. A relatively new technique for the production of ethanol from forest biomass is *enzymatic hydrolysis*. Cellulose is converted to glucose in high yield by the aid of organisms and then fermented to ethanol.

Enzymatic hydrolysis produces sugars from forest biomass with the aid of microorganisms. Several enzyme systems, based on the use of organisms such as *Trichoderma viride* or *Aspergillus niger*, have been shown to transform cellulosic feedstocks into sugars. Compared to acid hydrolysis, process conditions are mild, decomposition of the product sugars is not a serious problem, formation of harmful by-products from sugar degradation does not occur, and corrosion is minimal. However, biomass must be pretreated to make it accessible to enzymatic attack. Among the economic obstacles are the high cost of enzyme production and insufficient enzyme activity, resulting in a low rate of hydrolysis. No industrial processes based on enzymatic hydrolysis of forest biomass are in practical use.

Intense efforts have been put into research to improve the performance and economy of various processes of acid hydrolysis. The main disadvantage is generally the high cost of plant investment. The only process presently used on an industrial scale, the dilute acid percolation process, suffers from many technoeconomic deficiencies. Among them are high temperature and pressure, slow reaction, noncontinuous batch process, and problems in acid regeneration and corrosion. According to White (1983 a), glucose yields from several processes vary within a wide range but in most reports are 50–70% of the theoretical yields. In a Brazilian pilot plant using the Scholler-Madison-Soviet concentrated-acid process, the ethanol yield from dry wood was 182 l/t for eucalypt and 205 l/t for pine.

In addition, charcoal was produced as a by-product from the hydrolysis residue, the yields being 129 and 145 kg/t dry wood, respectively (Tomaselli 1982). Despite its higher sugar yield and favorable energy balance, the competitiveness of the concentrated acid process is low, mainly because of difficulties involved in corrosion and recovery of the acid.

A state of the art survey in cellulose conversion to ethanol indicated that the approach is technically feasible. However, competing sources of ethanol from sugar and starch fermentation, or synthetic alcohol from petroleum, continue to make cellulose an uneconomical raw material for ethanol. The survey concluded that recent technological advances and continuing efforts may eventually provide a price-competitive ethanol product from forest biomass. An economic analysis indicated a cost advantage for acid hydrolysis over enzymatic hydrolysis, but it was too early in the development of advanced processes to draw final conclusions. It was seen as necessary to achieve higher yields and, on the other hand, to convert all constituents of biomass feedstocks, such as hemicelluloses and lignin, to usable or salable products (White 1983a).

According to Zerbe (1985), the most likely wood-to-liquid fuel plant to compete in today's market would be a dilute acid hydrolysis plant on a scale to use 1000 dry tons of wood per day to produce 100 million liters of ethanol per year. However, in most countries, it would be impossible for such a plant to compete without a subsidy. Another option would be indirect production of methanol through gasification.

8.5 Use of Residual Forest Biomass for Fodder

Since farms are located in rural areas, generally not far from forests, it is natural to try to find local agricultural uses for unmerchantable components of forest biomass. Bark and foliage from whole-tree chips can be used to improve the tilth of light sand and heavy clay soils. They can also be used for composting, mulching, and landscaping purposes and as a growing medium in nurseries, and can serve as deep litter material for cattle and poultry as a substitute for straw or sawdust. The Forestry Sciences Laboratory of the U.S. Forest Service at Houghton, Michigan, is investigating the possibilities of using whole-tree chunks as a local forest-road construction material where gravel is not available (Hodek et al. 1988).

The above-mentioned agricultural and related use applications of residual forest biomass have only limited local significance, since none of them can absorb the cost of long-distance transport. In the long term, forest biomass may also be studied as a potential reserve for livestock fodder to satisfy mankind's ever-growing need for food. Use as animal feed might ultimately achieve great significance, although in the developed world as a whole surplus production rather than deficiency is currently the major concern in agricultural planning.

Even in developed countries, scarcity of farm products occurs locally and during times of crisis. Under such conditions, the possibility of using residual forest biomass as a supplement for traditional sources of animal feed could offer security and economic advantages. In developing countries where insufficient nourishment is a constant concern, any additional source of fodder would be welcome.

Four alternative fodder uses of residual forest biomass are discussed in the following sections:

1. Use of forest biomass as roughage for ruminants;
2. Use of forest biomass as energy for ruminants;
3. Use of foliage as a protein source;
4. Use of foliage as a biologically active fodder additive.

8.5.1 Wood and Bark as Ruminant Fodder

Wood and bark components of forest biomass can have two functions in livestock feeding. They may serve as dietary energy feed, and they may act as a pure roughage substitute with little or no nutritive value. Although the former function has a much greater potential in contributing to the fodder supply, roughage is a necessary component of ruminant feed.

Roughage is required in the ruminant ration to provide tactile stimulation of the rumen walls and to promote cud-chewing, which in turn increases salivation and the supply of a buffer for the maintenance of rumen pH. Common roughage materials are hay, corn cobs, oat hulls, etc., but, at times, the supply may be limited and costly. In such cases wood and bark of small particle size from hardwoods, especially aspen sawdust, could be used as a roughage substitute in a high-grain ration for lactating dairy cows (Fig. 8.18). However, since the digestibility of

Fig. 8.18. Whole-tree chips prepared from small-sized hardwoods can be used as a roughage extender in cattle feeding (courtesy of Hannu Kalaja)

untreated wood is poor, this type of roughage can contribute little to the dietary energy needs of ruminants. Therefore, cows fed with wood-containing rations tend to compensate for the lower digestibility by eating more. Whether high-producing cows already at maximum feed intake could do this is questionable. Consequently, using untreated sawdust as the only roughage in high-concentrate dairy rations cannot be recommended, but aspen sawdust could be used as a roughage extender. For beef cattle, depending on the other ration ingredients, concentrations of 5–15% screened sawdust appear practical (Baker et al. 1975).

Because carbohydrates are the primary compounds in wood and bark, the use of forest biomass as a *metabolizable energy feed* supply presupposes that its cellulosic component is digestible. Ruminants have an unusual advantage in being peculiarly adapted to digest cellulose. They are an outstanding example of a symbiotic arrangement consisting of the animal's rumen, a chamber preceding the true stomach, and the ability of the rumen to culture cellulolytic microorganisms, which enzymatically break down cellulose and hemicelluloses. Pure cellulose is fully digestible by ruminants and thus provides as much energy as the best feed grains. However, due to the existence of lignin-carbohydrate complexes and the crystalline nature of cellulose in wood, cellulose and hemicelluloses are generally unavailable as fodder to ruminants. If simple, economic ways could be found to make these constituents of tree biomass digestible, i.e., susceptible to enzymatic or microbiological attack, forest residue and unmerchantable tree species could supply large amounts of energy feed now provided by hay and grain (Scott et al. 1969).

Both in vivo and in vitro rumen tests indicate low digestibility for wood. Softwoods are essentially nondigestible and hardwoods only somewhat digestible. Bark generally has a higher digestibility than wood, but differences among species are large. Baker et al. (1975) investigated several softwood and hardwood species and found the highest digestibility values, 33 and 50%, for wood and bark of *Populus tremuloides*. All species remained considerably below the reported digestibility levels of more conventional animal feeds, such as about 60% for alfalfa and 90% cotton linters:

	In vitro digestibility of dry mass (%)	
	Wood	Bark
Acer saccharum	7	14
Betula alleghaniensis	6	16
Fraxinus nigra	17	45
Populus tremuloides	33	50
Tilia americana	5	25
Ulmus americana	8	27
Larix occidentalis	3	7

Hydrolysis and chemical pulping are the ultimate methods of releasing carbohydrates in wood and bark. Acid hydrolysis is applied on a small scale in the United States to make hardwood sawdust more readily digestible for beef cattle by producing simple sugars through degradation of the cellulose and hemicelluloses.

Chemical pulping, on the other hand, makes the cellulose available by removal of the lignin. This approach was actually applied in northern Europe during World War II to substitute softwood pulp for hay.

Delignification of wood by pulping produces materials with a high rumen digestibility, depending upon the extent but not upon the method of lignin removal. As delignification reaches completion, the proportion of digestible dry mass approaches 90% in hardwood pulps and 85% in softwood pulps. Process residues such as pulp fines can also serve as energy feed for ruminants (Baker et al. 1973).

However, the use of pulp for fodder is not feasible under normal economic conditions. In order to produce energy feed from forest biomass at high yield without a waste stream and at an acceptable cost, Baker et al. investigated the possibilities of the following pretreatment approaches:

1. It has been shown that fragmentation of wood to finer particles stimulates the bacterial degradation of wood in the rumen without lignin removal (Virtanen and Koistinen 1944; Virtanen and Nikkilä 1946). However, the response of various tree species to size reduction is quite selective. For example, Baker et al. found that the proportion of digestible sawdust in in vitro tests after a 30-min ball-milling treatment was 70% for *Populus tremuloides* but only 15% for *Alnus rubra.* The quantity, chemical nature, and distribution of lignin were assumed to be controlling factors. Due to the selective response, the ball-milling technique was concluded to be of limited value as a general means for upgrading the digestibility of wood residue as ruminant feed. Furthermore, it was questionable whether finely ground tree biomass will function as effectively in the ruminant as it does in an in vitro assay, since a very fine particle size tends to shorten the residence time in the rumen.
2. High-energy electron irradiation was an effective means of enhancing the digestibility of *Populus tremuloides* but did very little to improve digestibility of *Picea sitchensis.*
3. Treatment with gaseous or anhydrous liquid ammonia provided a substantial increase of in vitro digestibility of *Populus tremuloides* wood. A 30-min treatment with gaseous ammonia rapidly raised the digestibility to approximately that of hay. However, the response was weak for other tree species, resulting in 7–10% digestibility for *Picea sitchensis* and only 2% for *Quercus rubra* wood.
4. Treatment with dilute sodium hydroxide solution resulted in 55% digestibility for *Populus tremuloides* and *Tilia americana,* 11–41% for a number of other hardwoods, and 1–2% for *Picea sitchensis* and *Pseudotsuga menziesii.* Sodium hydroxide-treated wood has been used for livestock feeding in Europe during times of severe fodder shortages.
5. Treatment with sulfur dioxide vapor was found to disrupt the lignin-carbohydrate association in situ and to yield a product of high digestibility without physical removal of the lignin, which was extensively depolymerized and converted to soluble products. The 60–65% digestibility of treated hardwoods compares favorably with the digestibility of a high-quality hay or alfalfa. The digestibility of softwoods was somewhat lower, 46% for *Pseudotsuga menziesii* and 50% for *Pinus ponderosa.* Since the feed was accepted by ani-

mals during digestion trials, the treatment with gaseous sulfur dioxide was considered an interesting approach to make cellulosic compounds of residual wood metabolizable to ruminants.

6. Treatment with white-rot fungi is also a possible approach to improve the digestibility of wood. The fungus must have the ability to remove lignin faster than polysaccharides. In experiments with a number of fungi, lignin removal was always accompanied by removal of polysaccharides, thus resulting in a loss of usable components of the biomass.

Based on research originally carried out in the 1960s at the Eastern Forest Products Laboratory and the Animal Research Institute in Canada, a special process was developed in the late 1970s for upgrading the feed value of a wide range of lignocellulosic materials in terms of available energy. The process, called STAKE, is based on the thermochemical treatment of biomass by steaming at a high pressure and temperature in a continuous operation without chemical additives. The equipment operated by one man may be built stationary or mounted on a portable trailer, consisting of conveyors and feeding equipment, a stainless-steel horizontal digester, steam generator, and product bin. Aspen whole-tree chips with a particle size less than 25 mm are a suitable raw material, producing a brown, acid, fibrous material containing 47% crude fiber, 17% soluble carbohydrates, 1% crude protein, 1% crude fat, and 1% ash. The digestibility is 55%, which corresponds to the standard value for timothy hay but is lower than that for alfalfa. The product is accepted by ruminants of various types. For example, the mass gain of steers fed up to 30% treated aspen chips in their ration was as rapid as that of steers fed corn silage. A higher percentage of aspen reduced the rate of growth (Esdale 1977).

8.5.2 Foliage as a Source of Protein and Vitamins

The foliage component of forest biomass is for most utilization purposes only a harmful contaminant, since it has a chemical composition distinctly different from that of wood (Sects. 3.3.2.2, 3.3.3). It is, however, characterized by a large number of valuable feed ingredients such as protein, soluble nonstructural carbohydrates, pigments, vitamins, etc.

Since there is a constant shortage of fodder protein in the world, forest biomass can also be examined as a *source of protein* for livestock feeding. For example, spent sulfite liquor, wood hydrolyzates, and fines from chemical pulping can serve as a carbohydrate source for the production of single-cell protein or fodder yeast for use as an ingredient in feedstuffs. While ruminant animals, such as cattle and sheep, are able under certain preconditions to digest structural carbohydrates of forest biomass, nonruminants, such as pigs and chickens, can only digest protein and nonstructural carbohydrates.

Foliage can be regarded as a *protein reserve* of considerable dimensions. It exhibits both a high protein content and a favorable composition of essential amino acids. Table 8.13 shows an example of the occurrence of essential amino acids in the foliage of *Betula pendula* and *Pinus sylvestris* in young greenhouse-grown

Table 8.13. Proportion of essential amino acids in the leaf proteins of greenhouse-grown seedlings of *Betula pendula* and *Pinus sylvestris*, compared to the proteins of cereals cultivated outdoors (Sirén et al. 1970)

Compound	*Betula pendula* leaves	*Pinus sylvestris* needles	Wheat	Maize	Soya
Percent of total dry mass					
All proteins	20	12	10	10	40
Percent of proteins					
Theonine	5.0	4.8	3.3	3.7	3.9
Valine	7.4	7.6	4.3	5.3	5.3
Methionine	1.3	1.0	2.5	3.1	1.7
Isoleucine	6.2	6.0	4.0	6.4	6.0
Leucine	10.4	10.3	7.0	15.0	8.0
Phenylalamin	6.5	6.4	5.1	5.0	5.3
Lysine	5.4	7.5	2.7	2.3	6.8
Essential amino acids, total	42.2	43.6	28.9	40.8	37.0

seedlings. The proportion of essential amino acids in the total amount of proteins is about 43% or somewhat higher than in soybeans and corn, and significantly higher than in wheat (Sirén et al. 1970). However, since the protein content is effected by growth conditions, Table 8.13 does not represent the foliage of large forest-grown trees.

The protein content of foliage varies among tree species. In trees of the temperate and boreal forest zones it is usually higher in hardwood leaves than in conifer needles, although most conifers offer the advantage of being in leaf throughout the year. Analytical studies on North American conifers indicate protein contents of 5.9–7.3% for the needles of *Abies balsamea, Picea glauca, P. rubens, Pinus contorta,* and *P. taeda* (Barton 1981). The following data from Dickson and Larson (1977) show that the protein content of hardwood foliage is rather high from the viewpoint of fodder:

	Concentration in dry mass (%)		
	Conifer foliage	Hardwood foliage	Alfalfa meal
Crude protein	6–12	12–20	18
Cellulose	20–36	15–30	26
Fats	5–10	3–8	3

Even in late summer, hardwood foliage has a protein content as high as that in clover and other legumes cut during their early growth stage. The following table shows the protein content in the leaves of some Finnish hardwoods during July (Pohjonen and Näsi 1983):

	Concentration in leaf mass (%)	
	Crude protein	True protein
Alnus incana	17.7	17.3
Betula sp.	14.6	12.8
Populus tremula	16.8	15.9
Salix sp.	15.3	13.3
Sorbus aucuparia	10.6	10.1

The amount of crude protein in foliage can be estimated by multiplying the percentage of nitrogen by 6.25. Nitrogen fertilization increases the protein content of tree foliage, a fact that is in agreement with experience in agriculture. Foliage from *Populus* hybrids grown on fertilized nursery plots contained 3.1% nitrogen and 19.5% crude protein on a dry mass basis. It was concluded that such foliage could probably be used directly as a substitute for an equivalent amount of alfalfa meal or other high-quality forage (Larson et al. 1976; Dickson and Larson 1977).

Näsi and Pohjonen (1981) found the crude protein content in foliage from young *Populus* and *Salix* plantations to be as high as 20.4% and that of true protein 17.2%. The content of crude fiber was only 14.1%. The protein content was considerably higher and the fiber content less than half when compared to that of hay. The digestibility of leaves varied widely by tree species. Tannins in leaves reduced both digestibility and protein utilization.

The content of protein and amino acids is highest in hardwood leaves in early summer and in conifer needles in winter. However, if hardwood leaves are to be used for protein, the most advantageous harvesting time is the end of the growing season just before the start of translocation of important compounds from the leaves, since the total mass of protein is then greatest. Conifer needles should be harvested during winter, but compared to hardwood leaves, their protein content is significantly less (Sirén 1973).

Pohjonen and Näsi (1983) proposed two fractionation procedures to separate fodder from the fuel component of small-sized hardwood biomass. One method is treating the trees mechanically with a farm tractor-mounted, manually-fed flail delimber to segregate two products: a mixture of foliage and pieces of thin twigs for fodder, and roughly cleaned stems for fuel. The other method is dewatering fragmented biomass by pressure to produce a fibrous, low-protein solid fraction for ruminant roughage or fuel, and a virtually fiber-free, high-protein juice for subsequent processing as a protein concentrate for nonruminants. However, in preliminary laboratory experiments the crude-protein content in the juice was only 0.7% of the total mass of the liquid and 9% of its dry mass, which was actually less than in the native leaf biomass studied.

Pohjonen and Näsi stored fresh chips from 3-m-high willows in airtight bags using an urea-phosphate mixture as preservative. The proportion of urea-phosphate was 2.5% of the total mass of fresh chips. The product was palatable to cattle in preliminary feeding experiments. The concept was technically feasible, but the production cost of the fodder was not competitive.

The need for high-protein, digestible forage is a chronic concern throughout the tropics. In the seasonally dry areas of the tropics particularly, tree foliage may be of great significance as an additional source of fodder. An example of promising forage tree is *Leucaena leucocephala*, whose foliage is used in Southeast Asia as fodder for cattle, water buffalos, goats, and other domestic animals. The leaves of this nitrogen-fixing tree contain as much as 4.2% nitrogen and 25.9% crude protein. The carotene content is among the highest recorded in plant specimens. The leaves are readily digestible and serve as both energy feed and roughage, therefore making an excellent food, comparable to alfalfa. They may be used fresh, in the form of dried meal, or as ensilage. However, the leaves also contain an unusally large amount of mimosine-related compounds, which may be toxic, especially for nonruminants. Consequently, animals cannot be fed solely on *Leucaena* foliage (Leucaena ... 1977). A large number of other tropical tree species can also be used for animal food, but the potential of tropical forage trees as a whole is insufficiently known and utilized.

Tree foliage is not only a potential source of protein, but is also of great interest as a *biologically active fodder additive*. Foliage contains a large number of vitamins and pigments, which are important because they are vitamin precursors. Carotene, for example, is converted to vitamin A, chlorophyll appears to have a number of vitaminlike, growth-enhancing effects, and xanthophylls increase the yellow color of eggs and butter (Sect. 3.3.2.2).

The research on tree foliage as a source of vitamins for livestock originated in the Soviet Union in the 1930s, and is presently concentrated primarily in the Scientific and Industrial Research Association Silava in Latvia. The advantages of foliage as a fodder additive were initially assumed to be due exclusively to the high carotene and mineral element content. However, part of the physiological benefits is obviously due to many other, little understood, biologically active compounds.

When used in small proportions as a feed supplement in the Soviet Union, conifer foliage has been found to improve cattle productivity, stimulate reproduction, and reduce susceptibility to disease. It is emphasized that needles are not considered as feed, and their feed value is rarely evaluated. Their use is thus recommended only for their biological activity. In a few cases, mainly under winter conditions, this feed supplement may be given in the form of fresh chips reduced from technical foliage. More commonly, technical foliage is processed to dried "vitamin meal". The production of this fodder supplement, commonly called *muka* after its Russian name, in the early 1980s was more than 150000 t (Ievin et al. 1981). The needle content in muka should be no less than 70% of the dry mass, the remainder being bark and wood from thin twigs and shoots.

The manufacturing process comprises chipping of crown material, segregation of technical foliage from other components of crown mass (Sect. 8.6), rapid drying, grinding, and pelletizing. The energy for drying may be produced by direct combustion or gasification of the fuel component rejected in foliage upgrading (Ievin et al. 1973; Ievin and Daugavietis 1986). Muka with a slightly reduced content of carotene can also be produced from steam-distilled needle residue after the removal of essential oils. Once the foliage and thin sections of the twigs have been segregated from other components of the crown mass, the manufacturing procedure is essentially the same as that commonly applied for conventional forage feed.

Keays and Barton (1975) published a comprehensive bibliography of Russian research on foliage utilization. In the Soviet Union, both research and practical experience support the use of muka as a fodder supplement, up to biologically safe limits, due to the improved health, growth, and reproduction of animals. The amount of muka which can be used in a ration is limited by its growth-inhibiting properties. A correct proportion may be 2–6%, depending on the type and age of the animal, tree species, season, wood content of the meal, methods of processing and storage, and the type of basic feed to which it is added. Since the main value of muka lies in its carotene content, the proportion should be sufficient to compensate for the use of synthetic vitamin A in the fodder mixture. For muka made from technical foliage of pine and spruce, the carotene content is approximately 120 ppm.

Losses in the biochemical activity of foliage and its derivatives due to harvesting, preparation, and storage are serious considerations in operational planning. While most of the major components in foliage change rather slowly during storage, carotene decomposes rapidly, especially during summer and when the foliage is separated from the branches. In the Soviet Union it has been recommended that technical foliage, whether or not separated from its branches, should be stored at winter temperatures for not more than 20 days and at summer temperatures for not more than 6 days. Ideally, technical foliage should be produced and used in the first to third day after removal of the branches from the stem. However, whereas research results in the Soviet Union emphasize rapid losses in carotene content in foliage after felling, a Canadian study on *Pinus contorta* and *Picea glauca* showed, with the exception of essential oils, no appreciable loss of important organic constituents such as carotene, chlorophyll, or protein during 20 days of storage in summer and 49 days in winter (Barton and MacDonald 1978).

To avoid storage loss, foliage is dried in the Soviet Union rapidly to approximately 10% moisture content and converted to muka, using simple technology that allows local processing in rather small production units. The first muka plant, with a capacity of 100 t/year, went into production in Latvia in 1955. Several 100 plants, somewhat larger in size, are presently operating in the Soviet Union.

Outside the Soviet Union, muka and other corresponding products are used or experimented with on a limited pilot scale, for example in the German Democratic Republic (Rahmlow 1977) and in Czechoslovakia (Ilavský 1986). In the United States and Canada, results of feeding experiments with muka have not been as encouraging as in the Soviet Union. When corn was replaced in the ration at a 5% level by muka from spruce and fir needles, the mass gain of chickens was reduced. Experiments conducted at a 3% replacement level indicated that muka adds little if anything of value to the modern rations normally fed to broiler chickens in the USA (Gerry and Young 1977; Hunt and Barton 1978; Gerry et al. 1979). In an experiment with sheep, dilution of timothy hay with 25% fall-harvested spruce and fir muka did not appear to affect palatability of the fodder or daily feed consumption of the animals, but the resulting mixture was less digestible than timothy hay when fed alone.

In North America, feeding experiments have not shown the improved growth responses demonstrated by a number of studies and practical experience in the Soviet Union. This is probably due to the common use of artificial vitamin preparations and energy-rich basal diets that have been optimized for maximum growth. The digestibility of coniferous foliage products is lower than that of alfalfa and like feeds. Varying amounts of lignin and cellulose from twigs in the technical foliage affect the digestibility which approximates 53% found in low- and medium-grade hay (Barton 1981). More research is required before profitable utilization of tree foliage for animal feed can occur under the economic conditions prevailing in the industrialized Western countries.

8.6 Upgrading Residual Forest Biomass

Wood from residual trees and tree components generally differs from merchantable stem wood in technical properties such as cell dimensions, chemical composition, and basic density. For most processing purposes, wood properties in forest residues are inferior, and these differences may adversely affect the course of the process and the nature of the end product. However, the problems are not insurmountable and can be mastered. A much more serious problem is the presence of bark, foliage, and inorganic contaminants.

The composition of residual biomass varies irregularly depending on tree species, size, season, logging method, storage, etc. The smaller the trees and the wider the spacing, the lower is the content of wood. Table 8.14 shows the average distribution of the above-ground biomass of small trees from thinnings of Scots pine, Norway spruce, and birch in southern Finland and Latvia. These proportions can be used as a basis for general planning, but in individual cases they must be applied with caution, as the composition varies in practice over a broad range.

In all assortments of residual forest biomass the wood content generally increases slightly in the course of harvesting. A part of the branches, foliage, and bark falls off when trees are felled, moved, stored, and comminuted. In compari-

Table 8.14. Distribution of above-ground mass from thinnings of young stands of *Pinus sylvestris*, *Picea abies*, and *Betula* spp. into biomass components in Finland (Hakkila et al. 1979) and in Latvia (Daugavietis 1985)

Tree species	Proportion in dry mass (%)				
	Wood from stem and branches	Bark from stem and branches	Twigs	Foliage	Total
In Finland:					
Scots pine	75	15	4	6	100
Norway spruce	73	12	5	10	100
Birch in summer	72	18	6	4	100
Birch in winter	75	19	6	–	100
In Latvia:					
Scots pine	75	12		13	100
Norway spruce	68	15		17	100
Birch in summer	82	13		5	100
Birch in winter	86	14		–	100

son with standing trees, whole-tree chips from the same type of raw material indicate a higher content of wood and lower contents of bark and foliage when arriving at the mill. On the other hand, the content of inorganic contaminants tends to increase due to soiling of the biomass.

The wood content of whole-tree and residue materials is thus higher at the point of utilization than initially in the forest, but impurities still cause serious problems in industrial processes designed more or less for bark-free wood. Although it may be technically possible to make the quality of the end product from unsorted biomass conform to standards, process disturbances tend to make the use of such raw material uneconomical. Therefore, upgrading is frequently necessary in view of feasible resource utilization. The purpose of upgrading is to produce uniform, high-quality raw material from low-value mixtures of biomass. Three primary tasks can be distinguished:

1. Removal of poorly sized particles, such as fines and oversized, from acceptably-sized chips;
2. Segregation of bark, foliage, and inorganic components from wood;
3. Direction of each component of biomass to its most economical use.

Upgrading is thus *fractionation* of raw material components from one another. Fractionation is basically a two-stage separation-segregation process. First, the bond is broken between two integrated fractions, such as branch and stem, or bark and wood. This stage of biomass upgrading is called *separation*. Secondly, one or more fractions without bonds between them are removed from the mixture. This stage is called *segregation* (Erickson 1970, 1971, 1972c; Baadsgaard-Jensen 1985).

In principle, forest biomass may be divided in the upgrading process into three fraction categories: high-value wood for industrial use, high-value foliage for specific purposes such as chemicals extraction or use as biologically active fodder additive, and low-grade fuel rejects. The degree of purity and uniformity of particle size that should be aimed at in upgrading depends on the raw material characteristics, handling equipment, process technology, end products, and value ratios between industrial raw materials and fuels. Upgrading may take place prior to comminution (Sects. 8.6.1, 8.6.2), after comminution (Sect. 8.6.3), or in conjunction with comminution (Sect. 8.6.4).

8.6.1 Upgrading Measures on Site

Since upgrading of rough chips is difficult, it is feasible to keep residual biomass clean from inorganic contaminants and to remove as much as possible of the bark and foliage from the raw material prior to comminution. For example, *winter logging* makes it possible to recover biomass from deciduous trees without foliage. *Transpiration drying* of whole trees or crown mass leads to significant shedding of foliage within a few weeks, depending on season, weather, site conditions, and tree species. Partly dried foliage is shaken off relatively easily when these trees are moved and chipped (Sect. 8.1.4.1).

Simola and Mäkelä (1976) observed foliage alteration in small-sized trees during transpiration drying in Finland. In Scots pine and Norway spruce felled during early spring, drying of the needles and change in color proceeded from crown base to top. In Scots pine the needles began to turn brown at the crown base at the end of June. In August, the top fifth of the crown was still green, and the brown needles were attached, although they easily shed when the trees were moved. The change pattern was similar but more rapid in Norway spruce. At the end of August the crown was brown, and a light movement caused the needles to shed. *Betula* sp. felled in early May burst into leaf while lying on the ground, but the leaves did not reach their normal size. At the beginning of June they yellowed and began to fall. The leaves of birch trees cut in early July fell in the course of August. Changes were very similar in *Alnus incana*.

Transpiration drying offers a simple way of leaving foliage in the logging area, thus reducing nutrient loss from the forest soil. However, before the foliage is shed part of its nutrients is translocated into branches and stem. Adverse features of the transpiration drying method are slowing of the harvesting process, dust problems during subsequent chipping, an increase in the portion of fines in the chips and, mainly in the spring, the propagation medium that conifers lying on the ground provide for bark beetles and other insects.

The role of inorganic contaminants should not be ignored in system planning. If the utilization of biomass expands to include logging slash, the problem becomes truly serious, since stones tend to become mixed with the biomass and cause costly interruptions in the subsequent comminution. Sand and grit in raw material cause rapid wear and tear in harvesting and processing equipment (Sect. 8.2.6). Therefore, *the avoidance of soiling* should be given serious attention in operational planning. Load-carrying forwarders offer an advantage over skidders, which drag the trees on the ground. The following figures by Conner (1978) show very high contents of sand and grit in whole-tree material from southern pine thinnings when off-road hauling was carried out by skidding. Use of forwarders and delimbing reduces the amount of inorganic contaminants significantly:

Raw material	Sand and grit in raw material, percent of dry mass
Whole trees, skidded	1.22
Whole trees, nonskidded	0.81
Delimbed stems, skidded	0.49
Delimbed stems, nonskidded	0.18

The high cost of single-tree delimbing is becoming a serious constraint to the use of conventional logging methods in small-tree operations. If delimbing is to be accomplished, it must, in the future, be based on multiple-tree handling. Trees left on the site in small bunches by a feller-buncher can be roughly delimbed and topped motor-manually by a chain saw on the site, or it may become possible to use grapple processors converted for multiple-tree delimbing.

A feasible solution for multiple-tree treatment is offered by *flail delimbing*. The principle was first applied in North America by equipping a rotating shaft or drum with a number of steel chains and mounting the device on a skidder or wheeled loader. Trees are delimbed on the ground by driving the prime mover to the crown base of the trees, and bunch-delimbing the trees by driving over the crowns, usually from base to top. Stokes (1985) found the consumption of productive time only 1.8 min per average bunch of 38 trees, when 7- to 13-cm-diameter *Pinus taeda* trees were treated with a skidder-mounted Hydro-Ax LL64 flail delimber. The method has been adapted for large-scale operations requiring spacious landing areas, and does not allow the recovery of crown mass.

Another alternative is to link the flail treatment with off-road hauling. An experimental approach in Sweden employed a rotating horizontal shaft on the load space of a forwarder. The device was shuttled to-and-for in a sled above the load space independently of loading. Another prototype was based on a device consisting of two counterrotating verticale flail shafts placed in the front of the load space. In the latter the driver pulled each bunch of trees through the device when loading the forwarder, but due to the shortage of free operation space, pulling the bunches transversely to the load may cause difficulties under thinning conditions. Since the time consumption of loading increases and the load space diminishes, the method reduces the forwarding output. However, an improved solid content of treated trees compensates for the space loss.

Chippers also can be equipped with a flail delimbing possibility. For example, two counterrotating, vertical flail shafts can be mounted in front of the in-feed system of a chipper. Another solution is a rotating horizontal shaft equipped with parallel chains, both ends of which are fastened. If flail delimbing is integrated in chipping with mobile machines in the field, the rejected crown components with their nutrients are left on site. If flail delimbing is integrated in centralized chipping at a terminal or mill yard, the rejects can be recovered for fuel.

The result of flail delimbing is controlled primarily by treatment variables such as the feed speed of biomass and the mass and revolving speed of the flails. Light treatment is generally sufficient if the goal is foliage separation only. A

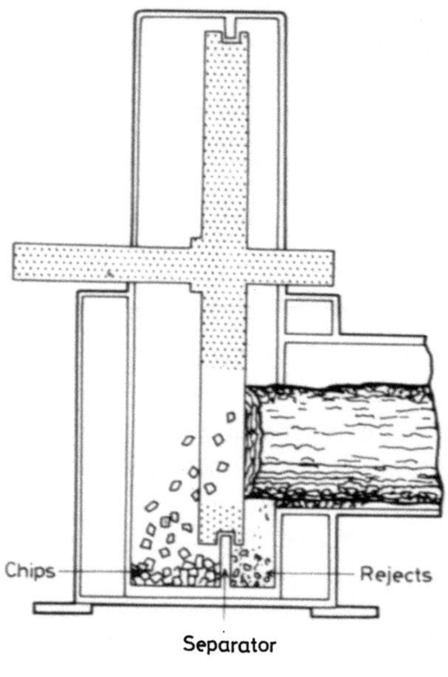

Fig. 8.19. A simplified illustration of the dirt separator of the Morbark disk chippers

more severe treatment with heavier flails and higher revolving speeds is required to remove branches (Helgesson 1977). A severe treatment also has a debarking effect. However, the more thorough the cleaning, the greater is the wood loss.

Sand, grit, snow, ice, foliage, pieces of twigs, and outer bark come loose from the feedstock through impact and vibration, when biomass is fed into a chipper. Part of the loose contaminants falls from the conveyor before reaching the in-feed opening of the chipper, but the majority is fed in. In a conventional disk chipper, these loose particles drop in front of the disk, pass the disk underneath, and are mixed with the chips produced. This mixing can be prevented with a built-in dirt separator, which was first developed by Morbark Industries, Inc. The disk and spout configuration is modified by separating the space in front of the disk from the space behind the disk (Fig. 8.19) and by discharging the rejects through a separate chute. According to the manufacturer, the amount of twig sections over 10 cm is reduced by 80% and that of sand, grit, and foliage by 50%. According to Wellwood (1979b), the content of sand and grit was 0.10% in the accepts and 2.17% in the rejects when a disk chipper was equipped with the dirt separator. The proportion of rejects varied between 0.5–5.0%.

8.6.2 Upgrading Tree Sections

Size reduction on a site or landing offers many advantages in recovery, handling, and transport of small-sized trees. However, there is a serious limitation to the generalization of whole-tree chipping due to the difficulty of subsequent separation and segregation of bark, foliage, and twigs from good-quality stem wood in the fragmented material. The problem of upgrading is greatly reduced when the raw material is bunch-delimbed, and possibly debarked, at a mill or terminal prior to comminution.

Multiple-stem delimbing can be accomplished with *flail delimbers* mounted on a chassis or attached to a high-speed boom capable of handling several trees or tree sections simultaneously. Although the machines can be made portable, the high capacity and the recovery goal of the residue restricts their use primarily to terminals or large landings. Figure 8.20 shows an example of a portable, heavy-duty flail-delimbing unit with an average processing capacity of 25 m^3 solid per work hour (Jonsson 1986b).

The stationary Manitowoc flail delimber consists of in-feed and out-feed rolls, a top flail drum, and a bottom flail drum. Both of the 39-cm-diameter, horizontal flail drums are equipped with a number of 55-cm chains in six rows. Each flail chain consists of both a base and working section. The base section is composed of three oval links and one round link, while the working section consists of six oval links and one quick-disconnect link. The chains are attached either in a straight or mesh configuration. For both configurations the six outer links are easily exchangeable by hand. Each drum, driven by a 93-kW electrical motor, rotates in the feeding direction at 300–525 rpm.

A centralized processing system of Weyerhaeuser Company in North Carolina for multiple-tree delimbing and debarking of *Pinus taeda* thinnings consists of a Manitowoc flail delimber-debarker, a chipper, a debris shredder, and associated

Fig. 8.20. A trailer-mounted flail delimber for mass handling of tree sections, as seen from behind (courtesy of Bruks Mekaniska AB)

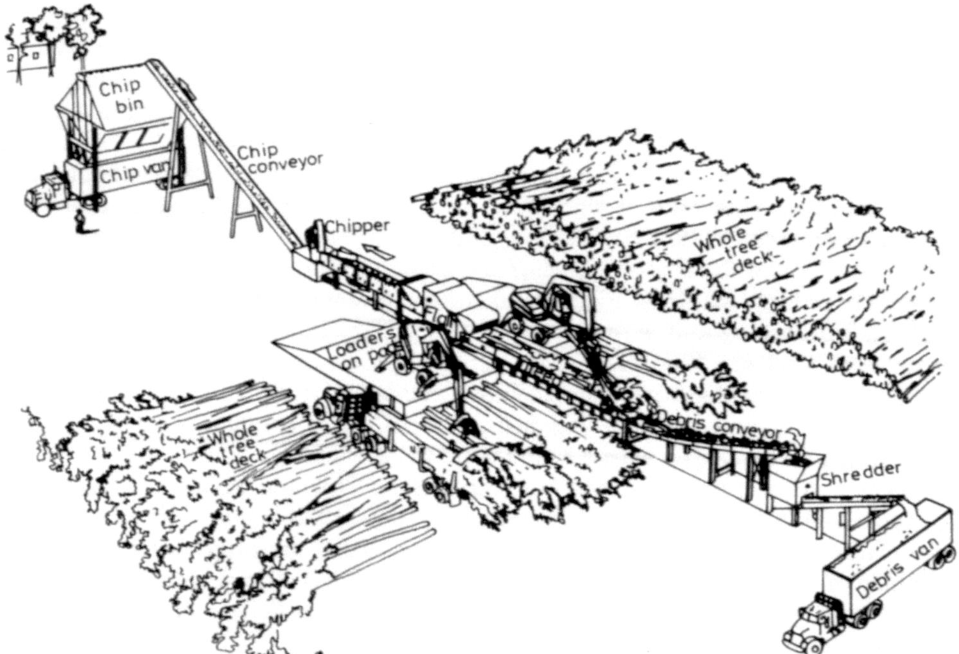

Fig. 8.21. A layout of a multiple-tree processing system for combined delimbing and debarking of southern pine in North Carolina, based on flail-delimbing treatment (courtesy of Weyerhaeuser Co.)

Fig. 8.22. A Rauma-Repola delimbing drum modified for tree sections; diameter 5.4 m, length 39.5 m (courtesy of Hannu Kalaja)

handling equipment (Fig. 8.21). The system is operated by seven men and is capable of a productivity of 530 m^3 per 8-h shift of whole trees of about 0.11 m^3 (Christensen 1986). After single-pass delimbing and debarking of small-sized southern pine by this equipment, the bark percentage in the chips was not more than 1.4%. However, some pieces of stem wood ended up as hog fuel, suggesting a less aggressive flail treatment. The debarking by beating chains left the stem with a somewhat hairy-looking surface, which may affect the fiber properties negatively (Jonsson and Nordén 1987).

Multiple-stem delimbing and debarking can also be integrated with *drum delimbers*. Although a rotating drum can be transported by mounting it on a trailer, the unit tends to become heavy from the standpoint of transport, and the productivity of such drum delimbers is hardly sufficient to cover the operating costs (Spångberg 1987). Nevertheless, high delimbing quality with little breakage and an output of 23.7 t per productive hour were obtained with the Cabro bunch-delimber prototype in Sweden in processing tree sections of Norway spruce. This 36-t, trailer-mounted unit uses a 600-cm-long, 100-cm-diameter drum equipped with 12 cutting augers on the inside. The revolving speed of the drum is 18–22 rpm and that of the cutters 110 rpm (Nordén 1987).

In Sweden there are several stationary drum-delimber terminals specially built for tree sections. The actual delimbing is done either by the abrasive action of the bolts or aggressive cutting edges on the inside of the rotating drum. The output is limited primarily by in-feeding and the slow flow of wood in the drum. The length of the special tree-section drums is at least 12 m. Lengthening of the drum

Fig. 8.23. A batch delimbing trough for tree sections (courtesy of AC Invest Product AB)

improves the delimbing result but increases the loss of stem wood to the energy fraction and slows down the process. Material smaller than the width of the drum slits tends to become stuck, breaks, and ends up as reject. In a study by Larsson et al. (1983), the net recovery of pulpable wood from tree sections was 84–104% compared to the original input of more than 5-cm-thick stem wood. Loss from breakage and worn-off wood was thus partly compensated and, in some cases, even overcompensated by the gain from tops and large branches.

A drum consists of one section or is divided into several sections which may rotate in alternate directions. The residue falls on a conveyor belt through the drum slits or slots in between the drum sections. In-feeding problems can be reduced and the wood flow increased by widening the diameter of the first section of the drum. The performance of such special delimbing drums is 50–80 m^3 solid wood per effective hour.

Some pulp mills in Sweden and Finland have modified their conventional stationary debarking drums for simultaneous treatment of conventional pulpwood and tree sections (Fig. 8.22). When the proportion of tree-section material increases, the capacity of the drum decreases, mainly due to in-feeding problems (Brattberg 1977). Conveyors to the drum must be enlarged because tree sections are entangled with one another and must be handled in bunches. The length of these drums may exceed 30 m and the diameter at the in-feed end 5 m. For example, the drum of the Billerud mill in Sweden is planned for an annual input of 500000 m^3 of tree sections to produce 300000 m^3 of debarked pulpwood and 200000 m^3 of bark and branch mixture for energy.

At small terminals and upper landings a more flexible alternative to stationary drums is a mobile *batch delimber trough* mounted on a trailer or truck. The basic components of the machine are a crane, operator's cabin, feed bin, delimbing trough holding 2–4 m^3 unlimbed, 6-m tree sections, and a belt for conveying the residue into a separate crusher or chipper (Fig. 8.23). Chain conveyors with pushing blades make the tree sections spin in the trough. Delimbing takes place partly through the friction between tree sections as they rotate around their own axis or around each other in the trough, and partly by delimbing rollers at the bottom of the trough, equipped with sharp cutting edges or impacting flails. When the treatment is completed, the batch of wood is lifted from the trough by the chain conveyor or crane, and replaced by a new batch dumped into the feed bin or by placing them there by a crane.

The output of a mobile batch delimber is 30–50 m^3 solid biomass per effective hour. If part of the raw material consists of heavier tree sections from larger pulpwood trees, then branch severing is faster, and productivity is higher. If the machine is used for combined delimbing and debarking, productivity slows by about 6 m^3 per productive hour. Stem wood losses to the energy fraction stay below 5%. In the delimbing of small trees from early thinnings, a net gain of pulpwood is possible due to the recovery from the tops (Scherman and Nordén 1983).

8.6.3 Upgrading Rough Chips

Although several mills in North America and Europe use or have used whole-tree chips for the production of pulp or composite board, use of whole-tree chips for processing purposes is not widespread. Application could be much larger if the quality of the chips could be improved.

The goal of chip upgrading is to improve the uniformity and the quality of comminuted biomass by removing the greatest possible proportion of contaminants and particles of improper dimensions with the least possible loss of wood. The upgrading result is influenced by variables such as tree species, size, season, method of comminution, condition of chipper knives, and duration of storage prior to upgrading. Particle size distribution of the chips, separation of bark from wood, disintegration of needle clusters, and microbial activity during storage are of great importance for the upgrading result.

Foliage and bark are rich in extractives, such as nitrogen compounds and nonstructural, easily soluble carbohydrates, all of which accelerate the biological reactions in chips during storage. These reactions result in high temperatures, risk of spontaneous combustion, a drastic increase in health-hazardous fungi, losses in dry matter and energy content, inferior quality of pulp, etc. (Sect. 8.1.4.2). Removal of foliage, bark, and fines immediately after chipping greatly reduces the deterioration rate during storage.

The effectiveness of any sorting process depends on the existence of some property or combination of properties that allows one material to be distinguished from another. Properties that are commonly used to purify or otherwise enhance a given commodity include size, shape, mass, color, and density. Unfortunately, the properties of wood chips and their contaminants do not differ sufficiently to provide a satisfactory degree of segregation. The problem is one of variability. Wood, bark, and to a lesser extent, foliage are each highly variable, possessing properties that span broad and overlapping ranges, thereby precluding the ready detection and segregation of one from the other. Accordingly, effective upgrading can be achieved only by inducing some singular difference that can then be exploited, and it is this approach that underlies many of the developments to date (Berlyn and Simpson 1988).

In practice, *screening* and possibly washing are usually the only upgrading measures for the quality improvement of rough chips from whole trees or logging slash. Because a large proportion of the bark and foliage separates from wood and crumbles during chipping (Erickson 1968), the wood content of the chips can be increased somewhat by removing the fine fraction in screening. Although mechanical screening per se is not effective in separating the various components into pure fractions, satisfactory initial separation of wood and needles can be effected (Hatton 1977a, b). The following example shows the average increase in wood content in flat-screening of whole-tree chips from small Scots pine trees in a sample collected from several Finnish mills. The composition of rough chips can be improved with conventional screen equipment. However, it will entail considerable loss of wood (Hakkila et al. 1979):

Removed fraction (mm)	Proportion in dry mass (%)				Recovery (%)	
	Wood	Bark	Foliage	Fines	Total mass	Wood
–	79.8	12.0	4.4	3.8	100	100
0–3	82.9	12.5	4.6	–	96	98
0–6	86.2	11.7	2.1	–	89	94
0–13	90.3	8.8	0.9	–	65	75

The usual vibratory and shaking screens of pulp and board mills are poorly adapted to handling chips that contain clusters of needles, slender twigs, and large branch pieces. They are unable to separate bark from wood by breaking the bond between them, and tend to become clogged when handling rough chips. Rotary drum screens and auger screens also do not give fully satisfactory results.

Better results have been achieved in segregating needle clusters, twigs, oversized particles, and even fine fractions with *disk screens*. Segregation takes place with toothed, rotary disks mounted side by side and one after the other. Disk screens are well suited to improving the handling properties of rough chips through segregation according to particle size. Their advantages are compactness, reliability, low capital and operating costs, high capacity, and a low level of noise and vibration (Fig. 8.24).

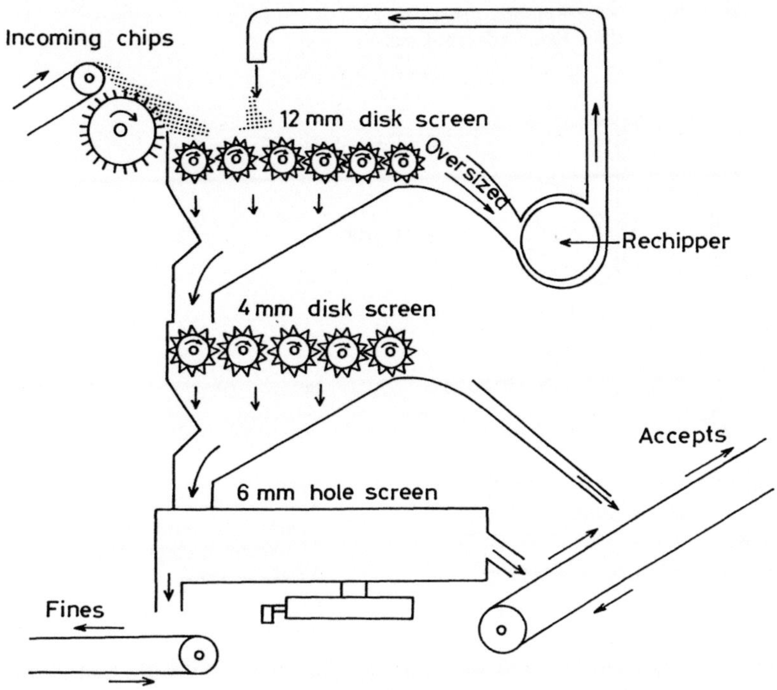

Fig. 8.24. A combination of two disk screens and a hole screen (D. M. Fisher 1976, personal communication)

As the disk screens are incapable of separating bark from wood, they cannot greatly increase the wood content of the chips. But because they do segregate oversized and elongated particles effectively from rough chips, they may be combined with conventional screening equipment in order to prevent the latter from blocking. Disk screen treatment often suffices to ensure uninterrupted in-feed of fuel in chip-fired power plants, and disk screening may also be sufficient in the wood panel industry. In the pulp industry disk screening offers a pretreatment alternative worthy of consideration.

Float-sink segregation employs an induced difference in mass density of wood and bark particles. If the density of a liquid lies between the densities of two biomass fractions, one fraction will float and may be skimmed off the surface, while the other will sink and may be collected from the bottom of the vessel. Researchers at the Battelle Memorial Institute in the United States found in the late 1950s that when vacuum was applied to the submerged mixture of bark and wood chips and then suddenly released, the wood fraction easily absorbed water and sank rapidly, whereas the vascular structure of the bark tended to collapse and became less susceptible to liquid penetration. Thereby the desired difference in mass density was obtained. The Vac-Sink process, based on this different behavior of wood and bark, was employed from the early 1960s on a full commercial scale for upgrading southern pine chips by a sulfate pulp mill in Savannah, Georgia. Performance suffered from the fact that those chips that were somewhat dry absorbed

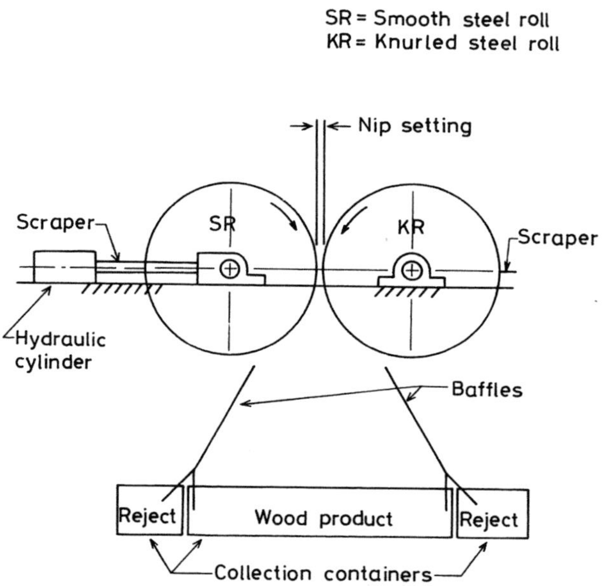

Fig. 8.25. Simplified schematic diagram of compression debarking of rough chips (Arola and Erickson 1973)

water less readily, tended to float, and were lost to the reject fraction. The plant is no longer operational (Baadsgaard-Jensen 1985; Berlyn and Simpson 1988).

If the bond between wood and bark is not broken, it is very difficult to produce clean chips from residual forest biomass without a considerable wood loss. An interesting approach is offered by the *compression-debarking* concept, which was first introduced by Hosmer Machine Co. and further developed by the Forestry Sciences Laboratory at Houghton, Michigan. The key part of the compression-debarking equipment consists of two horizontal counterrotating steel rolls (Fig. 8.25). In order to break the bond between wood and bark, the chips are forced through a narrow gap between the rolls as a continuous one-layer mat. This nip spacing between the rolls is somewhat narrower than average chip thickness so that the particles are subjected to compression. Some of the bark is fragmented into fines and some adheres to the surface of the rolls. The fragmented bark is removed with the fines by screening, and the adhering part of the bark is scraped or brushed off the rolls (Erickson 1970, 1971, 1972c, 1976a, b; Arola and Hillstrom 1972; Arola 1973, 1974; Arola and Erickson 1973; Mattson 1974a, b, 1975).

The main machine variables affecting the result of compression debarking are nip spacing, nip pressure, baffle position, and roll surface. The raw material variables affecting the result are the tree species, time of cutting and chipping, and storage. During the growing season bark is detached easily from wood and adheres readily to the rolls. Steaming before the compression treatment improves the debarking and eliminates the seasonal variation. It softens the inner bark and cambium, loosens the bark to some extent from the wood, and makes the bark tacky and more readily adherent to the surface of the compression rolls. If, after screening, the chip fractions with the highest bark content are sent to a hammer mill or for attrition treatment, the bark can be separated and segregated thoroughly. During the compression treatment, wood particles tend to open up, which facilitates the penetration of pulping chemicals. The cooking time is shortened when the chips are digested rapidly and evenly. However, the weak points of the method are low capacity and relatively high processing costs.

St. Anne-Nackawick Pulp and Paper in New Brunswick operated a compression-debarking pilot plant in the late 1970s. Steam preconditioning of the chips and close control of press variables and roll surface were necessary to maintain debarking efficiency. The initial bark content of hardwood whole-tree chips was reduced from 12.0 to 3.6%, and that of softwood chips from 20.0 to 5.9%. The overall 9.1% wood loss exceeded the goal of 5.0%. Pulping hardwood chips showed a 9.0% reduction in cooking time and a 3.7% lower alkali demand (Wawer and Misra 1977). Although the technical performance of the plant was satisfactory, due to the low output and high cost of debarking it proved more economical to mix the whole-tree chips unbarked with conventional chips, and the operation of the pilot plant was discontinued. A multiple-stage compression-debarking system using six subsequent pairs of compression rolls with various surface structures has also been investigated by the Wood Technology Institute in Poznan, Poland (Odolinska 1982).

Air segregation is based on the fact that chip fractions attain a different terminal velocity, which is defined as the maximum velocity a particle attains in free fall under the action of gravity. The terminal velocity of comminuted biomass depends primarily on the mass density and thickness of the particles. Layering of foliage and fines when blown through the chipper spout onto a truck load reflects differences in terminal velocity (Fig. 8.26).

Air segregation may be carried out by blowing air at the chips from the side as they fall freely, or the air can be blown from above or below as the chips are blown from the chipper. This concept was first tested for segregation of knot wood from stem wood (Enroth 1946), but it seems to be best suited for the removal of foliage, particularly from hardwood chips. Segregation of softwood foliage is more difficult since it contains two different components: single needles and needle clusters, including sections of twigs.

After several intermediate phases of investigation, the Forestry Sciences Laboratory at Houghton, Michigan, developed in the late 1970s the *vacuum-airlift seg-*

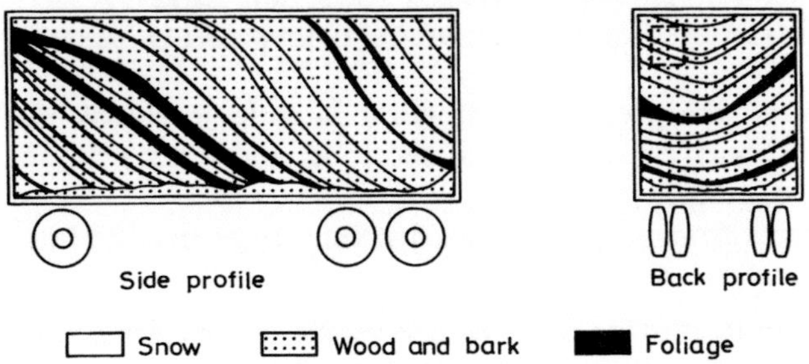

Fig. 8.26. Layering of needles in whole-tree chips loaded onto a truck trailer by blowing from a chipper (Verkasalo 1987)

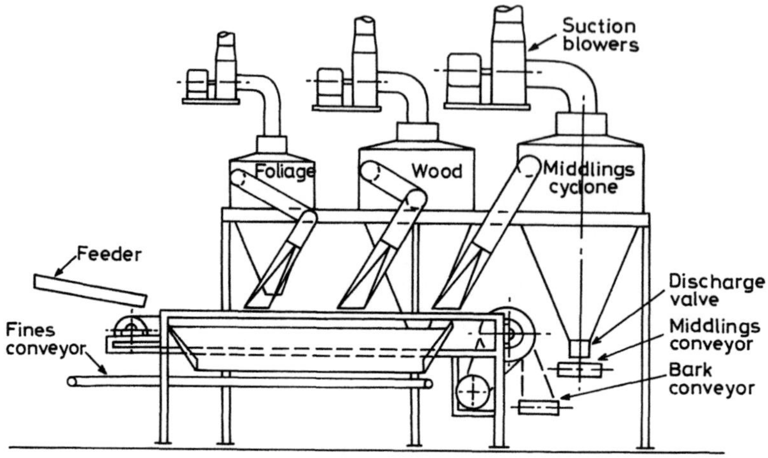

Fig. 8.27. Multiple-stage vacuum-airlift segregator for upgrading rough chips (Arola et al. 1976)

regator (Sturos 1972, 1973a, b, 1978; Sturos and Marvin 1978; Arola et al. 1976). The chips are scattered into a single layer on a wire-mesh conveyor belt which passes several fields of air current. First, some fines fall through the mesh belt. In each field an air current is directed onto the chips through the wire mesh from below, and the effect is augmented by the vacuum above. In the first field, foliage is removed from the chips. Good-quality wood particles are recovered in the second field. An unsorted, residual middlings fraction containing particles composed of both wood and bark is removed in the third field. Oversized particles, twigs, and bark are left on the belt and discharged as rejects to become fuel. The development goal has been an upgrading result in which 50–70% of the raw material is recovered as accepts and 5–10% as fuel rejects, whereas the remaining 20–45% is subsequently sent onto compression debarking (Fig. 8.27). Vacuum-airlift segregation and compression debarking complement each other. The Forestry Sciences Laboratory has proposed an upgrading system containing the following operations (Fig. 8.28):

1. Disk screening for removal and rechipping of oversized particles;
2. Vacuum-airlift segregation for removal of foliage, fines, and bark;
3. Presteaming the residual fraction of unsorted material in preparation for compression debarking;
4. Compression debarking of the presteamed fraction in order to break the bark/wood bond, remove bark by adhesion to the rolls, and reduce the remaining bark particles to smaller size;
5. Drum screening the accepts from compression debarking to remove the fragmented bark, residual foliage, and other fines.

The proposed process appears to provide chips of acceptable quality for sulfate pulping of southern pine. One of the primary advantages of integrating a vacuum-airlift segregator and compression debarker into a combined process is reduced capital costs. Alternate processes, with lower costs, can be used for indus-

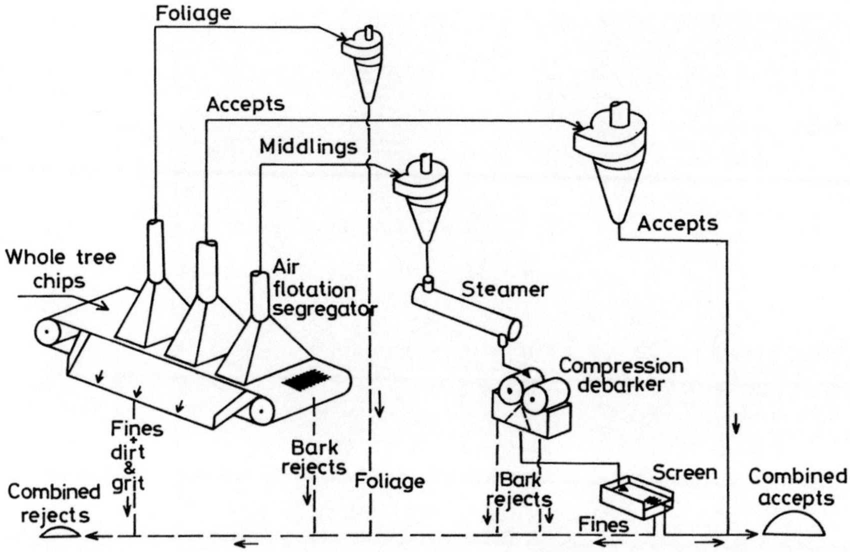

Fig. 8.28. A combined vacuum-airlift and compression debarking system for upgrading rough chips (Sturos 1978)

tries that can tolerate lower quality. The following example of southern pine whole-tree chips illustrates the removal of various chip components by vacuum-airlift segregation and subsequent compression debarking and drum screening (Sturos 1978):

Treatment	Mass removed (%)					
	Foliage	Bark	Grit	Fines	Wood	Input
Vacuum-airlift segregation	59.5	44.1	62.0	74.6	14.1	20.3
Compression debarking and screening	17.9	40.2	17.5	19.1	3.1	9.0
Total	77.4	84.3	79.5	93.7	17.2	29.3

The air segregation principle is widely applied in the Soviet Union. Chips reduced from conifer crowns are fractionated pneumatically to produce technical foliage, industrial chips primarily for board manufacturing, and fuel chips. Chips made from fresh crown mass of Norway spruce produce approximately the following fractions: technical foliage 46%, industrial chips 40%, and fuel chips 14% (Ilavský and Kevinš 1985). The most valuable product is technical foliage for the chemical industries and fodder. Two types of upgrading plants exist. The more common two-man-operated IPS-1 sorter plant has the capacity to treat 0.6–1.0 t of fresh crown mass per hour (Fig. 8.29), whereas the one-man-operated SIKO-2 is able to upgrade 2.5–3.2 t fresh mass per hour (Fig. 8.30).

Fig. 8.29. IPS-1.0-M upgrading plant for sorting chips reduced from coniferous crown mass in the Soviet Union (courtesy of Silava)

Ballmilling is a common refining technique in the mineral industries. Mattson (1974b) found ballmilling a feasible method of fragmenting residual bark after compression debarking of whole-tree chips to facilitate further upgrading by screening. Kone Osakeyhtiö in cooperation with Enso-Gutzeit Oy has built a pilot-scale ballmilling plant in Finland for upgrading 20000 m^3 solid of pine whole-tree chips annually. The raw material is first prescreened for the removal of fines and oversized fractions and then transported by a conveyor to a rotating drum, where the bark is separated from the wood and pulverized by attrition and hammering with iron balls. The ball diameter depends on the properties of the chips and the aim of upgrading, but may be 100 mm. After the drum treatment, the chips are screened for removal of oversized and fine fractions. The capacity of the plant and the debarking result are affected by the tree species, composition of chips, duration of the preceding storage, etc. Revolving speed and the number and mass of the iron balls are process variables. Bark content of chips is reduced from 12–14 to 3–4% with a wood loss of about 10%. The impact of the balls causes chips to break along the grain, resulting in an excessive amount of pin chips, which may cause problems in sulfate pulping.

The Pulp and Paper Research Institute of Canada (PAPRICAN) and the Forest Engineering Research Institute of Canada (FERIC) have jointly developed a chip upgrading method based on *agitation in water*. This Paprifer technique also depends on the fact that bark and foliage are inherently weaker than wood and can be fragmented without unduly damaging the wood. This method consists of three phases, conditioning, agitation, and segregation (Berlyn et al. 1979).

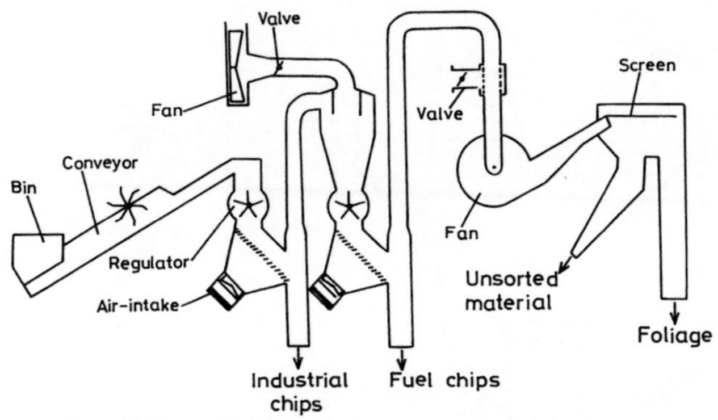

Fig. 8.30. A simplified schematic diagram of raw material flow in the pneumatic SIKO-2 chip sorter (Ievin et al. 1984)

Conditioning weakens bark and foliage as well as their bond to wood. It can be accomplished by storing the chips in a pile for about 6 weeks or with a 6- to 10-min exposure to steam. Agitating conditioned chips vigorously in water at a consistency of 13–19% breaks the bark free from wood and then reduces the bark and foliage to small particles. Virtually any standard of bark removal can be met, depending on the degree of conditioning and the energy applied to the slurry through the rotor, but the more demanding the standard, the greater is the wood loss as fines. Segregating the bark and foliage components from the wood chips entails washing over a screen plate. Dry methods are inappropriate, since the fine particles adhere to the wet chips.

A pulper is used to agitate the chips. The process can be set up in a variety of ways. In a simple batch operation one pulper is sufficient. If several pulpers are arranged in a series, the operation can be made fully continuous. Foliage, inner bark, juvenile bark, and soft, decayed wood are readily broken into small particles, whereas thick, corky tissues from mature, outer bark of some tree species can be a problem. Removal of inorganic contaminants is very effective. The bark and foliage content of moderately conditioned softwood whole-tree chips is typically reduced from 15–20 to 2–3% while incurring a wood loss in the range of 4–9%. In trials with hardwood whole-tree chips, the content of bark and foliage has been reduced typically from 15–20 to 4–5%. Wood losses are approximately 10%, of which 3% is from the fines generated by the chipper and 7% is process-induced. Adverse features of the method are slight breakage of chips along the grain, a high moisture content of the fuel rejects, and problems caused by the effluent (Berlyn et al. 1979; Berlyn and Simpson 1988).

8.6.4 Upgrading Stump and Root Mass

Fresh stump and root mass can be used in wood-deficit situations as marginal raw material for sulfate pulp. An established harvesting solution is to split the stump-root systems into three to five pieces on the site and then to season the raw materi-

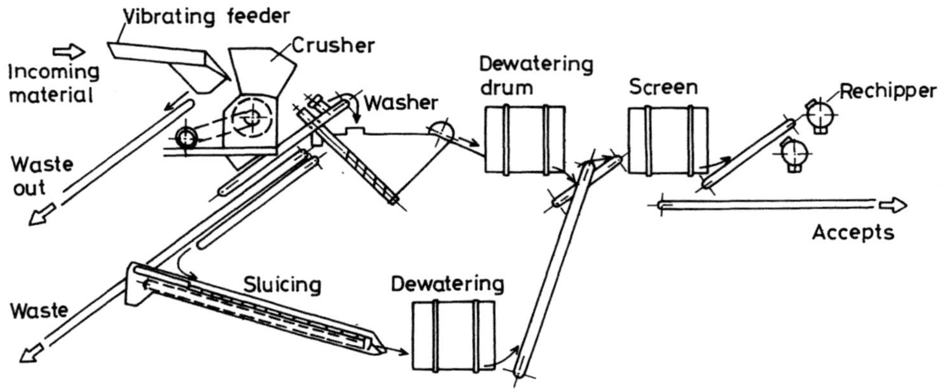

Fig. 8.31. The flow of stump and root mass at the processing plant of Joutseno Pulp Company (Blomqvist 1978)

al in piles for 6–12 months before transport to the mill or processing terminal. Large quantities of contaminants arrive at the point of utilization with these pieces of stump and root mass. It is necessary to upgrade the raw material by removing bark and soil. To protect the equipment from intolerable wear and tear, sand and stones should be segregated at an early phase of the process, preferably prior to comminution of the stump pieces to pulpable particles.

In 1975 Joutseno-Pulp in Finland was the first company to build a processing plant designed entirely for fresh stump and root mass. Although a considerable proportion of the inorganic contaminants was left at the landing due to long-term storage in piles, split pieces of stump and root mass still contained about 25% soil at the processing plant. Since comminution with sharp tools was not possible, the raw material was reduced in a stroke-beam crusher. The product was then washed by strong water jets, dewatered in a rotating drum, and screened. Wood material entering the waste conveyor was returned to be screened. Oversized particles were reduced by a drum chipper (Fig. 8.31). The product was sufficiently clean from the viewpoint of sulfate pulping, but the particle geometry was somewhat unfavorable and the content of fines and oversized fractions tended to remain excessive. The use of this marginal raw material was feasible as long as conventional pulpwood chips were in short supply (Blomqvist 1978). However, when the availability of normal wood raw material improved, the use of marginal stump and root wood was discontinued in the late 1970s after 5 years of operation.

A well-known application of a complete system for integrated comminution and upgrading of marginal stump and root mass into pulpable chips is the Mackmyra Cellulosaflis AB processing plant in Sweden (Fig. 8.32), which has been in operation since 1978. Although split pieces of stump and root mass may contain as much as 50% soil, the content of contaminants in the product from the plant is only 0.1%.

Stump mass is first cleaned in a rotary drum which separates soil, bark, and fine rootlets without water through slitlike openings in its shell. The pieces then

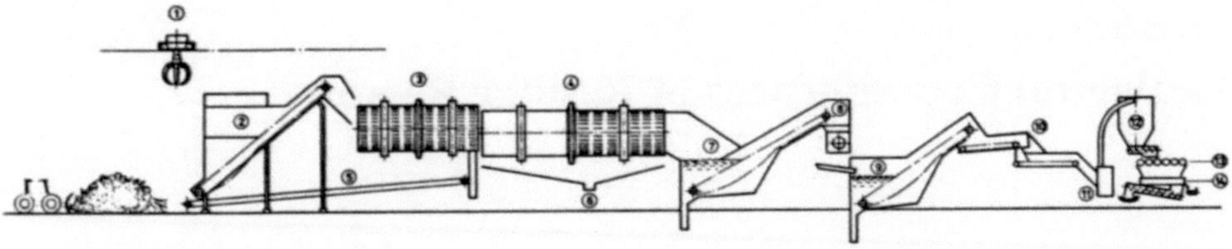

Fig. 8.32. The flow of stump and root mass at the processing plant of Mackmyra Cellulosaflis AB. *1* Loader; *2* screw and chain conveyor; *3* dry drum; *4* wet drum; *5* conveyor for dry rejects; *6* removal of wet rejects and wastewater; *7* stone trap; *8* wood splitter; *9* stone trap; *10* conveyors; *11* two chippers; *12* chip distributor; *13* screen; *14* screen (courtesy of Mackmyra Cellulosaflis AB)

proceed for washing to a wet drum, where small wood particles, rootlets, and dirty water are removed through drum slits. The cleaned stump material comes out through the outlet end of the wet drum along with large stones, which are now separated in a water tank by sinking. The stump pieces are transported by a water current to a chain conveyor and then to a splitter consisting of a powerful crusher mill with a few heavy hammers and no grate. When large pieces are broken apart, entrapped stones are released and segregated out in a subsequent water basin. The cleaned and reduced pieces are then transported to a chipper, and the chips are screened according to size. The screening is carried out in three stages, during which oversized chips and fines are removed. The oversized fraction is rechipped.

The dirty water from the wet drum and stone traps flows to a sedimentation basin to remove contaminants. Reject materials such as root pieces and bark are segregated, dewatered, and fragmented in a shredder for use as fuel together with screening fines. Loose soil is recovered as a humus product and used for soil improvement. Altogether, about 85% of the wood is recovered as a high-quality raw material for sulfate pulping, 10% is recovered as fuel, and 5% is lost. High-grade sulfate pulp can thus be produced from fresh coniferous stump-root systems.

Chapter 9

Ecological Consequences of Residue Removal

Because of the many technical, operational, and economical advantages offered by whole-tree logging technology and various forms of residue utilization, intensified biomass recovery will probably be applied more commonly in the future. Whole-tree utilization stimulates early thinnings and other silvicultural activities and thus benefits forestry and the forest industries as a whole. However, it also involves indisputable ecological and environmental risks.

Forest management always interferes with the functioning of the ecosystem, including soil, climate, and water relationships. In extreme cases, the exploitation may endanger the maintenance of a sustained yield of timber. The most significant and long-lasting ecological consequences of residue recovery obviously are the risks of excessive nutrient loss, and a subsequent decrease in the production potential of forest soil. Just as the productivity of a site can be improved through fertilization, it may also be impoverished by an improper harvesting practice. Although it is relatively easy to measure the nutrient removal from a site by various degrees of biomass recovery, it is much more difficult to estimate the ability of the site to sustain the loss.

Independent of the degree of biomass recovery, all forests must be managed for clean air, clean water, sustained yield of timber, and recreation. Therefore, the ecological consequences of intensive biomass removal must be known and taken into consideration in management planning. The present chapter deals with the effect of biomass removal on the nutrient balance of forest soil (Sect. 9.1), regeneration and growth (Sect. 9.2), and forest health and environment (Sect. 9.3). Finally, feasibility and techniques of returning nutrients to the forest are discussed (Sect. 9.4).

9.1 Effect of Residue Removal on Nutrient Balance

The problem of extended biomass removal, or redistribution in conjunction with the use of logging machines, is new and complex. Scientific knowledge is still inadequate to provide a satisfactory answer regarding the long-term consequences of residue utilization under various climatic, soil, and stand conditions. Great caution must be exercised in arriving at conclusions, or drawing up recommendations for lessening the impact of biomass removal on the ecosystem.

The scientific literature focusing on nutrient cycling in temperate forests has expanded tremendously in the past 2 decades. To facilitate research and application of nutrient cycling concerns in forest productivity, Kimmins et al. (1985) compiled an exhaustive bibliography of several 1000 references arranged by four subject categories: nutrient cycling, biomass, nutrient content of biomass, and computer simulation models.

Despite past efforts of scientists, there is an urgent need for decision-making tools, whereby harvest planners and managers can apply prescriptions that will maintain or improve site productivity. Nutritionally-oriented guidelines or laws regulating harvesting intensity are generally based rather on intuition than on strong scientific evidence. Computer simulation models to predict the effects of management on long-term productivity have been developed, but it is probably too soon to expect a high degree of accuracy of simulation models, as there are major gaps in our scientific understanding of forest ecosystems. Therefore, it is necessary to proceed empirically and conduct site-specific research to indicate the effects of various management practices on site productivity (Dyck et al. 1986).

9.1.1 Nutrient Budget of a Forest Ecosystem

A forest *nutrient budget* involves several biological cycles with geological and meteorological overprinting (Norton and Young 1976). In a forest ecosystem untouched by man, *the input* consists of particulate and dissolved elements in precipitation, the contribution from chemical weathering of the inorganic substrate, and biological fixation of atmospheric nitrogen. *The output* is the particulate and dissolved material that leaves the forest in groundwater and surface flow as well as the loss of nitrogen through denitrification and ammonia volatilization. Under natural conditions input and output are roughly balanced, and minor compared with the total amount of nutrients cycling within the system.

Switzer and Nelson (1972) categorized *the nutrient cycle* into three subcycles. The geochemical pathway of import-export to the forest ecosystem consists of mineral weathering, bulk precipitation, aerosol impaction, nitrogen fixation, and deep soil leaching. The biogeochemical pathway takes place between plant and soil and involves nutrient uptake, stem flow, litterfall, foliar leaching, root sloughing, root exudation, litter decomposition, and mineralization. The biochemical pathway involves the translocation of nutrients within the living plants to tissues as they are formed, while they are functioning, and from them before they are discarded.

The total nutrient supply of the forest soil is generally relatively large compared with the requirements of the trees. For example, in a mature Norway spruce stand on sandy moraine in central Sweden 9.3% of the nitrogen, 1.1% phosphorus, 0.3% potassium, and 1.3% calcium were bound in the biomass, and in a 29-year-old *Pinus radiata* stand in New Zealand the respective proportions were 11.2, 2.5, 0.4, and 0.2% (Table 9.1).

However, only a small part of the nutrients is in a form *available for the trees*. To become accessible to plants, the minerals require chemical weathering. Nitrogen, on the other hand, requires mineralization, i.e., the decomposition of organic material by microorganisms into inorganic ammonium (NH_4^+) or nitrate (NO_3^-) ions. According to Viro (1969), no more than about 1% of the total nitrogen supply in Finnish forest soil is mineralized. The mineralization process may be affected by management practice, but it is slow in any case. Hence, a large nutrient store as such is not a reliable index of soil fertility. A small active store might very well support a greater biomass than a large inactive one (Staaf and Björkroth 1981).

Table 9.1. The total supply of main nutrients in a mature *Picea abies* stand in Sweden and in a 29-year-old *Pinus radiata* stand in New Zealand (Messina et al. 1985)

Stand compartment	N	P	K	Ca
	(kg/ha)			
Picea abies stand in Sweden				
Trees	430	40	200	510
Forest floor	1100	60	80	260
Soil, 0–50 cm	3100	3600	64000	40000
Pinus radiata stand in New Zealand				
Trees	440	70	470	330
Forest floor	300	30	150	180
Soil	3050	2570	130720	170600

The nutrients that are readily available to the roots of the forest trees are thus much more important than the total amount. Unfortunately, quantifying the available nutrients is difficult. Chemical analysis should produce values similar to those the plant perceives as available, but this is rarely the case. Therefore, terms like available and exchangeable are simply chemical indices without direct biological significance (Messina et al. 1985).

Nutrient budgets alone do not provide reliable evidence of long-term depletion of *site productivity*, i.e., the capability of a site to produce biomass. To obtain better answers concerning the effects of timber removal on the production potential of a forest soil, the phenomenon must be understood in terms of *nutrient cycling*, in a dynamic sense. In principle, the development of nutrient availability is a combination of changes in turnover rates and the size of the nutrient stores. Removal of biomass might then be seen as the loss of a substrate that would supply nutrients to the subsequent tree stand during a phase of rapid growth (Staaf and Björkroth 1981).

The distribution of nutrients in an ecosystem varies depending on site fertility and the age and development of the stand. According to Switzer et al. (1978), the sustainable level of organic matter for the Gulf Coastal Plain of the southern United States totals about 360 t/ha, which is attained in pine-dominated ecosystems at around 65 years of age. From this time onward, about 70% of the organic matter is in the standing crop, 5% on the forest floor, and the remaining 25% within rooting depth in the soil. Table 9.2 shows the distribution of nutrients in old-field pine stands which have attained the sustainable or equilibrium level of biomass.

The nutrient store of most forest soils is thus rather large but firmly bound and inaccessible to plants. The trees obtain the main part of the nutrients from the humus layer. Rapid decomposition of organic matter and undisturbed cycling of nutrients are, therefore, of great importance for sustaining the fertility of the site.

To improve decision-making in forestry and to reduce the reliance on intuition, various approaches to assess the effects of harvesting on site productivity have been taken. Empirical research is often site-specific and is generally designed

Table 9.2. The distribution of total N, moderately available P, and exchangeable K, Ca, and Mg in old-field *Pinus taeda* stands on the Gulf Coastal Plain in the United States (Switzer et al. 1978)

Nutrient	Nutrients (kg/ha)				
	Stand compartment				Total
	Standing crop	Forest floor	Soil 0–20 cm	Soil 20–120 cm	
N	470	250	1820	4880	7420
P	57	15	230	1170	1472
K	330	23	190	2310	2853
Ca	650	180	700	4300	5830
Mg	114	24	155	5945	6238

to relate a dependent variable, such as growth, to independent variables, such as harvesting intensity or nutrition. Simulation methods also are generally based on empirical relationships, but are more flexible than empirical models since ecosystem processes are simulated. The model can make predictions outside the range of conditions for which the empirical relationships were derived, thus allowing greater applicability. Several ecosystem simulation models that incorporate nutrition, particularly nitrogen, have recently been developed for scientific purposes, such as teaching aids, or as forest management tools, but their accuracy is not yet sufficient (Dyck et al. 1986).

9.1.2 Nutrient Loss in Conventional Logging

Harvesting of timber interferes with the natural development of the ecosystem by accelerating the removal of organic matter and nutrients, which are lost not only through direct biomass removal but also through leaching due to soil disturbance and exposure of the forest floor, as well as by possible erosion. Loss of nutrients may lead to declining productivity, but the impact of harvesting on production capacity is also dependent on site characteristics.

Table 9.3 presents an example of nutrient loss after conventional stem-only logging in the thinning of young Scots pine (Mälkönen 1974) and birch (Mälkönen and Saarsalmi 1982) stands in southern Finland and in final cutting of an overaged Norway spruce stand (Kubin 1977) in northern Finland. The nutrient content is largest in birch. In fact, compared with conifers, wood and bark of deciduous trees generally contain more nutrients.

As long as the recovery of biomass is restricted to the stem of the tree, severe danger of impoverishment of site productivity does not seem to exist, excluding perhaps very poor soils (Rennie 1957) and certain fast-growing plantations in tropical and subtropical climates. The problem of second-rotation growth decline in some *Pinus radiata* stands in southern Australia and New Zealand may be at least partly a consequence of nutrient deficiencies (Keeves 1966; Whyte 1973).

Table 9.3. Nutrient loss as a result of conventional stem-only harvesting of *Pinus sylvestris* (Mälkönen 1974), *Picea abies* (Kubin 1977), and *Betula* spp. (Mälkönen and Saarsalmi 1982) stands in Finland

Tree species	Stand age (years)	Nutrient loss (g/m^3 of timber)			
		N	P	K	Ca
Pinus sylvestris	28	408	52	244	541
P. sylvestris	45	382	35	206	407
P. sylvestris	47	427	44	215	322
P. sylvestris average		406	44	222	423
Picea abies	135	396	44	308	1012
Betula spp.	20	1319	151	397	757
Betula spp.	40	979	91	337	712
Betula spp. average		1149	121	367	734

Table 9.4. Budget on nitrogen compounds for a 100-year-old *Picea abies* stand in central Sweden, and a 140-year-old *P. abies* stand in northern Sweden. The calculation covers a full rotation period under conventional harvesting practice (Nykvist 1977)

Nitrogen source	Central Sweden	Northern Sweden
	(kg N/ha/rotation period)	
Precipitation	+700	+280
Biological fixation	+280	+390
Total input	+980	+670
Denitrification	−100?	−140?
Leaching during		
Bare land period	−75	−75
Full stand period	−150	−170
Removal from		
Thinnings	−110	−130
Final cutting	−110	−130
Total output	−545	−645
Balance	+435	+25

The growth-limiting nutrient is commonly nitrogen, particularly in boreal forests. A nitrogen budget calculation for two mature Norway spruce stands in Sweden shows that input from precipitation and biological fixation is adequate to compensate the loss resulting from conventional timber harvesting. In fact, central Sweden seems to have a considerable nitrogen surplus (Table 9.4).

Atmospheric input of phosphorus in Sweden in the 1970s was ten times larger than before European industrialization, and large enough to compensate the loss from leaching and conventional timber harvesting. However, the input was not sufficient to balance the leaching of potassium, calcium, or magnesium. This leaching was probably offset by chemical weathering (Nykvist 1977).

After logging, decomposition of residue proceeds most rapidly in the foliage. Under Swedish conditions, needles lose 15–20% of their mass annually, although the rate tends to decrease with time. Nitrogen and phosphorus mineralization is initially slow, while potassium and magnesium are released quickly. Nitrogen release from needles is at its maximum after 3–6 years, after which it drops to almost zero at 10 years. Branches and twigs decompose more slowly, 5–15% per year for the fine material and 3–5% for thicker branch sections. Nitrogen and phosphorus are not released from the woody parts of the branches until after at least 10 years' storage. They accordingly constitute a minor but long-lasting nutrient source for young stands (Staaf 1984).

9.1.3 Nutrient Loss Caused by Intensive Biomass Removal

If the recovery of biomass is intensive and the operation is frequently repeated, harvesting will probably result in permanent impoverishment of the site. *Collection of litter* from forest soil, which was common practice in Germany from the 16th to the 19th century, is an extreme case. The organic matter was used for animal bedding and then as a soil improvement medium in farming. Over a period of 150–200 years as much as 200–300 t/ha of dry organic substance were removed. The loss of nutrients from the ecosystem was 500–600 kg/ha N, 50–100 kg/ha P, and 150–200 kg/ha K, respectively. A negative change gradually took place in the structure and composition of the forest soil, and the growth capacity of the ecosystem was seriously impaired. After the collection of organic substance ceased, the production rate of the soil started to improve slowly (Miljøministeriet 1985). The destructive effect of continued litter gathering on the productivity of forest stands has also been demonstrated by experiments (Ganter 1927; Kern et al. 1971). Regular litter collection must not be equated with timber harvesting, but these Central European experiences may nevertheless be seen as concrete proof of the potential danger inherent in uncontrolled recovery of forest biomass.

According to Messina et al. (1985), the extent of nutrient loss in connection with timber harvesting is controlled by several independent factors such as:

1. Intensity of harvest (clear-cutting versus partial logging; whole-tree versus stem-only harvesting);
2. Species composition (nutrient distribution in stand components and rates of cycling);
3. Site fertility (strategies of ecosystem stability against disturbance);
4. Stand age (differing nutrient efficiencies at different stand ages).

If the above-ground biomass of the tree were to be fully recovered, the *nutrient losses* would grow alarmingly compared with stem-only harvesting, as shown by Fig. 9.1 and Table 9.5 from Great Britain (Binns 1975), the Federal Republic of Germany (Kreutzer 1975), and Finland (Mälkönen 1975). The loss of nitrogen would increase roughly 2.5-fold in pine forests and fourfold in spruce forests. The figures for beech represent leafless trees during the winter season.

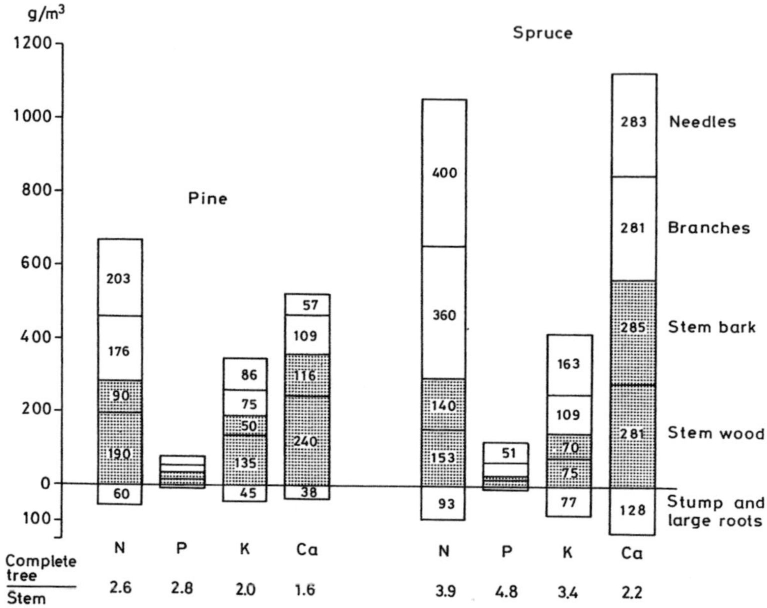

Fig. 9.1. The amount of nutrients, g/m^3 of unbarked merchantable stem wood, removed hypothetically in complete-tree logging of *Pinus sylvestris* and *Picea abies* in Finland (redrawn from Mälkönen 1975)

Table 9.5. The amount of various nutrients in whole-tree biomass as a ratio of that in the merchantable unbarked stem

Species	Country	Whole tree/stem, nutrient ratio			
		N	P	K	Ca
Pinus sylvestris and					
P. nigra var. *maritima*	Great Britain	2.2	3.3	2.9	2.2
P. sylvestris	Germany	2.3	2.4	2.1	1.5
P. sylvestris	Finland	2.6	2.8	2.0	1.6
Picea abies and					
P. sitchensis	Great Britain	4.3	5.2	3.1	2.5
P. abies	Germany	3.6	4.3	2.7	1.7
P. abies	Finland	3.9	4.8	3.4	2.2
Fagus sylvatica					
(leafless)	Germany	1.4	1.6	1.4	1.3

Another example of the concentration of nutrients in the above-ground biomass of a stand is given in Fig. 9.2 from a 22-year-old *Pinus radiata* plantation in New Zealand. Although stem wood consisted of 77% of the total above-ground dry mass, it contained only 20% of the nitrogen and phosphorus, 40% potassium, and 50% magnesium. At the other extreme, needles comprised only 3% of

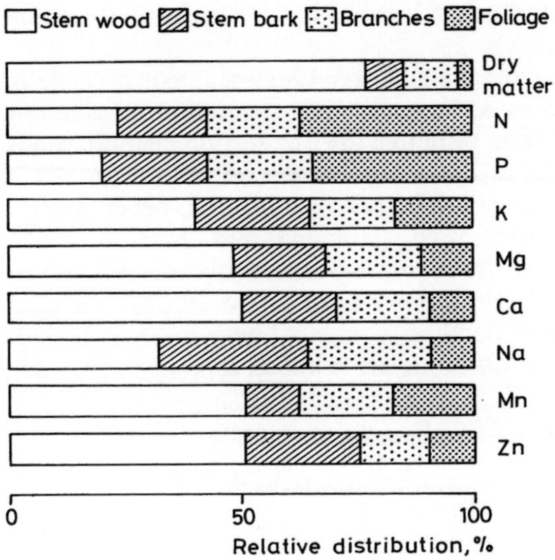

Fig. 9.2. The relative distribution of dry mass and eight nutrients in the above-ground components of a 22-year-old *Pinus radiata* stand in New Zealand (redrawn from Madgwick et al. 1977a)

the biomass but contained 10% of the calcium and zinc, 15–20% potassium, magnesium, and manganese, and 35% nitrogen and phosphorus.

As a rule of thumb, using stem-only harvesting as the basis for comparison, each percentage increase in biomass recovery is accompanied by about a 3% increase in nutrient removal. The effect of biomass removal on nutrient loss is of course specific for any particular nutrient and for those tissues that constitute basis of the yield increase (Switzer et al. 1978).

The effect of increased biomass removal on the *production potential of a stand* depends on the fertility of the site. Boyle et al. (1973) conducted theoretical studies on the impact of repeated whole-tree logging on the nutrient balance and wood production of a 40-year-old aspen-dominated hardwood stand growing on a loam and sandy loam site in Wisconsin. On this good site, reserves, inputs, and recycling of nitrogen, phosphorus, and potassium apparently would be able to sustain an infinite number of future crops of whole-tree chips, including foliage, while supplies of calcium might begin to limit growth rates and production after nine successive 30-year rotations. The authors emphasized, however, that similar harvest operations on sites of lower fertility or on those subject to severe erosion or leaching may result in significant reductions in the production capacity of the forest ecosystem even after a single crop.

In a study of whole-tree harvesting of young bottomland hardwood stands in Mississippi, dominated by *Fraxinus pennsylvanica*, Francis (1984) found that nutrients withdrawn by biomass removal exceeded the potential input through rainfall during an 11-year rotation period, but the net loss was small. The danger of depleting the fertile soil of its nutrients by intensive biomass recovery was considered minor. However, it was emphasized that this conclusion should not be extended to soils with low cation exchange capacity and low base saturation. Compared with summertime operations, winter logging resulted in about 20% reduction in nitrogen, phosphorus, and potassium loss but in 27% increase in calcium loss.

For several reasons, complete recovery of all tree biomass is not possible in practice. Nutrient loss is thus seldom as large as many theoretical calculations may suggest. Under Swedish conditions, although the tree-section method results in a radical reduction in the amount of slash, the biomass left on the ground is nevertheless 9–45% of that in conventional stem-only logging, depending on stand conditions and the equipment used (Holappa and Jonsson 1987). Shumakov (1971) studied real nutrient loss in connection with two alternative harvesting methods in final cuttings in the European part of the Soviet Union. Partial removal of crown mass in whole-tree skidding, compared with tree-length skidding, resulted in additional nutrient loss from the stand as follows:

Tree species	Additional nutrient loss resulting from removal of crown mass (%)			
	N	P	K	Ca
Pinus sylvestris and *Picea abies*	75	61	53	40
Betula, Populus, and *Alnus* spp.	37	29	40	37

Another example illustrates northern European whole-tree chipping practice, where small-sized birch is recovered, generally without leaves, and hauled by forwarders to the roadside for chipping. Assuming that the method increased biomass recovery by 30% compared with conventional logging, the removal of main nutrients increases per area by roughly 40 to 60% (Hakkila and Kalaja 1983):

Basis of calculation	Additional nutrient loss resulting from recovery of crown mass (%)			
	N	P	K	Ca
Per unit volume of timber	5	20	23	8
Per unit of area	37	56	60	40

When stump and root mass is harvested, root sections thinner than 2 to 5 cm are left on the site. Because the nutrient concentration is greatest in the fine roots, a large proportion of the nutrients in the below-ground parts of the trees will remain in the ground. The following table represents 5-year-old *Pinus elliottii* in Florida, but it reflects the nutrient ratios in the roots of larger trees as well (White et al. 1971):

	Concentration (%)			
	N	P	K	Ca
Under 5-mm fine roots	0.69	0.13	0.15	0.30
Lateral roots	0.28	0.09	0.12	0.14
Taproot	0.18	0.05	0.10	0.08

Table 9.6. Total nitrogen loss resulting from biomass removal, denitrification, and leaching during a 5-year period after clear-cutting in Appalachian oak ecosystems (Rauscher 1980)

Site index	Logging regime	Loss from biomass removal	Loss from leaching, etc.	Total loss
		(kg N/ha)		
40	Stem-only	238	303	541
40	Whole-tree	553	130	683
60	Stem-only	322	383	705
60	Whole-tree	746	150	896
80	Stem-only	425	507	932
80	Whole-tree	985	185	1170

Nutrients being released from slash in a clear-cutting area are partly leached. When slash is removed, the *leaching of nutrients* decreases compared with conventional logging. Rauscher (1980) pointed out that it is invalid to use nitrogen losses from conventional stem-only harvesting as a nondepletion standard with which more intensive harvesting is compared. When postharvest losses of nitrogen from denitrification and leaching are considered, the total losses for whole-tree harvesting in clear-cut areas are not very much larger than those for stem-only harvesting (Table 9.6). The effect of residue utilization on the nitrogen balance of an ecosystem is thus not necessarily as harmful as the nitrogen content of the biomass may suggest.

Recovery of forest residue may theoretically *accelerate soil acidification*. According to Nilsson (1984), the process is as follows: Tree roots take up nutrients as positively or negatively charged ions. The uptake of positive ions, which is coupled with a corresponding release of acid (H^+), dominates over the uptake of negative ions, resulting in gradual acidification of the forest soil during the development of a stand. However, the change is not permanent, because after the death of the trees, decomposition of the biomass starts a reverse process, releasing nutrients and neutralizing acids (Fig. 9.3). If the balance between accumulation and

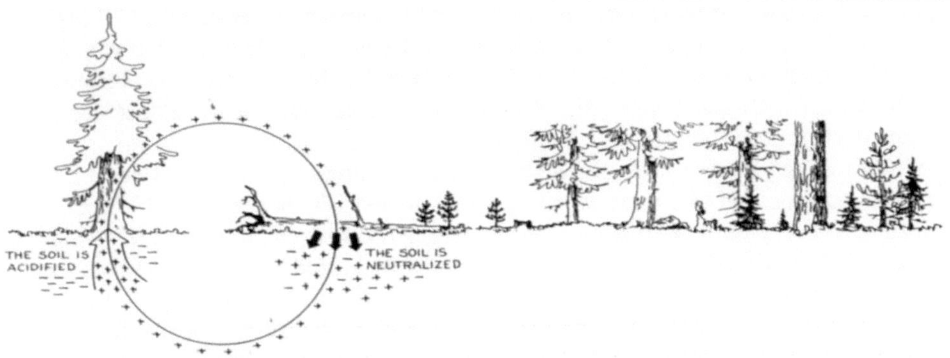

Fig. 9.3. During its lifetime the tree gives off H^+ ions thus acidifying the soil. After the death of the tree decomposing biomass reverses the process, and the soil is partly neutralized (Nilsson 1984)

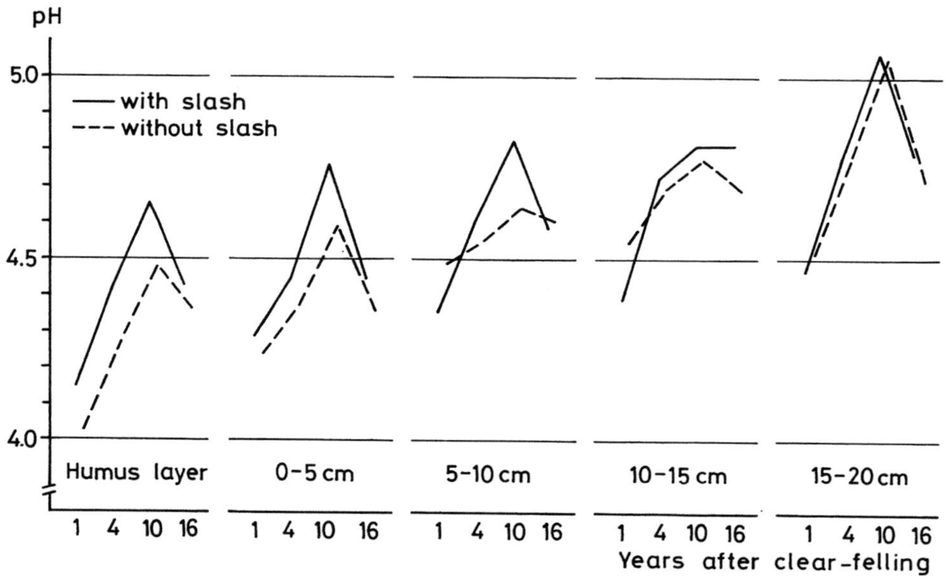

Fig. 9.4. The effect of slash removal on the pH of the humus layer and mineral soil after clear-cutting in Sweden (redrawn from Nykvist and Rosén 1985)

decomposition of organic matter during the life span of a stand is disturbed by intensive removal of biomass, acidification becomes dominant. Because clear-cutting results in an increase of the pH level in the organic humus layer, possible permanent acidification after slash removal will not show up until later. According to Nilsson, biological acidification in fast-growing stands may theoretically be even higher than that caused by the acid input from the atmosphere, commonly known as acid rain. Mahendrappa (1986) emphasized that the potential for acidification of forest soils depends on the site quality and tree species and, consequently, intensive biomass harvesting should not be practiced indiscriminately.

According to a Swedish study, the pH level of the humus layer increased significantly after clear-cutting, but the increase was 0.1 pH units less if slash was removed. Differences were also found in the pH level of the mineral soil, although they decreased with soil depth (Fig. 9.4). It was concluded that the effect might last up to 20 years (Nykvist and Rosén 1985).

Studies of the humus layer show evidence of an ongoing soil acidification in European forests, and one cannot exclude acid deposition as a responsible agent. In addition, however, it is important to consider the loss of basic cations when studying soil acidification. Consequently, intensive biomass recovery should be considered together with acid deposition (Andersson 1985). Generally, the need for research on the ecological impacts of whole-tree utilization is particularly urgent on poor and sensitive sites and in tropical and subtropical plantations, where fast growth and short rotations are accompanied by a high rate of nutrient accumulation.

9.1.4 Restricting Operations on Sensitive Sites

Intensified removal of biomass may have undesirable consequences in the forest ecosystem, as described in previous sections. Particularly on poor sites, compensation by fertilization may be needed in the future. Since trees recover commercial fertilizers rather ineffectively, application has to be in a suitable form and in amounts considerably larger than those removed with the biomass. However, the quantities required will not exceed the levels used by conventional forest fertilization.

Nutrient drain for hardwood stands in the southern United States logged by whole-tree methods is two to three times that in conventional stem-only harvesting, but the drain may be less critical than for pine stands because rotations are generally longer and soil nutrient reserves are often higher. Whole-tree harvesting has made many hardwood stands operable that previously were not, thereby increasing the silvicultural opportunity to improve the stands. Judged by these comparisons, in the southern United States whole-tree harvesting would appear to be more attractive, both economically and ecologically, in hardwood stands (Phillips and van Lear 1984).

On nutrient-poor and dry sites, rather than gradually creating a need for artificial fertilization, residue recovery should be controlled in such a way, and on such a scale, that the natural cycling of nutrients is not seriously interrupted. It is much easier to prevent an undesirable development than to restore sites damaged by a disruption of the nutrient cycle (Tamm 1975).

In order to prevent excessive biomass removal from biologically sensitive soils, the National Board of Forestry in Sweden has developed general guidelines for biomass harvesting. Two levels of intensity are acknowledged. Even when the aim is to harvest most of the crown mass, the actual assumption is that one-quarter of it remains in the stand. When the unmerchantable top with branches is cut off and left on the ground, or when foliage and thin twigs are removed on site with transpiration drying before haulage of whole trees from the site, the efficiency of residue recovery is estimated to be roughly 50%. The following restrictions are recommended (Skogsstyrelsen 1985):

1. As a rule, crown mass should not be harvested more than once during the rotation period of a stand.
2. Crown mass should not be removed at all from dry sites, from medium moist sites with coarse soil, where the ground water table is deeper than 1–2 m, or from poor organic peat sites.
3. Recovery of half the crown mass is acceptable from organic sites covered by herbal vegetation.
4. Most of the crown mass may be recovered from moist mineral sites and organic sites with tall herbal vegetation.

In addition, special restrictions are placed on certain sites and logging conditions. Table 9.7 shows an example of special restrictions on medium moist moss sites. They do not apply to first commercial thinnings. The Swedish guidelines have been criticized by forest owners and forest industries for inadequate scientific background, too detailed instructions, and difficulties in practical application.

Table 9.7. The recommendation of the National Board of Forestry in Sweden regarding the level of acceptable residue recovery under various geographic conditions on medium-moist moss sites

Geographic area of Sweden	Form of harvesting	Acceptable residue recovery (%)					
		Ground vegetation					
		Dwarf shrub			Blueberry		
		Coarse soil	Medium soil	Fine soil	Coarse soil	Medium soil	Fine soil
Areas with a high risk for early summer dry spells in the northeast	Stump harvest	–	–	50	–	75*	75
	First thinning	–	75	75	75	75	75
	Other harvest	–	–	50*	50	50*	50
Hard-to-renew areas in north	Stump harvest	–	–	50	–	75*	75
	First thinning	–	75	75	75	75	75
	Other harvest	–	–	50*	–	50*	75
Areas with considerable acid emission in the southwest	Stump harvest	–	–	50	–	75*	75
	First thinning	–	75	75	75	75	75
	Other harvest	–	–	–	50	50	75
The rest of Sweden	Stump harvest	–	50	75*	50	75*	75
	First thinning	–	75	75	75	75	75
	Other harvest	–	50*	75*	50*	75*	75

* Asterisk refers to additional specific restrictions

The National Board of Forestry in Denmark has also developed guidelines for intensive biomass utilization. Rather severe restrictions are recommended. Harvesting stump and root mass is only permitted in case of windfalls, and it is recommended that on poor sites even the above-ground residue be left in the forest. Whole-tree chipping of softwoods on sites with a production capacity lower than 9 m^3/ha/a should be restricted to the first and second thinnings only and compensated through fertilization. Whole-tree logging of hardwoods should be carried out when trees are leafless. Softwood needles should be gone before off-road haulage through at least 2 months of spring or summer transpiration drying. Slash removal is also widely restricted and should be carried out only when the needles have dropped. These and other ecological requirements would reduce the available forest biomass in Denmark to only 35–50% of the potential reserve (Miljøministeriet 1985).

In most countries, operation recommendations have not been published, but the importance of ecological considerations and of sustaining the production capacity of forest soil is recognized. It is commonly agreed that the ecological consequences of increased biomass utilization must be set as a charge against the benefits of the operation. According to Binns (1975), while it would not be possible on many British soils to accept complete utilization of all stages of the tree crop, whole-tree chipping in early thinnings of very closely planted stands could help the operations significantly without the drain of organic matter and nutrients becoming alarming. On the other hand, crown material should be left in the forest in later thinnings to give some structure to the soil to support machine operations, and to recycle nutrients. Harvesting tree tops with branches after final fellings could be allowed in order to facilitate the regeneration process.

9.2 Effect of Residue Removal on Regeneration and Growth

9.2.1 Logging Residue as a Work Difficulty Factor in Regeneration

Logging residue is often a serious hindrance to regeneration operations, especially after clear-cutting of unmanaged, overmature forests. On the other hand, residue left behind after final cuttings of repeatedly thinned forests is seldom heavy enough to totally prevent operation of site preparation machines or manual planting.

Compared with motor-manual logging, on-site logging machines tend to create piles of slash with some intermediate scattering, rather than an even distribution of residue over the entire site. Though the planting may then be easier in spots, a large proportion of the area may not be available for reforestation purposes, because the piles cannot be readily redistributed during site preparation and are often left in situ. Moreover, delimbers and slashers located at landings and along roadsides during whole-tree operations leave large volumes of debris, which generally prevent treatment of significant areas for stand establishment (Riley 1984).

Slash is one of the most important work difficulty factors of *mechanized site preparation*, adversely affecting its efficiency, especially if the residue is left partly in piles by logging machines. Under severe residue conditions, site preparation machines may jam with debris, resulting in lost production time and effectiveness of the operation. Even a 10- to 15-cm layer of fresh residue is too thick for good results with most scarification machines (Simonsson 1974). If residue is removed or crushed before site preparation, the area of uncovered mineral soil increases, though driving speed and areal output of the machine is barely affected (Gustafsson and Larsson 1981). Whole-tree harvesting thus eliminates certain slash-related regeneration problems. Integrating biomass recovery and site preparation can result in a credit to be applied to either the harvesting or site preparation operation (Stokes and Watson 1986).

Watson and Stokes (1984) studied mechanical site preparation following clear-cutting of southern pine stands with understory hardwoods in Alabama and Mississippi. The conventional stem-only harvesting averaged recovering 52% of the above-ground biomass while two whole-tree harvesting systems recovered 76 and 85% of the biomass, thus leaving the site relatively free of material to be removed during site preparation. As a result, the cost of performing an adequate site preparation treatment was at least US $ 137 less per hectare in conjunction with intensive biomass harvesting. This savings translated into a site preparation credit of at least US $ 3.20 per green tonne of fuel chips.

During *manual planting*, slash slows down the movements of a worker, and adds an entirely new phase to the work, i.e., removal of slash from the planting spot. However, planters are often reluctant to clear away debris from a planting site, but would rather select the next clear area. Slash accumulations thus contrib-

Table 9.8. The effect of slash on time consumption in manual planting (Hakkila 1973)

Work phase	Time consumption (cmin/seedling)		
	Slash abundant	Slash moderate	Slash removed
Moving from spot to spot	9.8	10.0	7.3
Removal of slash	9.9	6.7	0.0
Hoeing the planting hole	23.4	14.4	18.7
Planting the seedling	25.2	20.7	17.5
Total effective time	68.3	51.8	43.5

ute to clumped tree distribution. Abundant slash interferes with the preparation of the planting hole and the planting of the seedling. Accordingly, in a Finnish study 6 years after planting, Norway spruce seedlings had better developed root systems in areas which had been cleared from slash (Raivonen and Leikola 1980). Furthermore, seedlings planted in logging residue are more prone to mechanical damage such as broken tops and branches, pressed down stems, etc. (Björkroth 1983).

In addition to reducing work performance, residue seems to add to the mental and physical strain of the worker. The effect of slash removal on the structure and consumption of work time, output, and strain in manual planting with a semicircular hoe was studied by the Finnish Forest Research Institute. A clear-cut area of Norway spruce was divided into plots where the amount of residue was either abundant (stem-wood yield over 200 m^3/ha), moderate (yield 100 m^3/ha), or completely removed (Table 9.8). When the output of manual planting work in moderate residue was represented by the figure 100, abundant residue reduced relative output down to 76. Residue removal changed the structure of the work and raised average heart rates of two workers from 107 to 115 beats/min. Notwithstanding the rise in heart rate, the workers did not consider work in slash-free plots more strenuous and actually took shorter rest breaks than when working in the other two areas. Taking into consideration the spontaneous shortening of the rest breaks, slash removal raised relative output to 128.

In *mechanical planting*, abundant residue slows tractor movements, interferes with the making of furrows or planting holes and with the planting of seedlings. It may also cause failed planting attempts and mechanical breakdowns, and is a safety hazard to the planting machine operator. Consequently, residue affects the quality of the work and causes loss of productivity. In the southern United States, windrowing slash to facilitate regeneration operations is a common practice.

In Great Britain, the residue from pine is frequently crushed by a chopper and left as mulch to facilitate replanting, while residue from spruce is either left to decompose as such or piled into windrows, which may result in wide strips without trees. This creates pressure for greater utilization of the residue (Binns 1975).

9.2.2 Site Preparation Effect of Stump Removal

Any kind of soil preparation in a regeneration area, regardless of the level of residue recovery, means a drastic change in the decomposition conditions of the humus reserves, and may have other consequences as well. Extraction of stumps causes substantial *mixing of the soil.* This in turn leads to more rapid decomposition of organic matter, and increase in the loss of soil nutrients, nitrate nitrogen in particular, through faster leaching. The risk is greatest on dry soils with a coarse texture and a thin humus layer, and in climates with high temperatures. In contrast, moist, fine-grained soil with a nutrient-demanding herb cover may tolerate more soil mixing without significant effect on long-term productivity (Lundmark 1984).

The size of the soil area disturbed by stump uprooting is often sufficient for two or even three seedlings, and some *scarification effect* may also occur between the stumps. However, because 2000–3000 planting spots are required per hectare in North America and northern Europe, and much more in Central Europe, the site preparation effect is hardly adequate as such. Additional spots may be prepared in connection with harvesting, but this increases the costs of the operation. An excavator-mounted stump harvester requires 10–15 cmin for each extra spot prepared in conjunction with stump extraction (Nylinder 1976).

Site preparation with a conventional scarifier after stump removal is superior to normal site preparation. According to Österlöf (1979), stump extraction achieved an increase in:

1. The area of the scarified patch per planting spot, from 0.75 to 1.60 m^2;
2. The area of exposed mineral soil per planting spot, from 0.15 to 0.27 m^2;
3. The percentage of planting spots located in mineral soil, from 25 to 40%;
4. The productivity of a conventional scarifier of around 10%.

Harvesting stumps *facilitates mechanization* of planting, especially if both above-ground residue and stumps are removed (Gustafsson and Larsson 1981 a, b). Manual planting output increases by 10% after stump extraction (Österlöf 1979). In Central and Eastern Europe and the southern United States, stumps are sometimes removed from a regeneration area to expedite soil preparation and operation of planting machines, even if there is no demand for stump and root wood.

Stump harvesting may also improve the forest terrain for traffic. A significant *leveling of microtopography* was observed after recovering stump-root systems with an excavator-mounted harvester in Sweden (Kardell 1976). However, *deep holes* which may fill with water form occasionally. In Hungary, the depth of the holes of *Populus* spp. and *Robinia pseudoacacia* stumps is 20–50% greater than the stump diameter, for *Pinus nigra* var. *nigra* the depth is equal to the diameter, and for *Quercus* sp. the depth is about one-half of the stump diameter (Fig. 9.5). Holes can be partly filled up in conjunction with uprooting by vibrating, shaking, or pushing loose soil from the stump-root system back into place.

Pine sites in the southern United States harvested with a tree puller, which recovers the central stump-root system but leaves lateral roots in place in the soil, show only few and shallow holes. This is because the soil falls back into the cavi-

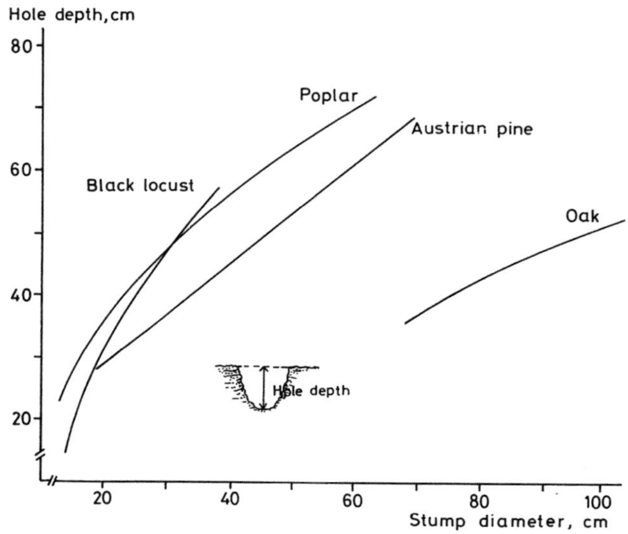

Fig. 9.5. The depth of the hole left after stump extraction as a function of stump diameter in Hungary (redrawn from Horvath-Laiko et al. 1970)

ties and leaves a hole less than 30 cm in depth. These holes are filled up after rains and movement of machines over them (Koch 1985 b).

Stump extraction may result in a general increase in soil moisture, which together with removal of thicker roots may *weaken the bearing capacity* of the ground. On poorly drained sites, stump harvesting may therefore have certain undesirable consequences (Y. Jonsson 1985). On erosion-prone sites, such as steep slopes, stump removal must be avoided.

9.2.3 Initial Development of a Stand After Residue Removal

Harvesting systems are an integral part of silviculture, even if logging is not (Armson 1976). Clear-cutting has extensive radical consequences for the biology of forest soil. These effects on the ecosystem include the loss of a large proportion of the biomass and an increase in nutrient mobilization. Leaching of nutrients is also stimulated as runoff from the regeneration area increases.

After conventional stem-only logging, a layer of residue more or less evenly covers the ground of a regeneration area. The decomposing biomass delivers energy to the organisms that release plant nutriment, thus serving both as a source of energy and as a *supply of nutrients.* Furthermore, residue prevents drying of the soil surface and reduces the variations in humidity and temperature (Björkroth 1983).

Possible growth reduction after residue removal in a regeneration area may actually be more closely associated with the *physical effects of slash,* rather than with the loss of nutrients. Of particular concern are those physical properties of the soil which are intimately associated with the decomposition and mineralization of organic matter, the prevention of nutrient leaching, and the maintenance of proper soil aeration and drainage (Switzer et al. 1978). Daily temperature variations, risk of night frost, and wind velocity at ground level, all increase in a regeneration area after residue removal. The moisture content in surface soil decreases, which may improve growth on clay soils, but reduces it on dry, sandy soils. The growth of ground vegetation increases, and the composition of the plant community changes (Söderström 1974).

In a Swedish regeneration experiment, the removal of residue did not affect the survival and height growth of Norway spruce seedlings during the first 3 years. However, thereafter growth was significantly faster in the slash-covered areas. It was possible to compensate the increment loss from residue removal by fertilization for a number of years, but by the tenth year after planting and fertilization, the trees in unfertilized, slash-covered areas were again growing better than those in fertilized, slash-free areas (Sirén 1975).

In another extensive experiment in Sweden, two kinds of plots were compared: those were the slash was removed completely, and those where it was left in place and further supplemented by addition of slash from cleared plots. After 14 years, in both cases the amount of organic substance was the same in the humus layer, but it was 9% higher in the mineral soil in the plots which had slash supplements. Moreover, the quantity of nitrogen as well as the proportion of mineralized nitrogen released through decomposition of organic matter was greater under a slash cover. The differences were more pronounced on poor soils in northern Sweden (Björkroth 1983).

As shown in Fig. 9.6, Norway spruce grew better while Scots pine showed a retarded height increment on slash-covered plots during the early years after planting. On the most northern plots, residue had a significantly positive effect

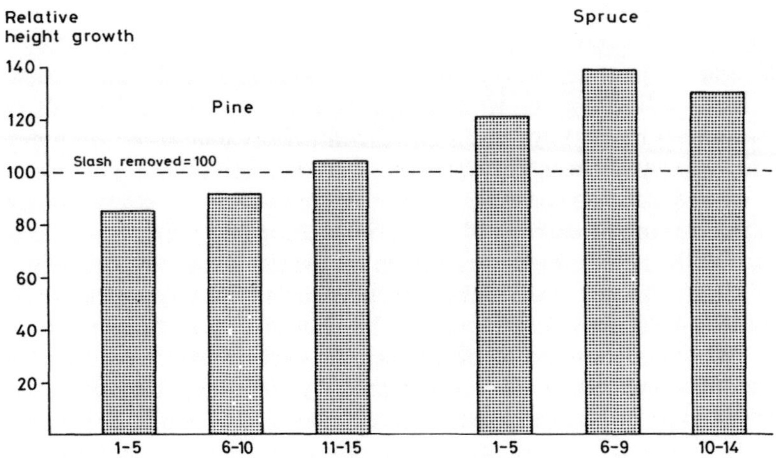

Fig. 9.6. Relative height growth of *Pinus sylvestris* and *Picea abies* on medium sites with a double amount of slash in Sweden. The height growth on plots without slash is defined as 100 (redrawn from Björkroth 1983)

on the diameter increment of pine. It was concluded that during the early years the residue affects the plants mainly in a mechanical way, favoring spruce and disfavoring pine. In a later phase the effect of slash on the nutrient conditions becomes more important, and by then both species show better growth on sites where slash remains (Björkroth 1983).

Raivonen and Leikola (1980) studied the development of Norway spruce seedlings on slash-covered and slash-free plots in central Finland. Six years after planting the mortality was lower on the plots where slash remained. Removal of the slash increased height growth slightly on slopes but decreased it on flat ground where the seedlings suffered more from frost damage. The proportion of weak seedlings was higher on slash-free plots. In northern Maine as well, young spruce-fir stands show best development where a well-distributed cover of slash remains (McCormack 1984).

9.2.4 Growth of a Thinned Stand After Residue Removal

The consumption of nutrients is distributed between the tree crop and the ground vegetation within limits determined by stand density. During the development of a tree crop from a seedling stand to a closed forest, ground vegetation diminishes. Nutrients are released to the trees at the very time when the need increases (Mälkönen 1977).

The nutrient demand of a stand is generally at its greatest soon after the first thinning. The increase in growth of the remaining trees after the thinning is actually due partly to a manuring effect of the residue. Therefore, even a relatively moderate removal of residue at this stage may reduce the growth increment of the stand (Mälkönen 1976). With the removal of about 30% of the trees, however, only one-tenth of the nitrogen and phosphorus contained in the tree biomass above ground is lost (Tamm 1969).

Experiments have been started recently in many countries to study the long-term effects of biomass removal on the growth of forest trees on different sites, but most of them are still too young to shed light upon the problem. One of the earliest experiments was carried out in an 88-year-old Scots pine stand on a dry *Cladonia*-type site in southeastern Norway in 1928. In repeated thinnings all slash was removed from one plot and spread onto another, the latter thus continuously receiving a double covering of crown and top residue compared with normal logging operations. The diameter growth was greater on the slash-covered plots while the effect on height growth was small and uncertain. During the 32-year trial period, the double slash cover resulted in a 0.3 m^3/ha/a or 20% growth increase. It was concluded that complete removal of all logging slash from poor pine sites may result in an increment reduction of approximately 10% (Brantseg 1962). In another Norwegian experiment, the losses of increment after complete slash removal from Scots pine and Norway spruce stands on fertile sites amounted to 1.5 m^3/ha/a or 11% (Tveite 1983).

A similar experimental design was used by Andersson (1984) on a 42-year-old Scots pine stand on a medium blueberry shrub-type site in Sweden. The stand was thinned in 1964 and again in 1974. Up to the time of the latest measurements in

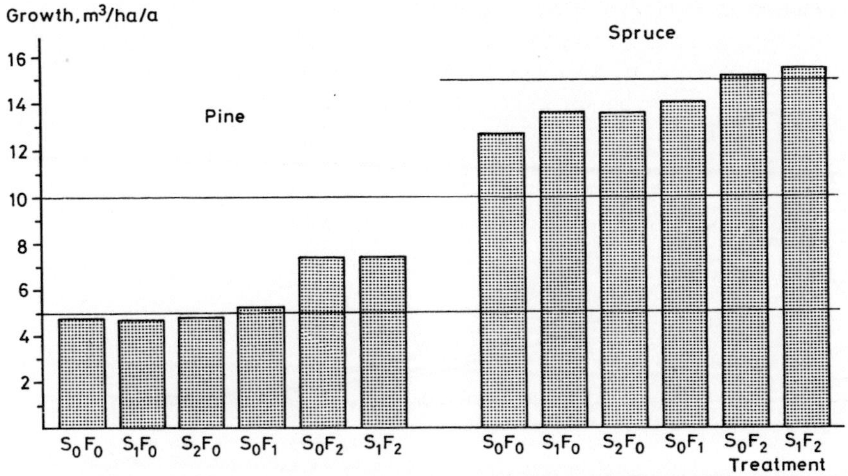

Fig. 9.7. Effect of slash removal and fertilization on the volume growth of *Pinus sylvestris* and *Picea abies* in southern Finland during the 5-year period following the first thinning. S_0 = slash removed; S_1 = slash left; S_2 = double amount of slash; F_0 = no fertilizers; F_1 = loss of N, P, and K compensated by fertilizing; F_2 = fertilization comprising 180 kg N/ha, 36 kg P/ha, and 72 kg K/ha (redrawn from Kukkola and Mälkönen 1985)

1980, the increment was greater in areas covered by a double amount of slash. The difference was 0.75 m^3/ha/a or approximately 10% over the follow-up period.

Kukkola and Mälkönen (1985) studied the effect of slash and fertilizers on the growth of 40- to 50-year-old Scots pine and Norway spruce stands after the first thinning in southern Finland. Removal, leaving, or doubling the residue had no effect on growth of Scots pine stands during the first 5-year period after thinning. The removal of the residue from spruce stands had a slight negative effect, although the differences were not statistically significant. Where the nutrient loss from slash removal was covered by an equivalent amount of nitrogen, phosphorus, and potassium through fertilizer compensation, the growth increment was better than with slash. Easily soluble fertilizers thus gave a quicker response than the slowly releasing nutrients from forest biomass. Compared with all other kinds of treatment, normal fertilization gave a statistically significant growth increase (Fig. 9.7). A period of 5 years only, however, is not adequate for determining the long-term effects of various types of treatment.

In most yield experiments, logging slash has been removed as completely as possible. However, since some of the crown and top mass is always left on site even during whole-tree logging, the reduction in growth, attributable to nutrient removal, is in reality less than indicated by these experiments.

In Sweden, Olsson (1985) assessed the total growth reduction caused by nutrient loss, logging damage to soil and trees, and strip roads in the first thinning of Scots pine on an average site on mesic soils. Three alternative thinning methods were compared. The estimated future timber yield of the two later thinnings and the final cutting was theoretically 522 m^3/ha in all, if the conventional motor-manual log-length method was applied during the first thinning. If a crane-

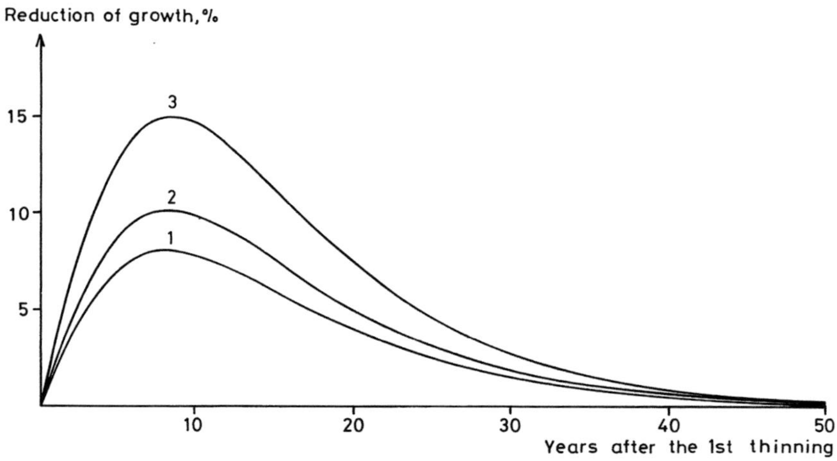

Fig. 9.8. Theoretical reduction of the annual growth due to soil and tree damage, nutrient loss, and area left under strip roads in the first thinning of a *Pinus sylvestris* stand. *1* Conventional motor-manual log-length method; *2* mechanized log-length method; *3* partially mechanized tree-section method (redrawn from Olsson 1985)

mounted, single-grip grapple harvester instead of a chain saw was used to prepare the timber, the future yield was reduced slightly to 518 m^3/ha due to increased logging damage. If the tree-section method was used and the timber was harvested with the branches intact, the yield was 507 m^3/ha or 3% less than after the motor-manual log-length method. In Olsson's opinion, the adverse effect of crown removal on the production capacity of medium-quality Scots pine sites is acceptable when weighed against the technical advantages of the tree-section system. The relative reduction in growth during a 50-year period after first thinning is shown in Fig. 9.8.

9.3 Effect of Residue Removal on Forest Health and Environment

9.3.1 Logging Damage to Soil and Trees

Logging operations result in damage to forest soil and remaining trees. The damage, which tends to be more severe when using mechanized methods, lowers timber quality and reduces tree growth. The risk is high for such species as spruces and many hardwoods, which have a superficial root system, grow on soft soils, and are sensitive to decay. The problem is less with pines, because they often have a deeply penetrating taproot with few superficial laterals, generally grow on dry, sandy sites, and are resistant to decay.

Whole-tree harvesting systems, compared with stem-only harvesting, have two distinguishing features. First, off-road haulage is carried out with branches still attached to the stem, and, secondly very little slash is left on the ground. The absence of protective slash has a definite influence on soil compaction, which also depends on the type of machine, amount and nature of applied force, and the strength and water potential of the soil (Grey and Jacobs 1985). The protective effect of slash may be largely due to reduced wear to the humus layer rather than improved carrying capacity of the ground (Kärkkäinen 1970).

Empirical studies have shown that on certain soil types compaction, as a result of machine traffic, can cause a decrease in tree growth. However, it is not really understood why growth is reduced, and results are therefore generally site-specific. A gap in our knowledge is understanding why compaction reduces growth and to what extent the effect is nutritionally based (Dyck et al. 1986).

The effect of slash cover on rut formation under two eight-wheel forwarders was studied under bare-ground conditions in Sweden. Even a 3–5-cm layer of slash significantly reduced rutting. However, deeper slash coverage seemed to give only slight additional protection to the soil (Hallonborg 1982a).

	Depth of ruts (cm)			
	Forest land, good bearing capacity		Farm land, poor bearing capacity	
	Forwarder I	Forwarder II	Forwader I	Forwarder II
No slash cover	6	7	7	11
Slash cover (3–5 cm)	8	4	7	6
Slash cover (10–15 cm)	4	5	5	7

The protective effect of slash is important on sites with poor bearing capacity (Fig. 9.9). Root damage caused by strip road-operating forest tractors or stand-

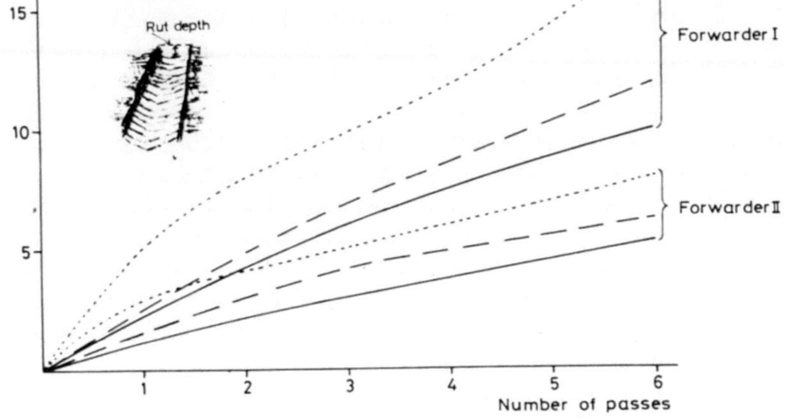

Fig. 9.9. Rut formation as a function of the number of passes with two different forwarders on poorly bearing farm land, depending on the thickness of the protective cover of slash (Hallonborg 1982a)

Table 9.9. Causes of damage to remaining trees during thinning of *Pinus sylvestris* and *Picea abies* stands using motor-manual and partly mechanized tree-section methods in Finland (Sirén 1986)

Cause of damage	Distribution of damage (%)			
	Damage point			Total
	Root	Root collar	Stem	
Wheel or track of forwarder	98	83	18	52
Boom of forwarder crane	2	17	38	23
Tree section during loading	0	–	25	14
Stakes of forwarder	–	–	11	6
Chassis of forwarder	–	–	6	3
Motor-manual cutting	–	–	1	0
Others and unknown	–	–	1	1
Total	100	100	100	100

operating feller-bunchers may thus be expected to increase in whole-tree thinning, due to the absence of the protective slash cover.

Skidding of whole trees from partial cuttings always results in excessive damage to remaining trees, regardless of the degree of delimbing. In mixed northern hardwood stands in Vermont, 31% of the remaining trees were injured in shelterwood areas and 44% in thinned areas, when trees were cut by a wheel loader-mounted feller-buncher and hauled to landing by a grapple or cable skidder. Most injuries occurred when whole trees were skidded against remaining trees. The majority of stem wounds were scrapes, where the bark was removed and the wood exposed but not broken. Direct root injury from skidder wheels was next in importance (Kelley 1983).

The damage can be essentially reduced by bucking trees with crown into shorter sections and hauling the tree sections with load-carrying forwarders. The proportion of damaged trees in Finland was measured in 24 thinned stands, where the trees had been cut motor-manually and hauled as tree sections with forwarders to the landing. During summertime operations, 1.8–3.3% of the remaining trees in Scots pine stands were damaged, and 4.8–8.4% in Norway spruce stands. Soil freezing and snow cover considerably improved the result during winter operations (Sirén 1986). The causes of damage are presented in Table 9.9.

A satisfactory thinning result during whole-tree logging requires careful timing and work organization. Slopes, soft soil, very narrow strip roads, overlong timber, and an unsuitable logging season have a great bearing on the amount of damage. However, despite the absence of a protective cover of slash, damage can be kept at an acceptable level through training by influencing worker attitudes, selection of equipment and methods, and planning of work.

9.3.2 Effect of Residue Removal on Pathogenic Fungi

A great number of fungi, some of them pathogenic, thrive in logging residue. Most fungi do not seem to have a direct influence in forestry, except that they decompose slash, thus increasing the rate of nutrient recycling. Few of the pathogenic fungi are known to spread from logging residue to surrounding trees. The most important of them is *Heterobasidion annosum* (*Fomes annosus*), which causes serious damage commonly known as root rot disease in coniferous forests in many parts of the world. Stumps of summer-felled trees play a central role in the life cycle of the fungus. The fungus spreads from tree to tree, generally by air, first to a stump cross-section of a freshly felled tree, then to roots, and via root connections further to neighboring healthy trees, both young and old.

Consequently, stump removal has been recommended as a feasible countermeasure to fight root rot disease in many countries, e.g., in Denmark (Yde-Andersen 1970) and Great Britain (Greig and Burdekin 1970). In a field test in Great Britain, only 46% of Scots pine seedlings were alive 11 years after planting in an infested regeneration area in control plots where the stump-root systems were left intact. In plots without stumps, 80% of the seedlings survived (Greig and Burdekin 1970). In a Swedish experiment, the stumps and roots were carefully removed from two heavily infected clear-cut areas of Norway spruce. After 25–28 years, the root rot disease had infected only 1–2% of planted spruce seedlings. On control plots, as many as 12–17% of the trees were infected. Disease risk correlated positively with the distance from infected stump-root systems (Stenlid 1987).

Fungi may exist in any part of the stump-root system. Hence, a basic requirement for stump removal is the extraction of as much of the wood material as possible, at least all roots more than 5 cm in diameter. Immediate removal of stump and root mass from the site is not necessary, as long as the woody material is left to dry with only a little earth around.

Honey fungus (*Armillaria mellea*) is another example of pathogenic fungi, of which the mycelia spread in a regeneration area from old stump-root systems to young seedlings. Although not as dangerous as the root rot fungus, combating the honey fungus by stump removal is also recommended in some cases.

9.3.3 Effect of Residue Removal on Insects

Logging residue is the habitat of many insects and other invertebrates. Some of these insects, such as *Hylobius abietis* (large pine weevil), and several species of bark beetles, may cause serious damage to living trees and unbarked timber.

Branches and stem parts under 6–7 cm in diameter are usually too small and thin-barked for pest beetles, but many of them live and breed in thicker stem parts and stumps of conifers. *Ips, Orthotomicus, Pityogenes,* and *Tomicus* are common northern European insects often found in residual tree tops, while *Hylastes, Hylobius, Pissodes, Tetropium, Tomicus,* and *Trypodendron* infest stumps (Löyttyniemi 1974).

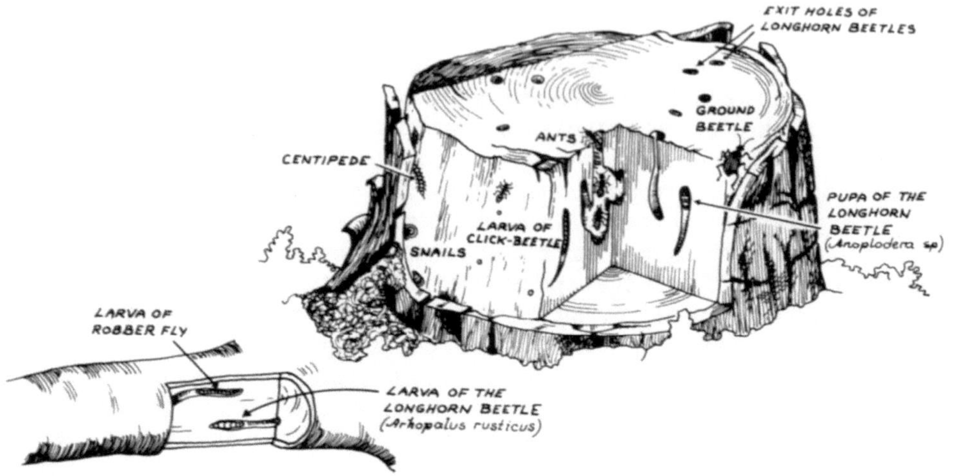

Fig. 9.10. A 10-year-old *Pinus sylvestris* stump, colonized by invertebrates (Ehnström 1984) (courtesy of the Swedish University of Agricultural Sciences)

The large pine weevil breeds almost exclusively in coniferous stumps in managed forests. According to a Swedish survey, a coniferous stump-root system contains an average of 300 larvae, 40% of which live in the stump proper and the remaining 60% in the roots (Skogsstyrelsen 1978). The adults attack young pine and spruce seedlings and may cause serious damage to a newly established plantation. Removal of stumps prevents a mass emergence of the pine weevil, which rarely breeds in the thinner roots left in the ground after stump extraction. If the seedlings are planted after stump removal, the damage can be avoided. The hazard from the black turpentine beetle in the southern United States could be correspondingly reduced by the removal of central pine stumps and their taproots (Coughran and Koch 1974).

Bark beetles infest stumps only during the years immediately after cutting, and subsequently leave behind a habitat for a variety of other animals and plants. At first the communities in stumps change frequently, but later the inhabitants may remain for several decades (Fig. 9.10). In addition to beetles, one finds robber, wood, and crane flies. The larvae of some flies are predators feeding on beetle larvae. When one species moves on, other species enter the existing tunnels and exit holes. Various species of ants are often among the first to colonize stumps. During later successive stages of decomposition, the stumps will also become a habitat for species which do not live off the wood but take shelter there to protect themselves from extreme drought or cold. Among these are terrestrial mollusks, which are extremely susceptible to desiccation. Several types of centipedes are almost exclusively found in old rotten stumps. A large variety of springtails also live in stumps (Ehnström 1984).

Large-scale stump extraction might thus reduce the populations of pest beetles by destroying their breeding grounds, but it would simultaneously weaken the conditions of existence of many other invertebrates. Vertebrates, e.g. the woodpecker, that feed on animals in wood would suffer accordingly.

Although modern logging practices under favorable market conditions generally offer insects a relatively poor breeding ground, residue recovery seems to further improve the hygiene of the forests. The long-term consequences of residue recovery for the populations of pest insects are not known. In the past, however, continued litter collection in Central Europe depleted the nutrient stores of forest soil from large areas, which resulted in increased insect damage to needles. Although residue harvesting is not directly comparable with continued litter collection, it may be assumed that recovery of logging slash may also increase the risk of needle damage on poor, dry pine sites (Mälkönen et al. 1985).

9.3.4 Effect of Residue Removal on Environment

The environmental impact which accompanies harvesting activities may take a wide variety of forms, depending on geology, topography, climate, etc. Destruction of the soil surface structure, increased sedimentation and disruption of sensitive limnological habitats, nutrient decline through erosion and biomass removal, and landslides due to a reduced contribution to soil strength from tree roots are only some examples (Grey and Jacobs 1985).

The environmental consequences of forest operations may affect the soil, the climate, and water relations in and outside the forests. The forest landscape and its suitability for multiple use and recreational activities are also affected. Here, the option between clear-cutting and thinning is generally much more important than the level of mechanization or the intensity of biomass recovery.

Adverse harvesting effects are often a result of the breaking or compaction of fragile soils, which may in turn result in *water contamination*. Reentering clear-cut areas to salvage living residue and ground slash may cause additional soil disturbance, because the protective slash cover is removed. Nevertheless, slash recovery obviously lowers the quality of water far less than operations such as soil preparation or prescribed burning. The consequences of stump removal, on the other hand, are more or less similar to those of site preparation.

The amount of rainfall reaching the soil surface in clear-cut areas of southern pine is affected only slightly by slash removal. Regardless of how much of the growing stock is cut, no more than about 5% of the surface absorption in the southern part of the United States is affected. Whether slash is left on the site or removed does not seem important in this respect, but recovery of slash leaves fewer nutrients to contribute to leaching. If nitrogen pollution of water is a problem, harvesting of slash from clear-cuts may reduce it (Ursic 1978).

Soil exposure, which is inconsequential on some sites, may be disastrous on others. On *erosion-prone sites*, the surface of the mineral soil needs protection against the impact of raindrops and other disturbances. Usually the litter layer on the forest floor is sufficient, but some of this may be scraped away during logging. Biomass residue can serve as a substitute mantle protecting the soil. The need usually occurs on steep slopes where disturbance is difficult to avoid and the potential for erosion is great. Consequently, part or all of the residue on such areas should be left to protect the soil (Ruth and Harris 1975).

Rosén (1984) studied the effects of clear-cutting at three small watersheds in Sweden. Runoff increased by nearly 100%. The annual *leaching loss* of nitrogen in a natural stand was less than 1 kg/ha, but 6 kg/ha after clear-cutting. When slash was removed, the leaching was somewhat smaller, i.e., 4.5 kg N/ha/a. Part of the nitrogen was thus derived from slash. When timber was prepared with logging machines, residue was concentrated in piles covering about 10–15% of the clear-cut area, and the concentration of nitrogen increased drastically in the soil water under piles (Rosén 1986).

Intensified recovery of forest residue favorably affects aesthetic and *recreational considerations*, especially if the foreground areas and landing sites along the travel routes are also kept clean. In fact, large whole-tree chippers were first developed in the United States in the beginning of the 1970s to keep parks and other recreational areas around population centers clear of dead trees and logging slash. The utilization of whole-tree chips for industrial processing or energy was originally of secondary importance.

In the short term, any clear-cutting defaces *the forest landscape*. Removal of logging slash cannot significantly change a depressing scenery. However, because walking becomes easier, it can make the area more accessible for recreation. It can also help a young plantation to achieve a natural appearance more quickly and

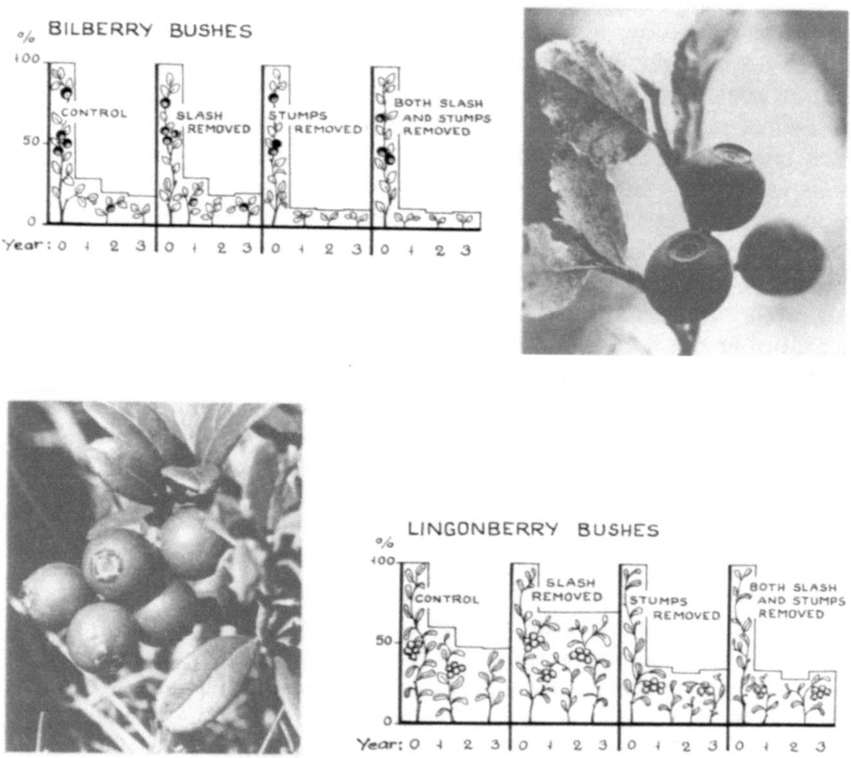

Fig. 9.11. The effect of slash and stump removal on the survival of bilberry and lingonberry after clear-cutting in Sweden (Kardell 1984) (courtesy of the Swedish University of Agricultural Sciences)

reduce fire hazards. Slash harvesting is therefore generally recognized as positive, but stump harvesting in the initial phase as negative.

Picking berries and mushrooms is a popular outdoor activity in the boreal forest zone. The extent of berry bushes is greatly reduced after final cutting. In a newly cut clearing, the biomass of lingonberry is reduced to two-thirds and that of blueberry or bilberry to only one-tenth of such biomass found in a natural forest. The situation is stabilized after 1 year, and then a slow recovery begins. Both species are assisted by slash removal and hurt by stump removal (Fig. 9.11). In the latter case, a great deal of the berry stems disappears along with the uprooting of stumps. On the other hand, raspberry bushes thrive after stump removal, but their growth is retarded by slash recovery (Kardell 1984).

Fresh logging residue, especially unlimbed hardwood tops, is winter forage for certain *wildlife* species such as deer and hare. On clear-cutting sites slash offers protection for birds. It may also offer protection for seedlings against cattle and big game. On the other hand, slash and stump removal accelerates the growth of grasses and herbs, thus benefiting many species. Taken all together, the intensity of biomass recovery does not seem to be of great importance for wildlife.

9.4 Returning Nutrients to Forest Soil

When harvesting biomass from sensitive soils, at least the main part of the foliage should be left on site. However, separating the foliage from the crown is often impractical and may make whole-tree logging methods unprofitable. An alternate solution, therefore, might be to return nutrient-rich parts of the biomass to the forest. Two possibilities are discussed here: either the foliage is separated from other components of the biomass at a terminal or point of utilization and returned to the forest (Sect. 9.4.1), or the biomass is used as fuel and the ash is returned after combustion (Sect. 9.4.2).

9.4.1 Returning Foliage

Logging residue after mechanized thinning tends to be concentrated on and alongside the strip roads. In clear-cutting areas, logging machines may collect a large part of the slash into piles.

The distribution of slash in connection with four Swedish thinning methods is estimated theoretically in Table 9.10. The distance between the strip roads is 25 m and the strip road width is 4 m. The following methods are compared: 1 = harvesting delimbed pulpwood with a forwarder-mounted grapple harvester; 2 = harvesting whole-tree chips using motor-manual felling and a terrain chipper equipped with a flail-type foliage separator; 3 = harvesting tree sections with motor-manual felling and a forwarder equipped with a grapple saw; 4 = harvesting tree sections with a small feller-buncher and a forwarder equipped with a grapple saw (Jonsson 1986a).

A thick cover of residue on the strip road protects the ground from machines, but uneven distribution is harmful from an ecological point of view. The nutrients

Table 9.10. Location of logging slash in a Norway spruce stand after four alternative thinning methods (Jonsson 1986a). For explanations of the logging methods, see the text

Logging method	Amount of logging slash, percent of original			
	On the strip roads	Alongside the strip roads	Between the strip roads	Average
1	310	110	30	100
2	60	80	10	40
3	30	30	30	30
4	20	20	20	20

are not equally available to all trees, and leaching from slash heaps contaminates water. These problems might be avoided if all the crown material is harvested, and the foliage later returned and spread evenly on the logging site.

Nitrogen availability is strongly affected by the C/N ratio in forest litter and soils. While intensive recovery of stem mass and large branches accelerates the total removal of nitrogen, excessive woody residue may, on the other hand, in fact cause a net reduction in nitrogen availability, because the nitrogen content of the stem is low compared with twigs and foliage. As removal of the woody parts of the residue increases, the C/N ratio of the remaining residue decreases, thus making nitrogen more easily available for trees. If only nutrient-rich foliage and small twigs are left on site after thinning, or alternately returned later, more nitrogen would probably be readily available for the trees (Jonsson 1986a).

Decomposition of organic matter in on-ground residue is accelerated after clear-cutting, because exposure of the forest floor to direct sunlight raises the soil temperature, and removal of trees reduces transpiration, thus increasing soil moisture. In clear-cutting areas, the nutrients are unfortunately released during a developmental phase when there are no trees to take them up, and therefore much of the released supply is lost through leaching. If, instead, foliage were to be returned in a later phase when the canopy of the young stand is closing, trees could recover most of the nutrients and leaching would be reduced radically. The overall nutrient economy of a stand would consequently improve.

Using theoretical nutrient balance calculations, Axelsson (1986) estimated the effect of foliage return on the total stem volume yield in three Swedish thinning stands after application of a motor-manual tree-section method. The growth increase after foliage return was highest in a Norway spruce stand with a large needle mass. According to the author, the field experiments show an essentially lower effect for double doses of foliage than the theoretical calculations in Table 9.11 would indicate.

If foliage could be separated from whole trees, tree sections, or whole-tree chips at a wood handling terminal, the return of needles and small twigs would be technically possible. Spreading could probably be carried out by forwarder-mounted equipment basically similar to that used for forest fertilization. However, the net profit from needle return should be weighed against the value of the foliage mass as fodder, biochemical feedstock, or fuel. The economic feasibility

Table 9.11. Effect of foliage return on total yields of stem wood after thinning using a motor-manual tree-section method, calculated for the whole rotation period in Sweden (Axelsson 1986)

Amount of foliage returned	Change in yield (m^3/ha)		
	Picea abies on a good site	Mixed stand on a medium site	*Pinus sylvestris* on a poor site
None	−27	−10	−5
All	−7	−4	−2
Double	+13	+3	+1

of returning foliage to forests seems only marginal, when compared with other utilization alternatives and the high costs in relation to the relatively low nutrient and solid contents of the load. Moreover, foliage could probably be used more profitably as a soil improvement agent in agriculture and horticulture than in forestry.

9.4.2 Returning Ash from Forest Biomass

Large amounts of residual forest biomass and bark from the forest industries are used for the production of energy. With the exception of nitrogen, almost all of the nutrients in the biomass remain in the ash after burning. The ash is generally deposited in landfill areas, resulting in extra transport and dumping costs. At the landfill areas, rainwater passes through the mounds of ash, whereby nutrients and heavy metals may leach into surface and ground-water. This is a poor use of scarse landfill space, for which operating costs are escalating rapidly.

Power plant ash can be used for various purposes. European countries utilize 40–50% of the ash from coal-fired power plants for cement production, concrete mixing, construction of road base, mineral filler in asphaltic mix, and ceramics. In the United States, less than 20% of all ash is utilized (Fulekar and Dave 1986). However, none of these applications take advantage of the versatile and natural combinations of nutrients which are typical of ash produced from biomass.

The nutrient flow from forest soils caused by intensive biomass recovery could be reduced by recycling the ash. The beneficial effect of ash in crop husbandry is well known throughout the world, but very few research results on recycling ash in the forest have been published. The majority of available information concerns peatland forests in Finland and Sweden (Bramryd and Fransman 1985; Silfverberg and Huikari 1985). Little research has been done on the suitability of ash for mineral soils in forestry and agriculture. Lack of information and workable recycling techniques prevent practical applications, but many producer plants, municipalities, forest industry enterprises, forest owners, and farmers are examining possibilites for utilizing the ash from wood, bark, and other clean biofuels.

9.4.2.1 Impurities in Power Plant Ash

In a furnace, part of the ash travels on air currents with the flue gases as *fly ash*, which is recovered from the combustion air with various types of emission control equipment. Some of the fly ash penetrates through the separation devices and escapes into the outside air. The type of separator affects the properties of the fly ash, since the finest fractions, transported with combustion gases, are important for the chemical composition of the ash residue. Part of the ash is too heavy to be carried in the flue gases and remains on the grate. The proportions of *grate or bottom ash* and fly ash are dependent on furnace type.

Ash is removed either manually, by stoker or scraper conveyors, by flushing with water, or pneumatically from the furnace. High temperatures and residual charcoal in unquenched ash result in danger of fire, unless conveyors and the ash silo are airtight. Handling dry ash also leads to problems with dust.

For these reasons, *ash is extinguished* by spraying the conveyor with water or by pouring the ash into a water tank. However, ash used for soil improvement should be fairly dry. Slight dampening to a moisture content of 10–20% binds the dust to some extent but does not make spreading more difficult. Wet sludge ash, on the other hand, is unfit for spreading. Nutrients also leach out of wet sludge ash.

Residual charcoal in power plant ash not only results in loss of energy, but also makes utilization of the ash more difficult and uneconomical. Transportation and spreading costs, calculated for pure ash, increase with the rise in the amount of charcoal. High-carbon ash may inactivate herbicides applied to agricultural land, particularly on no-till sites (Estes 1987).

Table 9.12 shows average proportions of soluble pure ash, insoluble sand, and unburnt charcoal in ash from 11 chip-fired, 8 bark-fired, and 6 milled peat-fired power plants in Finland. Naturally, the variations between plants and even within a single plant are wide, depending on fuel purity and combustion effectiveness.

The mineral compounds of the ash melt when exposed to high temperature and form a viscous fluid and ultimately a slag when cooled. *Slag formation* lowers

Table 9.12. Proportion of soluble pure ash, insoluble sand, and unburnt charcoal in the ash residue from power plants using wood, bark, and peat fuels in Finland (Hakkila and Kalaja 1983)

Type of ash	Fuel	Proportion (%)			
		Charcoal	Sand	Pure ash	Total
Bottom ash	Chips	8	25	67	100
Fly ash	Chips	22	19	59	100
Bottom ash	Bark	17	26	57	100
Fly ash	Bark	32	14	54	100
Bottom ash	Peat	28	62	10	100
Fly ash	Peat	7	56	37	100

Table 9.13. Content of heavy metals in wood ash residue in Swedish tests (Miljöeffekter ... 1983)

Element	Content (mg/kg)	
	Bottom ash	Fly ash
As	0.2–3	1–60
Cd	0.4–0.7	6–40
Co	0–7	3–200
Cr	–60	40–250
Cu	15–300	200
Hg	0–0.4	0–1
Mn	2500–5500	6000–9000
Ni	40–250	20–100
Pb	15–60	40–1000
Se		5–15
V	10–120	20–30
Zn	15–1000	40–700

the value of ash as a fertilizer and causes handling and spreading problems. The degree to which a fuel will form slag appears to be dependent primarily on the amount of minerals present and secondarily on the composition of the mineral matter. Fusion temperature decreases and slag formation increases with increased sand content up to about 40% SiO_2 (Osman and Goss 1983). Because wood fuels are relatively poor in silicon, however, slag formation seldom hampers the recycling of wood and bark ash. The silicon content of straw and peat ash is much higher.

The content of heavy metals varies within a wide range. The variation may be larger within a fuel type than between various biomass fuels. This is partly due to differences in combustion and gas cleaning efficiency in power plants. Metals often tend to become vaporized in the burning process and then condense around small particles in the combustion gases. The more effective the filtration of the gases, the larger is the proportion of heavy metals left in the fly ash (Naturvårdsverket 1983).

If ash is to be recycled, excessive contents of heavy metals in the ash residue are unacceptable. Generally, ash from pure wood-based fuels is harmless as far as heavy metals are concerned, if, for instance, 5 t of ash is spread per hectare. The heavy metal content rises if coal, oil, or impregnated wood is used for additional fuel. Nevertheless, the ash produced under the S. D. Warren bioash program from daily combustion of 2000 t of chips and 200 t of coal in Maine has a metal content far below the allowable maximum limits of the standards established by the Maine Department of Environmental Protection (Resource ... 1986). Table 9.13 shows the variation in heavy metal content of bottom and fly ash from several wood-fired plants in Sweden.

To solve the problems of ash recycling in an economical way, producers should pay increased attention to the following points:

1. Wood and bark ash should be kept separate from fossil fuel ash. Use of oil as a minor additional fuel is acceptable.

2. Wood and bark must be burnt as completely as possible. The charcoal content can be reduced by careful adjustment of the combustion process.
3. Ash cooling to reduce dust whirling must be done at the power plant with as little water as possible. When the moisture content exceeds 30%, ash spreading will be hampered, especially in temperatures below freezing.
4. Storage at the plant must be well arranged to allow easy loading and to avoid dust hazards. The storage capacity should be sufficient for a flexible delivery system.
5. In the long run, granulation technology should be developed for large ash producers.

9.4.2.2 Nutrient Content of Power Plant Ash

The nutrient content of power plant ash is much lower than that of pure wood and bark ash, because ash residue in practice contains a considerable amount of impurities. Moreover, *nutrients may leach* out in the process of handling and storing forest biomass. Floating, water storage, and wet-drum debarking of timber result in significant leaching of nutrients, which in turn results in a lowered nutrient content of the bark ash in forest industry power plants. The order in which nutrients leach out of biomass is potassium, phosphorus, and nitrogen. Because potassium compounds volatilize at about 1300 °C, some losses also occur during the combustion phase at the power plants. The most nutrient-rich ash is produced by chip-fired plants, as the fuel is only subject to a slight degree of nutrient leaching.

Unimproved wood and bark ash from power plants contains phosphorus and especially potassium in significantly lesser proportions than pure ash from wood or bark (Table 9.14). If charcoal and insoluble sand are removed, the nutrient contents increase, but remain even then below those in the ash of freshly felled forest biomass (Sect. 3.3.3).

Table 9.14. The proportions of some primary elements in dry ash residues from chip-, bark-, and peat-fired heating plants in Finland (Hakkila and Kalaja 1983)

Fuel	Treatment of ash	Proportion (%)			
		P	K	Ca	Mg
Chips	Unimproved	1.7	5.3	17.2	2.7
Chips	Charcoal removed	2.1	6.6	21.2	3.3
Chips	Sand and charcoal removed	2.7	8.6	27.7	4.3
Bark	Unimproved	1.1	3.7	19.6	1.7
Bark	Charcoal removed	1.5	5.1	27.3	2.4
Bark	Sand and charcoal removed	1.9	6.9	36.3	3.1
Peat	Unimproved	1.1	0.2	6.2	1.0
Peat	Charcoal removed	1.2	0.2	6.6	1.1
Peat	Sand and charcoal removed	3.2	0.5	16.7	2.6

For the above reasons, power plant ash contains lesser amounts of trace elements than pure ash. This is shown by the following figures, which are based on samples collected from six bark-fired power plants in Finland (Alestalo 1983, pers. commun.). The variation is wide and erratic, which complicates value estimation and utilization:

Element	Content in ash residue from bark-fired power plants
	(g/kg)
P	4–16
K	16–66
Ca	134–288
Mg	14–34
Mn	8–31
Fe	11–22
	(mg/kg)
Co	27–88
Cu	102–780
Zn	300–6000
V	20–470
B	67–358

9.4.2.3 Power Plant Ash as a Soil Improvement Agent

Ash affects soil both as a *primary* and as a *trace element nutrient*, and as a *neutralizing or liming agent*. Like manure, ash contains a rich variety of trace elements where nutrients, with the exception of nitrogen, are present in almost natural proportions.

Wood and bark ash is a low-cost alternative to commercial lime, which can be used to compensate for the acidity of farm lands, particularly since it contains a sufficient portion of magnesium in addition to calcium. For example, in Finland the important Ca/Mg relationship is a satisfactory 4.5–8.5 in tree ash, while the proportion in bark ash is slightly too high at 10–15 (Hakkila 1986). When ash is used for liming, secondary benefits in the form of phosphorus, potassium, and trace elements also accrue.

If ash is used as a neutralizing agent, the correct amount to apply depends on its calcium and magnesium contents. Liming materials are compared using their *equivalent neutralizing value* (ENV), which is the effectiveness in percent of a particular material relative to standard limestone with an ENV of 100. The ENV of a liming material is a function of the sum of the calcium and magnesium contents expressed as calcium carbonate and the fineness of the particles. In a laboratory study in New York State the ENV of boiler ash from a paper mill was 47. Therefore, double the amount of boiler ash was required, compared with standard limestone (Naylor and Schmidt 1985). According to a Finnish experiment, the liming effect of various kinds of ash, calculated both chemically and by the rise in the pH value of the soil, was as follows (Saarela 1982):

Liming agent	Proportional liming effect	
	Chemically	By pH value
Limestone powder	100	100
Wood ash	115	106
Bark ash	64	57
Peat ash	13	12
Straw ash	32	...
Coal ash	14	12

Crops favoring "sweet" soils, i.e., soils with a high pH value, benefit most from wood ash fertilization. Field experience shows significant yield increases of grass and hay crops, corn, alfalfa, and vegetables. Corn and alfalfa grew better in a greenhouse and laboratory study in Vermont when wood ash was used instead of lime as a neutralizing agent. Overliming injuries occurred when excess limestone was used, but generally not when excess ash was applied. The presence of boron and phosphorus in the ash was apparently great enough to mitigate possible detrimental effects of extra amendment, because the overdose injury in conjunction with liming could be partly overcome with boron and phosphorus additions. It was concluded that wood ash can be used safely as a liming material within the guidelines currently recommended for limestone.

In addition to its liming abilities, wood ash is a significant source of phosphorus, potassium, and other nutrients. If ash is used for *phosphorus and potassium fertilization*, the amount to be spread is determined primarily by the phosphorus content rather than the potassium content, since the former is lower compared with the needs of the trees. If, for example, the recommended amount of phosphorus is 50 kg/ha and the phosphorus content of ash is only 1%, then ash application should be 5 t/ha.

In 1983, a large-scale ash and sludge recycling program was started under a municipal regulatory program in the state of Maine by an S. D. Warren kraft pulp mill, burning 2000 t of chips and 200 t of coal daily. The "bioash" program is regulated by the Maine Department of Environmental Protection. Each site goes through a review process before it is approved for ash fertilization. A site management plan is developed, the soils are reviewed, and any potential environmental problems are identified. To prevent runoff, ash is not spread on snow-covered or frozen ground. A monitoring program over several years has revealed no impact on ground or surface waters (Resource ... 1986). According to experiments, a buildup of heavy metals in agricultural soils should not be a problem. Local landowners have accepted both ash and sludge as agricultural supplements, and more than 75% of all the ash and sludge generated is being beneficially utilized for improvement of farm lands. Recycling results in significant savings in landfill costs to the kraft mill, and in equally significant fertilization and soil improvement savings to the farms (Pepin and Coleman 1984). According to Bramryd and Fransman (1985), ash fertilization results in an increase in the concentration of heavy metals in soil and biomass, but as great or even greater increase is caused by application of commercial fertilizers.

Ash utilization in forestry has been studied in the Nordic countries ever since 1918, when the first ash fertilization experiment was conducted in Sweden (Malmström 1952). All of the old experiments are on peatlands. There are no published research results on mineral soils, where nitrogen is generally the limiting nutrient, and ash is considered to be of little value.

Most of the Finnish ash fertilization experiments are carried out on rather poor, originally treeless, drained bogs, where the growth reaction of Scots pine to an ash application of 5 t/ha has generally been very favorable. In many cases the increase in annual growth is 4 m^3/ha, and in one case as much as 8 m^3/ha. In the best stands, with a natural nitrogen content of peat as much as 2.5%, the growth is as fast after ash fertilization as in the fertile *Oxalis-Myrtillus* type forests in southern Finland. On the other hand, the growth response to ash fertilization is slow on sites where the total nitrogen content in the 20-cm surface layer of peat is only 1.0% (Silfverberg and Huikari 1985).

Some of the nutrients found in ash are directly utilized by trees. Some, on the other hand, will make *microbe activity* more vigorous. As a result of this activity, organic nitrogen reserves not normally available for trees are mobilized in a usable form. Nitrogen-rich peat containing 1.5–2.5% pure nitrogen by dry mass becomes an excellent growing substrate after ash fertilization.

The effect of ash on nitrogen-rich, drained peat lands is more permanent than that of artificial fertilizers. The following observations may be made in favor of ash as compared to commercial fertilizers (Reinikainen 1980):

1. Surface vegetation changes more permanently;
2. Peat decomposes more quickly and deeper;
3. Tree growth is less subject to disturbances; no symptoms of deficiencies can be seen;
4. The duration of growth improvement lengthens from 5–8 to 30–40 years.

Koch (1985b) proposed additions of nitrogen and phosphorus to biomass ash and application of such fortified ash to the slash-free forest floor. Followed by chopping with a drum roller, the treatment would likely afford better site preparation than the windrow and burn system which is commonly used in the southern parts of the United States. According to Koch, ash produced in a furnace, supplemented with needed nutrients, and evenly distributed on a relatively undisturbed site is much more valuable than incomplete ash from slash and soil burned in windrows.

9.4.2.4 Ash Recycling Techniques

Artificial fertilizers are spread in forests in fairly small amounts per hectare, making aerial spreading a viable method. Ash is used in amounts 10 or 20 times that of fertilizers per unit area, for example, 5 t/ha. The labor costs per hectare of aerial spreading of ash are therefore prohibitively high. Consequently, ash must be spread from the ground.

Since it is not economically feasible to put ash in sacks, transporting it into the forest for manual spreading is difficult. *Manual spreading* itself is expensive,

Fig. 9.12. Returning ash to farm land with conventional manure-spreading equipment

and health hazards are not inconsiderable. Small particles floating in the air may enter the respiratory system and from there other organs. Impurities on the skin cause irritation and may even penetrate through the skin. The high alkalinity of calcium oxide (CaO_2), the major component of wood and bark ash, causes irritation of the skin, mucous membranes, and respiratory passages. Silicate (SiO_2) in quartz form causes silicosis or gritty lung disease. Harmful trace elements may cause skin allergies and, in the long run, symptoms of poisoning (Juntunen 1982).

Therefore, manual spreading can be considered only under exceptional circumstances, such as establishing small-scale test plots for yield studies. Tractor-based spreading equipment seems to be the only feasible alternative for large operations under most forest conditions.

Even in *tractor spreading*, the driver is subject to serious dust hazards if the cabin is poorly insulated, but with careful measures the cabin can be protected against dust. It is essential that the cabin is tight and pressurized, and that incoming air is closely filtered. The dust problem may be minimized by increasing the moisture content of the ash (Juntunen 1982).

In agriculture, the work can be done with a tractor that drops the ash straight behind, making the width of the spreading swath of little significance. Smaller amounts can be spread with manure or lime spreading equipment of the farm (Fig. 9.12). In larger farm operations, ash can be topdressed or incorporated into the soil with contractor-owned heavier *lime spreading equipment* with only minor modifications. Ash loading from storage piles to a spreader can be done with a front-end loader, or with a boom loader with the timber grapple replaced by an ash scoop.

In the forest, spreading is done from logging strip roads used for thinning. The ash must be spread evenly to a distance of 12–15 m on either side of the ma-

Fig. 9.13. Returning ash to forest land with forwarder-mounted spreading equipment

chine. As the trees, particularly those with a long crown, further hamper spreading, the demands on the effectiveness of the machine are more critical in forests. Because the machine must be well fitted for forest terrain, *forwarder-mounted equipment* is the most natural solution for large operations (Fig. 9.13).

For ash spreading, a $10-12\,m^3$ ash bin is mounted in the load space of the forwarder. A screw, belt, or scraper conveyor located on the bottom of the bin transfers the ash to one or two slings that spread the ash on both sides of the strip road. Horizontal screws rotating above the conveyor, or a vibrator, may be used to prevent bridging of the ash. In cold weather, moist ash can be prevented from freezing by heating the sides of the bin with exhaust gas from the tractor. The timber grapple of the loader of the forwarder is replaced by an ash scoop. The output per effective hour of forwarder-mounted equipment in Finnish experiments was $22\,m^3$ or 10 t of dry ash, when the hauling distance from the roadside to the stand was 300 m (Kalaja 1986).

Ash is brought to the work site in trucks, for instance in a fuel chip truck and is transferred later from the ground into the tractor bin. The storage areas are often uneven, resulting in ash loss. Stones may also be picked up along with the ash from the ground, which may cause breakage of the conveyors and slings of the spreader. These problems could probably be avoided with *a system based on an on-off-road vehicle* which would be able to handle both highway transport and spreading in the forest. When traveling on highways the vehicle could haul an ash trailer equipped with a pneumatic or other unloading mechanism for moving the ash to the spreader on the forest landing. The prime mover, however, cannot match the performance of a forwarder in forest terrain.

Some of the technical problems with spreading might be avoided by *pelletizing ash*. Pellets would reduce the moisture content to an economical level, make storage easier, alleviate problems with dust, make the spreading result more uniform, and reduce clogging of the machinery. However, pellet production is not economically feasible for forestry applications at the present because it is technically difficult and requires heavy investment costs.

Scientific and Common Names of Tree Species

Conifers

Abies alba Mill. silver fir
A. amabilis Dougl. ex Forbes, Pasific silver fir
A. balsamea (L.) Mill. balsam fir
A. concolor Lindl. ex Hildebr. white fir
A. firma Sieb. & Zucc. Japanese fir, momi
A. grandis (Dougl. ex D. Don) Lindl. grand fir
A. lasiocarpa (Hook.) Nutt. subalpine fir
A. mariesii Mast. Maries's fir, todomatsu
A. nephrolepis Max. Sibirian white fir, Bunbi-namu
A. procera Rehd. noble fir
A. sachalinensis Mast. Sachalin fir
A. sibirica Lebed. Sibirian fir
A. veitchii Lind. Veitch's silver fir
Araucaria angustifolia O. Kuntze, Parañа pine
Cedrus deodara Loud. deodar cedar
C. libani A. Rich. cedar of Lebanon
Chamaecyparis lawsoniana (A. Murr.) Parl. Port-Orford-cedar
C. nootkatensis (D. Don) Spach, Alaska-cedar
C. obtusa Sieb. & Zucc. Hinoki-cedar
Cryptomeria japonica (L.f.) D. Don, Japanese cedar, sugi
Cunninghamia spp. China fir
Juniperus communis L. common juniper
Larix decidua Mill. European larch
L. laricina (Du Roi) K. Koch, tamarack, eastern larch
L. leptolepis Gord. Japanese larch, karamatsu
L. occidentalis Nutt. western larch
L. sibirica Lebed. Sibirian larch
Metasequoia glyptostroboides Hu & Chang, dawn redwood
Picea abies (L.) Karst. Norway spruce
P. engelmannii Parry ex Engelm. Engelmann spruce
P. glauca (Moench) Voss, white spruce
P. glehnii Mast. Sachalin spruce
P. jezoensis Carr. Yeso, Hondo spruce
P. mariana (Mill.) B.S.P. black spruce
P. pungens Engelm. blue spruce
P. rubens Sarg. red spruce
P. sitchensis (Bong.) Carr. Sitka spruce
Pinus banksiana Lamb. jack pine

Pinus contorta Dougl. lodgepole pine
P. contorta var. *contorta* Dougl. shore pine
P. contorta var. *latifolia* Engelm. lodgepole pine
P. densiflora Sieb. & Zucc. Japanese red pine, akamatsu
P. echinata Mill. shortleaf pine
P. elliottii Engelm. slash pine
P. maritima Mill. Corsican pine
P. montana Mill. mountain pine
P. monticola Dougl. ex D. Don, western white pine
P. nigra Arn. black pine
P. nigra var. *maritima* Mel. Corsican pine
P. nigra var. *nigra* Austrian pine
P. palustris Mill. longleaf pine
P. peuce Griseb. Macedonian pine
P. pinaster Ait. maritime pine
P. ponderosa Dougl. ex Laws. ponderosa pine
P. pumila Regel, Japanese stone pine
P. radiata D. Don, Monterey pine, radiata pine
P. resinosa Ait. red pine
P. strobus L. eastern white pine
P. sylvestris L. Scots (Scotch) pine
P. taeda L. loblolly pine
P. thunbergii Parl. Japanese black pine, kuromatsu
Pseudotsuga menziesii (Mirb.) Franco, Douglas-fir
P. menziesii var. *glauca* (Mayr) Franco, blue Douglas-fir
Sequoia sempervirens (D. Don) Endl. redwood
Thuja occidentalis L. eastern white-cedar
T. plicata Donn ex D. Don, western red-cedar
Thujopsis dolabrata Sieb. & Zucc. Japanese thuja
T. dolabrata var. *hondai* Makino Japanese thuja
Tsuga canadensis (L.) Carr. eastern hemlock
T. heterophylla (Raf.) Sarg. western hemlock
T. mertensiana (Bong.) Carr. mountain hemlock
T. sieboldii Carr. southern Japanese hemlock

Hardwoods

Acacia spp.
Acer negundo L. boxelder
A. pennsylvanicum L. stribed maple
A. platanoides L. Norway maple
A. rubrum L. red maple
A. saccharum Marsh. sugar maple
Alnus glutinosa (L.) Gaertn. black alder, European alder
A. incana (L.) Moench., grey alder
A. rubra Bong. red alder

Alnus rugosa (DuRoi) Spreng. speckled alder
Betula alleghaniensis Britt. yellow birch
B. ermanii Cham. Erman's birch
B. maximoviczìana Reg. Japanese birch
B. papyrifera Marsh. paper birch, white birch
B. pendula Roth, silver birch, European white birch
B. platyphylla Suk. var. *japonica* Hara, Japanese birch
B. populifolia Marsh. gray birch
B. pubescens Ehrh. hairy birch
Camellia japonica L. Japan rose
Carpinus betulus L. European hornbeam
Carya ovata (Mill.) K. Koch, shagbark hickory
C. tomentosa Nutt. mockernutt hickory
Castanopsis cuspidata Schott, Chinese chestnut
Celtis occidentalis L. hackberry
Cornus florida L. flowering dogwood
Corylus avellana L. filbert
Eucalyptus spp.
Fagus crenata Bl. Japanese beech
F. grandifolia Ehr. American beech
F. sylvatica L. European beech
Fraxinus americana L. white ash
F. excelsior L. European ash
F. nigra Marsh. black ash
F. pennsylvanica Marsh. green ash
Hevea braziliensis (Willd. ex A. Juss) Mull-Arg. Brazilian rubber tree
Juglans regia L. English walnut
Leucaena leucocephala de Wit, ipil-ipil
Liquidambar styraciflua L. sweetgum, redgum
Liriodendron tulipifera L. yellow-poplar, tulip tree
Machilus thunbergii Sieb. & Zucc. Chinese bandoline wood
Magnolia virginiana L. sweetbay
Nyssa aquatica L. tupelo-gum, water tupelo
N. sylvatica Marsh. blackgum, black tupelo
Populus deltoides Marsh. eastern cottonwood
P. grandidentata Michx. bigtooth aspen
P. nigra L. '*Italica*', Lombardy poplar
P. tremula L. European aspen
P. tremuloides Minchx. quaking aspen
P. trichocarpa T. & G. black cottonwood, balsam cottonwood
Prunus pensylvanica L. pin cherry
P. serotina Ehrh. black cherry
P. virginiana L. choke cherry
Quercus alba L. white oak
Q. borealis var. *maxima* (Marsh.) Ashe, northern red oak
Q. coccinea Muench. scarlet oak
Q. falcata Michx. southern red oak

Quercus falcata var. *falcata* southern red oak
Q. falcata var. *pagodaefolia* Michx. southern red oak, cherrybark oak
Q. glauca Thunb. Japanese oak, green oak
Q. laurifolia Michx. laurel oak
Q. marilandica Muench. blackjack oak
Q. mongolica Fisch. Japanese oak
Q. myrsinaefolia Blume shirakashi (Jap.)
Q. nigra L. water oak
Q. petrea Liebl. durmast oak, European oak
Q. phillyraeoides A. Gray, ubamegashi (Jap.)
Q. prinus L. chestnut oak
Q. robur L. English oak, European white oak
Q. rubra L. northern red oak
Q. shumardii Buckl. Shumard oak
Q. stellata Wang. post oak
Q. velutina Lam. black oak
Robinia pseudoacacia L. black locust
Salix caprea L. sallow willow
S. gracilis Anderss. slender willow
Sorbus aucuparia L. European mountain ash
Tilia americana L. basswood
Trema orientalis (L.) Bl. anabiong
Ulmus alata Michx. winged elm
U. americana L. American elm

Equivalents and Conversion Factors

Abbreviations

a = year
bd ft = board foot
atm = atmosphere
Btu = British thermal unit
°C = degree of Celcius scale
cal = calorie
cm = centimeter
cmin = centiminute = 0.01 min
d = day
F = degree of Fahrenheit scale
ft = foot
g = gram
gal = gallon
h = hour
ha = hectare
hp = horsepower
hph = (metric) horsepower hour
Hz = hertz (basic unit of frequency)
in = inch
J = joule (basic SI unit of energy) = 1 Nm
J/s = joule/second = Nm/s
kcal = kilocalorie
kg = kilogram
kgf = kilogram-force
kgf·m = kilogram-force·meter
kJ = kilojoule
km = kilometer
kN = kilonewton
kNm = kilonewtonmeter
kp = kilopond
kPa = kilopascal
kW = kilowatt
kWh = kilowatt-hour
l = liter = 1 dm^3
lb = pound
lbf = pound-force
lbf·ft = pound-force·foot
m = meter

mg = milligram
mi = mile
min = minute
MJ = megajoule
mm = millimeter
MPa = megapascal
MW = megawatt
N = newton (basic SI unit of force) = 1 kg m/s^2
Nm = newtonmeter = 1 J
Nm/s = newtonmeter/second = 1 J/s = 1 W
oz = ounce
Pa = pascal = 1 N/m^2
ppm = parts per million
rpm = revolutions per minute
s = second
sh tn = short ton (US) = 2000 lb
t = tonne (metric ton) = 1000 kg
W = watt (basic SI unit of power) = 1 Nm/s = 1 J/s
yd = yard
μg = microgram = 10^{-6} g
μm = micrometer = 10^{-6} m

Metric Prefixes with Exponent Values

Prefix	Symbol	Exponent
tera	T	10^{12}
giga	G	10^{9}
mega	M	10^{6}
kilo	k	10^{3}
hecto	h	10^{2}
deca	da	10^{1}
		10^{0}
deci	d	10^{-1}
centi	c	10^{-2}
milli	m	10^{-3}
micro	μ	10^{-6}
nano	n	10^{-9}

Metric Conversion Factors

Length

1 mm	=	0.0393701 in	1 in	=	25.400 mm
1 cm	=	0.393701 in	1 in	=	2.5400 cm
1 m	=	3.28084 ft	1 ft	=	0.3048 m
1 m	=	1.09361 yd	1 yd	=	0.9144 m
1 km	=	0.621371 mi	1 mi	=	1.60934 km

Area

1 mm^2	=	0.0015500 in^2	1 in^2	=	645.16 mm^2
1 cm^2	=	0.15500 in^2	1 in^2	=	6.4516 cm^2
1 m^2	=	10.763915 ft^2	1 ft^2	=	0.0929030 m^2
1 m^2	=	1.19599 yd^2	1 yd^2	=	0.836127 m^2
1 ha	=	2.47105 acres	1 acre	=	0.404686 ha
1 km^2	=	0.386102 mi^2	1 mi^2	=	2.58999 km^2

Volume

1 cm^3	=	0.061024 in^3	1 in^3	=	16.38706 cm^3
1 m^3	=	35.3147 ft^3	1 ft^3	=	0.0283168 m^3
1 m^3	=	1.30795 yd^3	1 yd^3	=	0.764555 m^3
1 m^3	=	0.353147 cunit	1 cunit	=	2.83168 m^3
1 m^3 stacked	=	0.275896 cord stacked	1 cord stacked	=	3.62456 m^3 stacked $\approx 2.5\,m^3$ solid
1 m^3	=	423.7759 bd ft	1 bd ft	=	0.002359738 m^3
1 l = dm^3	=	0.264172 gal (US)	1 gal (US)	=	3.785412 l

Mass

1 g	=	0.0352740 oz	1 oz	=	28.3495 g
1 kg	=	2.20462 lb	1 lb	=	0.453592 kg
1 t (tonne)	=	1.10231 sh tn (US)	1 sh tn (US)	=	0.907185 t (tonne)
1 t (tonne)	=	0.9841 long ton (UK)	1 long ton (UK)	=	1.0161 t (tonne)

Density

1 kg/m^3	=	0.062428 lb/ft^3	1 lb/ft^3	=	16.0185 kg/m^3

Force

1 N	=	0.10197 kgf (kp)	1 kgf	=	9.8066 N

Moment of force

1 Nm	=	0.10197 kgf × m	1 kgf × m	=	9.8066 Nm

Energy

1 J	= 0.94782×10^{-3} Btu	1 Btu	= 1.05506×10^{3} J
1 J	= 0.10197 kgf × m	1 kgf × m	= 9.8066 J
1 J	= 0.23885×10^{-3} kcal	1 kcal	= 4.1868×10^{3} J
1 J	= 0.27778×10^{-6} kWh	1 kWh	= 3.6×10^{6} J
1 J	= 0.37767×10^{-6} metric hph	1 hph	= 2.6478×10^{6} J

Power

1 W	= 1 Nm/s	1 Nm/s	= 1 W
1 W	= 0.10197 kgf × m/s	1 kgf × m/s	= 9.8066 W
1 W	= 0.23885×10^{-3} kcal/s	1 kcal/s	= 4.1868×10^{3} W
1 W	= 0.85985 kcal/h	1 kcal/h	= 1.163 W
1 W	= 1.3596×10^{-3} (metric) hp	1 hp (metric)	= 735.50 W
1 W	= 1.3410×10^{-3} (foot-pound system) hp	1 hp (foot-pound system)	= 745.70 W
1 W	= 3.4121 Btu/h	1 Btu/h	= 0.29307 W
1 kW	= 1.3596 metric hp	1 hp (metric)	= 0.7355 kW

Pressure

1 Pa	= 10×10^{-6} bar	1 bar	= 100×10^{3} Pa
1 Pa	= 10.197×10^{-6} kgf/cm^2	1 kgf/cm^2	= 98.066×10^{3} Pa
1 Pa	= 0.14504×10^{3} lbf/in^2 (psi)	1 lbf/in^2	= 6.8948×10^{3} Pa
1 Pa	= 9.8692×10^{-6} atm	1 atm	= 101.32×10^{3} Pa

Temperature

1 °C	= 0.555 Fahrenheit − 32	1 Fahrenheit	= 1.8 °C + 32

Miscellaneous forestry units

1 m^2/ha	= 4.35600 ft^2/acre	1 ft^2/acre	= 0.229568 m^2/ha
1 m^3/ha	= 14.2913 ft^3/acre	1 ft^3/acre	= 0.069972 m^3/ha
1 m^3/ha	= 0.1429 cunits/acre	1 cunit/acre	= 6.9973 m^3/ha
1 m^3 (stacked)/ha	= 0.111651 cord (stacked)/acre	1 cord (stacked)/acre	= 8.95647 m^3 (stacked)/ha ≈ 6.2 m^3 (solid)/ha
1 t/ha	= 0.4461 t/acre	1 t/acre	= 2.2417 t/ha

References

Abol P J 1984 Lage und Methoden der primären Bearbeitung des Rundholzes in den osteuropäischen Ländern. TIM/ECE/WP.1/R.50/Add. 1, 27 pp

Ageeva M I, Sarelainen A N, Levkina G M 1978 (Papermaking properties of pulp from young small-diameter wood obtained from forest thinning.) Fiz-Khim Issled Drev ee Kompleks Ispol'z 1978:106–120

Ager B H, Nilsson N-E, von Segebaden G 1964 Beskrivning av vissa skogstekniskt betydelsefulla bestånds- och trädegenskaper samt terrängförhållanden. (Description of some for logging operations important characteristics of forest stands, trees and terrain). Studia For Suec 20, 159 pp

Ahlgren T, Larsson M, Nordén B, Olsson K 1982a Motormanuell träddelshyggning i slutavverkning. (Motor-manual tree-section logging in final cutting). Skogsarbeten Mimeogr, 21 pp

Ahlgren T, Larsson M, Nordén B, Richardsson G 1982b Motormanuell och delmekaniserad drivning av träddelar i slutavverkning. (Motor-manual and partly mechanized harvesting of tree sections in cutting.) Skogsarbeten Mimeogr, 30 pp

Ahonen M, Mäkelä M 1972 Juurakoiden irrottaminen maasta pyöräkuormaajalla. (Extraction of stump-root systems by wheel loaders.) Fol For 140, 21 pp

Albrektson A 1980 Tallens biomassa. Storlek – utveckling – uppskattningsmetoder. (Biomass of Scotspine (Pinus sylvestris L.) Amount – development – methods of mensuration.) Sver Lantbruksuniv Inst Skogsskötsel Rapp NR 2-1980, 189 pp

Aldred A H, Alemdag I S 1988 Guidelines for forest biomass inventory. Can For Serv Inf Rep PI-X-77, 134 pp

Alemdag I S 1980 Manual of data collection and processing for the development of forest biomass relationships. Can For Serv Inf Rep PI-X-4

Alemdag I S 1982a Biomass of the merchantable and unmerchantable portions of the stem. Can For Serv Inf Rep PI-X-20

Alemdag I S 1982b Methods of estimating forest biomass from stand volumes: a case study with Ontario jack pine. Pulp Pap Can 83(9):41–43

Alemdag I S, Stiell W M 1982 Spacing and age effects on biomass production in red pine plantations. For Chron 58:220–224

Alestalo A, Hentola Y 1966 Sulfaattisellua havupuiden kuorellisista latvuksista, oksista ja kannoista. (Sulphate pulp from unbarked softwood tops, branches and stumps.) Pap Puu 48:737–742

Alexandersson H 1984 Chipping of tree residuals and small-dimension hardwood in southern Sweden. Skogsarbeten Results 4/1984, 4 pp

Alexandersson H 1985 Järnvägstransport av träddelar. (Tree-section transportation by railroad.) Skogsarbeten Redogör 9/1985, 23 pp

Alexandersson H, Carlsson T, Rådström L 1984 Transportalternativ för energiråvaror. (Haulage alternatives for energy raw materials.) Skogsarbeten Redogör 5/1984, 40 pp

Alexandersson H, Jonsson T 1985 Sönderdelning av trädrester med SIM-krossen. (Reducing residual wood with SIM crusher.) Skogsarbeten Results 6/1985, 4 pp

Alhojärvi P 1981 Hakkeen keinokuivaus. (Artificial drying of wood chips.) Työtehoseuran Metsätied 343, 6 pp

Altman J 1980 Whole-tree chipper survey – December 1975 and September 1979. APA Tech Rel 80-R-5, 3 pp

Altman J 1982 Pluses and minuses of using whole tree chips. APA Tech Rel 82-R-71, 2 pp

Andersson F 1985 Research in Scandinavia. Site productivity research. IEA/FE Project CPC-10, Rep 2:11–14

Andersson S-O 1984 Logging residue is good for forest. In: Andersson B, Falk S (eds) Forest energy in Sweden. Swed Univ Agr Sci:98–99

Anonym 1983 Débroussaillement et valorisation énergétiques. Le Bois National, Octobre 1983:28–29

Apgar W P, Dickey H C, Young H E 1977 Estimated digestibility of conifer muka fed to sheep. Univ Maine Orono, Res Life Sci 24(2), 6 pp
Armson K A 1976 Forest management in Ontario. Ontario Min Nat Resources Rep, 7 pp
Arola R A 1973 Compression debarked chips from a whole-tree chipper. USDA For Serv Res Note NC-147, 4 pp
Arola R A 1974 Comparisons of wet and ovendry analyses of compression debarking tests on wood chips. USDA For Serv Res Note NC-178, 4 pp
Arola R A 1976 Wood fuels – how do they stack up? In: Energy and the wood products industry. For Prod Res Soc Proc P-76-14:34–45
Arola R A 1983a Chunking of small trees – an alternative to chipping. For Prod Res Soc, FPRS Proc 7340:117–127
Arola R A 1983b Steep slope felling and bunching – The Menzi Muck. In: Timber harvesting in the central Rockies, Publ XCM-87, Colorado State Univ: 155–161
Arola R A 1984 Chunkwood – technologies for comminuting and utilizing forest biomass as an energy source. Mimeogr, 20 pp
Arola R A 1985 Comminuting trees from thinnings into chunkwood – a new alternative. Sver Lantbruksuniv Inst Skogsteknik Rapp Upps 42, pp 19
Arola R A, Erickson J R 1973 Compression debarking of wood chips. USDA For Serv Res Pap NC-85, 11 pp
Arola R A, Hillstrom W A 1972 Compression debarking of branchwood chips from Finland. USDA For Serv Res Note NC-143, 4 pp
Arola R A, Miyata E S, Sturos J A, Steinlib H M 1981 Felling and bunching small timber on steep slopes. U S For Serv Res Pap NC-203, 12 pp
Arola R A, Radcliffe R C, Winsauer S A 1985 Chunking bundled stems. For Prod J 35(4):40–42
Arola R A, Radcliffe R C, Winsauer S A, Matson E D 1982 A new machine for producing chunkwood. U S For Serv Res Pap NC-211, 8 pp
Arola R A, Sturos J A 1982 A portable chunking machine. U S For Serv Res Note WO-11, 4 pp
Arola R A, Sturos J A, Mattson J A 1976 Research in quality improvement of whole-tree chips. Tappi 59(7):66–70
Arola R A , Winsauer S A, Radcliffe R C, Smith M R 1983 Chunkwood production: a new concept. For Prod J 33(7/8):43–51
Art H W, Marks P L 1971 A summary table of biomass and net annual primary production in forest ecosystems of the world. In: Forest biomass studies. Univ Maine Coll Life Sci Agr 1–32
Artiano A 1986 Chip slicer – state of the art. Tappi 69(2):38–40
Assarsson A 1969 Some reactions during chip storage and how to control them. Pulp Pap Mag Can 70(18):74–79
Atmer B, Thörnqvist T 1982 Fiberegenskaper i gran (Picea abies Karst.) och tall (Pinus silvestris L.). (The properties of tracheids in spruce (Picea abies Karst.) and pine (Pinus silvestris L.) Swed Univ Agr Sci Dep For Prod Rep 134, 78 pp
Auchter R J 1972 Pulping without barking increases fiber yield. For Prod Lab Appr Tech Article, 2 pp
Aurell R, Hartler N 1965 Kraft pulping of pine. II. The influence of the charge of alkali on the yield, carbohydrate composition, and properties of the pulp. Svensk Papperstidn 68:97–102
Axelsson B 1986 Biomassaåterföring – biologiska aspekter (Returning biomass – biologic aspects). Mimeogr, 11 pp
Baadsgaard-Jensen J 1985 Micro-fractionation of whole-tree components. IEA-FEA-JAC 14 Rep, Dan Inst For Tech Copenhagen, 86 pp
Back E L, Lundqvist K E 1976 The use of bark for wood based panels – a critical evaluation. Symposium on extending the use of wood residues, Bucarest. TIM/SEM.5/R.26, 17 pp
Bailey G R 1969 An evaluation of the line intersect method of assessing logging residues. Can Dep Fish For For Prod Lab Inf Rep VP-X-23, 41 pp
Bailey G R 1970 A simplified method of sampling logging residue. For Chron 46:289–294
Bailey I W 1957 The structure of tracheids in relation to the movement of liquids, suspensions, and undissolved gases. In: Thimann K V (ed) The physiology of forest trees. Ronald New York, 71–82
Baker A J 1983a Wood fuel properties and fuel products from woods. In: Fuelwood management and utilization seminar: Proceedings, Michigan State Univ, East Lansing MI, 14–25
Baker A J 1983b Charcoal. In: Davis R C (ed) Encyclopedia of American forest and conservation history. Vol 1. Macmillan New York, 73–77

Baker A J, Millett M A, Satter L D 1975 Wood and wood-based residues in animal feeds. In: Turbak A F (ed) Cellulose technology research, A C S Symp Series 10, Am Chem Soc Washington D C, 75–105

Baker A J, Mohaupt A A, Spino D F 1973 Evaluating wood pulps as feedstuff for ruminants and substrate for Aspergillus fumigatus. J Animal Sci 37:179–182

Barclay H J, Pang P C, Pollard D F W 1986 Above ground biomass distribution within trees and stands in thinned and fertilized Douglas-fir. Can J For Res 16:438–442

Barefoot A C, Hitchings R G, Ellwood E L 1964 Wood characteristics and kraft paper properties of four selected loblolly pines. I. Effect of fiber morphology under identical cooking conditions. Tappi 47:343–356

Barefoot A C, Hitchings R G, Wilson E G, Kellison R C 1972 Related aspects of the morphology of loblolly pine and papermaking. In: Proc symp on the effect of growth acceleration of the properties of wood. U S For Serv For Prod Lab Madison 1971; K1–K33

Barton G M 1981 Foliage. In: Goldstein J S (ed) Organic chemicals from biomass, CRC Press, Boca Raton, 249–280

Barton G M, MacDonald B F 1978 A new look at foliage chemicals. Tappi 61(1):45–48

Baskerville G L 1965 Dry-matter production in immature balsam fir stands. For Sci Monogr 9, 42 pp

Baskerville G L 1966 Dry-matter production in immature balsam fir stands: roots, lesser vegetation, and total stand. For Sci 12:49–53

Beebe D E Jr 1980 Mechanical/thermal cogeneration in the forest products industry. In: Energy generation and cogeneration from wood. For Prod Res Soc Proc P-80-26, 90–93

Beemtsma J B 1979 Sampling Norway spruce needle by age class. In: Frayer W E (ed) Forest resource inventories II. Colorado State Univ, 792–798

Belin L 1984 Allergic alveolitis-fungal spores. In: Gislerud O, Heding N (eds) Storing, drying and internal handling of fuel chips. Dan Inst For Tech Copenhagen: 12–15

Bell G W 1985 A South African equipment manufacturer's approach to developing smallwood logging systems. In: For Prod Res Internat-Achievements and the future, Pretoria, 7:14.5, 6 pp

Benemann J R 1978 Biofuels: a survey. EPRI ER-746-SR, Palo Alto Ca, 130 pp

Bengtsson P, Myhrman D 1982 Gripsågar – en översikt. (Grapple saws – a review.) Skogsarbeten Results 27, 4 pp

Benson R E, Schlieter J A 1979 Residue characteristics in the northern Rocky Mountains. In: Harvesting and utilization opportunities for forest residues in the northern Rocky Mountains. U S For Serv Gen Tech Rep INT-110:33–43

Benson R E, Schlieter J A 1980a Volume and weight characteristics of a typical Douglas-fir/western larch stand, Coram experimental forest. Montana. U S For Serv Gen Tech Tep INT-92, 28 pp

Benson R E, Schlieter J A 1980b Logging residues in principal forest types of the northern Rocky Mountains. U S For Serv Res Pap INT-260, 14 pp

Bérard J A 1981 Energy biomass harvesting being developed in Canada. Am Soc Agr Eng Pap 81-1598, 6 pp

Bergman Ö 1972 Deterioration and protection of wood chips in outside chip storage. FAO/Norad symposium on production, handling and transport of wood chips, TWC/72/26, 33 pp

Bergman Ö, Nilsson T 1979 An experiment on outdoor storage of whole-tree chips. Sver Lantbruksuniv Inst Virkeslära Rapp R109, 21 pp

Berlyn R, Hutchinson J, Gooding R 1979 A method for upgrading chips from full trees. Pulp Pap Can 80(8):56–60

Berlyn R W, Simpson R B 1988 Upgrading wood chips: the Paprifer process. Tappi 71(3):99–106

Besley L 1960 Relationships between wood fibre properties and paper quality. Pulp Pap Mag Can 61(7):136–146

Besley L 1962 Quality control could start in the woods. Tappi 45(2):20A–32A

Bhat K M, Kärkkäinen M 1981a Variation in structure and selected properties of Finnish birch wood II. Observations of the anatomy of root wood. Silva Fenn 15:180–189

Bhat K M, Kärkkäinen M 1981b Variation in structure and selected properties of Finnish birch wood III. Proportion of wood elements in stems and branches in Betula pendula Roth. Silva Fenn 15:1–9

Bhat K M, Kärkkäinen M 1981c Variation in structure and selected properties of Finnish birch wood IV. Fibre and vessel length in branches, stems, and roots. Silva Fenn 15:10–17

Biblis EJ, Lee W-C 1978 Properties of particleboard from loblolly pine rootwood and above ground stumpwood. For Prod J 28(11):55–56
Binnis W O 1975 Whole-tree utilization – consequences for soil and environment – experience and opinion in Britain. Elmia 75, Konf Sk 2:18–25
Bjerkelund T C 1985 Central processing and associated transportation techniques: examples of state-of-the-art from eastern North America. In: Mitchell C P, Nilsson P O, Zsuffa L (eds) Research in forestry for energy. Sver Lantbruksuniv Inst Skogsteknik Rapp Upps 49:397–407
Björkroth G 1983 Hyggesavfall och barrträdens tillväxt. (Logging slash and the growth of conifers.) Skogsfakta 1:15–17
Blankenhorn P R, Bowersox T W, Kuklewski K M, Stimely G L, Murphey W K 1985 Comparison of selected fuel and chemical content values for seven Populus hybrid clones. Wood Fiber 17:148–158
Blankenhorn P R, Bowersox T W, Murphey W K 1978 Recoverable energy from the forests. An energy balance sheet. Tappi 61(4):57–60
Blasinska J 1977 (Study of the utilization of pine branchwood for pulp manufacture.) Przeglad Papierniczy 33(9):346–350
Blomqvist L 1978 An industrial plant for chipping stump and root wood for chemical pulping. In: McMillin C W (ed) Complete tree utilization of southern pine. For Prod Res Soc Madison: 294–297
Böhm W 1979 Methods of studying root systems. Springer Berlin Heidelberg New York, 188 pp
Boutelje J B 1966 On the anatomical structure, moisture content, density, shrinkage, and resin content of the wood in and around knots in Swedish pine (Pinus silvestris L.) and in Swedish spruce (Picea abies Karst.). Svensk Papperstidn 69:1–10
Boyle J R, Phillips J J, Ek A R 1973 "Whole tree" harvesting: nutrient budget evaluation. J For 71:760–762
Braekke F H 1986 Distribution on yield of biomass from young Pinus sylvestris and Picea abies stands on drained and fertilized peatland. Scand J For Res 1:49–66
Bramryd T, Fransman B 1985 Utvärdering av äldre gödlings- och kalkningsförsök med torv- och vedaska i Finland och Sverige. (Evaluation of old fertilization and liming experiments with peat and wood ash in Finland and Sweden.) Naturvårdsverket Rapp Snv Pm 1991, 60 pp
Brantseg A 1962 Irrigation and twig-covering experiments in Scots pine forests. Comm Inst For Fenn 55.7, 14 pp
Brattberg T 1977 Rensning av klena träd i barktrumma. (Cleaning small trees in a debarking drum.) Projekt helträdsutnyttjande Stockholm, PHU Rapp 33, 10 pp
Breloh P, Kirch C 1984 Waldschäden in der Bundesrepublik Deutschland. Allg Forstz 39:1265–1267
Brown J K 1974 Handbook for inventoring downed woody material. U S For Serv Gen Tech Rep INT-16, 24 pp
Browning B L 1967a Methods of wood chemistry. I. Interscience New York, 384 pp
Browning B L 1967b Methods of wood chemistry. II. Interscience New York, 498 pp
Bublitz W J 1976 Kraft pulping characteristics of chips from wood residues. Tappi 59(5):68–71
Bundesministerium für Ernährung, Landwirtschaft und Forsten 1985 Waldschadenserhebung 1985
Burger H 1939a Der Kronenaufbau gleichaltriger Nadelholzbestände. Mitt Schweiz Anst Forst Versuchwes 21:5–57
Burger H 1939b Holz, Blattmenge und Zuwachs. IV Mitteilung. Ein 80 jähriger Buchenbestand. Mitt Schweiz Anst Forst Versuchwes 21:307–348
Bylin C V 1982 Estimating dbh from stump diameter for 15 southern species. U S For Serv Res Note SO-286, 3 pp
Calvé L R, Shields J A, Gravel M C 1986 Maximizing aspen poplar residue utilization for waferboard production. For Prod J 36(5):39–45
Cannel M G R 1984 Woody biomass of forest stands. For Ecol Managem 8 (3/4):299–312
Carlsson T, Larsson M 1981 Komprimering av träddelar med Hydrovågs lastkomprimator. (Compacting tree sections with the Hydrovåg load compactor.) Skogsarbeten Mimeogr, 21 pp
Carlsson T, Rådström L 1984 Compaction of the load presses down the costs. In: Andersson B, Falk S (eds) Forest energy in Sweden. Swed Univ Agr Sci:43–45
Carpenter E 1980 Wood fuel potential from harvested areas in the eastern United States. U S For Serv Resour Bull NC-51, 14 pp

Carre J 1980 Technological value of whole-tree chips for the particleboard industry. Holz Roh- Werkst 38:337–344
Castaños L J, Arola R A 1982 Harvesting wood for energy in North America. In: Smith W R (ed) Energy from forest biomass. Academic Press, New York, 25–38
Castrén M, Lukkala K 1986 Hakelämmityksen ja työperäisten altisteiden vaikutus maaviljelijöiden hengityselinoireisiin. (The effects of heating with wood chips and occupational predispositions on farmers' respiratory symptoms.) Työtehoseuran Julk 282, 64 pp
Chaney W R, Kozlowski T T 1969 Seasonal and diurnal changes in water balance of fruits, cones, and leaves of forest trees. Can J Bot 47:1407–1417
Chase A J, Hyland F, Young H E 1971 Puckerbrush pulping studies. Life Sci Agr Exp Sta Univ Maine Tech Bull 49, 64 pp
Chase A J, Hyland F, Young H E 1973 The commercial use of puckerbrush pulp. Life Sci Agr Exp Sta Univ Maine Tech Bull 65, 54 pp
Chase A J, Young H E 1976a Corrugating pulp from puckerbrush. Res Life Sci Univ Maine Orono 23(10):1–9
Chase A J, Young H E 1976b The potential of softwood thinnings and standing dead softwood as a source of wood pulp. Life Sci Agr Exp Sta Univ Maine Tech Bull 82, 25 pp
Chen T-y 1979a Studies on the integrated utilization of China fir from thinning operation. B. Experiments on the manufacture of particleboard by making use of full-tree. Q J Chin For 12(1):65–86
Chen T-y 1979b Experiment on the manufacture of particleboard from full-tree of India-charcoal Trema. Bull Exp For NCHU 1/1979:42–59
Chen T-y, Paulitsch M 1974 Inhaltsstoffe von Nadeln, Rinde und Holz der Fichte und Kiefer und ihr Einfluss auf die Eigenschaften daraus hergestellter Spanplatten. Holz Roh-Werkst 32:397–401
Choong E T, Abdullah G, Kowalczuk J 1976 Mineral content in bark and wood of selected delta hardwoods. Louisiana State Univ Wood Util Notes 29, 4 pp
Christensen E 1986 Centralized processing. XVIII IUFRO World Congress, Ljubljana, 9 pp
Christopherson N S 1983a Mechanized harvesting of hardwood logging residue. Am Soc Agr Eng Pap 83-1602, 24 pp
Christopherson N S 1983b Dimensional characteristics of hardwood top and limb residue. U S For Serv Res Note NC-300, 5 pp
Christopherson N S, Barnett P L 1983 Harvesting hardwood top and limb residue for energy. The seventh international FPRS industrial wood energy forum '83, Proceedings 1:100–108
Clark A III 1978 Total tree and its utilization in the southern United States. For Prod J 28(10):47–52
Clark A III 1979 Suggested procedures for measuring tree biomass and reporting tree prediction equations. In: Frayer W E (ed) Forest resources inventories II. Colorado State Univ, 615–628
Clark A III, Taras M A 1976 Comparison of above ground biomasses of the four major southern pines. For Prod J 27(10):25–29
Clark J d'A 1969 Fibrillation, free water, and fiber bonding. Tappi 52:335–340
Colombo P, Gorbetta D, Pirotta A, Ruffini G 1964 The influence of thickness of chips on pulp properties in kraft cooking. Svensk Papperstidn 67:505–511
Conner A L 1978 Utilization of whole-tree chips in an unbleached kraft linerboard mill. In: McMillin C W (ed) Complete tree utilization of southern pine. For Prod Res Soc Madison, 268–273
Côté W A Jr, Kutscha N P, Simson B W, Timell T E 1968 Studies on compression wood. VI. Distribution of polysaccharides in the cell wall of tracheids from compression wood. Tappi 51:33–40
Coughran S J, Koch P 1974 Prototype trials of horizontal root shear and harvester of southern pines with taproot intact. Am Soc Agr Eng Pap 74-1536, 18 pp
Crossley T L 1938 Paper from fruit tree prunings and forest slash. Pulp Pap Mag Can 39:568–570
Crossley T L 1952 Multi-fibrous hardwood pulp. Pulp Pap Mag Can 53(7):126–130
Crow T R, Schlaegel B E 1988 A guide to using regression equations for estimating tree biomass. North J Applied For 5(1):15–22
Cummins W H 1941 A method for sampling the foliage of a silver maple tree. J For 39:382–384
Cunia T 1979a On tree biomass tables and regression: some statistical comments. In: Frayer W E (ed) Forest resources inventories II. Colorado State Univ, 629–642
Cunia T 1979b On sampling trees for biomass tables construction: some statistical comments. In: Frayer W E (ed) Forest resources inventories II. Colorado State Univ, 643–664
Curran C E 1938 Relation of growth characteristics of southern pine to its use in pulping. J For 36:576–581

Czereyski K, Galinska I, Robel H 1965 Rationalization of stump extraction. FAO/ECE/LOG/158, 29 pp
Dadswell H E, Wardrop A B 1955 The structure and properties of tension wood. Holzforschung 9:97–104
Dadswell H E, Wardrop A B 1959 Growing trees with wood properties desirable for paper manufacture. Appita 12:129–136
Dadswell H E, Watson A J, Nicholls J W P 1959 What are the wood properties required by the paper industry in the trees of the future? Tappi 42:521–526
Danielsson B-O, Marks J, Säll H-O 1977 Komprimering av klena träd och träddelar. (Compressing small trees and tree components.) Skogshögsk Inst Skogstekn Rapp Upps 119, 47 pp
Danielsson B-O, Nilsson P O 1977 Översiktlig beräkning av tillgänglig mängd skogsavfall. (Survey of availability of forest residues.) PHU Rapp 46, 36 pp
Daugavietis M O 1985 Basic principles of technology of whole tree biomass processing. In: Proceedings of IUFRO International Interdivisional Conference on Thinning, Moscow-Riga, 127–136
Davis W E 1971 Utilization of stumpwood saves 100000 cords/year of pulpwood. Pulp Pap Mag Can 72(12):111–112
de Montalembert M R, Clément J 1983 Fuelwood supplies in the developing countries. FAO For Pap 42, 125 pp
Débroussaillement et destruction de la végétation indésirable en forêt – inventaires des matériels et des techniques 1968. Cah Cent Tech Bois 78, 44 pp
Demin K A, Vasjukov V A, Kuznecova A I 1970 (Preservation of sapling stands in mechanized extraction of mature stumps.) Lesn Holzjajstvo 11/1970:45–48
Dickens C H, Edén P, Hogfors S, Persson J (eds) 1983 Wood fuel. Heat from the forest. Domänverket SSR Stockholm, 73 pp
Dickerhoof H E 1976 Particleboard production, markets, and raw materials in the United States. For Prod J 26 (10):16–20
Dickson R E, Larson P R 1977 Muka from Populus leaves: a high-energy feed supplement for livestock. Tappi For Biol Wood Chem Conf Madison, 95–99
Dickson R E, Larson P R, Isebrands J G 1976 Individual tree variation and age patterns in cell wall chemistry of rapidly grown cottonwood. Applied Polym Symp 28:1231–1238
Dietz P 1984 Zentrale Holzaufarbeitung als Betriebssystem der Fürstlich Fürstenbergischen Forstverwaltung. Allg Forstz 39:649–651
Dillner B 1972 Outside stored chips as raw material for pulping. FAO/Norad symposium on production, handling and transport of wood chips, Hurdal. TWC/72/7a, 14 pp
Dinwoodie J M 1961 Tracheid and fibre length in timber – a review of literature. Forestry 34:125–144
Dinwoodie J M 1965 The relationship between fibre morphology and paper properties: a review of the literature. Tappi 40:440–447
Dinwoodie J M 1966 The influence of anatomical and chemical characteristics of softwood fibers on the properties of sulfate pulp. Tappi 49:57–67
Drolet J C 1973 Stump volume of swamp-grown black spruce. For Managem Inst Inf Rep FMR-X-57, 9 pp
Dunfield J D, Silversides C R, Tsay T-B 1974 Engineering characteristics of trees – dimensions and weights of trees east of the Canadian Rocky Mountains. For Man Inst Inf Rep FMR-X-60, 77 pp
Dyck W J, Messina M G, Hunter I R 1986 Predicting the nutritional consequences of forest harvesting on site productivity. In: Ågren G I (ed) Predicting consequences of intensive forest harvesting on long-term productivity. Swed Univ Agr Sci Inst Ekol Miljövård Rapp 26:9–29
Dyer R F 1967 Fresh and dry weight, nutrient elements and pulping characteristics of northern white cedar. Agr Exp Sta Univ Maine Tech Bull 27, 40 pp
ECE/FAO 1986 European timber trends and prospects to the year 2000 and beyond. I. ECE/TIM 30, 323 pp
Edberg U, Engström L, Hartler N 1973 The influence of chip dimensions on chip bulk density. Svensk Papperstidn 76:529–533
Edlin H L 1965 The summerwood: springwood ratio in conifers. J Soc For Great Brit 38:91–112
Ehnström B 1984 Stumps and invertebrates – many species are threatened. In: Andersson B, Falk S (eds) Forest energy in Sweden. Swed Univ Agr Sci, 106–107
Elliott D R 1960 The transportation of pulpwood chips in pipelines. Pulp Pap Mag Can 61(5):170–175

Enčev E 1972 Untersuchung einiger physikalisch-mechanischer Eigenschaften der Äste von Nadelhölzern. Holztechnologie 13:232–236

Enroth A 1946 Kanto-flissorteraren och dess inverkan på cellulosans egenskaper. (Kanto chip classifier and its influence on the properties of the chemical pulp.) Papp Trävarutidskr Finland 28(11):172–181

Erickson J R 1968 Bark separation during chipping with a parallel knife chipper. U S For Serv Res Note NC-54, 3 pp

Erickson J R 1970 Bark removal after chipping – a progress report. Pulp Pap Mag Can 71(5):78–79

Erickson J R 1971 Bark-chip segregation: a key to whole-tree utilization. For Prod J 21(9):111–113

Erickson J R 1972a Field chipping: problems and solutions. FAO/Norad Symp on Production, Handling and Transport of Wood Chips, TWC/72/14a, 11 pp

Erickson J R 1972b The moisture content and specific gravity of the bark and wood of northern pulpwood species. U S For Serv Res Note NC-141, 3 pp

Erickson J R 1972c The status of methods for debarking wood chips. Tappi 55:1216–1220

Erickson J R 1974 Harvesting of forest residues. Reprint from AIChE Symp Series 71(146):27–29

Erickson J R 1976a Removing bark from southern hardwood whole-tree chips. For Prod J 26(2):45–48

Erickson J R 1976b Steaming chips facilitates bark removal. USDA For Serv Res Note NC-216, 4 pp

Erickson J R 1976c Exploratory trials with a spiral-head chipper to make hardwood "fingerling" chips for ring flakers. For Prod J 26(6):50–53

Esdale W J 1977 Nutritional release no. 4, project PT-021A. Stake Technology Ottawa, 1 p

Eskelinen A, Häggblom R, Peltonen J 1984 Osapuuna- ja metsähakkeenahankinnan kustannuskilpailukyky. Yritysmallitarkastelu. (A model for optimising the use of wood raw material) Metsäteho Rep 389, 31 pp

Eskilsson S 1969 Fiber properties in the spruce root system. Cellul Chem Technol 3:409–416

Eskilsson S 1972 Whole tree pulping. Part 1. Fibre properties. Svensk Papperstidn 75:397–402

Eskilsson S 1974 Whole tree pulping. Part 3. Pulp properties. Svensk Papperstidn 77:165–174

Eskilsson S, Hartler N 1973 Whole tree pulping. Part 2. Sulphate cooking. Svensk Papperstidn 76:63–70

Eskilsson S, Hartler N 1974 Whole tree utilization. Symp Int EUCEPA Madrid, 7, 24 pp

Estes G 1987 Agricultural use of wood ash: liming value, nutrient content and herbicide interactions. Research proposal. Typescr, 7 pp

European timber trends and prospects to the year 2000 and beyond 1986. ECE/TIM/30 Vol 1, 323 pp

Evelyn J 1729 Silva: or, a discourse of forest-trees, and the propagation of timber in His Majesty's dominions. The fifth edition. London, 363 pp

FAO 1982 World forest products demand and supply in 1990 and 2000. FAO For Pap 29, 346 pp

FAO 1986a Yearbook of forest products 1984. FAO For Ser 19, 400 pp

FAO 1986b Wood gas as engine fuel. FAO For Pap 72, 172 pp

Farrington A 1980 Wood and digester factors affecting kraft pulp quality and uniformity. Appita 34:40–46

Fayle D C F 1968 Radial growth in tree roots. Univ Toronto Tech Rep 9, 183 pp

Fegel A C 1941 Comparative anatomy and varying physical properties of trunk, branch, and root wood in certain northeastern trees. Bull N Y State Coll For Tech Publ 55, 22 p

Fellegi J, Janči J, Chovanec D 1974 Pulping and papermaking properties of softwood and hardwood branches. Symp Int EUCEPA Madrid 8, 11 pp

Fengel D, Grosser D 1975 Chemische Zusammensetzung von Nadel- und Laubhölzern. Holz Roh-Werkst 33:32–34

Filén H, Jantunen M, Salo K 1984 Kotimaisten polttoaineiden kaasutus. Osa I. Vastavirtakaasutus. (Gasification of indigenous fuels. Part I. Updraft gasification.) Tech Res Centre Finland Res Rep 250, 58 pp

Fiolek S 1961 Resistance stresses against the vertical force in extracting stumps to be used for industrial purposes. Transactions For Res Inst 218. Warsaw, 20 pp

Flinkman M, Thörnqvist T 1986 Lagring av buntade träddelar och hyggesrester. (Storage of bundled unlimbed pulpwood and logging residues.) Sver Lantbruksuniv Inst Virkeslära Rapp 180, 49 pp

Förslag till svensk standard SS 187106. (Proposal for a Swedish standard SS 187106.) 1986. SMS Remmiss 1438, 7 pp

Francis J K 1984 Yield and nutrient removal by whole-tree harvest of a young bottomland hardwood stand. U S For Serv Res Note SO-305, 4 pp
Fröding A 1987 Massaflismetoden – en studie av en ny metod för glenskogsgallring. (The pulp chip method – a study of a new method for first thinning.) Sver Lantbruksuniv Inst Skogstekn Res Upps 82, 51 pp
Fulekar M H, Dave J M 1986 Disposal of fly ash – an environmental problem. Int J Env Stud 26:191–215
Ganter K 1927 Streuversuchsflächen der badischen forstlichen Versuchsanstalt an der Universität Freiburg i. Br. Alg Forst-Jagdztg 103:353–358
Gärdenäs F 1986 Skogsbränslets inverkan på driftstörningar och pannförluster – två värmevärk. (The wood fuel's effect on breakdowns and energy losses – two heating plants.) Sver Lantbruksuniv Inst Virkeslära Rapp 178, 47 pp
Gardner J A F, Hillis W E 1962 The influence of extractives on the pulping of wood. In: Hillis W E (ed) Wood extractives and their significance to the pulp and paper industries. Academic Press New York, 367–403
Gardner R B 1977 Transporting wood chips by hydraulic pipeline. Reprint Proc Second PPI International Symp Rotterdam, 13 pp
Gardner R B, Schaffer E L, Erickson J R 1978 Converting forest residue to structural flakeboard – the fingerling concept. USDA For Serv Rep Pap INT-200, 31 pp
Gasslander J E, Mattsson J E, Sundberg U 1979 A pilot study on the energy requirements in forest logging. Royal Coll For Dept Oper Eff Rep 127, 33 pp
Geimer R L, Price E W 1978 Construction variables considered in fabrication of a structural flakeboard. In: Structural flakeboard from forest residues. USDA For Serv Gen Tech Pap WO-5:69–80
Gerry R W, Young H E 1977 A preliminary study of the value of conifer muka in a broiler ration. Univ Maine Orono, Res Life Sci. 24(1), 6 pp
Gerry R W, Young H E, Rowe R J 1979 A second look at conifer muka in broiler rations. Univ Maine Orono, Res Life Sci 26(4), 7 pp
Gilliusson R 1984 National research and development programmes in member countries of the IEA forestry energy agreement area "harvesting, on-site processing and transport". Swed Univ Agr Sci Bioenergy Progr Garpenberg, 47 pp
Gislerud O 1973 Heltreutnyttelse. III. Biomassetap under terrengtransport av tynningsvirke. (Whole-tree utilization. III. Biomass loss in off-road haulage of thinning wood.) NISK Skogstekn Avd, 10 pp
Gislerud O 1974 Heltreutnyttelse. II. Biomasse og biomasseegenskaper hos tynningsvirke av gran, furu, bjørk och or. (Whole tree utilization. II. Biomass and biomass properties of trees from thinnings of spruce, pine, birch, and alder.) Rapp Norsk Inst Skogforskn 6, 59 pp
Gislerud O 1982 Storing and drying of energy chips for small-scale chip heating. Mimeogr, 8 pp
Gislerud O 1983 Experimental drying and storing of fuel chips. IUFRO All-Division 5 Conf Madison, 5 pp
Gislerud O 1984 Small-scale storing and drying of fuelwood and fuel chips. In: Gislerud O, Heding N (eds) Storing, drying and internal handling of wood fuels. Dan Inst For Tech Copenhagen, 22–24
Gislerud O (ed) 1988 Production, storage and utilization of wood fuels. Proc IEA/BE Conf Task III/Activity 6 and 7. Sver Lantbruksuniv Inst Skogsteknik Upps Res 134, 247 pp
Gislerud O, Wilhelmsen G 1978 Fyring med ved, flis og trekull. (Heating with wood, chips and charcoal.) Norsk Inst Skogforsk Rapp 5/78, 82 pp
Gladstone W T, Barefoot A C Jr, Zobel B J 1970 Kraft pulping of earlywood and latewood from loblolly pine. For Prod J 20(2):17–24
Gleaton E N, Saydah L 1956 Fibre dimensions and papermaking properties of the various portions of a tree. Tappi 39:157A–158A
Glesinger E 1977 Forests and the limits of growth. Third Annual Complete Tree Lecture, Orono, 14 pp
Goldstein I S 1981 Composition of biomass. In: Goldstein I S (ed) Organic chemicals from biomass. CRC Press, Boca Raton, 9–18
Golob T B 1986 Analysis of short rotation forest operations. Nat Res Council Can Bioenergy Prog NRCC 26014, 127 pp
Golob T B, Silversides C R 1978 Study of a root-cutting shear. For Man Inst Inf Rep FMR-X-109, 35 pp

Golob T B, Tsay T-B, MacLeod D A 1976 Analysis of forces required to pull out stumps of varying age and different species. For Man Inst Inf Rep FMR-X-92, 28 pp

Grammel R 1978 Zentrale Aufarbeitung und Vermarktung von Rohholz – eine Antwort auf die ökonomische und ökologische Herausforderung der mitteleuropäischen Forstwirtschaft? Mitt Forst Versuchs Forschungsanst Baden-Württemberg, Heft 101:3–33

Grammel R 1985 Centralized wood processing. Mimeogr, 9 pp

Granskog J E, Anderson W C 1981 Dense undergrowth reduces feller-buncher productivity in shortleaf pine plantations. U S For Serv Res Note 50-274, 4 pp

Greene W D, Lanford B L, Mykytka E F 1987 Stand and operating effects of feller-buncher productivity in second thinnings of southern pine. For Prod J 37(3):27–34

Greig B J, Burdekin D A 1970 Control and eradication of Fomes annosus in Great Britain. Proc Third Internat Conf Fomes Annosus, 21–32

Grey D C, Jacobs E O 1985 The impact of mechanized harvesting systems on forest site quality. For Prod Res Internat – Achievements and the future, Pretoria, Vol 7:14.3, 18 pp

Grier C C, Waring R H 1974 Conifer foliage mass related to sapwood arc. For Sci 20:205–206

Groton E S, Brooks R T Jr, Waldrop T A 1985 Highlights of a woods biomass analysis for 13 southeastern states. TVA/ONRED/LER-85/35, Tech Note B56, 38 pp

Gugeleb S M, Emeljanov V M 1970 (Logging residues as a raw material for chips.) Lesn Promst 1/1970:19–22

Gunnarfelt B 1973 Grönriskvantiteter vid helträdslunning vintertid. (Quantity of branches and tops when skidding full trees under winter conditions.) Skogshögsk Inst Skogstekn Rapp Upps 60, 35 pp

Gustafsson J, Larsson M 1981a Markberedning efter sönderdelning och spridning av hyggesrester. (Soil preparation after smashing and spreading of logging residues.) Skogsarbeten Ekonomi 12/1981, 4 pp

Gustafsson J, Larsson M 1981b Markberedning efter uttag av stubbar och hyggesrester. (Scarification after removal of stumps and logging slash.) Skogsarbeten Ekonomi 11/1981, 4 pp

Haataja B A, Adams R D, Arola R A 1984 Total tree chunkwood as raw material for flake products. In: Corcoran T J, Gill D R (eds) COFE/IUFRO Proceedings Orono Maine, 157–163

Hägglund E 1925 Jämförande undersökningar över utbyte och engenskaper av cellulosa framställd enligt sulfitmetoden av rå och lufttork granved från olika delar av trädet. (Comparative investigations on yield and properties of cellulose prepared by the sulfite method from green and air-dry spruce wood from different parts of the tree.) Suom Pap Puutav 7:38–45

Hägglund E, Larsson S 1937 Om grankvistens kemiska sammansättning och dess förhållande vid sulfitkokningsprocessen. (On the chemical composition of knots in spruce and their behavior in the sulfite pulping process.) Svensk Papperstidn 40:356–360

Hakkila P 1962 Polttohakepuun kuivuminen metsässä. (Forest seasoning of wood intended for fuel chips.) Commun Inst For Fenn 54(4), 81 pp

Hakkila P 1966 Investigations on the basic density of Finnish pine, spruce and birch wood. Commun Inst For Fenn 61(5), 58 pp

Hakkila P 1967 Vaihtelumalleja kuoren painosta ja painoprosentista. (Variation patterns of bark weight and bark percentage by weight.) Commun Inst For Fenn 62(5), 37 pp

Hakkila P 1969 Weight and composition of the branches of large Scots pine and Norway spruce trees. Commun Inst For Fenn 67(6), 37 pp

Hakkila P 1971 Coniferous branches as a raw material source. Commun Inst For Fenn 75(1), 60 pp

Hakkila P 1972 Mechanized harvesting of stumps and roots. Commun Inst For Fenn 77(1), 71 pp

Hakkila P 1973 The effect of slash on work difficulty in manual planting. Commun Inst For Fenn 78(1), 36 pp

Hakkila P 1975 Kanto- ja juuripuun kuoriprosentti, puuaineen tiheys ja asetoniuutteitten määrä. (Bark percentage, basic density, and amount of acetone extractives in stump and root wood.) Fol For 224, 14 pp

Hakkila P 1978 Pienpuun korjuu polttoaineeksi. (Harvesting small-sized wood for fuel.) Fol For 342, 38 pp

Hakkila P 1979 Wood density survey and dry weight tables for pine, spruce and birch stems in Finland. Commun Inst For Fenn 96(3), 59 pp

Hakkila P 1984 Forest chips as fuel for heating plants in Finland. Fol For 586, 62 pp

Hakkila P 1986 Recycling of wood and bark ash. Metsäntut Tied 211, 44 pp

Hakkila P, Kalaja H 1974 Oksaraaka-aineen kasaus Melroe Bobcat M-600 kuormaajalla. (Bunching of branch raw material by Melroe Bobcat M-600 loader.) Fol For 204, 16 pp
Hakkila P, Kalaja H 1980 Harvesting fuel chips with the Pallari swath harvester. Fol For 418, 24 pp
Hakkila P, Kalaja H 1981 Kopo palahakejärjestelmä. (Kopo block chip system) Fol For 467, 24 pp
Hakkila P, Kalaja H 1983 Puu- ja kuorituhkan palauttamisen tekniikka. (The technique of recycling wood and bark ash.) Fol For 552, 37 pp
Hakkila P, Kalaja H, Mäkelä M 1975 Kokopuunkäyttö pienpuuongelman ratkaisuna. (Full-tree utilization as a solution to the problem of small-sized trees.) Fol For 240, 78 pp
Hakkila P, Laasasenaho J, Oittinen K 1972 Korjuuteknisiä oksatietoja. (Branch data for logging work.) Fol For 147, 15 pp
Hakkila P, Leikola M, Salakari M 1979 Production, harvesting and utilization of small-sized trees. SITRA Sarja B 46, 163 pp
Hakkila P, Mäkelä M 1973 Harvesting of stump and root wood by the Pallari Stumpharvester. Commun Inst For Fenn 77(5), 57 pp
Hakkila P, Mäkelä M 1974 Jatkotutkimuksia Pallarin kantoharvesterista. (Further studies of the Pallari Stumpharvester.) Fol For 200, 15 pp
Hakkila P, Mäkelä M 1975 Pallarin vesakkoharvesteri. (Pallari Busharvester.) Fol For 249, 18 pp
Hakkila P, Palenius I 1971 Vihreä hake massateollisuuden raaka-aineena. (Total chips as pulp raw material.) Pap Puu 53:645–651
Hakkila P, Panhelainen A 1970 On the wood properties of Pinus contorta in Finland. Commun Inst For Fenn 73(1), 43 pp
Hakkila P, Winter A 1973 On the properties of larch wood in Finland. Commun Inst For Fenn 79(7), 45 pp
Hakkila P, Wojcik T 1980 Thinning young pine stands with the Makeri tractor in Poland. Fol For 433, 29 pp
Halinen M 1986 Avohakkuualueille korjaamatta jäävän ainespuun osuus leimikon kokonaistilavuudesta Etelä-Suomessa. (Industrial roundwood waste on clear cutting sites.) Metsäteho Rep 397, 15 pp
Hall C A S, Cleveland C J, Kaufmann R (eds) 1985 Energy and resource quality. The ecology of the economic process. John Wiley New York, 577 pp
Hall D O, Coombs J 1987 Biomass energy in Europe. In: Biomassan uusia jalostusmahdollisuuksia. (New possibilities to process biomass.) VTT Symposium 75, Espoo, 27–49
Hallonborg U 1982a Ristäckets inverkan på spårbildningen. (The effect of slash cover on rutting.) Skogsarbeten Results 3/1982, 4 pp
Hallonborg U 1982b Terminaler – produktion och kostnader. (Terminals – production and costs.) Skogsarbeten Results 17/1982, 4 pp
Hallonborg U 1984 The optimum tree-section forwarder. In: Andersson B, Falk S (eds) Forest energy in Sweden. Swed Univ Agr Sci, 28–29
Hamilton D D 1984 A concept for recovering on-the-ground logging residues. In: Egnéus H, Ellegård A (eds) Bioenergy 84. Elsevier London, Vol II, 19–25
Hamilton J R 1961 Variation of wood properties in southern red oak. For Prod J 11:267–271
Hankala T, Toropainen M 1985 Puun energiakäytön laajentamismahdollisuudet. (The possibilities to expand the use of wood for energy.) In: Hakkila P (ed) Metsäenergian mahdollisuudet Suomessa (The potential of forest energy in Finland.) Fol For 624:60–70
Hannus K, Pensar G 1970 Extraktivämnen i barr. (Extractives in needles.) Åbo Akad Inst Träkemi Cellul B44, 41 pp
Hannus K, Pensar G 1973 Silvichemicals in technical foliage. I. Water steam distilled oil from pine material. Pap Puu 55:509–517
Hans R 1970 Die Zusammensetzung nichtcellulosischer Polysaccharide in verschiedenen Altersstufen des Holzes von Thuja occidentalis und Abies dahlemensis. Holzforschung 24:60–64
Hans R 1971 Die Polysaccharidzusammensetzung des Holzes verschiedener Altersstufen von Taxodium distichum. Holzforschung 25:15–18
Hansen R 1975 Lastbilstransport och komprimering av träddelar. (Truck transport and compaction of tree-sections.) Skogsarbeten Ekonomi 15/1975, 4 pp
Hansen R 1977 Lastbilstransporter av stubb- och rotved. (Trucking of stumps and rootwood.) Skogsarbeten Redogör 3/1977, 32 pp
Harpner J L 1967 A Darwinian approach to plant ecology. J Ecol 55(2):247–270

Harris A 1988 An economic comparison of covered vs. outside storage of whole-tree fuelwood chips. For Prod J 38(1):15–16
Harris R A 1984 Fuel values of stems and branches in white oak, yellow-poplar, and sweetgum. For Prod J 34(3):25–26
Harris R A, Phillips D R 1989 Bulk density of seven typical industrial wood fuels. For Prod J 39(1):31–32
Harris W F, Goldstein R A, Henderson G S 1973 Analysis of forest biomass pools, annual primary production and turnover of biomass for a mixed deciduous forest watershed. In: Young H E (ed) IUFRO biomass studies. Univ Maine Coll Life Sci Agr, 43–64
Hartig R 1896 Das Rothholz der Fichte. Forstl Naturwiss Z 5:96–105, 157–169
Hartig R 1901 Holzuntersuchungen. Altes und Neues. Springer Berlin Heidelberg New York, 99 pp
Hartler N 1976 Wood property requirements for paper. Proceed 16th IUFRO World Congr Oslo Div 5:64–75
Hartler N 1986 Chipper design and operation for optimum chip quality. Tappi 69:62–66
Hartler N, Lindström C, Stade Y 1977 Influence of stem, stump, root and branch bark in the kraft processing of pine, spruce and birch. Svensk Papperstidn 80:121–124
Hartler N, Stade Y 1977 Chipper operation for improved chip quality. Svensk Papperstidn 80:447–453
Hartler N, Stade Y 1979 Chip specifications for various pulping processes. In: Hatton J V (ed) Pulp and paper technology series. Joint Textbook Committee of the Paper Industry, 273–301
Harum J, Labosky P Jr 1985 Chemical constituents of five northeastern barks. Wood Fiber 17:274–280
Hatton J V 1977a Quantitative evaluation of pulpwood chip quality. Tappi 60:97–100
Hatton J V 1977b The potential of mechanical screening for separating components of chipped softwood branches and tops. Tappi 60:141–143
Hatton J V, Keays J L 1971 Laboratory precision studies. I. Lack of interaction in kraft pulping mixtures of western hemlock tree components. Svensk Papperstidn 75:128–131
Hatton J V, Keays J L 1972 Complete-tree utilization studies. III. Yield and quality of kraft pulps from the components of Pseudotsuga menziesii. Tappi 55:1505–1508
Haug B 1972 Field chipping: problems and solutions. FAO/Norad symposium on production, handling and transport of wood chips, TWC/72/22, 10 pp
Hedin T, Malm E, Wennerblom A 1963 Comparison of springwood and summerwood enriched dissolving pulps. Tappi 46:548–550
Heding N 1984 The critical moisture content. In: Gislerud O, Heding N (eds) Storing, Drying and internal handling of wood fuels. Dan Inst For Tech Copenhagen, 6–11
Heding N, Jakobsen H H 1985 Tørring af grønflis til fyringsformål. (Drying of green fuel chips.) Dan Inst For Tech Copenhagen, 2-1985, 41 pp
Heebink B G 1974 Particleboards from lodgepole pine forest residue. USDA For Serv Res Pap FPL 221, 14 pp
Heikka T, Piirainen K 1981 Pienhakkureiden voimankäyttö. (Power consumption of small chippers.) Fol For 496, 22 pp
Heimola A, Lindholm C A, Virkola N E 1980 Suitability of bleached pine branch wood sulfate pulp as raw material for fine paper. Pap Puu 62 (4a):177–178, 181–182, 187–188, 191–192
Helander A B 1933 Kuusen ja männyn vesisolujen pituusvaihtelut. (Variations in tracheid length of pine and spruce.) Found For Prod Res Finl Publ 14, 75 pp
Helge K 1976 The use of whole tree chips in the production of fibre building board. Symposium on extending the use of wood residues, Bucarest. TIM/SEM.5/R.3, 17 pp
Helgesson T 1977 Gisselkvistning, en buntkvistningsmetod för gallrings- och röjningsvirke. (Flail delimbing – a bundle delimbing method for wood removed in thinning and cleaning. STFI Meddelande Serie A 472, 21 pp
Hepp T E, Brister G H 1982 Estimating crown biomass in loblolly pine plantations in the Carolina flatwoods. For Sci 28:115–127
Heth H, Donald D G M 1978 Root biomass of Pinus radiata D. Don. South Afr For J 107:60–70
Heyward F 1933 The root system of longleaf pine on the deep sands of western Florida. Ecology 14:136–148
Hiett L A, Beers W L Jr, Zachariasen K A 1960 Relationship between wood density and other wood and pulp properties. Tappi 43:169–173

Hillis W E 1962 (ed) Wood extractives and their significance to the pulp and paper industries. Academic Press New York, 513 pp
Hillis W E 1987 Heartwood and tree exudates. Springer Berlin Heidelberg New York Tokyo, 268 pp
Hillis W E, Swain T 1962 The influence of extractives on the color of groundwood and newsprint. In: Hillis W E (ed) Wood extractives and their significance to the pulp and paper industries. Academic Press New York, 405–419
Hitchcock H C III 1979 Converting traditional CFI data into biomass values: a case study. In: Frayer W E (ed) Forest resources inventories II. Colorado State Univ, 596–614
Hitchcock H C III, McDonnell J P II 1979 Biomass measurement: a synthesis of the literature. In: Frayer W E (ed) Forest resource inventories II. Colorado State Univ, 544–595
Hodek R J, Arola R A, Bowman J K, Schulze G B 1988 Forest roads built with chunkwood. In: Gislerud O (ed) Production, storage and utilization of wood fuels. I. Production and Utilization. Sver Lantbruksuniv Inst Skogsteknik Upps Res 133:128–149
Hoffman B F 1986 Harvesting small trees for biomass. Maine Agr Exp Sta Bull 813, 30 pp
Hoffman B F, Mills J E 1983 The felling frame – a new tool for thinning sapling-pole stands. Univ Maine Coll For Res, Tech Notes 90, 5 pp
Holappa E-L, Jonsson L 1987 Inventering med tvärsnittslinjemetoden. (Inventory with line intersect method.) Sver Lantbruksuniv Inst Skogsskötsel, 38 pp
Holmes C A, Eickner H W, Brenden J J, White R H 1979 Fire performance of structural flakeboard from forest residue. USDA For Serv Res Pap FPL 315, 22 pp
Honer T G 1971 Weight relationships in open- and forest-grown balsam fir trees. In: Forest biomass studies. Univ Maine Coll Life Sci Agr, 65–78
Horn R A, Auchter R J 1972 Kraft pulping of pulpwood chips containing bark. Pap Trade J November/1972:55–59
Horntvedt R 1983 Seasonal variations in dry weight, chlorophyll and soluble carbohydrate content of Norway spruce needles (Picea abies (L.) Karst.). Medd Norsk Inst Skogforsk 38.2, 14 pp
Horvath-Laiko I, Kardas I, Maurer E 1970 (Results of a study on stump extracting machine K-2A.) Erdesz Kut 66:211–223
Hosia M, Kortelainen V-A 1971 Undersökningar rörande utnyttjandet av barrträdsgrenar som råmaterial för kemisk massa och hård fiberskiva II. (Investigations on utilizing softwood branches as pulp and hardboard raw material II.) Pap Puu 53:721–727
Hosia M, Lindholm C-A, Toivanen P, Nevalainen K 1971 Undersökningar rörande utnyttjandet av barrträdsgrenar som råmaterial för kemisk massa och hård fiberskiva. I. (Investigations on utilizing softwood branches as pulp and hardboard raw material. I.) Pap Puu 53:49–66
Howard E T 1973a Heat of combustion of various southern pine materials. Wood Sci 5:194–197
Howard E T 1973b Physical and chemical properties of slash pine tree parts. Wood Sci 5:312–317
Howard J O 1973 Logging residue. U S For Serv Resour Bull PNW-44, 26 pp
Howard J O 1981 Logging residue in the Pacific Northwest: characteristics affecting utilization. U S For Serv Res Pap PNW-289, 41 pp
Howard J O, Ward F R 1972 Measurement of logging residue – alternative applications of the line intersect method. U S For Serv Res Note PNW-193, 8 pp
Hunt J R, Barton G M 1978 Nutritive value of spruce muka (foliage) for the growing chick. Anim Feed Sci Technol 3:63–72
Hunt K, Hatton J V 1975 Full forest utilization: Part II. Quality and kraft pulp yield of eastern Canadian hardwoods. Pulp Pap Can 76(11):97–102
Hyland F 1974 Fiber analysis and distribution in the leaves, juvenile stems and roots of ten Maine trees and shrubs. Life Sci Agr Exp Sta Univ Maine Tech Bull 71, 59 pp
IEA Forestry energy summary report 1978–1986. 1987. National Energy Administration Sweden, Statens Energiverk R2, 53 pp
Ievin I K, Andreev V N 1985 Technology and machinery for forest improvement felling with logging of trees and tree lengths. In: Proceedings of IUFRO International Interdivisional Conference on thinning, Moscow-Riga, 44–51
Ievin I K, Berzinš G, Kevinš J, Ozolinš A 1984 (Processing and utilization of crown mass.) In: Harstela P, Savel'ev P, Sairanen P, Sirén M, Epalts A (eds) (Thinning operations. Results from a Finnish-Soviet joint research study.) Fol For 600:25–29
Ievin I K, Daugavietis M O 1986 Tree crown biomass as a source of biologically active substances and energy. Proceed 18th IUFRO World Congr Ljubljana Div 3:140–146

Ievin I K, Daugavietis M O, Polis O R, Deruma V J 1981 Tree verdure as a source of organic raw material. Am Soc Agr Eng Pap 81-1593, 7 pp

Ievin I K, Galvāns U I, Daugavietis M O, Kevinš J J 1973 Utilization of needles and leaves in the USSR. In: Young H E (ed) IUFRO biomass studies. Coll Life Sci Agr Univ Maine Orono, 469–475

Ilavský J 1986 Harvesting of tree biomass and its utilization for energetic and feeding purpose. Pap 18th IUFRO World Congr Ljubljana, 9 pp

Ilavský J, Kevinš J 1985 Technology and system of machines for tree crowns processing. In: Proceedings of IUFRO International Interdivisional Conference in thinning, Moscow-Riga, 149–157

Immonen K, Seppälä R 1984 Polttopuun ja palaturpeen alueittainen tuotanto, jakelu ja käyttö. (Regional production, distribution and usage of fuel wood and sod peat.) SITRA Sarja B 76, 164 pp

Ivanova I S 1960 (Sulfate pulp from logging wastes.) Bumazh Prom 2:13–15

Jääskeläinen P 1981 Betulinol and its utilization. Pap Puu 63:599–603

Jackson M, Megraw R A 1986 Impact of juvenile wood on pulp and paper products. In: Robertson D (ed) Juvenile wood. What does it mean to forest management and forest products? For Prod Res Soc Proc 47309:75–81

Jahkola A 1983 Burning of wood and wood wastes in industry. In: Proceedings of the Finnish Austrian Workshop on Energy from Wood, Vienna, 169–191

Jane F W, Wilson K, White D J B 1970 The structure of wood. Adam and Charles Black London, 478 pp

Jarceva N A 1969 (The chemical composition of Pinus sibirica.) Lesn Z Arhangel'sk 12(1):112–116

Jirjis R, Theander O 1986 Preliminära resultat av de kemiska huvudkomponenternas förändring i lagrad trädbränsleflis. (Preliminary results on chemical composition changes in stored forest fuel.) Sver Lantbruksuniv Inst Virkeslära Upps 161, 21 pp

Jonstone W D 1971 Total standing crop and tree component distributions in three stands of 100-year-old lodgepole pine. In: Forest biomass studies. Univ Maine Coll Life Sci Agr, 81–89

Johnstone W D, Keays J L 1973 Kraft pulps from the components of lodgepole pine. IUFRO Biomass Studies Vancouver, 477–494

Jones E, Hughes R E 1984 A note on the ascorbic acid content of some trees and woody shrubs. Phytochemistry 23:2366–2367

Jones K C 1981 Energy requirements to produce fuel from harvesting residues. In: Mattsson J E, Nilsson P O (eds) Proceedings of the international conference "Harvesting and utilization of wood for energy purposes". Inf Proj Forestry Energy 19:184–196

Jonsson T 1985 Survey of central processing systems in Scandinavia. In: Mitchell P, Nilsson P O, Zsuffa L (eds) Research in forestry for energy. Sver Lantbruksuniv Inst Skogsteknik Rapp Upps 49:408–429

Jonsson T 1986a Biomassaåterföring – kompensationsgödsling med aska eller barr och finkvist. (Returning biomass – compensation fertilization with ash or needles and twigs). Skogsarbeten Mimeogr, 66 pp

Jonsson T 1986b Kvistning av träddelar med Bruks buntkvistare. (Delimbing tree sections with Bruks bunch delimber.) Skogsarbeten Results 23/1986, 4 pp

Jonsson T 1986c Trädflis: flisning, barking, sållning. (Whole-tree chips: chipping, debarking, screening.) Forskningsstiftelsen Skogsarbeten, 60 pp

Jonsson T, Nordén B 1987 Multiple tree delimbing/debarking by the Manitowoc flail. Study of the Wayerhaeuser installation in Plymouth, N. C. Skogsarbeten Mimeogr, 33 pp

Jonsson Y 1978 Jämförelser av drivningssystem för stubb- och rotved. (Comparisons of harvesting systems for stump and root wood.) Skogsarbeten Redogör 5/1979, 24 pp

Jonsson Y 1984 Sound raw material but shaky profitability. In: Andersson B, Falk S (eds) Forest energy in Sweden. Swed Univ Agr Sci, 40–41

Jonsson Y 1985 Teknik för tillvaratagande av stubbved. (Techniques for recovering stump wood.) Skogsarbeten Redogör 3/1985, 35 pp

Jordan C F 1971 Productivity of tropical forest and its relation to a world pattern of energy storage. J Ecol 59:127–142

Jukola-Sulonen E-L, Mikkola K, Nevalainen S, Yli-Kojola H 1987 Havupuiden elinvoimaisuus Suomessa vuosina 1985–1986. (Vitality of conifers in Finland in 1985–1986). Metsäntut Tied 256, 92 pp

Juntunen M-L 1982 Tuhkan levityksen terveydellisten haittojen arviointi. (Evaluation of the health hazards of ash spreading.) Metsäntut Tied 62, 17 pp

Kahala M 1980 Puutavaran valmistus moottorisahalla. (Preparation of timber by power saw.) Metsäteho Rep 364, 19 pp
Kahala M 1984 Osapuunakorjuu eteläsuomalaisissa harvennusmetsäolosuhteissa. (Part-tree harvesting in thinnings in southern Finland.) Metsäteho Rep 386, 19 pp
Kalaja H 1978 Pienpuun korjuu TT 1000 F palstahakkurilla. (Harvesting small-sized trees with terrain chipper TT 1000 F.) Fol For 374, 27 pp
Kalaja H 1984 Varastointi- ja siirtolaitteiden toimintahäiriöt metsähaketta käyttävissä lämpölaitoksissa. (Operation disturbances in storage and handling equipment in forest chip-fired heating plants.) Metsäntut Tied 166, 24 pp
Kalaja H 1986 Tuhkan levittäminen metsätraktorilla. (Spreading ash with forest tractor.) Metsäntut Tied 230, 18 pp
Kaludin K 1964 Der Gehalt an Vitamin C, Karotin und Vitamin E in Nadeln von Pinus silvestris. Arch Forstwes 13:383–391
Kaludin K, Kaludin J 1967 (Investigations into the carotene in leaves of certain coniferous tree species.) Gorskostop Nauka 4(3):81–88
Kamin'ski E 1961 Swiez a karpina sosnowa-nowy surowiec do produkcji celulozy i kalafonii. Sylwan 105:41–49
Kanninen K, Uusvaara O, Valonen P 1979 Kokopuuraaka-aineen mittaus ja ominaisuudet. (Measuring and properties of whole-tree raw material.) Fol For 403, 48 pp
Karchesy J, Koch P 1979 Energy production from hardwoods growing on southern pine sites. U S Dep Agr For Serv Tech Rep SO-24, 59 pp
Kardell L 1976 Skogsvårds- och naturvårdsaspekter på stubbrytning. (Silvicultural and nature conservation aspects of stump harvesting.) Mimeogr, 10 pp
Kardell L 1984 Berries or energy? In: Andersson B, Falk S (eds) Forest energy in Sweden. Swed Univ Agr Sci, 108–111
Karepova Z A, Repjah S M, Levdikova V L 1978 (On the chemical composition of water-soluble extractives from needles.) Khim Prir Soedin 5:648–650
Kärkkäinen M 1970 Hakkuutähteiden merkityksestä puuston vaurioitumisen ja raiteenmuodostuksen kannalta. (On the significance of waste in thinnings as to scars and tracks.) Silva Fenn 4:155–171
Kärkkäinen M 1972 Havaintoja kuusen oksaisuudesta. (Observations on the branchiness of Norway spruce.) Silva Fenn 6:90–115
Kärkkäinen M 1975 Kantojen käytön kehittyminen Suomessa. (Development of stump utilization in Finland.) Silva Fenn 9:284–302
Kärkkäinen M 1976a Puun ja kuoren tiheys ja kosteus sekä kuoren osuus koivun, kuusen ja männyn oksissa. (Density of moisture content of wood and bark, and bark percentage in the branches of birch, Norway spruce, and Scots pine.) Silva Fenn 10:212–236
Kärkkäinen M 1976b Havutukkien kuoren tiheys ja kosteus. (Density and moisture content of bark in pine and spruce logs.) Commun Inst For Fenn 87(5), 23 pp
Kärkkäinen M 1981 Polttopuun rasiinkaadon ja muiden kuivatusmenetelmien perusteet. (Foundations of leaf-seasoning and other drying methods of fuelwood.) Fol For 459, 15 pp
Kärkkäinen M 1984 Puutavaran mittauksen perusteet. (The basis of timber measurement.) Helsingin Yliop Monistusp Helsinki, 252 pp
Kärkkäinen M 1986 Malli männyn, kuusen ja koivun puuaineen oksaisuudesta. (Model of knottiness of wood material in pine, spruce, and birch.) Silva Fenn 20:107–116
Keays J L 1968 Whole-tree utilization studies: selection of tree components for pulping research. Can Dep For Rural Devel For Prod Lab Inf Rep VP-X-35, 31 pp
Keays J L 1971a Complete-tree utilization. An analysis of the literature I: unmerchantable top of bole. Can For Serv For Prod Lab Inf Rep VP-X-69, 98 pp
Keays J L 1971b Complete-tree utilization. An analysis of the literature II: foliage. Can For Serv For Prod Lab Inf Rep VP-X-70, 94 pp
Keays J L 1971c Complete-tree utilization. An analysis of the literature III: branches. Can For Serv For Prod Lab Inf Rep VP-X-71, 67 pp
Keays J L 1971d Complete tree utilization. An analysis of the literature IV: crown and slash. Can For Serv For Prod Lab Inf Rep VP-X-77, 79 pp
Keays J L 1971e Complete-tree utilization. An analysis of the literature V: stump, roots and stump-root system. Can For Serv For Prod Lab Inf Rep VP-X-79, 62 pp
Keays J L 1973 Complete-tree utilization of mature trees. AIChE Symp Ser 70(139):67–75

Keays JL 1974 The biomass of forest residuals. Can For Serv West For Prod Lab Mimeogr, 29 pp
Keays J L 1975 Full-forest utilization history and future. Lect Univ Maine, Orono, 14 pp
Keays J L, Barton G M 1975 Recent advances in foliage utilization. Can For Serv West For Prod Lab Inf Rep VP-X-137, 94 pp
Keays J L, Hatton J V 1971a Complete-tree utilization studies. I. Yield and quality of kraft pulp from the components of Tsuga heterophylla. Tappi 54:99–104
Keays J L, Hatton J V 1971b Complete-tree utilization studies. II. Yield and quality of kraft pulp from the components of Picea glauca. Tappi 54:1721–1724
Keays J L, Hatton J V 1975 The implication of full-forest utilization on worldwide supplies of wood by year 2000. Pulp Pap Int 17, June, 49–52
Keen R E 1963 Weights and centres of gravity involved in handling pulpwood trees. Pulp Pap Res Inst Can Woodlands Res Index 147, 91 pp
Keeves A 1966 Some evidence of loss of productivity with successive rotations of Pinus radiata in the southeast of South Australia. Aust For 30:51–65
Kehr E 1976 Zur Erweiterung der Rohstoffbasis für die Herstellung von Spanplatten und Faserplatten. Holztechnologie 17:208–217
Kehr E, Drechsler W 1978 Dünnholz, ein wichtiger Rohstoff für die Faser- und Spanplattenproduktion. Holztechnologie 19:67–74
Kelley R S 1983 Stand damage from whole-tree harvesting in Vermont hardwoods. J For 81(2):95–96
Kellogg R M, Wangaard F F 1969 Variation in the cell-wall density of wood. Wood Fiber 1:180–204
Kellomäki S 1981 Effect of the within-stand light conditions on the share of stem, branch and needle growth in a twenty-year-old Scots pine stand. Silva Fenn 15:130–139
Kellomäki S, Väisänen H 1986 Kasvatustiheyden ja kasvupaikan viljavuuden vaikutus puiden oksikkuuteen taimikko- ja riukuvaiheen männiköissä. Malleihin perustuva tarkastelu. (Effect of stand density and site fertility on the branchiness of Scots pines at pole stage. A study based on models.) Commun Inst For Fenn 139, 38 pp
Kern K G, Moll W, Blum W E, Braun H J 1971 Ertragskundliche, bodenkundliche und genetische Untersuchungen in alten Kiefern-Krüppelbeständen und benachbarten Kiefern-Jungwüchsen des Haardtrandes. Allg Forst-Jagdztg 142:97–104
Kerruish C M 1975 Developments in harvesting technology relevant to short rotation crops. Mimeogr, 16 pp
Kibblewhite R P 1980 Radiata pine corewood and slabwood, and their interrelations with pulp and handsheet properties. N Z J For Sci 10:533–550
Kiiskilä E 1987 Biomass gasification in an Ahlström pyroflow gasifier. In: Biomassan uusia jalostusmahdollisuuksia. (New possibilities to process biomass.) VTT Symposium 75, Espoo, 76–89
Kimmins J P, Binkley D, Chatarpaul L, de Catanzaro J 1985 Whole-tree harvest-nutrient relationships: a bibliography. Can For Serv Petawawa Nat For Inst, 377 pp
Kirjavainen R, Lahdenperä T 1983 Hakkeen ja palaturpeen vastaanotto-, varastointi- ja kuljetinlaitteiden suunnitteluperusteet. (Planning grounds for receiving, storing and conveying devices of chips and peat.) Project on Development of Regional Usage of Peat and Wood Fuels, Res Rep 17. SITRA, 104 pp
Kiščenko T I, Unt V Ja, Komsilov N F, Gerasimov B S, Abol B I 1962 (Harvesting trees with stump-root systems.) Lesn Promst 40(10):13–16
Klausmeir W H 1985 The production of liquid fuels from wood. In: Robertson D, White J (eds) Seventh International FPRS Wood Energy Forum '83, Vol II, For Prod Res Soc Proc 47338:490–502
Kleinschmidt S, Baskerville G L, Solomon D S 1980 Foliage weight distribution in the upper crown of balsam fir. U S For Serv Res Pap NE-455, 7 pp
Klemola A 1976 Development and trends in industrial utilization of fuller forest biomass (branchwood, foliage and stumps) in the wood chemical industry. In: Symposium on the harvesting of a larger part of the forest biomass. ECE/FAO/ILO TIM/EFC/WP.1/SEM.3/3, Vol 1:56–63
Kleppe P J 1970 The process of, and products from, kraftpulping of southern pine. For Prod J 20(5):50–59
Klinga L O, Back E L 1964 Fiber building board variables influencing the wear of cutting tools. Svensk Papperstidn 67:309–316
Kluender R A, Farrar K D, Stuart W B 1983 An evaluation of logging systems designed to recover residues for energy. Tennessee Valley Authority, Tech Note B50, 46 pp

Knigge W 1960 Die Holzeigenschaften der Küstentanne (Abies grandis). Allg Forstz 15(7):94–100

Knigge W, Schulz H 1966 Grundriss der Forstbenutzung. Paul Parey Hamburg, 584 pp

Knutell H, Sundin T 1982 Felling and bunching small trees from thinnings with small scale equipment. Skogshögs Inst Skogstekn Stenc 177, 54 pp

Koch P 1972a Utilization of the southern pines. The raw material. U S Dep Agr For Serv, Agr Handbook 420(1):1–734

Koch P 1972b Utilization of the southern pines. Processing. U S Dep Agr For Serv, Agr Handbook 420(2):735–1663

Koch P 1976 New technique for harvesting southern pines with taproot attached can extend pulpwood resources significantly. Appl Polym Symp 28:403–420

Koch P 1985a Utilization of hardwoods growing on southern pine sites: the raw material. U S Dep Agr For Serv, Agr Handbook 605(1):1–1418

Koch P 1985b Utilization of hardwoods growing on southern pine sites: processing. U S Dep Agr For Serv, Agr Handbook 605(2):1419–2542

Koch P 1985c Utilization of hardwoods growing on southern pine sites: products and prospective. U S Dep Agr For Serv, Agr Handbook 605(3):2543–3710

Koch P 1987 Gross characteristics of lodgepole pine trees in North America. U S For Serv Gen Tech Rep INT-227, 311 pp

Koch P, Coughran S J 1975 Development of a puller-buncher for harvesting southern pines with taproot attached. For Prod J 25(4):23–30

Koch P, Nicholson T W 1978 Harvesting residual biomass and swathe-felling with a mobile chipper. In: McMillin C W (ed) Complete tree utilization of southern pine. For Prod Res Soc Madison, 146–154

Koch P, Savage T E 1980 Development of the swathe-felling with a mobile chipper. J For 78:17–21

Kofman P D 1985 A manual on felling and bunching small trees from thinnings with small scale equipment on gentle terrain. Final Rep IEA-FE-CPC7, 116 pp

Kohan S M 1985 Biomass thermal gasification applications in the electric utility industry. In: Robertson D, White J (eds) Seventh International FPRS Industrial Wood Energy Forum '83, Vol II, For Prod Res Soc Proc 47338:258–262

Koivulehto P 1969 Juurakoiden maasta irroittamisesta. (On the extraction of stumps and roots.) Fol For 73, 12 pp

Kollmann F 1951 Technologie des Holzes und der Holzwerkstoffe I. Springer Berlin Heidelberg, 1050 pp

Kollmann F F P, Côté W A Jr 1968 Principles of wood science. Springer Berlin Heidelberg New York, 592 pp

Komšilov N F, Kiščenko T I, Kyalina L V 1963 (Fresh pine stump wood – a promising raw material.) Bum Promst 7:11–12

Koning J W Jr, Skog K E 1986 Use of wood energy in the United States – an opportunity. Manuscript, 14 pp

Korotjaev L V 1984 (Mathematical model of the crown of northern spruce.) Lesn Zhur 1:46–54

Kozak A 1970 Methods for ensuring additivity of biomass components by regression analysis. For Chron 46:402–404

Kozlowski T T, Clausen J J 1965 Changes in moisture contents and dry weights of buds and leaves of forest trees. Bot Gaz 126:20–26

Kozlowski T T, Keller T 1966 Food relations of woody plants. Bot Rev 32(4):293–382

Kramer H 1986 Relation between crown parameters and volume increment of Picea abies stands damaged by environmental pollution. Scand J For Res 1:251–263

Kreutzer K 1975 The effect of economizing on nutrient economy in central European forests. Elmia 75, Konf Sk2:37–54

Kubin E 1977 The effect of clear cutting upon the nutrient status of a spruce forest in northern Finland (64°28'N). Acta For Fenn 155, 40 pp

Kubínec V, Šedivý M 1959 Studium obsahu a vlastonosti buničiny z lesního odpadu. (A study of the composition and properties of pulp from forest waste.) Drev Vysk 4:191–208

Kuitto P-J 1986 Välivarastohaketus ja metsähakkeen autokuljetus. (Chipping at upper landing and trucking of forest chips.) Metsäteho Mimeogr, 25 pp

Kuitto P-J 1987 Osapuiden metsäkuljetus. (Forest haulage of tree sections.) Metsäteho Kats 10/1987, 4 pp

Kukkola M, Mälkönen E 1985 Consequences of whole-tree harvesting in young coniferous stands. IUFRO Conf Thinning, Moscow-Riga, 8 pp
Kurelek J 1984 The feller forwarder concept and its application. In: Corcoran T, Gill D R (eds) Proceedings COFE/IUFRO, 191–194
Kurrle F L 1963 Nitric acid digestion of logging residues of red spruce. Pulp yield and physical properties. Tappi 46:267–272
Kuusela K, Nyyssönen A 1981 The forest biomass as an energy source: the role of forest inventory. Univ Helsinki Dep For Mensur Managem Res Rep 14, 11 pp
Kwasnitscka K 1978 Whole-tree utilization of Norway spruce in centralized timberyards. In: McMillin C W (ed) Complete tree utilization of southern pine. For Prod Res Soc Madison, 211–215
Kyttälä T 1981 Komprimering av träddelar från gallring och slutavverkning. (Compacting tree-sections from thinning and final cutting.). Skogsarbeten Mimeogr, 23 pp
Laasasenaho J 1975 Runkopuun saannon riippuvuus kannon korkeudesta ja latvan katkaisuläpimitasta. (Dependence of the amount of harvestable timber upon the stump height and the top-logging diameter.) Fol For 233, 20 pp
Laitakari E 1934 Koivun juuristo. [The root system of birch (Betula verrucosa and odorata).] Acta For Fenn 41.2, 216 pp
Lajunen P 1975 Kuitulevytehdaspään hakkuutähdeprojektin osatutkimus. (A sub-study of the logging residue project of the fiberboard industry.) Typescript, 11 pp
Larson P R 1966 Changes in chemical composition of wood cell walls associated with age in Pinus resinosa. For Prod J 16(4):37–45
Larson P R 1969 Incorporation of ^{14}C in the developing walls of Pinus resinosa tracheids: compression wood and opposite wood. Tappi 52:2170–2177
Larson P R, Dickson R E, Isebrands J G 1976 Some physiological applications for intensive culture. U S For Serv Gen Tech Rep NC-21:10–18
Larsson M 1981 Development of transportation systems for logging residues, stumps and trees in Sweden. In: Mattsson J E, Nilsson P O (eds) Proc Internat Conf "Harvesting and utilization of wood for energy purposes." Inf Proj For Energy 19:166–183
Larsson M 1986 Central processing and associated logging and transportation techniques – JAC 15. Mimeogr, 6 pp
Larsson M, Lidberg B, Nordén B 1983 Buntkvistning av träddelar i KMW-Mekans upparbetningssystem. (Bunch delimbing of tree-sections in KMW-Mekans processing system.) Skogsarbeten Results 26/1983, 4 pp
Larsson M, Nordén B 1982 Komprimering av träddelar från slutavverkning med en separat utrustning. (Compacting tree sections from final cutting with separate equipment.) Skogsarbeten Mimeogr, 22 pp
Lavoie J-M 1980 The transport of full trees from the stump area to the mill woodyard. FERIC Techn Rep TR-44, 36 pp
Law K N, Koran Z 1979 Utilization of white spruce foliage. Pulp characteristics. Wood Sci 12:106–112
Law K N, Koran Z 1982 Kraft pulping of white spruce bole and branch. Can J For Res 12:299–303
Legg G W, Hart J S 1960 Alkaline pulping of jackpine and Douglas-fir. The influence of sulphide and effective alkali charge on pulping rate and pulp properties. Pulp Pap Mag Can 61(6):T299–T304
Lehmann G, Luthardt H, Günther B, Bauer A, Wolff E 1981 Einsatz grüner Hackschnitzel in der Plattenindustrie. (The use of green chips in the board industry.) Beitr Forstwirt 15(1):44–48
Lehmann W F, Geimer R L 1974 Properties of structural particleboards from Douglas-fir forest residues. For Prod J 24(10):17–25
Lehtonen I 1980 Knots in Scots pine (Pinus sylvestris L.) and Norway spruce (Picea abies L. Karst.) and their effect on the basic density of stemwood. Commun Inst For Fenn 95.1, 34 pp
Lemke J 1978 Zmienność ciezaru igiel w okólkach koron sosen róznych klas wieku. (Variation in needle weight with crown whorl in Scots pines of different age classes.) Sylwan 122(2):13–21
Leppälahti J, Kurkela E, Salo K 1984 Kotimaisten polttoaineiden kaasutus. Osa 2. Kaasukäyttöinen lämpökeskus. (Gasification of indigenous fuels. Part 2. Gas-fired heating station.) Tech Res Centre Finland Res Rep 251, 44 pp
Leucaena 1977 Promising forage and tree crop for the tropics. Nat Acad Sci Wash D C, 115 pp

Levčenko V P 1969 (Physical and mechanical properties of the wood of pine knots.) Lesn Z Arhangel'sk 12(1):93–96

Levin E D, Repjah S M 1984 (On the utilization of tree foliage.) Lesn Promst 5:22–24

Lewis W C 1971 Board materials from wood residues. USDA For Serv Res Note FPL-045, 11 pp

Liiri O 1964 Lastulevyn puuraaka-aine. Tutkimuksia suomalaisen puun raaka-aineen eräiden ominaisuuksien merkityksestä valmistettaessa 3-kerroksista laakapuristettua lastulevyä. (Wood raw material of particleboard. Investigations on the significance of some properties of Finnish wood raw material in the manufacture of 3-layer flat-pressed particleboard.) State Inst Tech Res Publ 80, 181 pp

Liiri O, Kivistö A, Laine L 1972 Vihreä hake lastulevyn raaka-aineena. (Branches and whole trees as raw material for particleboard.) Pap Puu 54:593—609

Lönnberg B 1975a Short-rotation hardwood species as whole-tree raw material for pulp and paper. 1. General aspects. Pap Puu 57:443–447

Lönnberg B 1975b Short-rotation hardwood species as whole-tree raw material for pulp and paper. 2. Wood raw material. Pap Puu 57:507–516

Lönnberg B 1976a Short-rotation hardwood species as whole-tree raw material for pulp and paper. 3. Chemical pulping. Pap Puu 58:113–124

Lönnberg B 1976b Short-rotation hardwood species as whole-tree raw material for pulp and paper. 4. Effect of bark upon chemical pulping. Pap Puu 58(4a):181–186, 189–190, 195–197

Lönnberg B 1976c Short-rotation hardwood species as whole-tree raw material for pulp and paper. 5. Bleached pulps; fine papers. Pap Puu 58:455–460, 463–466, 469–472

Lönnberg B 1976d Short-rotation hardwood species as whole-tree raw material for pulp and paper. 6. Chemi-mechanical pulps. Pap Puu 58:630–632, 635–638, 648

Lönnberg B 1976e Short-rotation hardwood species as whole-tree raw material for pulp and paper. 7. Absorbent pulps. Pap Puu 58:857–860, 863–871

Lönnberg B, Palenius I, Saxena V B 1975 Short-rotation hardwood as raw material for absorbent pulp. Pap Puu 57:721–727

Lönner G 1985 Processing terminals for industry and energy wood raw material. Mimeogr, 13 pp

Loomis R M, Phares R E, Crosby J S 1966 Estimating foliage and branchwood quantities in shortleaf pine. For Sci 12:30–39

Love P, Overend R 1978 Tree power. An assessment of the energy potential of forest biomass in Canada. Energy Mines Resources Can Rep ER78-1, 31 pp

Löyttyniemi K 1974 Hakkuutähteiden talteenoton vaikutus tuhoeläimiin. (Effect of residue recovery on pests.). In: Hakkuutähteiden talteenoton seurannaisvaikutukset. (By-effects of the harvesting of logging residues.) Fol For 210:13–15

Lundmark J-E 1984 Soil mixing can give rise to loss of nutrients. In: Andersson B, Falk S (eds) Forest energy in Sweden. Swe Univ Agr Sci: 100–102

Madgwick H A I 1970 Caloric values of Pinus virginiana as affected by time of sampling, tree age, and position in stand. Ecology 51:1094–1097

Madgwick H A I 1976 Mensuration of forest biomass. In: Oslo biomass studies. Univ Maine Coll Life Sci Agr Orono, 13–27

Madgwick H A I, Jackson D S 1974 Estimating crown weights of Pinus radiata from branch variables. N Z J For Sci 4:520–528

Madgwick H A I, Jackson D S, Knight P J 1977a Above-ground dry matter, energy, and nutrient contents of trees in an age series of Pinus radiata plantations. N Z J For Sci 7:445–468

Madgwick H A I, Olah F D, Burkhart H E 1977b Biomass of open-grown Virginia pine. For Sci 23:89–91

Magnusson B, Nylinder M 1977 Virkeskvantiteten hos stubb-rotsystem. (Amount of wood in stump-root systems.) Projekt helträdsutnyttjande Mimeogr, 36 pp

Mahendrappa M K 1986 Potential acidification of black spruce (Picea mariana (Mill.) B.S.P.) sites due to intensive harvesting. In: Ågren G I (ed) Predicting consequences of intensive forest harvesting on long-term productivity. Swed Univ Agr Sci Inst Ekol Miljövård Rapp 26:165–171

Mäkinen V O 1965 Männyn juurakoiden kuutiomäärästä ja sen osuudesta vastaavan runkopuun kuutiomäärään. (On the volume of the stump-root systems of Scots pine.) Metsätal Aikak 85:75–76

Mälkönen E 1972 Hakkuutähteiden talteenoton vaikutus männikön ravinnevaroihin. (Effect of harvesting logging residues on the nutrient status of Scotch pine stands.) Fol For 157, 14 pp

Mälkönen E 1974 Annual primary production and nutrient cycle in some Scots pine stands. Commun Inst For Fenn 84(5), 87 pp
Mälkönen E 1975 Whole-tree utilization – consequences for soil and environment. Experiences from Finnish research and practice. Elmia 75, Konf Sk2:26–30
Mälkönen E 1976 Effect of whole-tree harvesting on soil fertility. Silva Fenn 10:157–164
Mälkönen E 1977 Annual primary production and nutrient cycle in a birch stand. Commun Inst For Fenn 91(5), 35 pp
Mälkönen E, Annila E, Kallio T 1985 Energiapuun talteenotto metsäekosysteemin kannalta. (Recovery of fuelwood from the forest ecosystem point of view.) In: Hakkila P (ed) Metsäenergian mahdollisuudet Suomessa. (The potential of forest energy in Finland.) Fol For 624, 86 pp
Mälkönen E, Saarsalmi A 1982 Biomass production and nutrient removal in whole tree harvesting of birch stands. Fol For 534, 17 pp
Malmström C 1952 Svenska gödlingsförsök för belysande av de näringsekologiska villkoren för skogsväxt på torvmark. (Swedish fertilization experiments to illuminate the nutrient-ecologic preconditions of forest growth in peatlands.) Commun Inst For Fenn 40(17), 27 pp
Manwiller F G 1972 Tracheid dimensions in rootwood of southern pine. Wood Sci 5:122–124
Manwiller F G 1974 Fiber lengths in stems and branches of small hardwoods on southern pine sites. Wood Sci 7:130–132
Manwiller F G 1979 Wood and bark specific gravity of small-diameter pine-site hardwoods in the south. Wood Sci 11:234–240
Marchand M H 1983 Balancing supply and demand worldwide by the year 2000 through resource management. UNIDO/FAO, First consultation on the wood and wood products industry ID/WG.395/6, 65 pp
Marklund L-G 1981 Skogsenergitillgångar i Sverige. Preliminära skattningar grundade på riksskogstaxeringens material. (Forest energy resources in Sweden. Preliminary results based on material from the national forest survey.) Sver Lantbruksuniv Inst Skogstax Rapp 32, 66 pp
Marklund L-G 1987 Biomass functions for Norway spruce (Picea abies (L.) Karst.) in Sweden. Sver Lantbruksuniv Inst Skogstax Rapp 43, 127 pp
Markwardt L J, Wilson T R C 1935 Strength and related properties of woods grown in the United States. U S Dep Agr Tech Bull 479, 99 pp
Martin A J 1976 Suitability of the line intersect method for sampling hardwood logging residues. U S For Serv Res Pap NE-339, 6 pp
Marton R, Amidon T E, Koeppicus R 1976 Origins of some problems in whole-tree pulping. Tappi 59:107–112
Matjusjkina A P, Ageeva M I, Skvorzova L V 1966 (Sulfate pulp from stump and root wood of pine.) Bum Promst 12/1966:10–11
Mattson J A 1974a Compression debarking of stored wood chips. USDA For Serv Res Note NC-161, 4 pp
Mattson J A 1974b Beneficiation of compression debarked wood chips. USDA For Serv Res Note NC-180, 4 pp
Mattson J A 1975 Debarking chips from whole trees in the Lake States. USDA For Serv Res Pap NC-115, 9 pp
Mattson J A 1982 The potential application of auger cutters to biomass harvesting. In: Robertson D (ed) The sixth international FPRS industrial wood energy forum '82. Proc, Vol 1:95–99
Mattson J A 1983 The mechanics of auger cutters for wood. For Prod J 33(2):31–38
Mattson J A 1985 Harvest system alternatives: the loggers toolbox. In: Martin A J (ed) Timber harvesting. The link between management and utilization. SAF 85-09:122–127
Mattson J A, Arola R A, Hillstrom W A 1978 Recovering and chipping hardwood cull trees having heavy limbs. In: McMillin C W (ed) Complete tree utilization of southern pine. For Prod Res Soc Madison, 120–129
Mattson J A, Arola R A, Karsky R 1984 Developments in wood chunking technology. In: Robertson D, White J (eds) Comminution of wood and bark. For Prod Res Soc Proc 7336:169–180
Mattson J A, Winsauer S A 1985 The potential for harvesting "puckerbrosh" for fuel. U S For Serv Res Pap NC-262, 6 pp
Maxwell W G, Ward F R 1976 Photo series for quantifying forest residues in the: coastal Douglas-fir-hemlock type and coastal Douglas-fir-hardwood type. U S For Serv Gen Tech Rep PNW-51, 103 pp

Mayer-Wegelin H 1955 Europäische und Japanische Lärche, ihre kennzeichnenden Holzeigenschaften. Holz-Zbl 81(145):1727–1729
McCormack M L Jr 1984 Interaction of harvesting and stand establishment in conifers in northeastern North America. In: Corcoran T, Gill D R A (eds) Proc COFE/IUFRO: 233–239
McLauchlan T A, Lapointe J A 1979 Production of chips by disc chippers. In: Hatton J V (ed) Pulp Pap Technol Ser. Joint Textbook Comm Pulp Industry, 15–32
McMillin C V 1968 Gross wood characteristics affecting properties of handsheets made from loblolly pine refiner groundwood. Tappi 51:51–56
McMillin C V 1969 Ash content of loblolly pine wood as related to specific gravity, growth rate, and distance from pith. Wood Sci 2:26–30
McNab W H 1980 A technique for inventorying volumes and weights of windrowed forest residues. U S For Serv Res Pap SE-215, 8 pp
McNatt J D 1975 Humidity effects on properties of structural particleboards from forest residues. USDA For Serv Res Pap FPL-267, 12 pp
McNatt J D 1978 Manufacture and performance of full-size structural flakeboards from Douglas-fir forest residues. In: Structural flakeboard from forest residues. USDA For Serv Gen Tech Rep WO-5:118–132
Megraw R A 1985 Wood quality factors in loblolly pine. Tappi Atlanta, 88 pp
Melkko M 1976 Hakkuutähteiden kasaus liukupuomikuormaimella. (Bunching of logging residues with slide boom loader.) Metsätehon Kats 5/1976, 4 pp
Mellström C, Thörlind M U 1981 Skotning av hyggesrester. (Forwarding of logging residues.) Skogsarbeten Ekonomi 1/1981, 4 pp
Meng W 1976 Der Transport von Ganzbäumen zur zentralen Aufarbeitung unter besonderer Berücksichtigung der dabei verwendeten Transportfahrzeuge. Allg Forstz 87:103–105
Messina M G, Dyck W J, Hunter I R 1985 The nutritional consequences of forest harvesting with special reference to the exotic forests of New Zealand. IEA/FE Project CPC-10, Rep 1, 57 pp
Metz L J, Wells C G 1965 Weight and nutrient content of the above-ground parts of some loblolly pines. U S Dep Agr For Serv Res Pap SE-17, 20 pp
Mihailov G M 1964 (Vibration compaction of chips made of residual wood.) Trudy CNIIME 57, 3 pp
Mikkola P 1969 Metsähukkapuun osuus hakkuupoistumasta Etelä-Suomessa. (Proportion of waste wood in the total cut in southern Finland.) Fol For 74, 20 pp
Mikkola P 1972 Metsähukkapuun osuus hakkuupoistumasta Suomessa. (Proportion of waste wood in the total cut in Finland.) Fol For 148, 15 pp
Miljøministeriet, Skovstyrelsen 1985 Økologiske konsekvenser af øget biomasseudnyttelse i skovene. (Ecologic consequences of increased biomass utilization in the forest.) København, 76 pp
Mingle J C, Boubel R W 1968 Proximate analysis of some western wood and bark. Wood Sci 1:29–36
Mitchell C P, Säll H O 1983 Biomass harvesting, transport and storage-needs, specifications and mechanization. In: Nilsson P O, Zsuffa L (eds) Short rotation forest biomass – production technology and mechanization. Swed Univ Agr Sci Dep Oper Eff Internal Rep 229, 69 pp
Mitchell H L 1961 A concept of intrinsic wood quality, and non-destructive methods for determining quality in standing timbers. U S For Serv For Prod Lab Rep 2233, 14 pp
Mitchell R R, Fischer K 1985 The processing of industrial residues and end use constraints on chipped material. In: For Prod Res Internat – Achievements and the future, Pretoria, Vol 7:14.9, 18 pp
Mittal R K 1979 Energy intensity of various transportation modes. Transport Res Board Natl Acad Sci, Transport Res Rec 689, 25–31 pp
Møller F 1975 Lastbiltransport af heltrae fra nåletraestyndinger. (Lorry transport of whole trees from softwood thinnings.) Skovtekn Inst, 27 pp
Moltesen Nielsen P 1940 Stødrydning. (Harvesting stumps.) Reprint from Dansk Skovforen Tidsskr 181–220
Monsi M, Saeki T 1953 Über den Lichtfaktor in den Pflanzengesellschaften und seine Bedeutung für die Stoffproduktion. Jap J Bot 14:22–52
Mork E 1928 Die Qualität des Fichtenholzes unter besonderer Rücksichtsnahme auf Schleif- und Papierholz. Papier-Fabr 26:741–747
Moslemi A A 1974 Particleboard. 1. Materials. Southern Illinois Univ Press, Carbondale, Ill, 244 pp
Müller-Stoll W R 1948 Photometrische Holzstrukturuntersuchungen. Part II: Über die Beziehungen der Lichtdurchlässigkeit von Holzschnitten zu Rohwichte und Wichtekontrast. Forstwiss Cbl 68:21–63

Münch E 1938 Statik und Dynamik des schraubigen Baues der Zellwand, besonders des Druck- und Zugholzes. Flora 32:357–424

Murakami K, Hirako H, Araya K, Imamura R 1979 Fundamental study on whole-tree pulping of Pinus densiflora. Bull Kyoto Univ For 51:267–273

Murto J O, Kivimaa E 1951 Selluloosapuun lastutus II. (Chipping of pulpwood II.) Teoll Keskuslab Tied 66–68, 33 pp

Muszynski Z, McNatt J D 1984 Investigations on the use of spruce bark in the manufacture of particleboard in Poland. For Prod J 34(1):28–35

Myhrman D 1983 Klippmyran – teknisk beskrivning. (Cutting Ant – a technical description.) Skogsarbeten Results 25, 4 pp

Najdenov V I, Koperin I F, Purim V R 1982 (Physical and chemical properties of the ash of waste wood). Lesn Zhur 2:115–117

Näsi M, Pohjonen V 1981 Green fodder from energy forest farming. J Sci Agr Soc Finland 53(3):161–167

Naturvårdsverket 1983 Miljöeffekter av ved- och torvförbränning. (Environmental effects of wood and peat combustion.) Medd SNV PM 1708, 232 pp

Naylor L M, Schmidt E J 1985 Agricultural use of wood ash as a fertilizer/liming material. 39th Ann Northeast Reg Meet Nat Coun Pap Ind, Boston, 22 pp

Nielson R W, Dobié J, Wright D M 1985 Conversion factors for the forest products industry in western Canada. Forintek Can Spec Publ SP-24 R, 92 pp

Nikiforov G V, Iljuhin J M, Makienko V K 1973 (Conifer vitamin flour made from extracted fir foliage.) Lesn Hozjajstvo 11:92–93

Nilsson I 1984 Whole tree harvesting – a new acidification problem? In: Andersson B, Falk S (eds) Forest energy in Sweden. Swed Univ Agr Sci, 103–105

Nilsson T 1987a Jämförande lagringstudie av småved och bränsleflis. (Comparison of storages of chunk wood and fuel chips.) Sver Lantbruksuniv Inst Virkeslära Rapp 192, 76 pp

Nilsson T 1987b Lagring av småved i mindre stackar. (Storage of chunk wood in smaller piles.) Sver Lantbruksuniv Inst Virkeslära Rapp 196, 57 pp

Nordberg P H, Meier H 1966 Physical and chemical properties of the gelatinous layer in tension wood fibres of aspen (Populus tremula L.) Holzforschung 20(6):174–178

Nordén B 1987 Produktionsstudie – Cabro buntkvistare. (A study on productivity – Cabro bunch delimber.) Skogsarbeten Mimeogr, 8 pp

Nordlinger B G 1976 Use of whole tree chips from the above ground biomass in the particle- and fiberboard industry. Typescript, 14 pp

Nordlinger B G, Tufvesson M I 1975 Helträdsutnyttjande för fiber- och spånskivor. (Whole-tree utilization for particle- and fiberboards.) Rapp Proj Helträdsutnyttjande, 42 pp

Nördlinger H 1878 Liegt an schiefen Bäumen das bessere Holz auf der dem Himmel zugekehrten oder auf der unteren Seite? Cbl Ges Forstwes 4:246–247, 494–495

Norton S A, Young H E 1976 Forest biomass utilization and nutrient budgets. Mimeogr, 17 pp

Novikov A L, Gunjačenko I V 1964 (On the vitamin content of the needles in selected local and introduced tree species.) Botanika 6:208–213

Nurmi J 1986 Chunking and chipping with conescrew chipper. Fol For 659, 23 pp

Nurmi J 1987 Polttohakkeen kuivatus traktorikonteissa. (Drying of fuel chips and chunks in wooden bins.) Fol For 687, 40 pp

Nyholm P 1976 Separate harvesting of fresh stumps and roots, handling of stumpwood at the mill. Symposium on the harvesting of a larger part of the forest biomass. ECE/FAO/ILO TIM/EFC/WP.1/SEM.3/3 Vol 1:121–134

Nykvist N 1977 Skogliga åtgärders inverkan på storlek och tillgänglighet av ekosystemets näringsförråd. (The effect of stand treatment on the quantity and availability of nutrient supply in an ecosystem.) Sver Skogsvårdsförb Tidskr 75:167–178

Nykvist N, Rosén K 1985 Effect of clear-felling and slash removal on the acidity of northern coniferous soils. For Ecol Managem 11:157–169

Nylinder M 1976 Drivning av stubbvirke. (Harvesting of stump wood.) Mimeogr, 12 pp

Nylinder M 1977 Upptagning av stubb- och rotved. (Extraction of stump and root wood.) Skogsarbeten Redogör 5/1977, 19 pp

Nylinder M 1979 Relationstal. Träbränslen – olja. (Conversion factors. Fuelwood – oil.) Sver Lantbruksuniv Inst Virkeslära Rapp 110, 51 pp

Nylinder P 1961 Om träd- och vedegenskapers inverkan på råvolymvikt och flytbarhet. I. Tall. [Influence of tree features and wood properties on basic density and buoyancy. I. Scots pine (Pinus silvestris).] Skogshögsk Inst Virkeslära Rapp Upps 35, 63 pp

Odolinska L 1982 Barking of chips from the forest small wood. Typescript, 10 pp

Olofsson L 1975 Värmevärden för olika delar av tall, gran och björk. (Heating values for different parts of pine, spruce and birch.) Skogshögs Inst Skogstekn Rapp Upps 90, 47 pp

Olsson P 1984 Cutters become felling specialists. In: Andersson P, Falk S (eds) Forest energy in Sweden. Swe Univ Agr Sci: 22–24

Olsson P 1985 Increased utilization from thinnings – some Swedish experience of tree-section logging. IUFRO Conference on thinning, Moscow-Riga, 14 pp

Osman E A, Goss J R 1983 Ash chemical composition, deformation and fusion temperatures for wood and agricultural residues. Am Soc Agr Eng Pap 83-3549, 16 pp

Österlöf P 1979 Markberedning vid stubbrytning. (Scarification aspects of stump harvesting.) Skogsarbeten Redogör 5/1979, 24 pp

Ostermeier D M, Young T M, Weaver L A 1988 Fossil fuel using industries: their awaraness and perceptions of wood fuel. For Prod J 38(10):53–57

Ostlie L D 1988a Artificial drying of whole trees for use as a fuel source for electrical generation. In: Gislerud O (ed) Production, storage and utilization of wood fuels. II. Drying and storage. Sver Lantbruksuniv Inst Skogsteknik Upps Res 134:120–149

Ostlie L D 1988b The whole-tree burner: a new concept in power generation. In: Lönner G, Törnqvist A Proceedings from the workshop "Economic evaluations of biomass oriented systems for fuel". IEA Bioenergy Task III/Activity 4:2–28

Overend R 1982 Wood gasificiation – review of recent Canadian experience. In: Smith R (ed) Energy from forest biomass. Academic Press New York, 171–207

Ovington J D 1957 Dry matter production by Pinus sylvestris L. Ann Bot 21(82):287–314

Page D H 1967 The collapse behaviour of pulp fibres. Tappi 50:449–455

Page I 1975 Stumping with a hydraulic excavator. N Z For Serv For Res Inst For Estab Rep 73, 7 pp

Pahlitzsch R, Sommer I 1965 Erzeugung von Holzschneidspänen mit einem Messerwellen-Spaner. Holz Roh- Werkst 23:403–412

Panshin A J, de Zeeuw C 1980 Textbook of wood technology. McGraw-Hill New York, 722 pp

Papworth R L, Erickson J R 1966 Power requirements for producing wood chips. For Prod J 16 (10):31–36

Patel R N 1965 A comparison of the anatomy of the secondary xylem in roots and stem. Holzforschung 19:72–79

Patel R N 1971 Anatomy of stem and root wood of Pinus radiata D. Don. N Z J For Sci 1:37–49

Paul B H 1939 Variation in the specific gravity of the springwood and summerwood of four species of southern pines. J For 37:478–482

Paulitzsch M 1977 Waldhackschnitzel für die Spanplattenherstellung. Eigenschaften von Spanplatten aus Buchen- und Eichenkronen-Waldhackschnitzeln. Holz-Zbl 103(15):233–234

Pellikka M, Kotimaa M 1983 Polttohakkeen käsittelystä aiheutuva ilman homepölypitoisuus sekä siihen vaikuttavat tekijät. (The mold dust concentration caused by the handling of fuel chips and its modifying factors.) Fol For 563, 18 pp

Pepin R G, Coleman P 1984 Paper mill sludge and ash as soil conditioner. Reprint from Biocycle May/June 1984, 3 pp

Petříček V 1985 Possibilities of mechanization in precommercial thinning with regard to utilization of biomass. IUFRO Conf Thinning, Moscow-Riga, 15 pp

Petíček V 1986 Možnosti zdokonalení motomanuálních technologií v prořezávkách. (Possibilities of improving the powersaw technologies of cleaning.) Lesnictvi 32(3):245–256

Pettersen R C 1984 The chemical composition of wood. In: Rowell R M (ed) The chemistry of solid wood. Am Chem Soc, Adv Chem Ser 207, 614 pp

Philippou J L, Collett B M 1978 Particleboard from young-growth white fir logging residues. For Prod J 28(12):35–41

Phillips D R 1977 Total-tree weights and volumes for understory hardwoods. Tappi 60:68–71

Phillips D R, Clark A III, Taras M A 1976 Wood and bark properties of southern pine branches. Wood Sci 8:164–169

Phillips D R, van Lear D H 1984 Biomass removal and nutrient drain as affected by total-tree harvest in southern pine and hardwood stands. J For 82:547–550

Pohjonen V, Näsi M 1983 Wet biomass as animal feed – fodder as a by-product from energy forests. IEA Forestry Energy Agreement Programme Group B Rep 1, 22 pp

Poller S 1978 Studie über die chemische Zusammensetzung von Wurzel-, Stamm- und Astholz zweier Kiefern (Pinus sylvestris L.) unterschiedlichen Alters. Holztechnologie 19:22–25

Poller S, Scharf P, Schultze-Dewitz C, Zenker R 1973 Untersuchungen über die stofflichen Eigenschaften des Stockholzes der Kiefer (Pinus sylvestris L.). Holztechnologie 14:92–99

Pottié M A, Guimier D Y 1985 Preparation of forest biomass for optimal conversion. FERIC Spec Rep SR-32, 112 pp

Pottié M A, Guimier D Y 1986 Harvesting and transport of logging residuals and residues. FERIC Spec Rep SR-33, 100 pp

Powell L N, Shoemaker J D, Lazar R, Barker R G 1975 Whole-tree chips as a source of papermaking fiber. Tappi 58:150–155

Preston B 1982 Intermediate technology in deriving energy from forest biomass. In: Smith R (ed) Energy from forest biomass. Academic Press New York, 95–107

Producer gas: another fuel for motor transport. 1983. Report of an ad hoc panel of the Advisory Committee on Technology Innovation. Natl Acad Washington D C, 101 pp

Projekt hälträdsutnyttjande 1977a Helträdsutnyttjande inom massaindustrin. Slutrapport från projektgrupp massa. (Whole-tree utilization in the pulp industry. Final report of project group pulp.) Stockholm, 64 pp

Projekt hälträdsutnyttjande 1977b Helträdsutnyttjande inom skivindustrin. Slutrapport från projektgrupp skivor. (Whole-tree utilization in the board industry. Final report of project group Boards.) Stockholm, 27 pp

Puuvaraselvitys 1976 – Suomen raakapuu-, teollisuus-, jätepuu- ja metsätähdetase vuoteen 1980 sekä metsäteollisuuden puuraaka-ainenäkymät vuoteen 2000. (Timber report 1976 – Finland's roundwood, industrial residue and forest residue balances by 1980 as well as the wood raw material situation of Finnish forest industries by the year 2000.) Silva Fenn 10:314–336

Quellet D 1983 Biomass production equations for twelve commercial species in Quebeck. Can For Serv Inf Rep LAU-X-62E, 34 pp

Rahmlow H 1977 Nadelgrüngut als Futterreserve. Sozial Forstwirtsch 27(4):106–107

Raivonen M, Leikola M 1980 The influence of the removal of logging waste on the initial development of planted Norway spruce seedlings. Fol For 429, 13 pp

Ramaker T J, Lehmann W F 1976 High-performance structural flakeboards from Douglas-fir and lodgepole pine forest residues. USDA For Serv Res Pap FPL-286, 20 pp

Rauscher H M 1980 An analysis of nitrogen removal from southern Appalachian oak ecosystems by harvesting. Virginia Pol Inst State Univ, 146 pp

Reinikainen A 1980 Tuhkalannoituksen ekologiaa. (Ecology of ash fertilization.) Metsäntut Muhoksen tutk tied 20:24–27

Renewable energy technology: issues in the transfer and development of biomass-ethanol technology. 1985. United Nations, UNCTAD/TT/64, 30 pp

Rennie P L 1957 The uptake of nutrients by timber forest and its importance to timber production in Britain. Q J For 51:101–105

Resource Conservation Services, Inc 1986 The S.D. Warren bioash program. Brochure, 8 pp

Rheborg H 1972 Kvistkvantiteter vid helträdslunning. (A study of quantities of branches and tops gathered at the landing when skidding full trees). Skogshögs Inst Skogstekn Rapp Upps 48, 33 pp

Ribe J H 1973 Puckerbrush weight tables. Life Sci Agr Exp Stat Univ Maine Misc Rep 152, 92 pp

Richardson R 1986 Evaluation of Bruks off-road chippers. FERIC Tech Rep TR-71, 36 pp

Riedl H 1937 Bau und Leistungen des Wurzelholzes. Jahrb Wiss Bot 85:1–75

Riekko R, Anttila E 1980 Nestemäisten tuotteiden valmistus puusta ja turpeesta. (Manufacturing of liquid products from wood and peat.) SITRA Sarja B 61, Helsinki, 153 pp

Riley R 1984 The interactions of harvesting and stand establishment operations: an overview and perspective. In: Corcoran T, Gill D R A (eds) Proceedings COFE/IUFRO, 47–51

Rodgers H W 1948 The wood chipper. Pap Ind 30(6):883–888

Rogers K E 1981 Preharvest drying of logging residues. For Prod J 31(12):32–36

Rosén K 1984 Less leaching after removal of slash. In: Andersson B, Falk S (eds) Forest energy in Sweden. Swed Univ Agr Sci, 94–95

Rosén K 1986 Increased nitrogen leaching under piles of slash – a consequence of modern logging systems. In: Ågren G I (ed) Predicting consequences of intensive forest harvesting on long-term productivity. Swed Univ Agr Sci Inst Ekol Miljövård Rapp 26:173–175
Routhier J-G 1982 Implications of full-tree harvesting for biomass recovery. ENFOR Project P-54, LAU-X-50E, 132 pp
Runkel R 1942 Zellwandforschung unter dem Blickpunkt des technischen Einsatzes. Zugleich IV. Holz Roh-Werkst 5:413–427
Ruth R H, Harris A S 1975 Forest residues in hemlock-spruce forests on the Pacific Northwest and Alaska. A state-of-knowledge review with recommendations for residue management. U S For Serv Gen Tech Rep PNW-39, 52 pp
Saarela I 1982 Tuhka maanparantajana. (Ash as a soil improvement agent.) Pellervo 1/1982:32–35
Sabalitschka T, Kleffner U 1946 Einfluss von Alter und Jahreszeit auf den Vitamin C-Gehalt der Nadeln von Fichte und Kiefer. Pharmazie 1:90–92
Säll H-O 1981 Development of harvesters for energy forest plantations. In: Mattsson J E, Nilsson P O (eds) Proceedings of the international conference "Harvesting and utilization of wood for energy purposes". Inf Proj Forestry Energy 19:118–131
Salmi J 1969 Lahovikaisuuden vaikutus koivupuun lämpöarvoon. (The influence of decay on the heat value of birch wood.) Pienpuu Toimik Julk 168, 26 pp
Sanio C 1860 Einige Bemerkungen über den Bau des Holzes. Bot Ztg 18:193–198, 201–204, 209–217
Sanio K 1873 Anatomie der gemeinen Kiefer (Pinus silvestris L.). Jahrb Wiss Bot 9:50–126
Sarkanen K V, Ludwig C H 1971 Lignins. Occurrence, formation, structure and reactions. Wiley New York, 916 pp
Sarvas R 1964 Havupuut. (The conifers.) Werner Söderström Porvoo, 518 pp
Satonaka S 1982 Carbonization and gasification of wood. In: Smith R (ed) Energy from forest biomass. Academic Press New York, 147–154
Satoo T 1968 Primary production relations in woodland of Pinus densiflora. In: Young H E (ed) Symposium on primary productivity and mineral cycling in natural ecosystems. Univ Maine Press Orono, 51–80
Satoo T 1974 Primary production relations of a young stand of Metasequoia glyptostroboides planted in Tokyo: materials for the studies of growth in forest stands. 13. Bull Tokyo Univ For 66:153–164
Satoo T, Madgwick H A I 1982 Forest biomass. Martinus Nijhoff/Dr. W. Junk, The Hague Boston London, 152 pp
Saucier J R, Phillips D R 1985 Storing whole-tree fuelwood chips for maximum energy. For Prod J 35(6):53–56
Schaafsma A H, Hoksbergen F T J 1980 De motorkettingzaag met velframe. (Motor chain saw with felling device.) Ned Bosbouw Tijdschr 52(10):257–262
Scherman S, Nordén B 1983 Kvistning av träddelar med AC Invests (Hydrovågs) buntkvistare. [Delimbing tree sections with the batch delimber of AC Invest (Hydrovåg).] Skogsarbeten Results 27, 4 pp
Schertz F M 1929 Seasonal variation of chloroplast pigments in several plants on the Mall at Washington D. C. Plant Physiol 4:135–139
Schiefelbein G F 1985 Overview for DOE's biomass thermochemical conversion program. In: Robertson D, White J (eds) Seventh International FPRS Industrial Wood Energy Forum '83, Vol II, For Prod Res Soc Proc 47338:263–273
Schiess P, Schuh D, Miyata E S, Mann C N 1983 Concept evaluation of a walking feller-buncher, the Kaiser X5M Spyder. First Tech Conf Timber Harv Central Rockies, Colorado State Univ, 50 pp
Schulz H 1959 Güteklassen des Stammholzes und ihre Abgrenzung gegeneinander. Holz Zbl 85(57):753–754
Scott R W, Millett M A, Hajny G J 1969 Wood wastes for animal feeding. For Prod J 19(4):14–18
Sellaeg M 1974 Grønn flis fra grantynninger i trefiberplateindustrien. (Green chips of spruce thinnings for use in the fiberboard industry.) Driftstek Rapp 12:103–116
Seth R S, Page D H 1988 Fiber properties and tearing resistance. Tappi 81:103–107
Shanks D, Parsons L, Quick G R 1979 Rome tree extractor. Logging Res Rev 2/1979, 4 pp
Shumakov V S 1971 (Application of mineral fertilizers in forests.). Lesovedenie 4:71–79
Siemon G R, Wood G B, Forrest W G 1976 Effects of thinning on crown structure in radiata pine. N Z J For Sci 6(1):57–66
Silfverberg K, Huikari O 1985 Tuhkalannoitus metsäojitetuilla turvemailla. (Wood-ash fertilization on drained peatlands.) Fol For 633, 25 pp

Silversides C R 1981 Innovative transportation in the 2000s. In: Weyerhaeuser Sci Symp Forest-to-Mill Challenges of the Future, 23–48
Simola P 1977 Pienikokoisen lehtipuuston biomassa. (The biomass of small-sized hardwood trees.) Fol For 302, 16 pp
Simola P, Mäkelä M 1976 Rasiinkaato kokopuiden korjuussa. (Leaf-seasoning method in whole-tree logging.) Fol For 273, 18 pp
Simonsson N 1974 Avverkningsavfallets fördelning efter mekaniserad slutavverkning. (Distribution of logging residues after mechanized final cutting.) Skogsarbeten Ekonomi 5/1974, 4 pp
Sinclair A W J 1984 Recovery and transport of forest biomass in mountainous terrain. Can For Serv Inf Rep BC-X-254, 31 pp
Sirén G 1973 Protein ur skogsträd. (Protein from forest trees.) Svensk Naturvetenskap 26:41–46
Sirén G 1975 Hyggesavfallets borttagande och skogsodlingsresultat. (Removal of logging residues and the regeneration result.) Skogen 1/1975:30–33
Sirén G, Blombäck B, Aldén T 1970 Proteins in forest tree leaves. Skogshögsk Inst Skogsföryngring Rapp Upps 28, 22 pp
Sirén M 1986 Puuston vaurioituminen karsimattomien puiden ja puunosien korjuussa. (Stand damage in logging of undelimbed trees and tree parts.) Fol For 645, 7 pp
Sirjaeva G A 1967 (Seasonal changes in the carotenoid contents of Scots pine in different habitats.) Dokl Akad Nauk SSSR (1966) 172:733–736
Sirois D L 1981a Some engineering aspects of the Nicholson-Koch mobile chipper. Am Soc Agr Eng Summer Meeting Orlando, 11 pp
Sirois D L 1981b A mobile harvester for utilization of weed trees and residues. In: Weed control in forest management. Proc 1981 John S Wright For Conf, Purdue Univ, 177–191
Sjöholm R T 1977 Acetone-water extractives from the stump and root softwood and the influence of the removal of these through pre-extraction on the pulping. Acta Acad Aboensis Ser B 37(1), 130 pp
Sjöström E 1981 Wood Chemistry. Fundamentals and applications. Academic Press New York, 233 pp
Skalskij M U 1960 (The use of vibration method in stump extraction.) Lesn Hoz 12/1960
Skogsstyrelsen 1978 Om skydd mot insektsskador på skogsplantor. (On the protection against insect damage to forest tree seedlings.) Stockholm, 161 pp
Skogsstsyrelsen 1985 Skogstryrelsens almänna råd om begränsning vid uttag av träddelar utöver stamvirke på skogsmark. (General directions of Forest Service concerning restrictions in connection with harvesting of tree parts in addition to stem timber from forest lands.) Mimeogr, 6 pp
Sloan E 1958 The seasons of America past. Wilfred Funk New York, 150 pp
Snell J A K, Brown J K 1978 Comparison of tree biomass estimators – dbh and sapwood area. For Sci 24:455–457
Snell J A K, Brown J K 1980 Handbook for predicting residue weights of Pacific Northwest conifers. U S For Serv Gen Tech Rep PNW-103, 43 pp
Söderström V 1974 Hur inverkar hyggesavfallets borttagande på planteringsresultatet. (Effect of slash removal on regeneration result.) Inst Skogsteknik Rapp Upps 76:94–116
Sondell J 1984 Removal and transport of logging residue. In Andersson B, Falk S (eds) Forest energy in Sweden. Swed Univ Agr Sci, 38–39
Spångberg E 1987 Buntkvistning av träddelar i Carboverkens buntkvistare. (Processing tree sections with Carboverken bunch delimber.) Skogsarbeten Mimeogr, 10 pp
Spiers J J K 1984 Tree harvesting. In: Corcoran T, Gill D R A (eds) Proceedings COFE/IUFRO, 39–40
Springer E L 1979 Should whole-tree chips for fuel be dried before storage? U S For Serv For Prod Lab Res Note FPL-0241, 5 pp
Springer E L 1980 Potential problems in storing whole-tree chips. Proceed annual meeting southeastern section. For Prod Res Soc Savannah, Proc P-81-30:31–35
Springer E L, Zoch L L Jr, Feist W C, Hajny G J, Hemingway R W 1978 Storage characteristics of southern pine whole-tree chips. In: McMillin C W (ed) Complete tree utilization of southern pine. For Prod Res Soc Madison, 216–223
Sproull R C, Parker R B, Belvin W L 1957 Whole tree harvesting. For Prod J 7:131–134
Spurr S H, Hsiung W Y 1954 Growth rate and specific gravity in conifers. J For 52(3):191–200
Spurrell R M, Thorn P A, Adams T N 1987 Factors affecting combustion efficiency in wood-fired boilers. Tappi 70:100–106

Staaf H 1984 Harvesting of logging slash on highly productive sites. Buffering of negative effects. In: Andersson B, Falk S (eds) Forest energy in Sweden. Swed Univ Agr Sci:96–97

Staaf H, Björkroth G 1981 Complete tree utilization and soil fertility in Swedish forests. In: Mattsson J E, Nilsson P O (eds) Proc "Joint IEA/IUFRO Forestry Energy Workshop" in Garpenberg, Sweden, October 2, 1980:82–99

Stade Y 1976 The economics of using different biocomponents from the complete tree as raw material for kraft pulp. Symposium on extending the use of wood residues, Bucarest. TIM/SEM.5/R.26, 9 pp

Stamm A J 1969 Correlation of structural variations of lignins with their specific gravity. Tappi 52:1498–1502

Stamm A J, Sanders H T 1966 Specific gravity of the wood substance of loblolly pine as affected by chemical composition. Tappi 49:397–400

Standish J T, Manning G H, Demaerschalk J P 1985 Development of biomass equations for British Columbia tree species. Can For Serv Pac For Res Cent Inf Rep BC-X-264, 47 pp

Stanek W, State D 1978 Equations predicting primary productivity (biomass) of trees, shrubs, and lesser vegetation based on current literature. Can For Serv Pac For Res Cent Inf Rep BC-X-183, 58 pp

Stanković S, Senić R 1954 [About quantitative changes of essential oil, carotene and l-ascorbic acid in white pine (Pinus silvestris L.) and black pine (Pinus nigra Arn.) needles in a year.] Glas Sumarskog Fakulteta 8:291–312

Starck P, Palenius I 1970 Kuorineen ja neulasineen haketetun harvennuspuun käyttökelpoisuus sulfaattimassan valmistukseen. (Suitability of timber from thinnings for the production of sulfate pulp when chipped with bark and crown intact.) Keskuslab Sel 1005, 53 pp (unpubl)

Statkov N, Mateev A, Marinov T, Stoykov H 1969 (Timber waste in felling and possibilities for its use in industry.) Gorskostop Nauka 6(6):63–76

Statkov N, Mateev A, Marinov T, Stoykov K 1970 (Logging residues and their utilization possibilities.) Zemizdat Sofia, 146 pp

Statkov N, Mateev A, Marinov T, Stoykov K 1976 (Felling waste and possibilities for its use in industry.) In: Symposium on the harvesting of a larger part of the forest biomass. TIM/EFC/WP.1/SEM.3/3 Vol 1:49–55

Steinhilb H M, Erickson J R 1970 Weights and centers of gravity for quaking aspen trees and boles. USDA For Serv Res Note NC-91, 4 pp

Steinhilb H M, Erickson J R 1972 Weights and centers of gravity for red pine, white spruce, and balsam fir. USDA For Serv Res Pap NC-75, 7 pp

Stenlid J 1987 Controlling and predicting the spread of Heterobasidion annosum from infected stumps and trees of Picea abies. Scand J For Res 2:187–198

Stephenson E H 1986 Bell feller buncher saw. APA Tech Rel 86-R-53, 2 pp

Stevens D J 1985 Liquid fuels from biomass: an overview of research activities funded by DOE's biomass thermochemical conversion program. In: Robertson D, White J (eds) Seventh International FPRS Wood Energy Forum '83 Vol II, For Prod Res Soc Proc 47338:433–441

Stockman L 1962 The influence of some morphological factors on the quality of spruce sulphite and pine sulphate pulp. Svensk Papperstidn 65(23):978–982

Stocks B J 1980 Black spruce crown fuel weights in northern Ontario. Can J For Res 10:498–501

Stokes B J 1985 Flail-delimbing of loblolly pine – a case study. U S For Serv Res Note SO-315, 7 pp

Stokes B J, Lanford B L 1985 Prebunching and skidding functions in thinnings. Am Soc Agr Eng Pap 85-1594, 9 pp

Stokes B J, Sirois D L 1986 Evaluation of chipper-forwarder biomass harvesting concept. In: Rockwood D L (ed) Proceedings of the 1985 southern forest biomass workshop, Inst Food Agr Sci Univ Florida, 23–26

Stokes B J, Watson W F 1986 Integration of biomass harvesting and site preparation. In: Rockwood D L (ed) Proceedings of the 1985 southern forest biomass workshop, Inst Food Agr Sci Univ Florida, 62–67

Stout B F 1968 Root Systems of deciduous trees. Am Nurseryman 127(1):13, 120–122

Stuart W B, Walbridge T A Jr, King F, Farrar K D 1981 Economic feasibility of modifying conventional harvesting systems to recover residue in bale form. In: Stumbo D A (ed) Utilization of low-grade southern hardwoods. Feasibility studies of 36 enterprises. For Prod Res Soc Madison, 57–60

Sturos J A 1972 Determining the terminal velocity of wood and bark chips. USDA For Serv Res Note NC-131, 4 pp
Sturos J A 1973a Segregation of foliage from chipped tree tops and limbs. USDA For Serv Res Note NC-146, 4 pp
Sturos J A 1973b Predicting segregation of wood and bark chips by differences in terminal velocities. USDA For Serv Res Pap NC-90, 8 pp
Sturos J A 1978 Bark, foliage, and grit removal from whole-tree chips – results and economics. In: Tappi Pulping Conf Proc New Orleans, 121–134
Sturos J A 1984 Characterization and air drying of chunkwood and chips. USDA For Serv Res Note NC-308, 6 pp
Sturos J A 1989 Airflow resistance of chunkwood. For Prod J 39(3):19–24
Sturos J A, Coyer L A, Arola R A 1983 Air drying of chunkwood and chips. USDA For Serv Res Note NC-293, 4 pp
Sturos J A, Marvin J L 1978 A process for removing bark, foliage and grit from southern pine whole-tree chips. In: McMillin C W (ed) Complete tree utilization of southern pine. For Prod Res Soc Madison, 277–293
Sundberg U 1981 Joule, labour and dollar. In: Mattsson J E, Nilsson P O (eds) Proceedings of the international conference "Harvesting and utilization of wood for energy purposes". Inf Proj For Energy 19:197–200
Sundberg U, Silversides R 1988 Operational efficiency in forestry. Vol 1: Analysis. Kluwer Academic Dordrecht, 219 pp
Sutherland B J 1985 Brush harvester development and field test. Am Soc Agr Eng Pap 85-1628, 23 pp
Switzer G L, Nelson L E 1972 Nutrient accumulation and cycling in loblolly pine (Pinus taeda L.) plantation ecosystems: the first twenty years. Proc Soil Sci Soc Am 36:143–147
Switzer G L, Nelson L E, Hinesley L E 1978 Effects of utilization on nutrient regimes and site productivity. In: McMillin C W (ed) Complete tree utilization of southern pine. For Prod Res Soc Madison, 91–102
Tadaki Y 1976 (Biomass of forests with special reference to the leaf biomass of forests in Japan.) J Jap For Sci 58:416–423
Tamm C O 1969 Site damages by thinning due to removal of organic matter and plant nutrients. In: Thinning and mechanization, IUFRO meeting, R Coll For Stockholm, 175–179
Tamm C O 1975 Losses of plant nutrients and other adverse effects of full tree logging. Pap Elmia 1975, 10 pp
Tamminen Z 1962 Fuktighet, volymvikt, m.m. hos ved och bark. I. Tall. (Moisture content, density and other properties of wood and bark. I. Scots pine). K Skogshögsk Inst Virkeslära Rapp Upps 41, 46 pp
Tamminen Z 1964 Fuktighet, volymvikt, m.m. hos ved och bark. II. Gran. (Moisture content, density and other properties of wood and bark. II. Norway spruce). Skogshögsk Inst Virkeslära Rapp Upps 47, 56 pp
Taras M A 1978 Biomass of southern pines and their product yields. In: McMillin C W (ed) Complete tree utilization of southern pine. For Prod Res Soc Madison, 15–23
Tasman J E, Berzins V 1957 The permanganate consumption of pulp materials. III. The relationship of the Kappa number to the lignin content of pulp materials. Tappi 40:699–704
Tauber B A 1979 (The development prospects of all biomass of the tree.) Lesn Promst 60(6):5–6
Taylor F W 1968 Variations in the size and proportions of wood elements in yellow-poplar trees. Wood Sci Technol 2:153–165
Taylor F W 1977 A note on the relationship between branch- and stemwood properties of selected hardwoods growing in the mid-south. Wood Fiber 8(4):257–261
Taylor F W 1979 Property variation within stems of selected hardwoods growing in the mid-south. Wood Sci 11:193–199
Technical and economic evaluations of biomass utilization processes. 1980. DOE/ET/20605-T4, Tech Rep 1, 352 pp
Temler J 1980 Hardwood and softwood whole tree chips in different pulping processes. Pulp Pap Can 81(3):122–125
The nation's unused wood offers vast potential energy and product benefits. Comptroller General report to the Congress of the United States. 1981. EMD-81-6, 115 pp

Thorbjörnson B 1922 Några synpunkter beträffande sulfitkokning. (Some viewpoints concerning sulphite cooking.) Svensk Papperstidn 25:196–199, 252–254, 307–310, 346–351

Thörnqvist T 1982 Torkning av bränsleflis med solfångaruppvärmd luft. (Drying of fuel chips with solar panel warmed air.) Sver Lantbruksuniv Inst Virkeslära Rapp Upps 126, 54 pp

Thörnqvist T 1984 Storing of slash. Moisture content decisive. In: Andersson B, Falk S (eds) Forest energy in Sweden. Swed Univ Agr Sci, 76–79

Thörnqvist T 1985a Trädbränslekvalitet – vad är det? (Wood fuel quality – what is it?) Sver Lantbruksuniv SIMS Upps 14, 18 pp

Thörnqvist T 1985b Storing of forest residues and comminuted fuels from forest residues. Sver Lantbruksuniv Inst Virkeslära Rapp Upps 153, 4 pp

Thörnqvist T 1985c Drying and storage of forest residues for energy production. Sver Lantbruksuniv Inst Virkeslära Rapp Upps 155, 10 pp

Thörnqvist T 1986a Projekt storskalig säsonglagring av träbränsle – en sammanfattning av etapp 1. (Large-scale seasonal storage of tree fuel project – summary of phase 1.) Sver Lantbruksuniv Inst Virkeslära Rapp Upps 188, 50 pp

Thörnqvist T 1986b Trädbränslet förändras mellan skog och panna. (Wood fuel changes from forest to boiler.) In: Bäckström P-O Skogen i energiförsörjningen. (Forest in the energy supply.) Sver Lantbruksuniv Skogsfakta 10:73–77

Thörnqvist T 1987 Bränder i stackar med sönderdelat trädbränsle. (Spontaneous combustion in piles with comminuted wood fuel.) Sveriges Lantbruksuniv Inst Virkeslära Rapp Upps 163, 17 pp

Thörnqvist T, Gustafsson G 1983 Kalluftstorkningens betydelse för bränsleflisens lagrinsbarhet. (The importance of air drying for the storing property of fuel chips.) Sveriges Lantbruksuniv Inst Virkeslära Rapp Upps 142, 70 pp

Timell T E 1964 Wood hemicelluloses I. Adv Carbohyd Chem 19:247–302

Timell T E 1965a Wood hemicelluloses II. Adv Carbohyd Chem 20:409–483

Timell T E 1965b Wood and bark polysaccharides. In: Côté W A Jr (ed) Cellular ultrastructure of woody plants. Syracuse Univ Press Syracuse, 127–156

Timell T E 1967 Recent progress in the chemistry of wood hemicelluloses. Wood Sci Technol 1:45–70

Timell T E 1969 The chemical composition of tension wood. Svensk Papperstidn 72:173–181

Timell T E 1973 Studies on opposite wood in conifers. II. Histology and ultrastructure. Wood Sci Technol 7:79–91

Timell T E 1981 Recent progress in chemistry, ultrastructure, and formation of compression wood. The Ekman-Days 1981, Stockholm SPCI Rep 38 Vol 1:99–147

Timell T E 1982 Recent progress in the chemistry and topochemistry of compression wood. Wood Sci Technol 16(2):83–122

Timell T E 1986a Compression wood in gymnosperms 1. Bibliography, historical background, determination, structure, chemistry, topochemistry, physical properties, origin, and formation of compression wood. Springer Berlin Heidelberg New York Tokyo 1–706

Timell T E 1986b Compression wood in gymnosperms 2. Occurrence of stem, branch, and root compression woods, factors causing formation of compression wood, gravitropism and compression wood, physiology of compression wood formation, inheritance of compression wood. Springer Berlin Heidelberg New York Tokyo, 707–1338

Timell T E 1986c Compression wood in gymnosperms 3. Ecology of compression wood formation, silviculture and compression wood, mechanism of compression wood action, compression wood in the lumber and pulp and paper industries, compression wood induced by the balsam woolly aphid, opposite wood. Springer Berlin Heidelberg New York Tokyo, 1339–2150

Tomaselli I 1982 Liquidification of wood. In: Smith R (ed) Energy from forest biomass. Academic Press New York, 209–220

Tomčuk R I, Ryzhilo L E 1968 (Carotene content of spruce and fir needles.) Lesn Hozjaistvo 21(5):22–23

Tomčuk R I, Tomčuk G N 1966 (Tree foliage and its utilization.) Moscow, Izdatel'stvo Lesnaja Promyshlennost, 241 pp

Toropainen M 1982 Kotimaisten polttoaineiden käyttöön siirtymisen kannattavuus ja julkinen rahoitustuki. (Profitability of the use of indigenous fuels and public funding.) Metsäntut Tied 54, 111 pp

Travers W 1976 The challenge of using the whole tree. Symposium on extending the use of wood residues, Bucarest. TIM/SEM.5/R.26, 9 pp

Tree biomass – a state-of-the-art compilation. 1981. U S For Serv Gen Tech Rep WO-33, 34 pp
Trendelenburg R 1939 Das Holz als Rohstoff. Lehmanns München, 435 pp
Trendelenburg R, Mayer-Wegelin H 1955 Das Holz als Rohstoff. Hanser München, 541 pp
Tryon T, Edson D 1979 Forest biomass inventories in the northeastern United States – state of the art. In: Frayer W E (ed) Forest resources inventories II. Colorado State Univ, 730–735
Tsoumis G 1968 Wood as raw material. Source, structure, chemical composition, growth, degradation and identification. Pergamon Oxford, 276 pp
Turkia K, Kellomäki S 1987 Nuorten mäntyjen oksien läpimitan riippuvuus kasvupaikan viljavuudesta ja puuston tiheydestä. (Influence of the site fertility and stand density on the diameter of branches in young Scots pine stands.) Fol For 705, 16 pp
Tveite B 1983 Heiltretynning – fare for tillvekstap. (Whole-tree thinning – danger of growth loss.) Aktuelt fra statens fagtjeneste for landbruket 3:98–105
UNIDO 1983 First world-wide study of the wood and wood processing industries. Sectorial Studies Series 2, UNIDO/IS. 398, 195 pp
United Nations 1985 The forest resources of the ECE region (Europe, the USSR, North America). ECE/TIM/27, 223 pp
Ursic S J 1978 Effect of harvesting systems on water quality and supply. In: McMillin C W (ed) Complete tree utilization of southern pine. For Prod Res Soc, Madison, 103–107
U. S. Department of Agriculture Forest Service 1978 Final report – Forest residues energy program. U S Dept Agr For Serv, North Central For Exp Stat, 297 pp
Uusvaara O 1984 Hakepuun kosteuden alentaminen ennen haketusta korjuuseen ja varastointiin liittyvin toimenpitein. (Decreasing the moisture content of chip wood before chipping; harvesting and storage measures.) Fol For 599, 30 pp
Uusvaara O, Verkasalo E 1987 Metsähakkeen tiiviys ja muita teknisiä ominaisuuksia. (Solid content and other technical properties of forest chips.) Fol For 683, 53 pp
Van Wagner C E 1968 The line intersect method in forest fuel sampling. For Sci 14:20–26
Vasil'ev I, Demin K, Man'kovskij D 1970 (Mechanized harvesting of mature stumps.) Lesn Promst 7/1970:23–24
Verkasalo E 1987 Metsähakkeen kosteuden ja kuivamassan mittaus kuormaotantamenetelmällä. (Measurement of moisture content and dry weight of forest chips by load sampling methods.) Fol For 694, 35 pp
Vidrine C G, Woodson G E 1982 Bulk densities of selected materials from pine-site hardwoods. For Prod J 32(7):21–24
Virkola N-E 1976 Development and trends in industrial utilization of fuller forest biomass (branchwood, foliage and stumps) in traditional wood industry. In: Symposium on the harvesting of a larger part of the forest biomass. ECE/FAO/ILO TIM/EFC/WP.1/SEM.3/3 Vol 1:31–48
Viro P 1969 Prescribed burning in forestry. Com Inst For Fenn 67.7, 49 pp
Virtanen A I, Koistinen O A 1944 Fermentation of the native cellulose and pentosans in wood. Svensk Kem Tidn 56:391–400
Virtanen A I, Nikkilä O E 1946 Cellulose fermentation in wood dust. Suomen Kemistilehti 19B: 3–4
Virtanen P 1963 Fuel properties of barking refuse from Finnish tree species. Pap Puu 45:313–330
Visser J 1987 Evaporation as parameter to measure mass loss in biological drying pine trees. In: Malinovski J R (ed) Simpósio sobre exploracao transporte, ergonomia e seguranca em reflorestamentos. (Meeting on harvesting, transport, ergonomics and safety in plantation forestry.) Curitiba, Brazil, 386–401
Voipio R 1987 Puiden biomassan vitamiinipitoisuus. (Vitamin content of tree biomass.) Fol For 682, 30 pp
Von Schantz M, Hiltunen R 1983 Familie Pinaceae (Coniferae). Comment Physico-Math 57, 149 pp
Voronicyn K I, Pavlov V P 1968 (Use of logging residues in the production of fiberboards.) Symposium on the Mechanization of Harvesting of Small-Sized Wood and Logging Residues. FAO/ECE/LOG/208, 6 pp
Vurdu H, Bensend D W 1979 Specific gravity and fiber length in European black alder roots, branches and stems. Wood Sci 12:103–105
Wahlman M 1967 Importance of chip thickness in alkaline pulping, and chip thickness analysis. Pap Puu 49:107–110

Walbridge T A Jr, Stuart W B 1981 An alternative to whole tree chipping for the recovery of logging residues. In: Mattsson J E, Nilsson P O (eds) Proceedings of the international conference "Harvesting and utilization of wood for energy purposes". Inf Proj Forestry Energy 19:132–148

Waldschäden in der Bundesrepublik Deutschland. Ergebnisse der Waldschadenserhebung 1987. 1987. Schriftenreihe des Bundesministers für Ernährung, Landwirtschaft und Forsten, Reihe A 349, 89 pp

Walker R F 1976 Stump extraction and utilization – a literary review with regard to conditions in New Zealand. N Z For Serv For Res Inst For Estab Rep 93, 25 pp (unpubl)

Wallden J 1982 Riksskogstaxeringens inventering av virkesspill 1980. (Survey of logging residue by the National Forest Survey of 1980.) Sver Lantbruksuniv Inst Skogstaxering Rapp 33, 48 pp

Waring R H, Schlesinger W H 1985 Forest ecosystems. Concepts and management. Academic Press London, 340 pp

Warren W G, Olsen P F 1964 A line intersect technique for assessing logging waste. For Sci 10:267–273

Watanabe H, Tsutsumi J, Kanagawa H 1962 Properties of branch wood; especially on specific gravity, tracheid length, and appearance of compression wood. Bull Kyushu Univ For 35:91–96

Watson A J, Dadswell H E 1962 Influence of fibre morphology on paper properties. Appita 15:116–128

Watson W F, Miller D E, Stokes B J, Broussard M L 1986b Energy budget for an energywood harvesting system. In: Proceedings of the 1986 southern forest biomass workshop, Tennessee Valley Authority, 103–105

Watson W F, Sabo R F, Stokes B J 1986c Productivity of in-woods chippers processing understory biomass. In: Improving productivity through forest engineering. Counc For Engineer Mobile Alabama, 69–72

Watson W F, Stokes B J 1984 Evaluating the opportunity for reducing site preparation costs by increasing the utilization at final harvest. Final report. FS-SO-3701-832-2, 26 pp

Watson W F, Stokes B J, Savelle J W 1986a Comparisons of two methods of harvesting biomass for energy. For Prod J 36(4):63–68

Wawer A 1975 Bark in hardwood chips – effect on mill operations. Pulp Pap Can 76(7):51–54

Wawer A, Misra N 1977 Compression debarking whole tree chips. Pulp Pap Can 78(11):93–99

Wegelius T 1939 The presence and properties of knots in Finnish spruce. Acta For Fenn 48(1):1–191

Wegelius T 1946 Det finska granvirkets egenskaper och kvalitetsvariationer. (Properties and variations in quality of Finnish spruce wood.) Svensk Papperstidn 49:51–61

Welch R L 1980 Living residues in the South Atlantic States. For Prod J 30(6):37–39

Wells G C 1980 The Bell logger. LIRA Mach Eval 5(3), 4 pp

Wells G C, Metz L J 1963 Variation in nutrient content of loblolly pine needles with season, age, soil, and position on the crown. Soil Sci Soc Am Proc 27:90–93

Wellwood R W 1979a Complete-tree utilization. An analysis of literature (1970–1978). I. Biomass of forest tree components. ENFOR P-15, 89 pp

Wellwood R W 1979b Complete-tree utilization. An analysis of literature (1970–1978). II. Recovery of biomass. ENFOR P-15, 150 pp

White E H, Pritchett W L, Robertson W K 1971 Slash pine root biomass and nutrient concentrations. In: Forest biomass studies. Univ Maine Coll Life Sci Agr, 165–176

White H M 1983a Progress in the production of ethanol from cellulose. DOE/CE/50305-1, Washington D C, 36 pp

White H M 1983b A survey of current biomass gasification processes for methanol production. DOE/CE/50305-2, Washington D C, 24 pp

White M S, Argent R M, Sarles R L 1986 Effects of outside storage on the energy potential of hardwood particulate fuels: Part III. Specific gravity, ash content, and pH of water solubles. For Prod J 36(4):69–73

White M S, Curtis M L, Sarles R L, Green D W 1983a Effects of outside storage on the energy potential of hardwood particulate fuels: Part I. Moisture content and temperature. For Prod J 33(6):31–38

White M S, Curtis M L, Sarles R L, Green D W 1983b Effects of outside storage on the energy potential of hardwood particulate fuels: Part II. Higher and net heating values. For Prod J 33(11/12):61–65

Whyte A G D 1973 Productivity of first and second crops of Pinus radiata on the Mountere gravel soils of Nelson. N Z J For 18:87–103

Wiant H W Jr 1979 Estimating woody biomass using 3P and point sampling. In: Frayer W E (ed) Forest resource inventories II. Colorado State Univ, 757–762
Wiklund M 1970 Provsågning av timmer innehållande stubbvirke. (Test sawing of timber containing stump wood.) Svenska Träforskn Redogör 710204/963/50, 39 pp
Wiklund M 1971 Ökad råvarutillgång genom sänkt stubbhöjd. (Increased raw material availability through lower stump height.) Sågverken/Trävaruindustrin 10/1971:54–60
Williams D F 1976 Forest fuels in unthinned radiata pine stands. Aust For 39:238–244
Wilson B F 1964 Structure and growth of woody roots of Acer rubrum L. Harward Univ For Pap 11, 14 pp
Winsauer S 1984 Simulation of mechanized felling in dense softwood plantations. In: Corcoran T, Gill D R (eds) Proceedings COFE/IUFRO, 175–180
Wippermann H-J 1985 Central processing and associated transportation techniques in central Europe. In: Mitchell P, Nilsson P O, Zsuffa L (eds) Research in forestry for energy. Sver Lantbruksuniv Inst Skogstekn Upps Res 49:430–445
Wisherd K D, Wilson J B 1979 Bark as a supplement to wood furnish for particleboard. For Prod J 29(2):35–39
Wokes F, Melville R 1948 Vitamin C in the walnut. Biochem J 43(4):585–592
Wood J R, Goring D A I 1971 The distribution of lignin in stem wood and branch wood of Douglas-fir. Pulp Pap Mag Can 77(3):61–68
Worster H E, McCandless D L, Bartels M 1976 Kraft pulping of whole-tree chips from southern pine for linerboard. Tappi 59:64–66
Worster H E, Vinje M G 1968 Kraft pulping of western hemlock tree tops and branches. Pulp Pap Mag Can 69(7):T308–T311
Yde-Andersen A 1970 Fomes annosus in conifer stands of first and second rotation. Proc Third Internat Conf Fomes Annosus, 137–148
Yonezawa Y, Itho A, Kikuchi F, Usami K, Takano I, Takamura N 1962 On the branches and top part of tree as the raw materials for pulp and fiberboard. Bull For Exp Sta Meguro 113:145–152
Young H E 1964 The complete tree concept – a challenge and an opportunity. Reprint from Proc Soc Am For Denver, 3 pp
Young H E 1968 Challenge of complete tree utilization. For Prod J 18(4):83–86
Young H E 1975 Utilization of forest residues – a segment of the complete tree concept. 75th Nat Conv Soc Am For Washington D C, 6 pp
Young H E 1976 A summary and analysis of weight table studies. In: Oslo biomass studies. Univ Maine Coll Life Sci Agr Orono , 251–282
Young H E 1977a Hardwoods within the complete forest concept. For Chron 53:204–207
Young H E 1977b Forest biomass inventory: the basis for complete tree utilization. Joint For Biol/Wood Chem Conf TAPPI Madison, 12 pp
Young H E 1978a "Forest biomass inventory". In: McMillin C W (ed) Complete tree utilization of southern pine. For Prod Res Soc Madison, 24–28
Young H E 1978b Engineering the complete forest concept. National Off-Highway Vehicle Meeting and Exhibition, Soc Aut Eng Milwaukee, 17 pp
Young H E 1980a Puckerbrush power. IEA Meeting on Biomass Energy Rhinelander, 13 pp
Young H E 1980b Biomass utilization and management implications. In: Weyerhaeuser Sci Symp 3, Forest-to-Mill Challenges of the Future, 65–80
Young H E 1981 The complete tree concept – a fifteen-year progress report. In: Stumbo D A (ed) Utilization of low-grade southern hardwoods. For Prod Res Soc, 3–8
Young H E, Carpenter P M 1967 Weight, nutrient element, and productivity studies of seedlings and saplings of eight tree species in natural ecosystems. Maine Agr Exp Stat Tech Bull 28, 39 pp
Young H E, Chase A J 1965 Fiber weight and pulping characteristics of the logging residue of seven tree species in Maine. Maine Agr Exp Sta Tech Bull 17, 44 pp
Young H E, Hatton J V 1976 Chipping of tops and branches: a comparison of a drum- and a disk-chipper. Univ Maine Orono Res Life Sci 23(9):1–11
Young H E, Hoar L, Ashley M 1965 Weight of wood substance for components of seven tree species. Tappi 48:466–469
Young H E, Ribe J H, Wainwright K 1980 Weight tables for tree and shrub species in Maine. Univ Maine Orono Life Sci Agr Exp Sta Miscell Rep 230, 84 pp

Young H E, Strand L, Altenberger R 1964 Preliminary fresh and dry weight tables for seven tree species in Maine. Maine Agr Exp Sta Techn Bull 12, 76 pp

Young H E, Tryon T C 1978 A national forest biomass inventory. IUFRO S4.01 For Invent Meet Bucarest, 11 pp

Zavitkovski J 1971 Dry weight and leaf area of aspen trees in northern Wisconsin. In: Forest biomass studies. Univ Maine Coll Life Sci Agr, 193–205

Zerbe J 1982 Wood production of liquid or gaseous fuels. Paper for seminar on reducing forest biomass losses in logging operation. Moscow, 13 pp

Zerbe J 1985 Current developments in the conversion of wood to liquid fuels. Forest Products Research International – Achievements and the Future, Pretoria, Vol 5:13.22, 10 pp

Žerebov L P 1946 (Mechanical functions of the chemical constituents of wood.) Bumazh Prom 21(3/4):14–26

Zhitkov A V 1958 (Bark and its properties.) Bum Promst 33(2):4–6

Zimmermann MH 1983 Xylem structure and the ascent of sap. Springer Berlin Heidelberg New York, 143 pp

Zimmermann M H, Potter D 1982 Vessel-length distribution in branches, stem and roots of Acer rubrum L. IAWA Bull 3:103–109

Subject Index

MIX
Papier aus verantwortungsvollen Quellen
Paper from responsible sources
FSC® C105338

If you have any concerns about our products,
you can contact us on
ProductSafety@springernature.com

In case Publisher is established outside the EU,
the EU authorized representative is:
Springer Nature Customer Service Center GmbH
Europaplatz 3, 69115 Heidelberg, Germany

Printed by Libri Plureos GmbH
in Hamburg, Germany